Environmental
Risk Analysis
for Chemicals

Environmental Risk Analysis for Chemicals

Edited By

Richard A. Conway, P.E.

Corporate Development Fellow
Union Carbide Corporation
South Charleston, West Virginia

Van Nostrand Reinhold Environmental Engineering Series

VNR VAN NOSTRAND REINHOLD COMPANY
NEW YORK CINCINNATI TORONTO LONDON MELBOURNE

Manufactured in the United States of America

Published by Van Nostrand Reinhold Company
135 West 50th Street, New York, N.Y. 10020

Van Nostrand Reinhold Limited
1410 Birchmount Road
Scarborough, Ontario M1P 2E7, Canada

Van Nostrand Reinhold Australia Pty. Ltd.
17 Queen Street
Mitcham, Victoria 3132, Australia

Van Nostrand Reinhold Company Limited
Molly Millars Lane
Wokingham, Berkshire, England

15 14 13 12 11 10 9 8 7 6 5 4 3 2 1

Library of Congress Cataloging in Publication Data
Main entry under title:

Environmental risk analysis for chemicals.

(Van Nostrand Reinhold environmental engineering
series)
    Includes index.
    1. Chemicals—Environmental aspects.
2. Environmental impact analysis. I. Conway,
Richard A.
TD196.C45E58   363.7'3'0287   81-386
ISBN 0-442-21650-5   AACR1

Van Nostrand Reinhold Environmental Engineering Series

THE VAN NOSTRAND REINHOLD ENVIRONMENTAL ENGINEERING SERIES is dedicated to the presentation of current and vital information relative to the engineering aspects of controlling and monitoring and protecting man's physical environment. Systems and subsystems available to exercise control of both the indoor and outdoor environment continue to become more sophisticated and to involve a number of engineering disciplines. The aim of the series is to provide books which, though often concerned with the life cycle—design, installation, and operation and/or maintenance—of a specific system or subsystem, are complementary when viewed in their relationship to the total environment.

The Van Nostrand Reinhold Environmental Engineering Series includes books concerned with the engineering of systems designed (1) to control the environment within structures, including those in which manufacturing processes are carried out, and (2) to protect the exterior environment through control of waste products expelled by inhabitants of structures and from manufacturing and other processes and (3) to predict, measure, and monitor environmental quality. The series includes books on heating, air conditioning and ventilation, control of air and water pollution, control of the acoustic environment, sanitary engineering and waste disposal, illumination, and piping systems for transporting media of all kinds. Books on environmental assessment are included to point out control needs.

Van Nostrand Reinhold Environmental Engineering Series

ADVANCED WASTEWATER TREATMENT, by Russell L. Culp and Gordon L. Culp

ARCHITECTURAL INTERIOR SYSTEMS—Lighting, Air Conditioning, Acoustics, John E. Flynn and Arthur W. Segil

SOLID WASTE MANAGEMENT, by D. Joseph Hagerty, Joseph L. Pavoni and John E. Heer, Jr.

THERMAL INSULATION, by John F. Malloy

AIR POLLUTION AND INDUSTRY, edited by Richard D. Ross

INDUSTRIAL WASTE DISPOSAL, edited by Richard D. Ross

MICROBIAL CONTAMINATION CONTROL FACILITIES, by Robert S. Rurkle and G. Briggs Phillips

SOUND, NOISE, and VIBRATION CONTROL (Second Edition), by Lyle F. Yerges

NEW CONCEPTS IN WATER PURIFICATION, by Gordon L. Culp and Russell L. Culp

HANDBOOK OF SOLID WASTE DISPOSAL: MATERIALS AND ENERGY RECOVERY, by Joseph L. Pavoni, John E. Heer, Jr., and D. Joseph Hagerty

ENVIRONMENTAL ASSESSMENTS AND STATEMENTS, by John E. Heer, Jr. and D. Joseph Hagerty

ENVIRONMENTAL IMPACT ANALYSIS: A New Dimension in Decision Making by R. K. Jain, L. V. Urban and G. S. Stacey

CONTROL SYSTEMS FOR HEATING, VENTILATING, AND AIR CONDITIONING (Second Edition), by Roger W. Haines

WATER QUALTIY MANAGEMENT PLANNING, edited by Joseph L. Pavoni

HANDBOOK OF ADVANCED WASTEWATER TREATMENT (Second Edition), by Russell L. Culp, George Mack Wesner and Gordon L. Culp

HANDBOOK OF NOISE ASSESSMENT, edited by Daryl N. May

NOISE CONTROL: HANDBOOK OF PRINCIPLES AND PRACTICES, edited by David M. Lipscomb and Arthur C. Taylor

AIR POLLUTION CONTROL TECHNOLOGY, by Robert M. Bethea

POWER PLANT SITING, by John V. Winter and David A. Conner

DISINFECTION OF WASTEWATER AND WATER FOR REUSE, by Geo. Clifford White

LAND USE PLANNING: TECHNIQUES OF IMPLEMENTATION, by T. William Patterson

BIOLOGICAL PATHS TO SELF-RELIANCE, by Russell E. Anderson

HANDBOOK OF INDUSTRIAL WASTE DISPOSAL, by Richard A. Conway and Richard D. Ross

HANDBOOK OF ORGANIC WASTE CONVERSION, by Michael W. Bewick

LAND APPLICATIONS OF WASTE (Volume 1), by Raymond C. Loehr, William J. Jewell, Joseph D. Novak, William W. Clarkson and Gerald S. Friedman

LAND APPLICATIONS OF WASTE (Volume 2), by Raymond C. Loehr, William J. Jewell, Joseph D. Novak, William W. Clarkson and Gerald S. Friedman

STRUCTURAL DYNAMICS: THEORY AND COMPUTATION, by Mario Paz

HANDBOOK OF MUNICIPAL WASTE MANAGEMENT SYSTEMS: PLANNING AND PRACTICE, by Barbara J. Stevens

INDUSTRIAL POLLUTION CONTROL: ISSUES AND TECHNIQUES, by Nancy J. Sell

WASTE RECYCLING AND POLLUTION CONTROL HANDBOOK, by A. V. Bridgwater and C. J. Mumford

ENVIRONMENTAL RISK ANALYSIS FOR CHEMICALS, edited by Richard A. Conway

WATER CLARIFICATION PROCESSES: Practical Design and Evaluation, by Herbert E. Hudson, Jr.

HANDBOOK OF NONPOINT POLLUTION: Sources and Management, by Vladmir Novotny and Gordon Chesters

NATURAL SYSTEMS FOR WATER POLLUTION CONTROL, by Ray Dinges

Dedication

To the early leaders in the field of environmental risk analysis, including J. R. Duthie and A.W. Maki of Procter and Gamble Company; G. Baughman, R. Lassiter, D. Mount, J. Falco, D. Hansen, and C. E. Stephan of the U. S. Environmental Protection Agency; G. F. Lee of Colorado State University; K. L. Dickson of North Texas State University; J. Cairns, Jr., of Virginia Polytechnic Institute and State University; J. H. Smith and T. Mill of SRI, International; R. Kimerle and W. E. Gledhill of Monsanto Industrial Chemicals Co.; E. E. Kenaga, D. Branson, B. Neely, and H. C. Alexander of Dow Chemical Co.; J. L. Hamelink of Lilly Research Laboratories; A. K. Ahmed of the Natural Resources Defense Council; R. L. Metcalf of the University of Illinois; C. H. Ward of Rice University; J. B. Sprague of the University of Guelph; K. J. Macek of Bionomics; F. L. Mayer of U. S. Department of the Interior; F. Korte and W. Klein of the Gesellschaft für Strahlen und Umweltforschung in Munich; and many others, including the contributors to this book.

CONTRIBUTORS

A. Karim Ahmed, Natural Resources Defense Council (*Chapter 16*)

Bernard D. Astill, Eastman Kodak Company (*Chapter 12*)

Lowell H. Bahner, U. S. Environmental Protection Agency (*Chapter 14*)

Stephen L. Brown, SRI International (*Chapter 13*)

David R. Cogley, GCA Corporation (*Chapter 15*)

Richard A. Conway, Union Carbide Corporation (*Editor, Preface, Chapters 1, 3*)

Jack C. Dacre, U. S. Army (*Chapter 15*)

Wendell L. Dilling, The Dow Chemical Company (*Chapter 5*)

George Dominguez, CIBA-GEIGY Corporation (*Chapter 16*)

Patrick R. Durkin, Syracuse Research Corporation (*Chapter 10*)

Cleve A. I. Goring, The Dow Chemical Company (*Chapter 6*)

Walter J. Hansen, Union Carbide Corporation (*Chapter 3*)

Stephen E. Herbes, Oak Ridge National Laboratory (*Chapter 4*)

Buford R. Holt, U. S. Geological Survey (*edited Chapter 17*)

Philip H. Howard, Syracuse Research Corporation (*Chapter 10*)

R. Anne Jones, Colorado State University (*Chapter 17*)

Dennis A. Laskowski, Dow Chemical U. S. A. (*Chapter 6*)

G. Fred Lee, Colorado State University (*Chapter 17*)

Shonh S. Lee, SRI International (*Chapter 7*)

Haines B. Lockhart, Jr., Eastman Kodak Company (*Chapter 12*)

Donald Mackay, University of Toronto (*Chapter 2*)

Philip J. McCall, The Dow Chemical Company (*Chapter 6*)

Jay B. Moses, Eastman Kodak Company (*Chapter 12*)

Ahmed Nasr, Eastman Kodak Company (*Chapter 9, 12*)

Jerry L. Oglesby, University of West Florida (*Chapter 14*)

Benjamin R. Parkhurst, Oak Ridge National Laboratory (*Chapter 4, 11*)

(continued)

Parmely H. Pritchard, U. S. Environmental Protection Agency (*Chapter 8*)

Robert L. Raleigh, Eastman Kodak Company (*Chapter 12*)

David H. Rosenblatt, U. S. Army (*Chapter 15*)

Joseph Santodonato, Syracuse Research Corporation (*Chapter 10*)

George R. Southworth, Oak Ridge National Laboratory (*Chapter 4*)

R. L. Swann, The Dow Chemical Company (*Chapter 6*)

Clarence J. Terhaar, Eastman Kodak Company (*Chapter 12*)

Shan-Ching Tsai, Oak Ridge National Laboratory (*Chapter 4*)

Frank C. Whitmore, VERSAR, Inc. (*Chapter 3*)

Preface and Acknowledgments

There is a rapidly increasing need to know more about the impact of chemicals in the environment. Regulations promulgated in the United States under the Toxic Substances Control Act, the Clean Water Act, the Federal Insecticide, Fungicide, and Rodenticide Act, the Clean Air Act, and the Resource Conservation and Recovery Act make this knowledge essential. Other countries have similar legislation. Customers demand this environmental information, as do prudent managers of manufacturing operations. We should be able to predict to which environmental compartments a chemical will be transported, to what extent it will be transformed, and what effects it will have.

Since collection of this information for all chemicals cannot be rapidly achieved, a systematic approach to establish priorities for information gathering is needed. This book is planned to provide the basis for such an orderly process for use by those concerned with analysis of enivronmental impact, be they in industry, commerce, government, universities, or private research institutions.

The emphasis of the book is on the *process* of environmental risk analysis as opposed to a collection of test results with specific chemicals and situations. Topics include:

Reasons to study environmental impact
Entry of materials to the environment
Basic physical-chemical tests
Concerns in aquatic, atmospheric, and terrestrial environments
Mathematical model building for predictions
Model ecosystems
Human health considerations
Case studies in environmental risk analysis

Guides are provided to set up and carry out programs in this area. This book is not a symposium proceedings, but rather a planned coverage of the basics of environmental risk analysis for chemicals.

The purpose of the book is to provide rationales and protocols for those who make decisions involving environmental risks. A classic example would be a producer of a new industrial chemical who must decide whether adverse environmental effects will constrain or preclude successful introduction of the product. Producers and users of certain existing products have similar concerns. Those deciding the extent to which an aqueous or gaseous emission or a solid waste needs to be treated also need to assess environmental risks in a systematic manner.

Regulating agencies and public interest groups share such concerns with material generators. University courses in this area are expected.

A tiered approach to decision making, along with necessary tools and techniques, is presented in this book. Both basic principles and case studies are included.

For the purpose of this book the definitions of Sanders (Ref. 1-3) are used to distinguish between analysis and assessment. "Analysis" describes the collection and examination of technical and scientific data, whereas "assessment" includes evaluation of both the scientific data and the social, economic, and political factors that must be considered to reach an ultimate decision on the use, prohibition, or restrictive control of chemicals in the environment. Since considerable information exists in the general area of assessment, this book is primarily limited to the use of analysis in determining the probability and magnitude of an environmental risk.

Chapter Summaries. The first chapter defines risk analysis as "the collection and examination of technical and scientific data to measure the probability and severity of adverse effects." Alternative approaches are described for analyzing risks associated with entry of chemicals into the environment. Tiered testing protocols are provided, as well as techniques for modeling the data collected. Pertinent legislation and other driving forces are also described in this chapter by the Editor.

Those properties of a new chemical compound that a manufacturer can be reasonably expected to provide and methods for measuring or estimating them are the subjects of Chapter 2. Included are properties of a purely physical nature, properties affecting chemical conversion, and, to a lesser extent, biological properties. Complicating factors and issues are presented, as well as data sources. Collection of these basic data is viewed as a starting point for the in-depth evaluations described in Chapters 4 to 6. The considerable experience in this area of Donald Mackay of the University of Toronto is reflected in this chapter.

Entry of chemicals into the environment is discussed in Chapter 3, the purpose being to establish the initial concentration of a material in compartments of concern. This concentration is needed as a starting point in fate/effects studies discussed in subsequent chapters. Manufacturing sources are emphasized. Checklists and guides are provided to analyze each unique entry situation. A model is described for estimating the injection rate and accumulation of a pollutant; the PCB situation is developed as an example. The chapter is co-authored by Frank C. Whitmore of VERSAR, Inc., and Walter J. Hansen and the Editor, both of Union Carbide Corporation.

The factors contributing to a chemical being a local, regional, or global aquatic hazard are delineated in Chapter 4 by George R. Southworth and his coworkers at Oak Ridge National Laboratory. Many types of fate studies applicable to the aquatic environment are described. Guides are presented for aquatic effects testing, both acute and chronic.

The transport and transformations of chemicals in the atmospheric environment are discussed in Chapter 5 by Wendell L. Dilling of The Dow Chemical Company. Basic data on atmospheric properties and composition are related to a wide range of expected reactions. Test methods for persistence and fate prediction are described in detail; projected transformation schemes are also presented. Validated examples are included. Theoretical considerations and rate data presented should provide a starting point for those undertaking work in this area.

The environmental chemistry of a terrestrial environment involves a mixture of processes existing in water, air, and soil. In Chapter 6, Dennis A. Laskowski and his coworkers at The Dow Chemical Company discuss the physical, chemical, and biological aspects of terrestrial environments. Data requirements and methods for environmental fate analysis in soil are specified. Original means of data interpretation, an often neglected subject, are presented.

Mathematical modeling for prediction of chemical fate is covered in Chapter 7 by Shonh S. Lee, formerly of SRI International. Pertinent emission and environmental factors are identified. Equations representing important transport and transformation mechanisms in aquatic systems are presented. Methods of treating the environment as a series of completely mixed compartments is described.

The design and applicability of microcosms or model ecosystems in risk analysis is discussed thoroughly in Chapter 8 by P. H. Pritchard of EPA's Gulf Breeze Laboratory. The advantages and disadvantages of the microcosm approach are first delineated. A more exact definition of a microcosm study is proposed; it requires use of natural assemblages, standardized conditions, individual process studies, and mathematical modeling. System design is discussed in terms of scale, trophic level, innoculum, and operation; some limitations of synthetic communities are identified. Ways of calibrating against field conditions are specified. Fate and transport studies in microcosms and related systems are emphasized; however, emerging tests for toxicity are also addressed. Aquatic systems are emphasized, but soil systems are also described.

The foregoing chapters primarily involve ecological matters. However, there is also present an underlying concern for human health. To bring this into focus, human diseases caused by chemicals are discussed by Ahmed Nasr of Eastman Kodak Company in Chapter 9. The philosophy and teachings of occupational health are emphasized because occupational exposures are higher, the population is more readily observable, and responses are more easily measured. This is a useful point to begin considering the problems that result from exposure to chemicals in the environment. Of direct interest are modes of exposure, types of diseases, chemical carcinogens and cancer risks, toxicity testing, and epidemiology.

As the first in a series of eight case studies, the Syracuse Research Corporation's approach to chemical risk analysis is presented in Chapter 10 by Philip H.

Howard, Joseph Santodonato, and Patrick R. Durkin. This serves to tie together many of the concepts presented in earlier chapters. Key information sources and testing procedures are specified. A general format and approach are presented for organizing and interpreting results and for identifying higher risk chemicals. Sample information for several chemical classes is included. The approach has been used in the preparation of a series of EPA reports for various chemicals by reviewing and integrating available information on production and use, monitoring, environmental chemistry and fate, and toxicity. The concepts and procedural details can also be applied to in-company studies.

An application of environmental risk analysis to coal conversion wastes is discussed in Chapter 11. The environmental fate and effects of key classes of contaminants, such as polycyclic aromatic hydrocarbons, are determined. Procedures by which wastewater from individual processes have been fractionated and tested for toxicity are also described. The author of this chapter is Benjamin R. Parkhurst of the Oak Ridge National Laboratory.

A quantitative systematic scheme developed by Eastman Kodak Company for collecting, organizing, relating, and evaluating information concerning chemical hazards to health and the environment is described in Chapter 12 by Bernard D. Astill and coworkers. They identify the types of information needed to evaluate potential hazards and provide the user with required tests. They offer a numerical rating system for determining which of four levels of testing to perform. It can be used in guiding product development as well as in providing information for premanufacture notification.

The development of a dossier system on potentially problem chemicals is described in Chapter 13. The categories and extent of information needed at different levels of appraisal is specified, along with an extensive list of information sources. Particular emphasis is given to the order and format of information display that most effectively support environmental decisions by chemical producers and regulatory agencies. A sample dossier on carbon tetrachloride is presented. This study, directed by Stephen L. Brown, was done by SRI International for the United Kingdom Department of the Environment. Also discussed is the general approach of the U. K. in the area of environmental risk analysis for chemicals.

As a case study extending the principles discussed in Chapter 7, non-linear regression models for predicting bioaccumulation and other ecosystem effects of kepone, endrin, and textile effluents are described in Chapter 14 by Lowell H. Bahner and Jerry L. Oglesby, under the auspices of the U. S. Environmental Protection Agency. They consider the uptake of kepone from water, food, and sediments. Stimulation or inhibition of algae by textile effluents and kepone are also modeled, as is cumulative spawning of grass shrimp exposed to endrin. Also included is a short discussion of a complete ecosystem model, using mass–balance kinetic equations to compute pollutant distribution in the James River.

An approach to set limits for pollutants in soil that is based on projected transport at a given site to a receptor organism (in this case, man) is given in Chapter 15. The work was done by D. H. Rosenblatt, J. C. Dacre, and D. R. Cogley under U. S. Army auspices on problems related to arsenal wastes. The various modes of intake by man and alternative environmental pathways are covered, as are detailed means for estimating acceptable daily dosages. Numerical values for estimating purposes are derived; examples are provided. It is possible to extend the model to aquatic, plant, and wildlife toxicity.

A consensus position among conservationists, the industrial community, and academics regarding testing of new products is described in Chapter 16. It was developed under the auspices of the Conservation Foundation and under the leadership of George Dominguez and A. Karim Ahmed. Especially well discussed is short-term testing to predict health effects.

In Chapter 17, G. Fred Lee and R. Anne Jones present a risk analysis approach for application to solid wastes. Waste-specific, site specific studies are advocated, as opposed to a standard-leaching-test approach. Studies with dredging waste are described, as are concentration-exposure relationships as they affect impact.

Acknowledgments. The contributors are experts in the field and were selected to obtain a thorough coverage of this discipline. Their willingness to share their knowledge is much appreciated, as are their efforts to prepare chapters that are thorough, accurate, and useful.

I also acknowledge the encouragement James B. Johnson and others in Union Carbide Corporation provided me in serving as Editor. The review of the introductory material by William E. Gledhill is appreciated. I am indebted to Mrs. Delores Quesenberry for continuing support in assembling the overall manuscript and to Mrs. Karen McClung and Mrs. Regina Pauley for their typing of portions of the manuscript. Miss Sherry Williams drafted the figures for several chapters. As indicated in the Dedication of this book, I also am indebted to early leaders in this field whose original thinking and hard work provided a basis for our current efforts.

Richard A. Conway
Union Carbide Corporation
South Charleston, West Virginia

Contents

Environmental Risk Analysis for Chemicals

1

Introduction to environmental risk analysis

Richard A. Conway, P. E.
Corporate Development Fellow
Research and Development Department
Union Carbide Corporation
South Charleston, West Virginia

1-1 THE PROBLEM

Regulatory and other pressures have made it necessary to know the fate and effect of chemicals entering the environment. Chemicals enter the environment during manufacturing operations, product distribution, and product use. Other sources of chemicals include mineral extraction, petroleum refining, power generation with fossil-fuels, coal conversion, and natural emissions. It is necessary to develop methodology that will make it possible to predict where in the environment a chemical will be transported, the rate and extent of its transformation, and its effect on organisms and environmental processes at expected ambient levels. The process of making such predictions is termed environmental risk analysis.

Some of the factors contributing to this greatly increased need for environmental risk analysis for chemicals include the following:

- Discovery that certain chemicals persist in the environment and could have adverse effects at higher

1

levels, e.g., polychlorinated biphenyls (PCBs), dioxins, and fluoro-carbons.

- Rapid growth of the chemical industry (Fig. 1-1) and the many other human activities (e.g., fuel production and use) that introduce chemicals into the environment.
- The Federal Toxic Substances Control Act (TSCA), which requires information on ecological fate and effects for new products, significant-new-use products, priority commercial materials, and chemicals suspected of having substantial risks.
- The Federal Resource Conservation and Recovery Act (RCRA), which requires information regarding the fate and effects of leachates from land disposal of hazardous solid wastes.

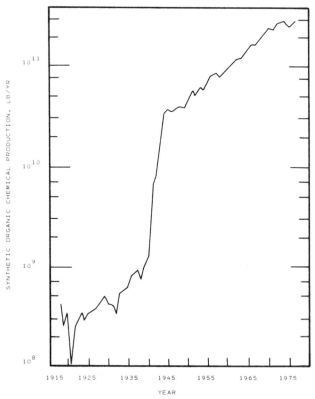

Fig. 1-1.   Growth of the synthetic organic chemical industry. *Source: First Annual Report to Congress by the Task Force on Environmental Cancer and Heart and Lung Disease*, U. S. Environmental Protection Agency, Washington, D. C., August 7, 1978.

- The Federal Clean Water Act Amendments, which require that receiving water quality criteria be developed for toxic pollutants.
- Studies of hazardous air pollutants under the Federal Clean Air Act.
- The need for managers of manufacturing and distribution operations to know the environmental impact of materials they handle so that they are able to make decisions, especially in emergencies.
- Customers' demand for environmental fate-and-effects information from their suppliers.
- The need for product development to be guided by a combination of information on efficacy, economics, health effects, and environmental effects.

The legislation cited above is that in the United States on a Federal level; however, similar pressures exist or are developing elsewhere. More detail on these driving forces is given in Section 1-6.

Collection of detailed information on all chemicals entering the environment is not practical as a short-term goal. Hutzinger points out that a chemical compound is more likely to be a severe pollutant if it fulfills the following criteria (Ref. 1-1):

- Large production volume.
- Environmental leakage likely.
- High dispersion tendency.
- Pronounced persistence.
- Tendency to bioaccumulate.
- High toxicity.

Systems for developing priorities for information gathering based on projected risks are required and are discussed in Section 1-4 and in various case studies in the latter chapters of this book.

Screening-type information is first needed regarding a chemical's chances of entering the environment, the transport or partitioning of the chemical within and between environmental compartments, the persistence of the chemical, and the effect of the chemical on the environment or on organisms in the environment. Based on screening results, it may be established that there is a need to conduct longer-term, more detailed tests that will provide results with higher degrees of confidence.

This book will stress the process of environmental risk analysis, i.e., the basic tests to be conducted on all chemicals, their modes of entry into the environment, fate and effects testing in the aquatic, atmospheric, and terrestrial environments, use of mathematical and ecosystem models, and human health considerations. Following the nine basic chapters is a section on case studies that demonstrates the use of principles of environmental risk analysis.

1-2 BASIC CONSIDERATIONS IN RISK ANALYSIS

The purpose of this book is to provide approaches and tools for analyzing the risks associated with the presence of chemicals in the environment. What does "risk analysis" mean?

- Risk is a measure of the probability and severity of adverse effects (Ref. 1-2).
- Analysis describes the collection and examination of technical and scientific data (Ref. 1-3).

Risks are established with varying degrees of confidence, according to the importance of the decision involved. This is done by first using simple screening tests on decisions having small potential impact and then applying tests of increasing complexity and accuracy as decisions having large impact are reached.

The foregoing is termed "tier testing." Early decisions may be whether to pursue further synthesis efforts with a new compound, while a decision with large environmental impact potential might be the inclusion of a new material in a high-volume product to be used by consumers and passed to the environment, e.g., detergents and aerosols.

Although this book deals with the analysis or determination of risk by examining the fate and effects of chemicals in the environment, the acceptability of a determined risk is the logical next step. The "assessment" of risk involves an evaluation of both scientific data and the social, economic, and political factors that must be considered in reaching an ultimate decision on the prohibition, control, or management of chemicals in the environment (Ref. 1-3). This includes the worth of the product, i.e., probability and intensity of beneficial effects (Ref. 1-2).

Final decision making involves the scientific measurement of risk and then a social judgment of whether the benefits of the product outweigh the risk. An analogous risk–benefit case can be developed for discharges associated with a manufacturing operation versus the benefit to the economy. Means for evaluating the acceptability of risk have been proposed by Kazmarek (Fig. 1-2, Ref. 1-4) and Comongis (Fig. 1-3, Ref. 1-5).

Threshold or no-effect levels are often used to set regulations for toxic effects, except in the cases of carcinogenesis and mutagenesis (Fig. 1-2). If no threshold is assumed, the risk is determined. Kazmarek recommends either the Mantel–Bryan log probit model or the linear model, with the former preferred in most cases. Conservative extrapolation should be used but it should not be so conservative to severely overestimate the risk. If the existing safety factor is clearly adequate, e.g., more than several thousandfold, or if the risk is vanishingly small, then no further action should be necessary. If the safety factor or risk is not indisputably adequate, other factors are considered such as technical feasibility of control, cost of control, and benefits of the action creating the risk.

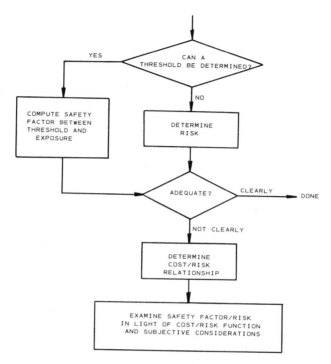

Fig. 1-2. Acceptability of Risk. *Source*: Kazmarek, E. A., (Georgia-Pacific Corp., Atlanta, GA), "Evaluation of Reasonable Risk," Presented at Annual Conference of Water Pollution Control Fed., Anaheim, CA, October, 1978.

Other criteria for the reasonableness of risk have been cited by Daniels (Ref. 1-2), based on a publication by Lourance (Ref. 1-6):

1. Toxicologically insignificant levels.
2. No detectable adverse effect.
3. Degree of necessity or benefit.
4. Best available practice, highest practicable protection, lowest practical exposure.
5. The threshold principle.
6. The Delaney principle regarding animal carcinogens in food.
7. Prevailing professional practice.
8. Custom of usage.
9. Exposure relative to natural background.
10. Occupational exposure precedent.
11. Public referenda and polling.
12. Comparison with accustomed hazards.

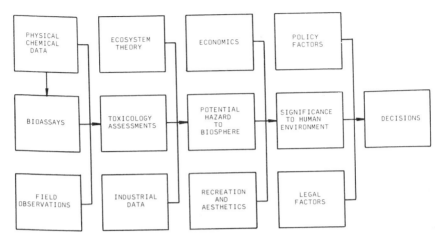

Fig. 1-3. Hazard evaluation and decision making chain. *Source*: Comongis, G., "Hazard Evaluation," presented at 3rd Annual ASTM Sym. on Aquatic Toxicity, New Orleans, LA, October, 1978.

"Nothing is wholly safe or dangerous *per se*. It is the quantity involved, the manner and conditions of use, and the susceptibility of the organism which deter—mine degree of hazard or safety. The perspective is that there is no escape from all risk no matter how remote. There are only choices among risks (Ref. 1-2)." Table 1-1 lists risks of various types, some of which are environmental (Ref. 1-7).

1-3 ALTERNATIVE TECHNICAL APPROACHES

There are at least four major approaches to environmental risk analysis:

- The "stochastic-statistical" approach, in which large amounts of data are obtained under a variety of conditions, and the correlations are established between the input of a certain material and its observed concentrations and its effects in various environmental compartments. An example of this is a lake study correlating phosphate input, phosphorous concentrations, and algal blooms.
- The model ecosystem ("microcosm") approach, in which a physical model of a given environmental situation is constructed, dosed with a chemical, and the fate and effects of the chemical are observed. An example would be a balanced aquarium with plants, animals, and protista (organisms without tissue differentiation).
- The "deterministic" approach, which uses a simple mathematical model to describe the rates of individual transformations and transports of the

chemical in the environment. Only the dominant mechanisms, grouped as much as possible, are studied to arrive at an estimate of Expected Environmental Concentration (EEC). Separate determinations of toxicity with indicator organisms are made. This is the approach proposed in the TSCA premanufacture-notice testing guidelines. Figures 1-4 and 1-5 (Ref. 1-8) depict this approach.

- The "baseline chemical" approach, in which transformations, transports, and effects are measured as in the deterministic approach, but the results then are compared with data on chemicals of known degrees of risk.

TABLE 1-1. Actions Increasing Risk of Death by One in a Million.

Action	Nature of Risk
Smoking 1.4 cigarettes	Cancer, heart disease
Drinking 0.5 liter of wine	Cirrhosis of the liver
Spending 1 hour in a coal mine	Black lung disease
Spending 3 hours in a coal mine	Accident
Living 2 days in New York or Boston	Air pollution/heart disease
Traveling 6 minutes by canoe	Accident
Traveling 10 miles by bicycle	Accident
Traveling 30 miles by car	Accident
Flying 1000 miles by jet	Accident
Flying 6000 miles by jet	Cancer caused by cosmic radiation
Living 2 months in Denver on vacation from New York	Cancer caused by cosmic radiation
Living 2 months in average stone or brick building	Cancer caused by natural radioactivity
One chest x-ray taken in a good hospital	Cancer caused by radiation
Living 2 months with a cigarette smoker	Cancer, heart disease
Eating 40 tablespoons of improperly stored peanut butter	Liver cancer caused by aflatoxin B
Drinking heavily chlorinated water (e.g., Miami) for 1 year	Cancer caused by chloroform
Drinking 30 12-oz. cans of diet soda	Cancer caused by saccharin
Living 5 years at site boundary of a typical nuclear power plant in the open	Cancer caused by radiation
Drinking 1000 24-oz. soft drinks from recently banned plastic bottles	Cancer from acrylonitrile monomer
Living 20 years near PVC plant	Cancer caused by vinyl chloride (1976 standard)
Living 150 years within 20 miles of a nuclear power plant	Cancer caused by radiation
Eating 100 charcoal broiled steaks	Cancer from benzopyrene
Risk of accident by living with 5 miles of a nuclear reactor for 50 yr	Cancer caused by radiation

Source: Wilson, R., "Analyzing the Daily Risks of Life", *Technology Review* (M.I.T.), 41–46, February, 1979.

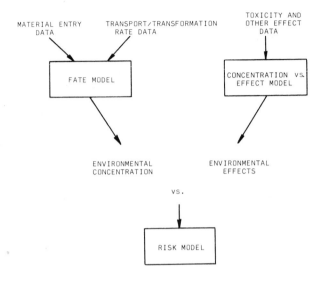

NOTE: MODELS CAN BE SIMPLE MATHEMATICAL EXPRESSIONS OR
YES-NO COMPARISONS WITH PRE-ESTABLISHED LEVELS AND
RATIOS THAT TRIGGER CERTAIN ACTIONS.

Fig. 1-4. Deterministic approach to environmental risk analysis.

Each of these approaches has merit, and the choice for any given situation depends on the need for extrapolation of results, the desired accuracy, the relative costs, and the available time and resources. The deterministic modeling approach will be emphasized in this book; however, Chapter 7 is devoted to the use of microcosms for parallel and confirmatory studies.

1-4 TIERED TESTING PROTOCOLS

In risk analysis schemes, information of the following nature is collected:

- Levels and distribution of input of material into the environment ("entry").
- Transport and transformation in the environment ("mobility" and "persistance" leading to "fate").
- Toxic concentrations ("effects").

First screening information is collected. Materials with high levels of input and/or certain distributions require extensive testing, as do products that persist (resist transformation) and are toxic at low levels. The estimates of ambient

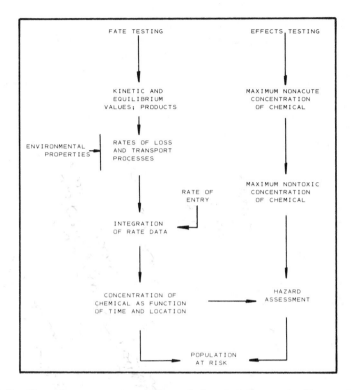

Fig. 1-5. Hazard assessment: relation of chemical fate and effects. *Source*: Mill, T., "Data Needed to Predict the Environmental Fate of Organic Chemicals," presented at Symposium on Environmental Fate and Effects, American Chemical Society, Miami, Florida, September, 1978. Proceedings to be published by Ann Arbor Science.

environmental concentration and toxic-effect concentration become increasingly more accurate as more testing is conducted.

The entry of synthetic organic chemicals, fuels, and other materials into the environment results from production, manufacturing, distribution, use, and disposal. Subsequent transport is associated with partitioning between the air, water, sediment, and biota. Transformation is related to factors such as biodegradation, chemical reactions, and photochemical degradation. Toxicity is determined by various bioassay procedures.

Many tier testing schemes have been proposed. These provide qualitative and/or quantitative guides in decision making regarding the need for further testing. Maki has analyzed the decision-making criteria in several of these schemes (Ref. 1-9):

- American Institute of Biological Sciences report on pesticide testing (Ref. 1-10) specifies that additional testing is required under these circumstances:

  Acute toxicity: Lethal concentration to 50% of the population (LC_{50}) < 1.0 mg/liter
  Estimated environmental concentration: (EEC)>LC_{50}/100
  Observed mammalian/avian reproductive effects
  Solubility <0.5 mg/liter, octanol–water partitioning >1000
  Half life >4 days
  Broad, large volume use pattern

- Eastman Kodak Company's quantitative rating scheme combines human health and ecological effects. Chapter 12 describes this tier testing based on quantity discharged, discharge distribution, octanol–water partitioning, biodegradation, waste-treatment effects, and fish toxicity.

- Monsanto Company's tier testing scheme with trigger levels between tiers is depicted in Fig. 1-6 (Ref. 1-11); tests are listed in Table 1-2. The associated administrative aspects are given in Fig. 1-7. The scheme makes four levels of testing. The ratio of maximum allowable toxicant concen-

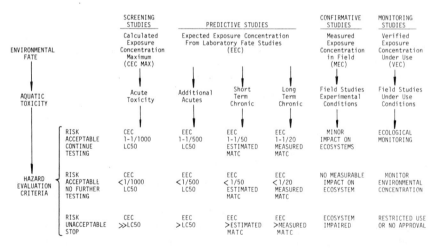

Fig. 1-6. Trigger levels in Monsanto Company's hazard evaluation procedures. *Source*: Kimerle, R. A., Gledhill, W. E., and Levinskas, G. J., "Environmental Safety Assessment of New Materials," Reprinted with permission, from STP 657, "*Estimating the Hazard of Chemical Substances to Aquatic Life.*" Copyright American Society for Testing and Materials, 1916 Race St., Philadelphia, PA 19103, 1978.

TABLE 1-2. Monsanto Tests for Assessing Aquatic Risks.

I.   Screening Tests

    A. Environmental Fate
       1. Transport
          a. Volatility
          b. Adsorption (soil)
          c. Solubility (aqueous)
          d. Partition coefficient (octanol–water)
       2. Persistence
          a. Microbial degradation
             (1) Soil
             (2) Activated sludge
             (3) River water
             (4) Ultimate (CO_2 evolution)
          b. Photochemical
             (1) Oxidation–Reduction
             (2) Photolysis
          c. Hydrolysis

    B. Aquatic Toxicity (LC_{50}/EEC)
       1. Algae
       2. *Daphnia*
       3. Fish

II. Predictive Studies[a]

    A. Environmental Fate
       1. Transport
       Follow up of adsorption screening tests
       2. Persistence
          a. Microbial
             (1) Anaerobic systems
             (2) Soils
             (3) Microcosms
             (4) Follow up of activated sludge test (metabolism)
          b. Photochemical
          Degradation products identity and as a next step their fate
       3. Chlorination/ozonation response

    B. Aquatic Toxicity
       1. Acute Tests
          a. Consideration of various water quality parameters
          b. Time/dose/mortality relationships
       2. Subchronic Tests
          a. 30–60 day growth/survival
          b. Egg/fry
          c. Physiological/behavioral
          d. Bioconcentration/Bioaccumulation
          e. Microcosm
       3. Chronic Test
       Fish and Invertebrates

(continued)

TABLE 1-2 (Continued)

III.   Confirmation Studies

Using results from screening and predictive tests as guides, environmental fate and aquatic toxic effects are measured using a full-scale activated sludge plant, a pond ecosystem, or a small stream ecosystem.

IV.   Monitoring

As the product is in manufacture and use, appropriate monitoring is conducted to check the certainty of prior limited-scale studies.

---

[a]Consider impurities, plant wastes, and specific environments involved, as well as follow-up work on product itself.

Source: Kimerle, R.A., Gledhill, W. C., and Levinskas, G. J., "Environmental Safety Assessment of New Materials." Reprinted, with permission, from STP 657, *Estimating the Hazard of Chemical Sustances to Aquatic Life.* Copyright American Society for Testing & Materials, 1916 Race St., Philadelphia, PA 19103, 1978.

tration (no-effect level) to estimated environmental concentration (MATC/ EEC) is used in deciding upon further testing.

- Unilever Research has developed a plan somewhat specific to detergent compounds (Ref. 1-12); trigger levels cited include:

    "EEC<0.01 LC_{50}" for most sensitive species and "EEC = 0.01 – 0.1 LC_{50}" indicate no further testing. Chronic tests are needed for slowly degrading materials or if LC_{50} data differ by an order of magnitude.

- The American Society for Testing and Materials' Subcommittee E-47.01 on Safety to Man and the Environment has drafted a standard practice to evaluate the hazard of a substance to aquatic organisms. Guides are presented for decisions regarding product acceptance, further testing, terminating the product, or use with monitoring. Figure 1-8 (Ref. 1-13) summarizes the test sequence; more details are in the draft ASTM procedure (Ref. 1-14).

- Other less quantitative tier test schemes include (Ref. 1-9): Stufer plan (a German plan based largely on product volume), Salmon & Freshwater Fisheries Laboratory plan, Commission of the European Communities Plan, FIFRA guidelines for registering pesticides (Ref. 1-15), Conservation Foundation's recommendations to TSCA (Chapter 16), Dow Chemical's methodology for testing new chemicals (Table 1-3), Soap and Detergent Industries of the UK, and, of course, TSCA itself (Table 1-4).

The philosophy cited in the TSCA guidelines endorses flexibility in testing according to estimated exposure and predicted toxic effects, but it remains to be seen whether or not most materials will have to be subjected to a complete listing of tests. It is most important, from cost-benefit and resource-allocation viewpoints that the concept that prevails is a flexible base set of data, with follow-up tests only in problem areas.

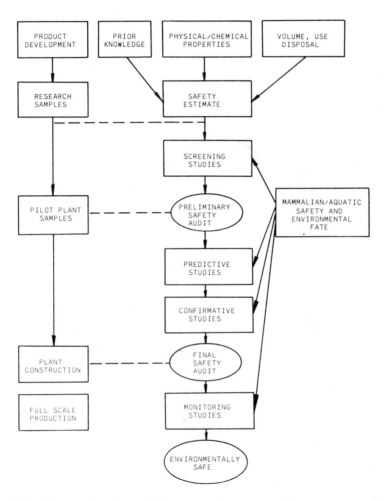

Fig. 1-7. Managerial aspects of Monsanto Company's risk assessment and commercialization of a new product. *Source*: Kimerle, R. A., Gledhill, W. E., and Levinskas, G. J., "Environmental Safety Assessment of New Materials. Reprinted with permission, from STP 657, *"Estimating the Hazard of Chemical Substances to Aquatic Life."* Copyright American Society for Testing and Materials, 1916 Race St., Philadelphia, PA 19103, 1978.

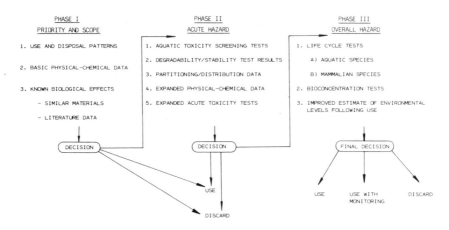

Fig. 1-8. ASTM aquatic hazard evaluation scheme *Source*: Maki, A. W., and Duthie, J. R., "Summary of Proposed Procedures for the Evaluation of Aquatic Hazards," Reprinted with permission, from STP 657, *"Estimating the Hazard of Chemical Substances to Aquatic Life."* Copyright American Society for Testing and Materials, 1916 Race St., Philadelphia, PA 19103, 1978.

1-5 MODELING AND DATA COLLECTION CONSIDERATIONS

Data on environmental input (entry), transport (mobility), and transformation (persistence) need to be collected and interpreted in order to estimate what is an effective environmental concentration. This concentration is then compared to toxicity information in order to analyze environmental risk. Often this is done best by means of simple mathematical models. However, in the absence of rate data, screening tests can provide a basis for yes–no decisions regarding likely fate of the chemical in various compartments without constructing a mathematical model (Fig. 1 in Ref. 1-16).

A detailed discussion of modeling is found in Chapter 7. A listing of 27 environmental fate models is given in Ref. 1-17.

Neely (Ref. 1-18), of Dow Chemical Company, has described a model for estimating the compartmental (air, water, sediment, fish) distribution of material (Fig. 1-9). For such a model the following equation was developed by Branson (Ref. 1-19, with modification):

$$V \frac{dCw}{dt} = k_0 - k_1 ACw - k_2 VCw - k_3 FCw + k_4 FC_f - k_5 SCw + k_6 SC_s \qquad \text{(Eq. 1-1)}$$

where

V = volume of water,
A = surface area,

TABLE 1-3. Dow Chemical Company Methodology for Testing New Compounds.

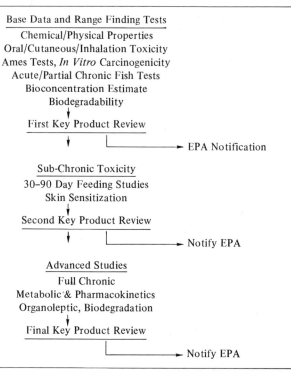

Base Data and Range Finding Tests

Chemical/Physical Properties
Oral/Cutaneous/Inhalation Toxicity
Ames Tests, *In Vitro* Carcinogenicity
Acute/Partial Chronic Fish Tests
Bioconcentration Estimate
Biodegradability

First Key Product Review

→ EPA Notification

Sub-Chronic Toxicity
30–90 Day Feeding Studies
Skin Sensitization

Second Key Product Review

→ Notify EPA

Advanced Studies
Full Chronic
Metabolic & Pharmacokinetics
Organoleptic, Biodegradation

Final Key Product Review

→ Notify EPA

Source: Maki, A., "An Analysis of Decision-Making Criteria in Environmental Hazard Evaluation Schemes", *Analyzing the Hazard Evaluation Process,* K. Dickson, A. Maki, and J. Cairns (eds.), published by Water Quality Section of the American Fisheries Society, Washington, D. C., 1979.

TABLE 1-4. Candidate Tests for Screening Ecological Impact of New Products.

I.   Chemical Fate (Transport, Persistence)

    A. Transport
        1. Adsorption isotherm (soil)
        2. Partition coefficient (water/octanol)
        3. Water solubility
        4. Vapor pressure

    B. Other Physical–Chemical Properties
        1. Boiling/melting/sublimation points
        2. Density
        3. Dissociation constant
        4. Flammability/explodability
        5. Particle size
        6. pH
        7. Chemical incompatibility

(continued)

TABLE 1-4. (Continued)

8. Vapor-phase UV spectrum for halocarbons
9. UV and visible absorption spectra in aqueous solution

  C. Persistence
    1. Biodegradation
      a. Shake flask procedure following carbon loss
      b. Respirometric method following oxygen (BOD) and/or carbon dioxide
      c. Activated sludge test (simulation of treatment plant)
      d. Methane and CO_2 productions in anaerobic digestion
    2. Chemical degradation
      a. Oxidation (free-radical)
      b. Hydrolysis ($25°C$, pH 5.0 and 9.0)
    3. Photochemical transformation in water

II. Ecological Effects

  A. Microbial Effects
    1. Cellulose decomposition
    2. Ammonification of urea
    3. Sulfate reduction

  B. Plant Effects
    1. Algae inhibition (fresh and seawater, growth, nitrogen fixation)
    2. Duck weed inhibition (increase in fronds or dry weight)
    3. Seed germination and early growth

  C. Animal Effects Testing
    1. Aquatic invertebrates (*Daphnia*) acute toxicity (first instar)
    2. Fish acute toxicity (96 hr)
    3. Quail dietary LC_{50}
    4. Terrestial mammal test
    5. *Daphnia* life cycle test
    6. *Mysidopsis bahia* life cycle
    7. Fish embryo–juvenile test
    8. Fish bioconcentration test

Source: "Toxic Substances Control; Discussion of Premanufacture Testing Policy and Technical Issues; Request for Comment" *Federal Register,* 44, 53, Part IV, pp. 16239–16292, March 16, 1979.

F = fish mass,
S = sediment mass,
Cw = chemical concentration in water,
k = rate constants (see Fig. 1-9),
C_f = chemical concentration in fish,
C_s = chemical concentration in sediment.

In this equation k_2 relates to hydrolysis; when related to biodegradation, the concentration of active bacteria would replace the volume term. Neely developed a decision tree to use this distribution in guiding testing (Fig. 1-10).

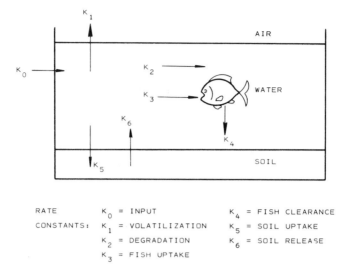

RATE	K_0 = INPUT	K_4 = FISH CLEARANCE
CONSTANTS:	K_1 = VOLATILIZATION	K_5 = SOIL UPTAKE
	K_2 = DEGRADATION	K_6 = SOIL RELEASE
	K_3 = FISH UPTAKE	

Fig. 1-9. Compartmental model showing the movement and distribution of a chemical in an aquatic ecosystem. *Source*: Neely, B., "An Integrated Approach to Assessing the Potential Impact of Organic Chemicals in the Environment," *Analyzing the Hazard Evaluation Process*, K. Dickson, A. Maki, and J. Cairns (eds.), published by Water Quality Section of the American Fisheries Society, Washington, D. C., 1979.

SRI International developed simple models for use in integrating rate data for transport and transformation phenomena for the EPA Environmental Research Laboratory in Athens, GA (Ref. 1-20). Means of estimating transport rate constants for intercompartmental flows in a model developed by A. D. Little, Inc., (Fig. 3-1) are given in Ref. 1-17.

Howard *et al.* have published useful papers on determining the *fate* of chemicals (Refs. 1-21, 1-22). Some of the important transport–degradation mechanisms cited are presented in Table 1-5. From the original entry point, the chemical may:

- Be retained at its original location or be transported throughout an eco-system component.
- Remain unaltered and cycle throughout the various components of the ecosystem.
- Be strongly adsorbed or absorbed and remain in some remote location.
- Be degraded.

There is described a test methodology to be used in screening environmental fate and in establishing quantitative data for the important mechanisms. Biochemical

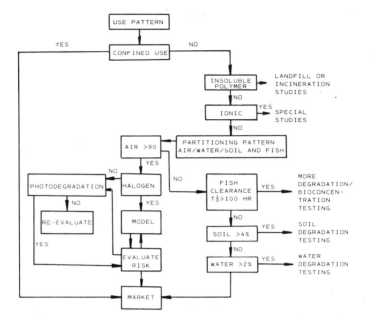

Fig. 1-10. A decision tree for designing an appropriate environmental testing program. *Source:* Neely, B., "An Integrated Approach to Assessing the Potential Impact of Organic Chemicals in the Environment," *Analyzing the Hazard Evaluation Process*, K. Dickson, A. Maki, and J. Cairns (Eds.), published by Water Quality Section of the American Fisheries Society, Washington, D. C., 1979.

TABLE 1-5. Mechanisms Affecting Environmental Fate of Chemicals.

I.   Original Input to Environment During Manufacturing, Distribution, Use, and Disposal

    A. To Water
       1. Treatment plant effluents at manufacturing and/or formulating plants
       2. Spills during manufacturing (original and formulating) and distribution
       3. Disposal after use

    B. To Soil
       1. Direct applications as an agricultural chemical or for vegetation control
       2. Land disposal, e.g., landfill or land cultivation operations
       3. Spills

    C. To Atmosphere
       1. Stack emission during manufacture
       2. Fugitive volatilization losses, e.g., from leaks, storage tank vents, and waste-water treatment
       3. Losses during use and subsequent disposal

(continued)

and photochemical (air and water) degradation procedures are detailed. Mill has elucidated the important environmental properties affecting each process (Table 1-6).

Table 1-5 (Continued)

II. Mechanisms for Transformation and Transport Within and Between Environmental Compartments

  A. Water
  1. Transport from water to atmosphere, sediments, or organisms
      a. Volatilization
      b. Adsorption onto sediments; desorbtion
      c. Absorption into cells (protista, plants, animals); desorption
  2. Transformations
      a. Biodegradation (effected by living organisms)
      b. Photochemical degradation (nonmetabolic degradation requiring light energy), direct or via a sensitizer
      c. Chemical degradation (effected by chemical agents), e.g., hydrolysis, free-radical oxidation

  B. Soil
  1. Transport to water, sediments, atmosphere, or cells
      a. Dissolution in rain water
      b. Adsorbed on particles carried by runoff
      c. Volatilized from leaf or soil surfaces
      d. Taken up by protista, plants, and animals
  2. Transformation
      a. Biodegradation
      b. Photodegradation on plant and soil surfaces

  C. Atmosphere
  1. Transport from atmosphere to land or water
      a. Adsorption to particulate matter followed by gravitational settling or rain washout
      b. Washout by being dissolved in rain
      c. Dry deposition (direct absorption in water bodies)
  2. Transport within atmosphere
      a. Turbulent mixing and diffusion within troposphere
      b. Diffusion to stratosphere
  3. Atmospheric transformations
      a. Photochemical degradation by direct absorption of light, or by accepting energy from an excited donor molecule (sensitizer), or by reacting with another chemical that has reached an excited state
      b. Oxidation by ozone
      c. Reaction with free radicals
      d. Reactions with other chemical contaminants

Based in part on: Howard, P. H., Saxena, J., and Sikka, H., "Determining the Fate of Chemicals", *Environmental Science and Technology,* 12, 4, 398–407 (1978).

TABLE 1-6. Environmental Processes and Properties.

Process	Key Environmental Property[a]
Physical transport	
Meteorological transport	Wind velocity
Bio-uptake	Biomass
Sorption	Organic content of soil or sediments, mass loading of aquatic systems
Volatilization	Turbulence, evaporation rate, reaeration coefficients, soil organic content
Runoff	Precipitation rate
Leaching	Adsorption coefficient
Fall out	Particulate concentration, wind velocity
Chemical	
Photolysis	Solar irradiance, transmissivity of water or air
Oxidation	Concentrations of oxidants and retarders
Hydrolysis	pH, sediment or soil basicity or acidity
Reduction	Oxygen concentration, ferrous ion concentration, and complexation state
Biological	
Biotransformation	Microorganism population and acclimation level

[a] At constant temperature.

Source: Mill, T., "Data Needed to Predict the Environmental Fate of Organic Chemicals," presented at Symposium on Environmental Fate and Effects, American Chemical Society, Miami, Florida, September, 1978. Proceedings to be published by Ann Arbor Science.

In regard to *effects* in aqueous and soil systems a typical screening study might include:

- Fathead minnow (*Pimephales promelas*) acute toxicity.
- *Daphnia* acute toxicity.
- *Mysidopsis bahia* acute toxicity (if salt water exposure is projected).
- *Daphnia* reproduction and growth.
- Seed germination and early growth.
- Algal inhibition.
- Soil respiration inhibition.

Any problems detected in the screening tests would be followed by more definitive, chronic studies. For compounds having low water solubility and high octanol–water partition coefficients, bioconcentration tests may be needed. The lists in Tables 1-2 through 1-4 provide other effects testing alternatives. A National Academy of Sciences publication in this area is useful (Ref. 1-23).

An in-depth analysis of application factors (AF) for converting acute toxicity data (LC50) to chronic data (MATC) has been conducted by Kenaga (Ref. 1-24). Fish, *Daphnia,* and mammals are considered. MATC data can be best derived in fish from critical life-stage tests (egg-fry) or in *Daphnia* reproduction tests.

Besides actual test work to determine transport, transformation, and toxicity, there are many calculative approaches available. Physical properties and some partitioning rate constants can be estimated; important properties have been listed by Lyman (Table 1-7, Ref. 1-17). A first reading on toxicity can be obtained by structural activity relationships (SAR); this approach is either based on a list of constituents that make a chemical suspect (Appendix II of Ref. 1-16) or on the structure and properties of a chemical, as in the Hanch model (Ref. 1-1). Verschueren has compiled extensive environmental data on organic chemicals in a useful format (Ref. 1-25).

The extent of the research and testing effort for chemical fate and effects prediction in the various environmental compartments has centered largely on the aquatic area, with lesser efforts devoted to the terrestrial and atmospheric areas. The initial testing guidelines suggested under TSCA also reflect this emphasis.

However, the atmosphere is a vital concern and will be recognized as such. For example, a sizable fraction of the PCBs in the Great Lakes reportedly enter by atmospheric pathways. Moreover, some of the major chemical losses are to the atmosphere, both product losses and residue-combustion end products.

Transformation and transport tests that might be conducted for atmospheric considerations include:

Gas-phase UV spectra
Direct photolysis
Photo-induced, free-radical oxidation (for chemicals with C–H or C=C bonds)
Oxidation by ozone
Reactions of particulates
Reactions in aerosols
Aerosol formation
Washout
Dry deposition
Sedimentation

See also Tables 1-5 to 1-8, as well as Chapter 5. Atmospheric *effects testing* could include the potential to react photochemically in producing ozone, to persist and react with ozone itself in the upper atmosphere, to cause damage to plants, to be odorous, and to cause human health problems through inhalation.

TABLE 1-7. Properties Important for Chemical Fate and Runnoff/River Modeling of Organic Chemicals.

Adsorption coefficient, K or K_{OC} (at equilibrium), for soils and sediments
   Octanol–water partition coefficient
   Rate of adsorption[a]
   Rate of desorption[a]
   Desorption coefficient (at equilibrium) when hysteresis is apparent[a]

Bioconcentration factor (at equilibrium) for aquatic life (especially fish)
   Rate of uptake[a]
   Rate of clearance[a]
   Octanol–water partition coefficient
   Adsorption coefficient for adsorption on microbial populations ([a]?)

Rate of volatilization from surface waters and soils
   Henry's Law constant (H)
   Solubility in water (fresh water and salt water)
   Vapor pressure and/or partial pressure
   Gas and liquid phase mass transfer coefficients ([a]?)
   Diffusion coefficients in air and water
   Apparent diffusion coefficient in unsaturated soils or other systems[a]
   Molar refractivity radius

Rates of various chemical reactions
   Hydrolysis (in water)
   Photolysis (in water and, less importantly, in air)[a]
   Oxidation, e.g., with RO (in water)[a]
   Reaction with hydroxyl radical (in air)[a]

Rate of biodegradation (in water, soil, and treatment plants)[a]
   Biochemical oxygen demand (or rates of related reactions)[a]
   Rate of metabolism by aquatic life[a]

Dissociation constant

Rate constant for transfer from air to water or land[a]

Rate constant for transfer from troposphere to stratosphere[a]

"Conductivity parameter" for chemical in soil[a]

Mean lifetime of chemical in any major compartment or overall rate constant for "decay" in the compartment[b]

---

[a]Properties for which (1) estimation methods appear to be of either limited applicability or uncertain reliability at present, and/or (2) no information on any estimation method is currently available.

[b]An estimation method to determine the mean atmospheric residence time has been proposed, but its utility has not yet been assessed.

Source: Lyman, W. J., Harris, J. C., Nelken, L. H. *Research and Development of Methods for Estimating Physicochemical Properties of Organic Chemicals of Environmental Concern,* Final Report Phase 1, February, 1979, Contract No. DAMD17-78-C-8073, prepared for U. S. Army Medical Research and Development Command by A. D. Little, Inc., Cambridge, MA.

Although biological effects are emphasized when analyzing environmental risks of chemicals, nonbiological effects should be considered, e.g.:

- Aesthetic effects, such as odors, fish-flesh tainting, and decreased attractiveness of recreational areas.
- Corrosion to structures or natural formations.
- Weather modification, depletion of the stratospheric ozone layer, or other adverse atmospheric modifications.

These effects can be important and need to be considered for persistent materials with high entry potential to the compartment of concern.

The use of model balanced ecosystems (often termed microcosms) is another basic approach in predicting environmental fate and effects of a product or effluent. A microcosm is designed to behave either like a specific (geographical) natural system or like a generic model of a natural system (Ref. 1-26). Examples that have been developed are various artificial streams, aquaria, suspended bags, and so on, involving soil, water, sediment, plants, and/or animals. It is possible to create assemblages of plants and animals, which sometimes function as a food chain (Ref. 1-27), or a natural assemblage can be taken from nature and brought into the laboratory, e.g., a soil core or pond water with bottom sediment. The latter approach seems to produce more readily interpretable data, as discussed in Chapter 8.

Microcosms are tools to be kept in mind, along with study of individual fate mechanisms and of individual effects. Aside from cost and time considerations, the choice largely depends on which provides better extrapolation of results for a given environmental situation. Is it more logical to (1) use the mathematical combination of rate constants derived for the individual transport/degradation mechanisms to predict an ambient environmental concentration of pollutants against which individually measured aquatic effects are compared, or (2) use the natural integration of transport, transformations, and effects through applying the pollutant to a given confined ecosystem? Microcosms also can be used in verifying mathematical models.

1-6 PERTINENT LEGISLATION AND OTHER DRIVING FORCES

The necessity for industry and others to have vital programs in environmental risk analysis (i.e., fate/effect testing) for chemicals is a result both of legislation and internal needs. Legislative emphasis is discussed in detail in this section for the four major federal acts and to a lesser extent for five others; Daniels (Ref.

TABLE 1-8. Intoxicating Lists.

Products	Water
Drugs (FDCA)	Water quality criteria (CWA)
Pesticides (FIFRA)	Drinking water standards (SDWA)
Food additives (FDCA)	Primary
Cosmetics (FDCA)	Secondary
Toxic substances (TSCA)	Synthetic organic chemicals
Existing commercial chemicals	Coagulant aids (flocculants)
New chemicals	Underground injection control (SDWA)
Testing priorities	
Candidates for regulation	**Wastewater (CWA)**
Unreasonable risks	Pretreatment standards
Safety standards (CPSC)	Pollutants
Transportation (HMTA)	Priority
	Toxic
Workplace (OSHA)	Conventional
Criteria documents	Nonconventional
Health standards	Hazardous substances (spills)
Emergency temporary health standards	Effluent limitations
Carcinogens	
Recognized	**Solids (RCRA)**
Suspected	Solid wastes
Potential occupational	Hazardous wastes
	Processes
Air (CAA)	Chemicals
Ambient air quality standards	Sludges
Primary	
Secondary	
Hazardous emission limits	
Known pollutants	

Source: Daniels, S. L., "Environmental Evaluation and Regulatory Assessment of Industrial Chemicals," presented at 51st Annual Conference of Water Pollution Control Federation, Anaheim, CA, October 3, 1978.

1-2) has summarized these in Table 1-8. Customer inquiries, marketing questions, plant needs, and ethics also dictate such an effort. Table 1-9 summarizes the types of studies that might be conducted in response to the various driving forces.

TSCA. The more well-known aspects of the Toxic Substances Control Act (TSCA) concern human health. However, of considerable significance are provisions related to fate and effects of products in the environment:

- Section 5 of TSCA includes the chemical fate and ecological effects portion of the premanufacture notification required for new chemicals and

TABLE 1-9. Types of Environmental Fate/Effects Studies on Products and Effluents.

Product orientation

Testing of new products and expanded-use products for premanufacture notification (PMN) purposes under TSCA.

Consortium testing of commercial materials identified as requiring further testing under TSCA Section 4 (e) Priority List.

Testing of materials identified in-house as potentially having substantial risk under TSCA.

Testing of materials for which product-development or sales/marketing personnel need ecological data.

Effluent orientation

Definitive laboratory fate/effects testing of materials in discharges to watercourses for aiding NPDES negotiations and pollutant-control priority setting.

Testing of solid wastes to determine whether hazardous, or, if on hazardous list, to determine qualification for a variance.

Testing of materials that might be considered hazardous pollutants discharged to the atmosphere.

Screening of periodic effluent samples for acute fish toxicity under existing or proposed permits at various plants.

Use of balanced ecosystems (microcosms) for on-site evaluation of effluent impacts.

Downstream testing of sediments or benthos or use of periphyton growth boxes for in-stream assessments (Ref. 1-30).

chemicals that are considered to be of "significant new use". A "new chemical" is any material to be manufactured or imported that is not listed on EPA's inventory of existing products. Table 1-4 lists candidate tests initially suggested in the proposed regulations. The extent of testing to be required depends upon the volume of material to be produced, its likelihood of reaching the environment, and indications of persistence or adverse effects from key screening tests. The required tests for some materials will cost a few thousand dollars, while an extreme case like a major detergent builder, having indications of persistence, may require extensive testing that could cost over a million dollars.

• Section 4(e): the Interagency Testing Committee (ITC) must recommend to the EPA a list of commercial chemicals (or classes of chemicals) to be tested for reasons including environmental effects. The ITC can designate up to 50 of these chemicals at any one time for priority consideration.

- Section 8(e): If a firm has reason to believe that one of their materials poses a substantial risk, it must report that risk. At times, test work may be needed to establish the nontriviality of a suspected risk.

Federal Water Pollution Control Act or Clean Water Act (CWA). Permits for plant wastewater under the National Pollutant Discharge Elimination System (NPDES) often include provisions for determining the risk of aquatic toxicity. In addition, if the wastewater components not specifically covered by guidelines can be shown through testing to be nonpersistent and not highly toxic, the permitted discharge of pollutants will be higher in some plant situations as a result of cost-benefit considerations.

One can argue that, under the Clean Water Act, knowing fate and effects of pollutants will not have a major impact on permitted discharge levels: (1) for toxic substances listed in the amended law, technology-based discharge levels will be set, as will quality standards for receiving waters, (2) for conventional pollutants (e.g., BOD and suspended solids) discharge levels are set for many processes on a kg/Mg product basis, and (3) for water-quality-limited situations, allocations for permitted discharge levels will be made according to the quality of the receiving water. However, these are still several potentially important reasons for doing the fate-and-effects work:

- New findings on listed toxic compounds could influence water quality standards (Ref. 1-28).
- All manufacturing units at a plant are not covered with regard to conventional pollutants by specific discharge guidelines. Allowable levels from noncovered units are negotiable.
- For nonconventional, nontoxic pollutants (e.g., dissolved solids, ammonia, and many residual organics) any data on fate and effects should be beneficial, especially if such parameters were not in the water-quality limitations for the receiving water.

The fate and effects of materials that have a high potential for spills also should be known.

Resource Conservation and Recovery Act (RCRA). Two mechanisms are proposed under RCRA regulations to determine whether a solid waste is hazardous: a list of characteristics of hazardous wastes, and lists of particular hazardous wastes and of processes that generate them.

- Included in the characteristics is the chemical content of an extract of the solid waste. Tests for bioaccumulation and toxicity to aquatic animals (life-cycle *Daphnia*) and terrestrial plants (radish seed germination) also may be required at some time.

- Extensive fate-and-effects testing is required to gain a variance for a specific solid waste falling under the generic listings of hazardous wastes in the mutagenic, bioaccumulative, or toxic organic categories.

Clean Air Act (CAA). These aspects regarding the CAA might warrant experimental fate-and-effects testing:

- Criteria Pollutants for which ambient air standards are set, e.g., for ozone. One might show that certain "hydrocarbons" do not photochemically react (or only do so very slowly) to form ozone; consequently, these chemicals would be exempt.
- National Emission Standards for Hazardous Air Pollutants (NESHAPs). The transformation of materials being considered for listing as hazardous air pollutants could be studied; perhaps products of a lesser hazard are formed.

Other Legislation. Several other pieces of federal legislation bear upon this matter of ecological impact, although to a lesser extent than those described above:

- The National Environmental Policy Act (NEPA) (1969) requires, for new sites, predictions of the impact of proposed discharges on all aspects of the ecology.
- The Marine Protection, Research, and Sanctuaries Act concerns materials that might be lost at sea during barging and tanker operations.
- The Federal Insecticide, Fungicide, and Rodenticide Act (FIFRA) applies to the agricultural products in the manner that TSCA applies to other chemicals.
- The Hazardous Material Transportation Act (HMTA) concerns materials that might be spilled.
- The Safe Drinking Water Act (SDWA) provides for standards for inorganic and organic materials; these standards affect the quantities that can be discharged, considering transport (partitioning) and degradation.

Customer Inquiries. For the past several years users of products have required information from suppliers concerning the environmental impact of the purchased materials, especially data on biodegradability and fish toxicity and sometimes on odor level and treatability. This activity is growing and appears to be vital to many sales efforts. Data also are needed for legal purposes.

Product Development Guidance. The environmental fate and effects of materials should be an important factor in deciding which materials to produce and favor (Fig. 1-7). For example, the photoactivity of a material governs its suitability as a solvent in some areas of the United States. These factors should be considered at an early stage in the synthesis effort, not as afterthoughts. Ecological effects are taking their place beside human effects, economics, and efficacy. Products can be changed in molecular structure or dropped early in their development as a result of health and environmental research (Ref. 1-29).

Operations Control. Managers of manufacturing and distribution facilities need information on the ecological fate and effects of the materials handled. This allows decisions to be made concerning priority of pollutant reduction efforts. In addition, proper reaction to accidential releases is facilitated.

Ethics. Whether or not pressure from regulations or other external sources exists, manufacturers should know what happens to their effluents and products when they reach the environment and what effect they have. Of course, this ethical need has to be balanced against other needs when competing for limited man-power resources. However, it is unlikely that many will dispute the fact that some information gaps concerning major discharges and major product areas should be filled each year even in the absence of external pressures.

REFERENCES

1-1. Hutzinger, O., Tulp, M. Th. M., and Zitko, V., "Chemicals with Pollution Potential," *Aquatic Pollutants, Transformation and Biological Effects,* O. Hutzinger, L. H. Van Lelyveld, and B. C. J. Zveteman (Eds.), Permagon Press, New York, 1978.

1-2. Daniels, S. L. (Dow Chemical Co., Midland, MI) "Environmental Evaluation and Regulatory Assessment of Industrial Chemicals," presented at 51st Annual Conference Water Pollution Control Federation, Anaheim, CA, October 3, 1978.

1-3. Sanders, W. M., III, "Exposure Assessment, A Key Issue in Aquatic Toxicology," *Aquatic Toxicology,* L. L. Marking and R. A. Kimerle (Eds.), STP 667, Am. Soc. for Testing and Materials, Philadelphia, PA, 1979.

1-4. Kazmarek, E. A. (Georgia-Pacific Corp., Atlanta, GA), "Evaluation of Reasonable Risk," presented at Annual Conference of Water Pollution Control Federation, Anaheim, CA, October, 1978.

1-5. Comongis, "Hazard Evaluation", presented at 3rd Annual ASTM Symposium on Aquatic Toxicity, New Orleans, LA, October, 1978.

1-6. Lourance, W. W., "Of Acceptable Risk – Science and the Determination of Safety," William Kaufman, Inc., Los Altos, CA, 1976.

1-7. Wilson, R., "Analyzing the Daily Risks of Life," *Technology Review* (M. I. T.), 41–46, February, 1979.

1-8. Mill, T., "Data Needed to Predict the Environmental Fate of Organic Chemicals," presented at Symposium on Environmental Fate and Effects, American Chemical Society, Miami, Florida, September, 1978. Proceedings to be published by Ann Arbor Science.

1-9.   Maki, A., "An Analysis of Decision-Making Criteria in Environmental Hazard Evalu-
       ation Schemes," *Analyzing the Hazard Evaluation Process,* K. Dickson, A. Maki, and
       J. Cairns (Eds.), published by Water Quality Section of the American Fisheries Soci-
       ety, Washington, D.C., 1979.

1-10. "Criteria and Rationale for Decision Making in Aquatic Hazard Evaluation (3rd
       Draft) – Aquatic Hazard of Pesticides, Task Group of the Am. Inst. of Biological
       Sciences," *Estimating the Hazard of Chemical Substances to Aquatic Life,* Publica-
       tion STP 657, Am. Soc. for Testing and Materials, Philadelphia, PA, 1978.

1-11. Kimerle, R. A., Gledhill, W. E., and Levinskas, G. J., "Evnironmental Safety Assess-
       ment of the New Materials," *Estimating the Hazard of Chemical Sustances to Aquatic
       Life,* Publication STP 657, Am. Soc, for Testing and Materials, Philadelphia, PA, 1978.

1-12. Lee, C. M. (Unilever), "Determination of the Environmental Acceptability of Deter-
       gent Compounds," *Analyzing the Hazard Evaluation Process,* K. Dickson, A. Maki,
       and J. Cairns (Eds.), published by Water Quality Section of the American Fisheries
       Society, Washington, D. C., 1979.

1-13. Maki, A. W., and Duthie, J. R., "Summary of Proposed Procedures for the Evaluation
       of Aquatic Hazard," *Estimating the Hazard of Chemical Substances to Aquatic Life,*
       Publication STP 657, Am. Soc. for Testing and Materials, Philadelphia, PA, 1978.

1-14. Duthie, J. R., "Standard Practice to Evaluate the Hazard of a Substance to Aquatic
       Organisms," Task Group Draft, ASTM Sub-Committee D-35.23 on Safety to Aquatic
       Organisms, January, 1979.

1-15. Ackerman, J. W., and Coppage, D. L., "Hazard Assessment: A Regulatory Viewpoint,
       "Analyzing the Hazard Evaluation Process, K. Dickson, A. Maki, and J. Cairns (Eds.),
       published by Water Quality Section of the American Fisheries Society, Washington,
       D. C., 1979.

1-16. *Prescreening for Environmental Hazards – A System for Selection and Prioritizing
       Chemicals,* EPA/560/1-77-002 (NTIS PB 267093), April, 1977, A. D. Little, Inc.,
       20 Acorn Park, Cambridge, MA 02140.

1-17. Lyman, W. J., Harris, J. C., and Nelken, L. H., *Research and Development of Methods
       for Estimating Physicochemical Properties of Organic Chemicals of Environmental
       Concern,* Final Report Phase 1, February, 1979, Contract No. DAMD17-78-C-8073,
       prepared for U. S. Army Medical Research and Development Command by A. D.
       Little, Inc., Cambridge, MA.

1-18. Neely, B., "An Integrated Approach to Assessing the Potential Impact of Organic
       Chemicals in the Environment," *Analyzing the Hazard Evaluation Process,* K. Dickson,
       A. Maki, and J. Cairns (Eds.), published by Water Quality Section of the American
       Fisheries Society, Washington, D. C., 1979.

1-19. Branson, D. R., "Predicting the Fate of Chemicals in the Aquatic Evnironment from
       Laboratory Data", *Estimating the Hazard of Chemical Substances to Aquatic Life,*
       J. Cairns, K. Dickson, and A. Maki (Eds.), Am. Soc. for Testing and Materials, Phila-
       delphia, PA, 1978.

1-20. Smith, J. H., *et al.,* "Environmental Pathways of Selected Chemicals in Freshwater
       Systems," prepared by SRI, Int., for U. S. Environmental Protection Agency ORD,
       Athens, GA, EPA-600/7-77-113, October, 1977 (NTIS PB 274 548).

1-21. Howard, P. H., Saxena, and Sikka, H., "Determining the Fate of Chemicals", *Envi-
       ronmental Science and Technology,* 12 (4), 398–407 (1978).

1-22. Howard, P. H., Saxena, J., Durkin, P. R., and Ou, L. T., *Reveiw and Evaluation of Available Techniques for Determining Persistence and Routes of Degradation of Chemical Substances in the Environment.* EPA-560/5-75-006, U. S. Environmental Protection Agency, Washington, D. C.; W. S. Natl. Tech. Inform. Serv. PB 243 825/7WP (1975).

1-23. *Principles for Evaluating Chemicals in the Environment,* Lib. Congress Cat. Card No. 74-31482, National Academy of Sciences, Washington, D. C., 1975.

1-24. Kenaga, E. E., "Aquatic Test Organisms and Methods Useful for Assessment of Chronic Toxicity of Chemicals," *Analyzing the Hazard Evaluation Process,* K. Dickson, A. Maki, and J. Cairns (Eds.), published by Water Quality Section of the American Fisheries Society, Washington, D. C., 1979.

1-25. Verschueren, K., *Handbook of Environmental Data on Organic Chemicals,* Van Nostrand Reinhold Co., New York, 1977.

1-26. Hart, J., *et al.*, "Making Microcosms in Effective Assessment Tool," Presented at Sym. on Microcosms in Ecological Research, Sponsored by Savannah River Ecology Laboratory, Augusta, GA, November, 1978.

1-27. Metcalf, R. L. (U. Ill.), "Biological Fate and Transformation of Pollutants in Water," *Fate of Pollutants in the Air and Water Environments, Part 2,* I. H. Suffet (Ed.), John Wiley & Sons, New York, 1977.

1-28. "Water Quality Criteria," *Federal Register,* **44,** 52, 15925–15981, March 15, 1979.

1-29. *Chem. Week,* p. 76, December 13, 1978.

1-30. Cairns, J., Jr., and Dickson, K. L., "Field and Laboratory Protocols for Evaluating the Effects of Chemical Substances on Aquatic Life," *J. of Testing and Evaluation,* **6,** 2, pp. 81–94, 1978.

PART 1
PRINCIPLES OF
ENVIRONMENTAL
RISK ANALYSIS

2

Basic properties of materials

Donald Mackay
Department of Chemical Engineering
and Applied Chemistry
University of Toronto
Toronto, Ontario, Canada

2-1 INTRODUCTION

It is self-evident that the transport, transformation, accumulation, and biological effects of environmental contaminants are strongly influenced, if not controlled, by these substances' physical and chemical properties. As our understanding of the dynamics of environmental systems improves, there is emerging a capacity to predict the environmental fate of newly introduced chemicals by means of techniques such as the evaluative models originally devised by Baughman and Lassiter (Ref. 2-1) and described in Chapter 7 by Lee. Such techniques predict the likely compartments of the environment into which the contaminants will flow and accumulate, thus exposing biota and humans to toxic effects.

It is reasonable that those responsible for introducing a new chemical compound to the marketplace and, hence, into the environment, should be required to provide physical, chemical, and biological property data that can be inserted into some analytical and assessment system to predict the compound's fate. Perhaps a protocol can be developed to identify those substances that would prove to be particularly troublesome environmentally, thus making it possible to better assess whether their social benefits exceed their environmental disbenefits. At least, if the substance will have some adverse environmental effect it is prudent to have advance warning so that appropriate control measures can be exerted.

In this chapter we address the question of what properties of a new chemical compound we can reasonably expect the manufacturer to provide and how these properties can be measured or estimated. It is recognized that many of these properties will be measured for commercial application purposes but it is likely that other data will be required for environmental purposes.

Some physical properties are obviously important, or even essential, in determining environmental fate. Examples are volatility, aqueous solubility, and extent of sorption. Some, such as viscosity, surface tension, or flammability limits, may be less significant but often quite useful. In Section 2-2 we examine these physical properties and subjectively assign them priorities.

In Section 2-3, we consider the more complex set of properties that control the extent to which the molecule may undergo some chemical conversion in the environment. Included are the tendencies to ionize, oxidize, and photolyze. Unfortunately the rates of these processes depend on a number of conditions, including temperature, pH, solar insolation, and the presence of other compounds; thus these properties are less determinate. They are, however, exceedingly important because chemical conversion processes may play a key role in destroying the substance in the environment, thus alleviating its adverse effects and possibly transforming it into other compounds that may have quite different properties and toxicities.

The third set of properties, considered in Section 2-4, are the biological effects, which are even less determinate than the chemical properties. The evaluation of the toxicity of environmental contaminants is exceedingly complex because of the many factors involved, such as the large number of species present that may be vulnerable to the toxic substance, the dependence of susceptibility on life stage and temperature, and the possibility of synergism or antagonism. Toxic substances may have sublethal effects on reproductive activity, may affect behavioral patterns, such as interference with mating or predator–prey relationships, or may cause changes in habitat or to food supply. There may be a long latent period before adverse effects become apparent. It should be clear that no simple test, or set of tests, is likely to convince the environmental biologist that all possibilities have been fully examined. It does seem, however, that the best approach is to set up a number of biological tests, with the aim of screening out the most potent environmental chemicals and thus identifying future problems, rather than waiting for adverse environmental reactions. Admittedly, the tests will be imperfect but progress can only be made by trial, error, and improvement.

The general approach taken in these sections is to discuss and justify parameters of interest and outline experimental techniques by which these data can be obtained. In some cases, there are emerging techniques that permit properties to be calculated from fundamental molecular properties rather than measured, thus permitting considerable economies of effort.

Section 2-5 reviews a number of complicating factors and issues, including (i) the possible presence, in commercial products, of trace chemicals that may have more adverse environmental effects than the primary chemical, (ii) the difficulties of dealing with mixtures of toxic substances, (iii) hazards posed during manufacture, transportation, and use of these products, and (iv) the possibility that the substance may be transformed into another more toxic compound.

Finally, Section 2-6 is an inventory of basic data sources that could act as a starting point in the search for published data.

Obviously, acquisition of such property data is only a start to a larger evaluative process. The nature of the sources of the substance and its physical properties result in partitioning, that is, their concentration in selected environmental compartments. Freons are of atmospheric concern, whereas mercury is a sediment problem. Detailed further consideration must therefore be given to those environments and processes, as reviewed in Chapters 4, 5, and 6 for aquatic, atmospheric, and terrestrial environments, respectively.

It is emphasized that the set of tests suggested should not be regarded as the last word. The determination of the environmental toxicity of chemical species is still in its infancy and considerable progress will undoubtedly be made in determining protocols for prior assessment of industrial and other chemicals. The set of tests and their assigned relative priorities reflect the author's judgment at the time of writing. Tiering of tests and decisions is also discussed in Chapter 1. It is unlikely that we can ever be totally confident that the set of tests will cover all eventualities. Perhaps the best test that screening mechanisms can be subjected to is to apply them in hindsight to chemicals that have proved to be most troublesome. Obvious examples are DDT, mirex, polychlorinated biphenyls, mercury, lead, and arsenic. If the protocol would have given advance warning of these substances, then we can have some confidence in the probability of its exposing new contaminants. Undoubtedly, however, the environment in its complexity has surprises in store for us, transporting or transforming contaminants in unexpected ways, leading to situations in which there may result adverse effects to ecological systems or human health. It is in the conviction that our present knowledge is inadequate that much of the fascination and incentive for environmental science lies.

2-2 PHYSICAL PROPERTIES

2-2.1 Pure Component Phase Equilibrium Data

These data quantify the relationships between state (solid, liquid, or vapor) of the compound, pressure, and temperature. They include melting point, normal

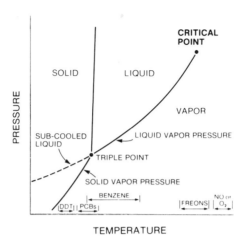

TEMPERATURE

Fig. 2-1. Typical pressure, temperature, state diagram for a pure substance showing the location of the environmental temperature range for selected compounds.

boiling point, triple point, critical point, and vapor pressures. These data characterize the pressure-temperature–state diagram illustrated in Figure 2-1. Obviously for environmental purposes we are concerned only with that part of the diagram that lies in the range of environmental temperatures. The location of this temperature range relative to the phase boundary lines is given for selected substances.

The melting and triple temperatures are usually very close and can be assumed to be equal. The temperature and pressure at the critical point can be very useful in correlating physical properties of the substance with those of similar substances, by means of the theory of corresponding states. In essence, this theory is based on the observation that similar compounds have similar properties when conditions are defined as a ratio with the critical properties. There are a number of empirical correlations that can provide estimates of physical properties if critical conditions are known. Critical information is relevant only if the substance is gaseous or very volatile.

For liquids it is essential to be able to deduce (from the melting point) if there is a possibility of solidification when the material is exposed to low environmental temperatures. This is obviously relevant to materials handling practices but the effect may be more subtle since, as is discussed later under solubility, there is an increased risk of precipitation from solution at temperatures below the melting point.

Volatility, as characterized by vapor pressure, is obviously important in controlling the extent to which the substance may become volatile. Volatilization can be surprisingly fast even for solid substances that are normally regarded as involatile. It is now recognized, for example, that DDT has been globally transported by volatilization. For most liquids, vapor pressure can be adequately characterized by measuring boiling point at several pressures and may be conveniently correlated by the Antoine Equation in either its 2- or 3-constant form, in which P is vapor pressure, T is absolute temperature, and A, B, and C are constants.

$$\log_e P = A - B/T \quad \text{or} \quad A - B/(T + C)$$

This equation incidentally provides an estimate of the enthalpy of vaporization ΔH since it can be shown that, when $(T + C)$ is absolute temperature, B is equivalent to $\Delta H/R$, where R is the gas constant, in appropriate units. The equation can be used for extrapolation purposes, provided, of course, that one recognizes the attendant hazards of any such procedure, especially if some phase transition occurs.

Measurement of boiling point or vapor pressure of fairly volatile compounds presents little difficulty. Methods involve the use of isoteniscopes or precision pressure gauges such as helical quartz gauges. The principal difficulties arise when the material has a high boiling point or low volatility. Examples are the series of polychlorinated biphenyls or organo-halide pesticides for which data may be available only at elevated temperatures. Extrapolation to environmental temperatures may introduce serious error, and it is preferable to measure the vapor pressure directly.

Hamaker and Kerlinge (Ref. 2-2) have reviewed methods of determining the vapor pressure of pesticides and concluded that such static methods as the use of gauges or the measurement of boiling point under reduced pressure are suitable when the vapor pressure exceeds 0.1 mm Hg. As a method of measuring boiling point, they describe the use of a differential thermal analyzer operating at reduced pressure. For lower pressures a dynamic method is preferred. Examples are the saturation of a gas stream with vapor and subsequent analysis of the vapor (Spencer and Cliath, Ref. 2-3), evaporation from dishes, or the more sophisticated Knudsen effusion or jet effusion techniques. A simple approach is to measure evaporative mass loss from dishes at controlled temperature, air turbulence, and surface area. If two liquids are evaporated from dishes under conditions of equal area and temperature and with identical air flows sufficiently vigorous to remove the vapors without, however, inducing excessive rippling or splashing, then the ratio of the loss rates in moles per unit time will approximately equal the ratio of the vapor pressures. A slight discrepancy is attributable to differences in molecular diffusivity in the gas phase, but this is usually small under turbulent conditions when molecules of nearly equal molecular weight are volatilizing. The method can also be used for solids, in which case smooth cast solid disks are the preferred configuration. Probably the most convenient reference liquids are the normal alkanes for which extensive data are available

(Ref. 2-4). Naphthalene is a convenient solid. Hartley (Ref. 2-5) has described such a system.

It is also possible to obtain an approximate indication of volatility by comparing retention time of the substance with those for reference substances on a non-polar gas chromatographic column.

In certain situations where a substance is volatilizing from water at a temperature below its melting point it is useful to have both the solid and subcooled liquid vapor pressure estimates, the latter being obtained by extrapolation of the liquid vapor pressure through the triple point.

Particularly difficult is a situation (such as the case with PCBs) in which the product is a complex mixture of isomers or related compounds, each with its individual properties. Although the vapor pressure of the mixture is probably acceptable for cursory evaluation purposes, for detailed fate analyses individual properties may be necessary.

Finally, for many substances it should be noted that critical or high-temperature conditions (and hence properties) are inaccessible because of the onset of decomposition or oxidation. An indication of the temperature (and other conditions) at which this occurs can be very useful in elucidating environmental fate in flue gases or during chemical analyses by gas chromatography at elevated temperatures.

2-2.2 Aqueous Solubility

Aqueous solubility is important not only in that it provides an upper limit to the extent of incorporation of the substance in the aquatic environment, but it also indicates hydrophobicity and thus, at least for organics, the potential for transfer into lipid phases of aquatic organisms. To state, as older handbooks often did, that a substance is "insoluble" or "sparingly soluble" is no longer acceptable for environmental assessment. All substances containing discrete molecules of reasonably low molecular weight demonstrate some aqueous solubility. The theory underlying aqueous solubility of nonelectrolytes has been reviewed by Prausnitz (Ref. 2-6); the same for the electrolytes has been performed by Butler (Ref. 2-7).

Established methods used to measure solubility normally involve contacting water and an excess of the substance isothermally for a time sufficient to achieve equilibrium. The aqueous phase is usually analyzed by a convenient conventional technique determined by the characteristics of the substance. For low solubility compounds (especially solids) it is essential to demonstrate that adequate equilibration time has been allowed, as such compounds tend to dissolve very slowly. Dissolving the substance in a volatile solvent followed by evaporative "plating" on the walls of the flask is often a preferred technique for exposing a large area to water during equilibration. An attractive method of preparing saturated solutions of slightly soluble materials is to flow water through a long column containing glass beads coated with the materials (May and Wasik, Ref. 2-8). Saturation can be confirmed by varying column length or water flowrate.

If the compound is even slightly volatile, extreme care is necessary in order to avoid evaporative loss from solution. Many organic compounds tend to form micelles of particle size about 1 μm at concentrations of approximately 1 mg/liter; this may obscure the true (lower) solubility. Filtration, centrifuging, or, preferably, avoidance of formation, is necessary. The range of environmental interest is 0 to 30°C; some indication of sensitivity of solubility to temperature is useful.

When measuring aqueous solubilities of sparingly soluble compounds, it is essential to ensure that the material is not contaminated with a small amount of a more soluble compound. This can easily result in spurious conclusions.

Environmental waters differ considerably from the distilled water used in physical chemical determinations. The most important difference is the presence of electrolytes, which occur in 35 parts per thousand in sea water. A 3.5% by weight sodium chloride solution is usually adequate to simulate sea water, although "standard sea waters" are available. Most organic compounds are "salted out" (have a lower solubility) to the extent of 20 to 30% in sea water. Rarely does the electrolyte content of inland waters significantly affect the aqueous solubility, although in some wastewaters solubility may be affected.

Solubilities may also be affected by the presence of other dissolved species, notably common ions (including H^+ and OH^-), organics, surfactants, and by suspended natural or anthropogenic organic or mineral matter. Some of these effects are discussed later under sorption. Rather than generalize it is probably more economical to suggest that if there is reason to believe that the substance is likely to be present in an aquatic environment containing an unusual assembly of other compounds, then the sensitivity of solubility to the presence of these compounds should be explored.

It would be very useful if techniques could be devised to predict the aqueous solubility of organics from readily accessible properties, such as molecular structure, molar volume and melting points. Such techniques would reduce the need for experimentation, would provide a check on the "reasonableness" of data and thus expose "bad" data. There are indications that this capacity is emerging, at least for homologous series of organics such as aromatic hydrocarbons (Ref. 2-9).

2-2.3 Octanol–Water Partition Coefficient

The work of Hansch and Leo has shown that the octanol–water partition coefficient K_{ow} is a very useful predictive tool in many situations where a substance's behavior is influenced by its partition into a lipid phase (Ref. 2-10). There are correlations available for predicting K_{ow} from molecular structure, as are extensive tabulations of data. Recent work has shown that K_{ow} correlates with bioaccumulation and other properties (Refs. 2-11, 2-12, 2-13). The rationale for this lies in the supposition that octanol is similar in polarity–hydrophobicity characteristics to the glyceryl esters that comprise the principal components of cell membranes. Presumably compounds that preferentially partition out of water into these membranes (and thus have high K_{ow} values) will exert greater disruptive influence on membrane processes or will accumulate more when the cell is exposed to a contaminated aqueous environment.

Karickhoff et al. (Ref. 2-14) have shown that K_{ow} is a good predictor of sorption, which provides an additional incentive for its measurement.

Experimentally, Leo's technique (Ref. 2-10) is normally employed to determine K_{ow}: the solute is dissolved in purified octanol and a volume of this solution contacted with various volumes of water. After equilibration and centrifuging, the two phases are sampled and analyzed. Care must be taken to avoid evaporative loss of solute. More recently, Chiou et al. (Ref. 2-15) presented a useful discussion of techniques and relationship to bioaccumulation.

An interesting recent development is the use of high-pressure liquid chromatography (HPLC) to determine K_{ow}. This is analogous to determining the activity coefficients of solutes in solvent liquids by measuring the retention time of the solute on a gas chromatography column with the solvent as liquid phase. The HPLC approach may be useful for measuring K_{ow} for components of mixtures (for example in wastewaters). Descriptions of the technique have been given by McCall (Ref. 2-16), Carlson (Ref. 2-17), and Mirrlees (Ref. 2-18).

2-2.4 Henry's Law Constant

The Henry's Law Constant or air–water partition coefficient H is an indication of the likely partitioning behavior of the solute between the atmosphere and water bodies. Essentially, it is the ratio of solute vapor pressure to aqueous solubility and may be expressed in units such as atm/(g mol/m^3), i.e., atm·m^3/g mol or as a concentration ratio in (g/liter)/(g/liter) or even as a Bunsen coefficient. Ionic substances such as NaCl are usually considered to have a negligible vapor pressure and, H, unless a nonionic, volatile form is in equilibrium, as in the case of SO_2 or NH_3.

When reliable solubility S (mol/m^3) and vapor pressure P (atm) data are available at the same temperature, H (atm·m^3/g mol) can be calculated directly from

$$P = HS$$

The value of H is of course temperature dependent, generally increasing as temperature increases. It may also be concentration dependent, although for non-associating solutes of low solubility the effect is likely to be slight. Substances that may associate in both water and air phases, such as carboxylic acids, may exhibit unusual concentration effects, exemplified by variable H values. Any factor that modifies solubility, such as pH, will modify H correspondingly.

A degree of caution is required in calculating H values for mixtures and for substances close to their melting point. The solubility and vapor pressure must be of the same material in the same physical (liquid or solid) state. For example, published PCB vapor pressure data are for liquid mixtures in which many of the components are in a subcooled but stable liquid form below their individual melting points. The published solubility data are for individual isomers, many of which are solid. From a physical–chemical viewpoint the only reliable method of obtaining H is to determine the ratio of pure component same-state vapor pressures and solubilities. For a mixture it is conceivable that the volatile components controlling the vapor pressure may not be the most soluble components; thus H may have spurious or unrealistic values for mixtures. The importance of reliable thermodynamically meaningful data can not be overemphasized.

As the molecular size of organic compounds increases, vapor pressure tends to decrease (often to unmeasurable levels) and solubility again decreases to unmeasurable levels. The result is that, in calculating H for substances such as polynuclear aromatics, the ratio is between two very small and unreliable numbers. It is then clearly preferable to measure H directly. In theory, this can be accomplished by measuring an aqueous concentration and an air phase concentration in equilibrium, i.e., essentially performing a head space analysis. In practice, this is experimentally very difficult and a preferable approach is to measure a volatilization rate from an aqueous solution under controlled conditions. Such a system is illustrated in Figure 2-2. A mass balance on the volatilizing solute shows that

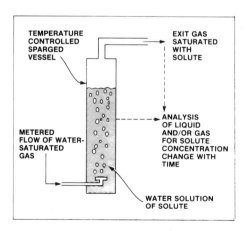

Fig. 2-2. Schematic diagram of apparatus for determining Henry's law constant of a solute between air and water.

for a gas rate G (air plus water vapor) of m^3/sec into a water volume $V (m^3)$ in which the solute concentration C is mol/m^3 and C_0 at time $t = 0$ (sec)

$$V \, dC/dt = -GHC/RT$$

thus

$$\ln(C/C_0) = -GHt/RTV$$

where H has units of $atm \cdot m^3/mol$, R is the gas constant (82×10^{-6} $atm \cdot m^3/mol$ K), and T is absolute temperature (K). This assumes that the exit gas is in equilibrium with the liquid, which is valid (i.e., >99%) provided that the gas is sparged into small bubbles through a sintered disk and the water depth exceeds 30 cm. The slope of the graph of $\ln(C/C_0)$ thus gives H directly and is a useful indicator of the likely evaporative behavior of the substance (Refs. 2-19, 2-20).

It should be noted that organic substances of very low vapor pressure may partition appreciably from aqueous solution into the atmosphere since their hydrophobic nature results in low values of S as well as H. In addition, only very little of a slightly soluble substance need evaporate to alter substantially the aqueous concentration. This is now well recognized for substances such as DDT (Ref. 2-21).

In some cases with new compounds there may be doubt as to the relevance of air–water partition measurements, since the compound is very involatile. A simple test is to measure its volatility relative to water from aqueous solution. A dilute subsaturated aqueous solution is prepared and batch distilled until about half the water has evaporated. The solute concentration is then measured and compared with the original value. If the concentration decreases, volatility is likely to be important. If it remains constant or increases then volatilization will not be significant in depleting the solute from water bodies; however, the substance may, of course, still have adverse effects in the atmosphere.

2-2.5 Sorption

Sorptive properties are of critical importance in determining environmental fate and especially in determining which environmental "compartments" the substance may accumulate. The ratio of the sorbed and "solution" forms of the substance is usually expressed as a sorption constant K_p in units of, for example, (μg solute/g sorbant)/(μg solute/g solution), or ppm/ppm. A high value of K_p implies preferential sorption. In most cases K_p is concentration dependent and several sorption isotherm equations, such as the Langmuir or Freundlich equations, are used to characterize this dependence. At very low solute solution concentrations, i.e., well below the solute solubility, K_p is often independent of concentration and a linear isotherm applies.

For environmental assessment purposes a knowledge of K_p is invaluable. For example, substances of high K_p value may preferentially sorb on particulates in the atmosphere or on suspended solids in the water column on in bottom sediments in lakes or they may remain in soils. Essentially the role of sorption is to reduce chemical potential or fugacity and, hence, the tendency to diffuse away by "trapping" high concentrations of the solute in an non- or less-available form.

Unfortunately there are no "standard" sorbants, although there are some likely candidates. Soils or sediments characterized for particle size distribution, organic carbon content, and mineral nature, and for which simultaneous measurement is made for partition properties of known related "reference" solutes, should form useful sorbants. Recent work by Karickhoff, Brown, and Scott (Ref. 2-14) has shown that sorption of organic hydrophobic compounds on a wide range of sediments correlates well with organic carbon content and also with the solute's octanol–water partition coefficient; thus the capability of making generalizations about sorptive behavior is emerging.

As with sediments, it appears that the organic carbon content of the soil is responsible for most of the sorptive activity and that sorption coefficients calculated on an organic carbon basis, are remarkably constant for one solute. Kenega and Goring (Ref. 2-22) have reviewed work in this area recently.

At this stage possibly the best approach is to measure sorption constants (K_p) for the substance for soils and sediments thought to be relevant to the nature of the solute and its mode of entry into, and transport through, the environment and simultaneously report the sorptive behavior of "reference" solutes of related structure about which more environmental data are available. Obviously if sorption is judged to be insignificant on the basis of such tests, extensive data are not required; however, if sorptive behavior in soils, for example, is deemed to be critically important more extensive data may be required.

Experimentally, the procedure first involves characterization of the sorbant, followed by measurement of partition. Characterization procedures are described in soil science texts and should include pH, particle size information, organic carbon content, and cation exchange capacity. Measurement of partition is normally done by contacting solute, solution and sorbant for about 24 hr, followed by separation by centrifuging and analysis of each phase. It is important to ensure that the solute is fully extracted from the sorbant. Difficulties are encountered with volatile and highly insoluble solutes, which require modifications in technique.

2-2.6 Other Properties

A number of other properties may be useful in facilitating the assessment of environmental dynamics and hazards. Among these is viscosity, which may control the spreading and penetration characteristics of accidental spills. Surface tension can also influence spill behavior. Flash point, fire point, and flammability or explosive limits (if any) may be important for volatile or ignitable liquids. For purposes of chemical analysis some data on solubility or miscibility with common solvents can assist in the selection of suitable extractants. A general description of the compound (colorless free-flowing liquid or green crystalline powder) may also be useful.

Such data, if available, assist in establishing a general profile of properties, but there is probably little justification for requiring data acquisition solely for environmental purposes.

2-2.7 Relationships between Physical Properties

It is not always appreciated that vapor pressure, aqueous solubility, octanol–water partition coefficient, and Henry's Law Constant are related thermodynamically. Indeed, they are experimental manifestations of three more fundamental thermodynamic properties: solute fugacity f, aqueous phase activity coefficient γ_a, and octanol phase activity coefficient γ_o. Briefly, the relationships are

Vapor pressure = f
Aqueous solubility (in mole fraction) = $1/\gamma_a$ for liquid solutes
$$\text{or} = f_s/\gamma_a f_{sc} \text{ for solid solutes}$$
Octanol water partition coefficient
  expressed as mole fraction ratio = γ_a/γ_o or
  expressed as molarity ratio = $\gamma_a v_a/\gamma_o v_o$
Henry's Law Constant $H = f_{sc}\gamma_a v_a$

where f_s is the solid fugacity (or vapor pressure) (in atm), f_{sc} is the subcooled liquid fugacity (or vapor pressure) (in atm), and v_a and v_o are the molar volumes of water and octanol (in m^3/g mol). Justification for these equations can be found in texts such as that by Prausnitz (Ref. 2-6).

These relationships can be useful in three respects. They can be used to avoid experimental measurement; for example, if solubility and Henry's Law Constant are known, the vapor pressure can be calculated. They provide a check on the reasonableness of the data; for example, two structurally similar liquids with solubilities of 50 and 25 μg/liter should have octanol–water partition coefficients that differ by the same ratio, (e.g., 40,000 and 80,000) because it is likely that they have similar γ_o values. Finally, there are emerging techniques for correlating and predicting γ_a from fundamental molecular properties. If successful, these could result in considerable economies of effort. To date they have been applied only to homologous series but the potential exists for wider application.

The topic of the relationship between these properties and the broader relationships to soil sorption and bioaccumulation has been reviewed recently by Kenega and Goring (Ref. 2-22).

2-2.8 Summary

In summary, the minimum desirable physical property information is a statement of the following type, backed up with a description of the techniques used and the likely error:

Compound—Benzene, molecular weight 78.
Melting Point 5.5°C
Normal Boiling Point 80.1°C
Vapor Pressures 40 mm Hg at 7.6°C
100 mm Hg at 26.1°C

(If volatilization proves to be an important pathway more detailed information may be justified.)

Aqueous solubility 1780 mg/liter at 25°C

(Solubility data at other temperatures may be useful in some situations.)

Octanol–Water Partition Coefficient at 25°C
135 $(mol/m^3) / (mol/m^3)$

Sorption Partition Coefficient K_p measured on a Black River sediment (pH 6.8, 11.1% organic carbon, cation exchange capacity 7.0 meq/100 g, particle size distribution 10% <50 mesh, 20% >10 mesh) at aqueous phase concentration of 890 mg/liter (50% of saturation) and sorbant concentration of 1.0 g/liter was 5.3 (μg/ml solution). Under similar conditions toluene was found to have a K_p of 12.4

(For situations in which sorption is significant several K_p determinations should be made and a linear or other isotherm equation obtained. It may also be desirable to test several sediments.)

Note on Accuracy and Precision. Where several determinations are made it is important that the results be reported directly and completely, to convey an impression of precision, rather than presented as means, even with standard deviations. The simultaneous determination and reporting of properties of compounds with similar but well-established properties is highly recommended as a means of indicating accuracy.

2-3 CHEMICAL REACTION PROPERTIES

2-3.1 Aqueous Ionization

The environmental behavior of such compounds as the carboxylic or phenolic acids may be substantially modified by ionization in aqueous solution and salt formation. Dissociation data at environmental temperatures are therefore essential and can be obtained by standard physical–chemical techniques. Basic organic compounds such as amines may display ionic behavior in the presence of acids. Such properties as well as unusual chelating capacity should be documented.

2-3.2 Hydrolysis

A substance, depending on its chemical nature, may be susceptible to hydrolysis in acid, alkaline, or neutral conditions. One approach is to follow the procedures established for pesticides and undertake hydrolysis experiments at three pH levels (3.0, 6.0, and 9.0) and three temperatures (25, 35, and 45°C). Judgment should be used to select additional conditions, such as lower temperatures or lower or higher pHs, if these data provide confirmation or elucidation of trends or tendencies detected under "standard" conditions.

   The usual experimental procedure is to prepare a dilute solution of the solute in water with a noncatalytic buffering agent (usually phosphates and borates) and monitor the solute concentration versus time under dark conditions at constant temperature. A range of pH and temperature is studied. The results are normally expressed as first-order rate constants. An example of such a study is that on Captan by Wolfe et al. (Ref. 2-23).

   For organic compounds Mabey and Mill (Ref. 2-24) have provided a critical review of hydrolysis data that can be used to check the experimental data with those of similar compounds. Correlations are also possible between molecular structure and hydrolytic rate constants.

   If hydrolysis is a significant process it is desirable to identify the products, as they, too, may have environmental effects.

2-3.3 Photolysis

It is only recently that the full significance of environmental photolytic reactions has been fully recognized. A complete study of the photodecomposition of a substance is usually impossible and unnecessary, but some indication of the probability of photolysis being an important transformation process is useful, especially if other transformation processes are slow. Comparison of the energy necessary for bond breakage, the energy available in photons of various wavelength and the incident solar spectrum shows that only the near ultraviolet, i.e., <290 nm, is likely to induce substantial photolytic reactions. Obviously a "simulated sunlight" lamp is preferable to the more artificial low-pressure mercury lamp. Choice of reaction conditions and media is complicated by the fact that the medium (e.g., water) may absorb radiation and, in the natural environment, sensitizers or quenchers may be present.

For aquatic studies a solution of the substance in distilled or natural water is placed in a quartz or Pyrex cell and exposed to radiation of known intensity, usually from a medium pressure mercury lamp. A calibration using standard actinometric techniques is essential. Alternatively, the cells can be exposed to natural sunlight for several days. Temperature and pH measurement or control is necessary. An estimate can be made of photolytic rates by analysis of the contents of the exposed cells and control cells that have been kept under identical dark conditions.

If photolysis is determined to be significant, it may be desirable to undertake more detailed studies varying pH, temperature, wavelength, presence of potential sensitizers or quenchers and to obtain data on the reaction products.

For atmospheric systems a similar approach is used; here the solute vapor is contained in a chamber, irradiated, and the concentration change monitored. Again a dark "blank" is essential. Dilling (Ref. 2-25) has described such a study.

It may also be desirable to test the photolytic decomposition of the substance in a soil medium, especially if the substance is to be applied to soils.

Translation of laboratory photolytic data to environmental conditions is very complex because of time, latitude, and water depth variation of exposure; its discussion is beyond our scope here. The best approach may be to devise laboratory conditions in which environmentally realistic exposures are obtained, with possibly some accelerated or high exposure testing for substances found to be only slowly subject to photolytic decay. An example of such a study is that of Zepp et al. (Ref. 2-26) for 2,4-D esters in surface waters. A review of enivronmental photochemistry has been published by Wolfe et al. (Ref. 2-27). Smith et al. (Ref. 2-28) have provided a useful brief review of photochemical and other environmentally significant transformation processes.

2-3.4 Chemical Oxidation

Many organic compounds are susceptible to free radical oxidation, in which a radical such as $RO\cdot$ or $RO_2\cdot$, produced from some other reaction (possibly photolytic in origin) extracts hydrogen, forming ROH or $ROOH$ and leaving the compound in a radical form that can then undergo subsequent reaction. It is significant not only that the molecule has changed in structure but that it will often now have very different properties, notably solubility and reactivity, both chemical and biological. Free radicals may be generated in the laboratory by decomposition of azo compounds at predictable concentrations. Comparison can then be made between observed laboratory rates and likely rates in the environment, provided that there are available estimates of prevailing radical concentrations in the environment.

A study of this type on a series of organic solutes has been described by Smith *et al.* (Ref. 2-28).

2-3.5 Summary

It will be apparent that characterization of chemical reactivity is considerably more complex and less precise than that of physical properties. Perhaps there will emerge a series of standard laboratory tests that can characterize dissociation, hydrolysis, photolysis and oxidation, such that estimates can be made of the expected environmental "half lives" or rate constants. Probably the best available model at present is the SRI study (Ref. 2-28), in which 11 chemicals were subjected to such tests, the objective being to identify and quantify the dominant processes.

The following is a typical chemical reaction profile.

Compound: An alkyl phenol
Dissociation Constant at $25°C$ $K_a = 6 \times 10^{-4}$
Hydrolysis ([a]) $K_H = 0.01$, half life = 70 hr.
Photolysis ([a]) $K = 0.1$ hr^{-1}, half life = 6.9 hr.
Oxidation ([a]) $K < 0.01$ hr^{-1}, half life > 100 hr.
([a]) Then follows a statement of reaction conditions including temperature, pH, radiation intensity, radical system used, etc.

In this case since photolysis is the fastest process it may be desirable to amplify the data, although a compound with such a low persistence is likely to be of little concern unless it is highly toxic.

2-4 BIOLOGICAL PROPERTIES

2-4.1 Introduction

There are two distinct biological concerns — the effect of the prevailing biotic community on the substance in determining that substance's fate, and the effect of the substance on the exposed biotic community. This community normally contains a vast number of species with varying and poorly understood interdependencies and includes man as the ultimate predator and possibly the ultimate victim. It is completely beyond our scope to provide a meaningful discussion of biodegradation processes, bioaccumulation, food web biomagnification, acute and sublethal toxicology to various species, and the complex ecosystem reverberations that occur naturally and may be induced by chemical contamination;

indeed it will obviously prove a fruitful and possibly everlasting study area for generations of environmental biologists.

In this section we attempt to suggest a series of test procedures that yield quantitative data that may relate to the environmental behavior of a new substance. The tests are far from satisfactory — they are incomplete and may even be misleading — but they represent, in the author's subjective judgment, a reasonable set from which improvements can be made. It will be many years before there can be devised a test protocol that can give a reasonable certainty of avoiding false negative or false positive results. The emphasis here is on the test principles rather than on the detailed laboratory procedures. Recent publications by ASTM on aquatic toxicology and hazard evaluation contain compilations of papers reviewing various aspects of this issue (Refs. 2-29, 2-30). Notable in the latter reference is the comprehensive review by Stern and Walker (Ref. 2-31). Draft and ultimately final test protocols are being produced by many government agencies, including the U.S. Environmental Protection Agency (Ref. 2-32).

2-4.2 Biodegradation

The objective of a biodegradation test is to bring the substance and a population of viable microorganisms into contact under conditions such that the conversion of the substance by biological (enzymatic) processes can be detected and measured. A number of schemes are possible, varying from the simple 5-day biochemical oxygen demand (BOD) test to the Warburg test to the more sophisticated continuous chemostat systems. The principal problems are the selection of a source of microorganisms and the determination of the contacting proportions, conditions, and time.

A favorite source of microorganisms is an activated sludge from a typical sewage treatment plant, preferably one receiving a variety of industrial effluents. If, for example, the substance is a hydrocarbon, it is desirable to select a sludge from an oil refinery in which the microbial population has become acclimatized to such compounds and in which hydrocarbon degrading organisms have flourished.

The simplest contacting arrangement is the standard 5-day BOD test, which has considerable merit because of its simplicity and wide use. A small inoculum of sludge is brought into contact with the substance in solution in aerated water and the depression in dissolved oxygen level is measured after sealed exposure for 5 days. A blank and control test are also done. For substances that biodegrade appreciably this procedure is excellent and sufficient. Analysis of the contents can confirm that the substance has been degraded.

For sparingly soluble or slowly degradable substances it is unlikely that the BOD test will be conclusive. Most of the substance may merely sorb into the microorganisms. If the BOD test proves inconclusive, more favorable conditions must be sought.

One approach is to arrange for an aqueous solution of the solute to be kept aerated, either by gentle shaking in a flask, or by bubbling air through the vessel, in the presence of inocula and having provided a basal salts medium and a substrate on which the microorganisms can grow (e.g., glucose and peptone). The system can be operated in batch fashion but is probably more realistic if operated continuously with a constant feed and overflow of settled supernatant liquid. Essentially this becomes a model wastewater treatment plant. A check should be made to ensure that volatilization losses are either kept to a minimum or are quantified. The system can be run for several days before the contents are subjected to analysis to establish a mass balance. It is important to ensure that extraction of the solute from the microorganisms is complete.

The general procedure is to subject the substance to an increasingly favorable biodegradation environment until either degradation is observed or it fails entirely. If degradation is observed, documentation of the degrading conditions (time, cell counts, temperature, etc.) could allow the statement of kinetic expressions that can be used as a basis for extrapolation to environmental conditions. If no degradation is observed, the most favorable degradation conditions should be reported as being ineffective. The result of such a test is an estimate of the "recalcitrance" of the substance to biodegradation. Clearly the most recalcitrant substances are of greatest concern, especially if chemical degradation routes are inoperative.

If biodegration is significant, an investigation of breakdown products may be justfied.

In this discussion we have ignored degradation under anaerobic conditions and by fungi and other organisms. The rather unsatisfactory justification for this omission is that if aerobic bacterial degradation is operational the others probably need not be considered; if it is not operational the others are also unlikely to be effective and it would probably be unwise to depend on them for appreciable environmental degradation.

In summary, the aim is to state the conditions under which a substance is observed to undergo a stated biodegradation. By comparing the test conditions with those in typical environments an environmental half-life may be inferred. It should be emphasized that this half-life may be in error by an order of magnitude but even this poor accuracy is preferable to no estimate or to such qualitative statements as "slowly biodegradable."

2-4.3 Bioaccumulation

If a substance is absorbed by an organism in food, during respiration, or by physical contact, and it has chemical properties that favor its being retained in the organism (usually in a lipid phase), it will tend to accumulate in increasing concentration. Small organisms may quickly reach a constant level but larger organisms such as fish may grow over a period of years and build up increasing concentrations of the substance. Clearly the most vulnerable species are those at higher trophic levels which ingest food consisting of biota that have already accumulated high concentrations. Fish-eating birds and humans with certain diets are most vulnerable in this regard.

The most realistic and practical test for bioaccumulation is probably one in which fish are exposed to the substance for a period of time (several weeks) under defined conditions well below the toxicity level. They are then analyzed to determine the retention of the substance. One well-documented approach is described by Neely (Ref. 2-33): selected fish were exposed to a known aqueous concentration of the substance and the uptake, and indeed an entire substance mass balance was measured. Comparison of the uptake data with those for other compounds provides an indication of the significance of this process.

Writing a simple equation to describe intake (I) and depuration (release) (D) rates of a toxicant by a fish of volume V resulting in a fish concentration change dC in time dt gives (in appropriate consistent units)

$$V\,(dC/dt) = I - D$$

The accumulation behavior thus depends on the relationship between intake and depuration rates — fairly high intake rates can be tolerated if they are associated with high depuration rates. It is, therefore, very useful to arrive at an estimate of the relative magnitudes of I and D by conducting separate experiments. For example, after the initial rate of increase of C with time, information about I may be obtained by subjecting a fish to a change from clean to contaminated environment, whereas placing the contaminated fish in clean water and measuring the drop in C gives information about D. These rates can then be combined to give an estimate of the potential for accumulation. Short-term tests based on this principle would be invaluable.

Perhaps the problem of assessing bioaccumulation will best be solved by establishing good correlations between bioaccumulation and physical-chemical parameters such as the octanol–water partition coefficient, taking into account the biodegradability of the substance. The review by Branson (Ref. 2-34) suggests that this may be feasible.

The test results should be in the form of a description of conditions (type, number and weight of fish, temperature, volume, concentration, time, etc.), statement of the analytical results, and a mass balance. Simultaneous testing with a standard substance may be desirable.

2-4.4 Toxicity Testing or Bioassays

The obvious approach is to select one or more species, expose them to the substance for a selected time and concentration, and measure the toxic effect of the substance either as death or as some more subtle physiological effect. This apparently simple task is rendered almost impossible by the complexities arising from the selecting the species, a specific life stage of the species, the concentration, time, and nature of the exposure, and the nature of measurable effects. Short-term, high-concentration (acute) exposures may give quite different results from long-term, low-concentration (chronic) exposures. The toxic effect may not be to kill the organism, but rather to prevent it from reproducing or eating or avoiding predators, all of which are indirect death sentences. Of particular human concern is carcinogenicity (induction of cancerous growths), mutagenicity (induction of a chromosomal change), or teratogenicity (induction of a defect in embryos). These latter effects are subtle and may take a decade or more to develop in humans. They have the common feature of being associated with genetic changes and there is therefore some hope that they can be identified by a common test.

In order of increasing complexity and cost a series of such tests might be as follows (details of such tests are given in Chapter 4-6):

Microorganisms. The standard Warburg analysis in which the respiration rate of sewage sludge is measured can be used to test toxicity to microorganisms by adding various dosages and determining the effect on respiration. An alternative is to measure respiration behavior in a soil microcosm or to measure the effect of the toxicant on the performance of a specific microorganism that is normally capable of a particular biochemical transformation that can be readily analyzed. Examples given in Ref. 2-31 include nitrite oxidation, sulfate reduction, and cellulose oxidation. In all cases, the objective is to measure how much toxicant is necessary to significantly deactivate the microorganism, the implication being that the most highly toxic compounds produce an effect at lower concentrations.

Primary Producers. A photosynthetic toxicity can be measured by subjecting algae to the substance and measuring carbon fixation by ^{14}C uptake. Alternatively, a simple determination of increase in dry weight can be used as a measurement of growth inhibition.

Fish and Aquatic Invertebrates. Because of its availability, *Daphnia* is a commonly used organism for invertebrate toxicity testing. A static 48 to 96 hr acute toxicity test is normally used, although continuous flow configurations are possible. For marine testing, shrimp is a suitable organism.

A standard procedure has been formulated in which rainbow trout are subjected to a 96 hr static test and a LC_{50} (lethal concentration to 50% of the population) is determined. A continuous flow version may also be employed; this is less subject to errors arising from evaporation or sorption of the toxicant. Unfortunately, the more meaningful results are costly. Other species such as fathead minnows may be used. It should be emphasised that a chemical can be harmful without killing the test organism. It may prevent reproduction, alter behavior, or have a subtle effect at a particularly sensitive life stage. Sublethal or subacute tests to elucidate such effects are neccessarily more complex, lengthy, and costly. Very subtle changes in feeding or growth rates or reduced responses to stimulants or predators can be detected at remarkably low concentrations. Such tests are certainly of considerable biological and ecological interest, but it remains to be seen whether or not they can be used for screening or testing toxic chemicals. Ultimately, a subjective judgment is necessary in deciding if a given toxic effect is significantly harmful.

Details of these tests are given in standard test descriptions, such as Ref. 2-35 and in various U.S. Environmental Protection Agency reports. The review by Stern and Walker (Ref. 2-31) lists many such reports. Undoubtedly, standard test methods will continue to be developed for some years.

Plants. The effects of the chemical on growth and reproduction of aquatic plants such as duckweed (*Lemma*) and other species can be conducted fairly quickly. If there is reason to believe that the compound may be dispersed on crops, it may be desirable to document the phytotoxicity or herbicidal activity using standard procedures. There is considerable information on techniques for contacting plants with air-borne contaminants, largely as a result of concerns about air pollutant effects on crops. A seed germination and early growth test has also been suggested (Ref. 2-32).

Mammals, Birds, and Human Health. Birds are often among the most vulnerable species to chemical contamination because they ingest both seed, containing pesticide, and fish, containing persistent hydrophobic compounds such as PCBs. Small mammals, such as mice, rats, and rabbits, are the traditional victims of toxicity testing, in which various dosages are applied and toxic effects measured. Mink and raccoon reproduction tests have also been proposed. Tests on these mammals are necessarily expensive and are used only in exceptional circumstances, for example, where occupational exposure is expected. When a significant human health threat is possible, it may be desirable to undertake tests on dogs or primates (e.g., monkeys), but such tests are not environmentally motivated. A dietary LC_{50} test on quail has been suggested (Ref. 2-32).

SUMMARY

It is obviously impractical to submit all chemicals to even a significant proportion of these tests. The first two, however, are fairly inexpensive and fast and could be suggested as mandatory (along with a mutagenicity test described later). Fish bioassays are justified if there is reason to suspect that the substance is toxic and will be present in aquatic environments. More complex mammal tests should be restricted to cases of unusual toxicity or when high human exposure is expected.

For carcinogenicity, mutagenicity, and teratogenicity testing the obvious first approach is the "Ames" test, in which a strain of a microorganism is exposed to the substance under conditions that allow mutants induced by the substance to be readily detected and measured. There is a sufficiently good correlation between this simple and inexpensive mutagenicity test and the more elaborate, time consuming, and expensive carcinogenicity tests with small mammals (and with epidemiological evidence from humans) that one or more tests based on this principle should be mandatory for new substances. Since the test has been developed recently, it is likely that in the near future, there will be improvements that will increase the reliability and permit more quantitative estimates to be obtained. In the meantime, it is prudent to treat the test results with some caution and scepticism.

In undertaking any toxicity assessment it is obviously desirable to be guided by the existing literature on compounds similar in chemical structure. Some literature sources are given in Section 2-6.

In summary, it is clear that some indication of toxicity and mutagenicity is essential. The depth of testing is a separate issue and presumably depends on factors such as the amount of material produced, the nature of its environmental dispersal, its likely dynamics and tendency to accumulate. At present there seems little alternative to informed subjective judgment on a case-by-case or compound-by-compound basis. It is hoped that, as fundamental understanding of toxic effects improves, it may be possible to predict toxicity to one species based on that measured for another. Considerable economies of effort would result in this way, and only the most useful test organisms would need to be used. A review of this topic by Kenaga suggests that already some extrapolation is possible and that the three species that could emerge as the best indicators are the rat, a species of fish, and *daphnia* (Ref. 2-35).

2-5 MISCELLANEOUS ASPECTS

2-5.1 Trace Contaminants

Production of a chemical compound, particularly an organic compound, almost inevitably results in coproduction of other compounds that, prior to marketing, may not be entirely separated from the principal product. The trace compound may be more environmentally significant or toxic. The best approach to identifying such problems is to subject the actual product to the test, with the recognition that it may be impure. Subjecting a specially purified batch of the product to toxicity testing could result in failure to detect such compounds. Because process conditions, raw materials, and product separation processes change, it may be desirable to subject the product to periodic testing to identify such changes, should they occur. The best safeguard may be detailed consideration of the chemistry of the synthesis by a competent and perceptive chemist who is aware of the possibilities of toxic reaction byproducts.

2-5.2 Mixtures

Many commercial products are mixtures of isomers or near-isomers, which thus have rather indeterminate properties reflecting some form of "average" behavior. Examples are PCBs and most petroleum products. For mixtures that are of little environmental concern, the "average" data are adequate, but for others it may be desirable to obtain more detailed single-component data, at least for the principal constituents. There are dangers in treating the properties of mixtures as the properties of a pure substance, particularly if the substance is to be subjected to extensive evaluation.

2-5.3 Production and Transportation Hazards

The greatest exposure to a toxic substance is likely to occur in the chemical plant where it is manufactured, in regions subjected to its air, water or solid waste effluents, or in regions where accidental spills occur during transportation or storage. Particular care is thus needed in these locations. The occupational health issue is beyond our scope here, but it is worth noting that the environmental data provide useful occupational information. Similarly, the physical–chemical and biodegradation data are invaluable in suggesting effluent treatment methods such as bio-oxidation, activated carbon treatment, or sedimentation. There exist established contingency plans that provide procedures to be adopted after transportation accidents involving hazardous materials, and, clearly, it is essential to incorporate appropriate data into the plan for large tonnage products. This includes data such as toxicity, volatility, explosion hazard, and suitable clean-up or neutralizing agents or methods.

Finally, it should be recognized that the chemical plant is likely to emit quantities of the product and byproducts in its effluent. There may be unexpected synergisms or even reactions that could cause unexpected toxic conditions. The best approach is to monitor the waste by chemical analysis and for toxicity to ensure that no unexpected problem exists or arises. As mutagenicity tests improve, it may also be desirable to use them to identify possible problems.

2-5.4 Transformation Products

The classic case of a transformation product proving to be the principal toxicant is the conversion of elemental mercury to its organometallic form. Depending on their valence, metal ions such as chromium display very different toxicities. Organic compounds may degrade into toxic metabolites or oxidation products. To cover all eventualities is clearly impossible, but it is essential to keep in mind the possibility of unexpected "downstream" conversion and resultant toxicity.

For metals, it is obviously important to know if the metal is capable of forming toxic organic forms, possible in anaerobic sediments, or if other valency states are differently toxic. If so, it becomes appropriate to assess carefully on paper and even by experiment the likelihood of such transformations occurring in the environment. Fortunately, the toxicology literature is sufficiently abundant to allow a good probability of identifying such possibilities.

For organics, a competent chemist or biochemist can probably speculate with fair accuracy about the degradation mechanisms and products, and identify troublesome products. Such time could prove to be very well spent in identifying potential problems.

2-5.5 Calculation of Environmental Fate and Effects

This chapter has been devoted primarily to the first, and easiest, stage of hazard assessment, namely determining relevant properties. The second stage is to combine these data to calculate likely environmental partitioning, persistence, transport, transformation and ultimately predict exposure concentrations. From such exposures it may then be possible to predict effects or proximity to effect levels. Finally a judgment can be made about the need for, and extent of, control. This four-stage process presents formidable problems in which scientific knowledge severely lags behind regulatory needs. Work has only recently started on the second stage, which is essentially an environmental modeling effort. Approaches to this have been described by Baughman and Lassiter (Ref. 2-1) and using the versatile "fugacity approach" by Mackay (Ref. 2-36). It is clear that significant advances in the chemistry, physics, biology and especially toxicology of industrial chemicals are needed before the regulatory need is fully satisfied.

2-6 LITERATURE SOURCES

The following is a list of literature sources that may form a useful starting point for searches for physical property or toxicity data. There is, however, no substitute for the well-planned literature search using such existing abstracting systems as ACS Chemical Abstracts, the US National Technical Information Service, etc.

Weast, R. C., *Handbook of Chemistry and Physics*, 52nd ed., CRC Press, Cleveland, OH, 1972.

Perry, J. H., and Chilton, C. H., *Chemical Engineers' Handbook*, 5th ed., McGraw-Hill Book Co., New York, 1973.

Reid, R. C., Prausnitz, J. W., and Sherwood, T. K. *The Properties of Gases and Liquids*, 3rd ed., McGraw-Hill, New York, 1977.

Boublik, T., Fried, V., and Hala, E., *The Vapor Pressures of Pure Substances* Elsevier, Amsterdam, 1973.

Linke, W. F., and Seidell, A., *Solubilities*, Amer. Chem. Soc., Washington, D. C., Vols. I and II, 1958.

Stephen, H., and Stephen, T., *Solubility of Inorganic and Organic Compounds*, Pergammon Press, New York, 1964.

Zwolinski, B. J., and Wilhoit, R. C., *Handbook of Vapor Pressures and Heats of Vaporization of Hydrocarbons and Related Compounds*, API-44-TRC Publications, 1971.

American Petroleum Institute, *Technical Data Book* (2 Volumes).   API Washington, D. C., (Continuously updated).

Herbicide Handbook, Weed Science Society of America, Champaign, IL., 1977.

Martin, H., *Pesticide Manual*, 3rd ed., British Crop Protection Council, London, 1972.

Leo, Hansch, C., and Elkins, D., *Chem. Reviews* 71, 525 (1971).

Bondi, A., *Physical Properties of Molecular Crystals, Liquids, and Glasses*, Wiley, New York, 1968.

C.A.Ch.E. *Physical Properties Data Book*, National Academy of Engineering, Washington, D. C., 1972.

American Chemical Society, *Physical Properties of Chemical Compounds*, Adv. in Chem. Series, Vols. 15, 22, and 29.

Matheson Gas Data Book, 5th ed., 1971.

Kirk Othmer Encyclopedia of Chemical Technology, 2nd ed., Wiley Interscience, New York, Vols. 1–22, 1966.

Bond, R. G., Straub, C. P., Prober, R. (eds.), *Handbook of Environmental Control*, CRC Press, Cleveland, OH, Vols. I to IV, 1972.

Sax, N. I., *Dangerous Properties of Industrial Materials*, 4th ed., Van Nostrand-Reinhold Company, New York, 1975.

Gleason, M. N., Gosselin, R. E., Hodge, H. C., and Smith, R. P., *Clinical Toxicology of Commercial Products*, 3rd ed., Williams and Wilkins, Baltimore, 1969.

Sunshine, I., *Handbook of Analytical Toxicology*, CRC Press, Cleveland, OH, 1969.

Christensen, H. E., Luginbyhl, T. T., Carroll, B. S., *The Toxic Substances List*, 1974 Edition, U. S. Department of Health Education and Welfare, Brockville, MD, 1974.

Toxic and Hazardous Industrial Chemicals Safety Manual for Handling and Disposal with Toxicity and Hazard Data, International Technical Information Institute, Tokyo, Japan, 1975.

Windholy, M., *et al., The Merck Index*, 9th ed., Merck & Co. Rahway, NJ, 1976.

Plunkett, E. R., *Handbook of Industrial Toxicology*, Chemical Pub. Co., New York, 1976.

"Chemical Hazards Response Information System," (CHRIS), (CG-446), Vol 2, *Hazardous Chemical Data*, U. S. Coast Guard, 1974.   Available from U. S. Govt. Printing Office, Washington, D. C., Stock No. 050-012-0094-8. *Registry of Toxic Effects of Chemical Substances*, U. S. National Institute for Occupational Safety and Health (1976). Available from U. S. Govt. Printing Office, Washington, D. C.

Toxicology of Metals, U. S. Environmental Protection Agency, Subcommittee on the Toxicology of Metals, National Technical Information Service, Springfield, VA, 1976.

Verscheuren, K., *Handbook of Environmental Data on Organic Chemicals*, Van Nostrand Reinhold Co., New York, 1977.

Chemical Restriction List, Aldrich Chemical Coy, Milwaukee, WI, 1977.

Manufacturing Chemists Association, Washington, D. C. Various publications including *Behavior of Organic Chemicals in the Aquatic Environment*, Vol. I (1966) and Vol. II (1968).

American Petroleum Institute, Washington, D. C. Various publications relating to hydrocarbon toxicity.

2-7 CONCLUSIONS

In conclusion, from a knowledge of a chemical's physical–chemical properties, its chemical conversion characteristics, its biological behavior, and its proposed production volume and patterns of use, it should be possible to make a subjective judgement about its overall "acceptability" and determine appropriate restrictions that may be placed on its use.

Undoubtedly, "scoring" or "rating" or "prioritizing" numerical schemes will be developed, allegedly to assist in identifying the most hazardous substances. It is the author's opinion that such schemes are usually misleading and are often pursued only because of intellectual laziness. There is no substitute for the careful gathering and assimilation of reliable physical–chemical, biological, and industrial data by a broadly experienced group of well-informed and well-intentioned individuals who can then make a balanced judgement, in which all the issues have been weighed subjectively. Indeed, it is likely that the greatest difficulties in controlling toxic substances in the environment will involve the mechanics of the decision-making process rather than the scientific difficulties of elucidating environmental fate and behavior.

It must be emphasized that the aim should not be to ban large numbers of toxic compounds. It should be to exert subtle and reasonable controls only if and when they are needed. It is likely that only a few substances will be judged to have a high environmental threat and low social necessity, resulting in a decision that society is better off without them. PCBs would probably have fallen in this category. A larger number may be deemed to be acceptable but require a level of control. Most are likely to be relatively harmless. Ideally, the screening process should give a reliable and early warning of hazard, not only to protect the environment but also to prevent waste of industrial effort on development of a product that is destined to be banned or restricted. There is little doubt that we now are close to having the tools with which to identify the most hazardous substances and it should be possible to permit even highly toxic compounds to be produced, used, and destroyed while maintaining a satisfactorily small risk to the environment and to human health.

REFERENCES

2-1. Baughman, G. L., and Lassiter, R. R., "Prediction of Environmental Pollutant Concentration," in ASTM STP 657 *Estimating the Hazards of Chemical Substances to Aquatic Life,* ASTM Philadelphia, p. 35, 1978.

2-2. Hamaker, J. W., and Kerlinge, H. O., "Vapor Pressure of Pesticides," *Advances in Chemistry* Series No. 86, American Chemical Society, Washington, D. C., pp. 39-54, 1957.

2-3. Spencer, W. F., and Cliath, M. M., *Environ. Sci. Technol.* 3, p. 670, 1969.

2-4. Zwolinski, B. J., and Wilhoit, R. C., *Handbook of Vapor Pressures and Heats of Vaporization of Hydrocarbons and Related Compounds,* API-44-TRC Publications, 1971.

2-5. Hartley, G. S., "Evaporation of Pesticides." *Advances in Chemistry* Series No. 86, American Chemical Society, Washington, D. C., pp. 115-134, 1969.

2-6. Prausnitz, J. M., *Molecular Thermodynamics of Fluid Phase Equilibrium,* Prentice-Hall, New York, 1973.

2-7. Butler, J. N., *Solubility and pH Calculations,* 2nd ed., Addison Wesley Pub. Co., Reading, MA, 1973.

2-8. May, W. E., and Wasik, S. P., "Determination of the Aqueous Solubility of Polynuclear Aromatic Hydrocarbons by a Coupled Column Liquid Chromatographic Technique," *Anal. Chem.,* 50, pp. 175-179, January 1978.

2-9. Mackay, D., and Shiu, W. T., "Aqueous Solubility of 32 Polynuclear Aromatic Hydrocarbons," *J. Chem. Eng. Data,* 22, p.4, 1977.

2-10. Leo, A., Hansch, C., and Elkins, D., *Chem. Rev.,* 71, 6, p. 525, 1971.

2-11. Neely, W., Branson, D., and Blau, G., *Environ. Sci. Technol.,* 8, p. 1113, (1974).

2-12. Kurihara, N., Uchida, M., Fujita, T., and Nakajima, M., *Pesticide Biochem. and Physi.* 2, 383, (1973).

2-13. Metcalf, R. L., Sanborn, J. R., Lu, P. Y., and Wye, D., *Arch. Environ. Contam. and Toxic.,* 3, 2, p. 151, 1975.

2-14. Karickhoff, S., Brown, D., and Scott, T., "Sorption of Hydrophobic Pollutants on Natural Sediment," *Water Research* 13, 241 (1979).

2-15. Chiou, C. T., Freed, W. H., Schnedding. D. W., and Kohnert, R. L., *Environ. Sci. Technol.* 11, p. 475, 1977.

2-16. McCall, J., *J. Med. Chem.,* 18, 6, p. 549, 1975.

2-17. Carlson, R., and Kopperman, H., J. Chromatog, 107, 213, (1975).

2-18. Mirrlees, M., Moulton, S., Murphy, C., and Taylor, P., *J. Med. Chem.,* 19, (5), p. 615, 1975.

2-19. Atkins, D. H. F., and Eggleton, A. E. J., Proc. Symp. Nuclear Techniques in Envir. Pollution, IAEA Vienna, 521, 1971.

2-20. Mackay, D., Shiu, W. Y. and Sutherland, R. P., "Determination of Air-Water Constants for Hydrophobic Pollutants," *Environ. Sci. Technol.* 13, 3, pp. 333-337 (1979). Henry's Law.

2-21. Mackay, D., Leinonen, P. J., *Environ. Sci. Technol.,* **9**, p. 1178, 1975.

2-22. Kenega, E. E., and Goring, C. A. I., "Relationship between Water Solubility, Soil-Sorbtion, Octanol-Water Partitioning and Bioconcention of Chemicals in Biota," ASTM 3rd Symposium on Aquatic Toxicity, New Orleans, LA, Oct. 17-18, (1978).

2-23. Wolfe, N. L., Zepp, R. G., Doster, J. C., and Hollis, R. C., "Captan Hydrolysis," *J. Agric. Food Chem.,* **24**, pp. 1041-1045, 1976.

2-24. Mabey, W., and Mill, T., "Critical Review of hydrolysis of organic compounds in water under Environmental Conditions," *J. Phys. Chem. Ref. Data,* 7, p. 383, 1978.

2-25. Dilling, W. L., Bredeweg, C. J., and Terfertiller, W. B., *Environ. Sci. Technol.,* **10**, p. 351, 1976.

2-26. Zepp, R. G., Wolfe, N. L., Gordon, J. A., and Baughmann, G. L., "Dynamics of 2, 4-D Esters in Surface Waters: Hydrolysis, Photolysis and Vaporization," *Environ. Sci. Technol.,* 9, pp. 1144-1150, 1975.

2-27. Wolfe, N. L., Zepp, R. G., Baughmann, G. L., Fincher, R. C., and Gordon, J. A., "Chemical and Photochemical Transformation of Selected Pesticides in Aquatic Systems," USEPA Report 600/3-76-067, 1976.

2-28. Smith, J. H., Mabey, W. R., Bahoros, N., Holt, B. R., Lee, S. S., Chou, T-W., Bomberger, D. C., and Mill, T., (SRI International), "Environmental Pathways of Selected Chemicals in Freshwater Systems," Part I, Background and Experimental Procedures, Report EPA 600/7-77-113, (1977) and Part II, "Laboratory Study," EPA 600/7-78-074 (1978).

2-29. ASTM STP 634 "Aquatic Toxicology and Hazard Evaluation," Philadelpha (1977).

2-30. ASTM STP 657 "Estimating the Hazards of Chemical Substances to Aquatic Life," Philadelphia, PA (1978).

2-31. Stern, A. M. and Walker, C. R., "Hazard Assessment of Toxic Substances: Environmental Fate Testing of Organic Chemicals and Ecological Effects Testing, in ASTM STP 657 "Estimating the Hazards of Chemical Substances to Aquatic Life," Philadelphia, PA, p. 81, 1978.

2-32. Draft Guidelines produced in connection with the administration of the Toxic Substances Control Act, USEPA Washington, (1978).

2-33. Neely, W. B., Blau, G. E., Alfrey, T., *Environ. Sci. Technol.,* **10**, p. 72, 1976.

2-34. Branson, D. R., "Predicting the Fate of Chemicals in the Aquatic Environment from Laboratory Data" in ASTM STP 657 "Estimating the Hazards of Chemical Substances to Aquatic Life" ASTM Philadelphia, PA, p. 55, 1978.

2-35. Kenega, E. E., *Environ. Sci. Technol.,* **12**, p. 1322, 1978.

2-36. Mackay, D. "Finding Fugacity Feasible", *Environ. Sci. Technol.,* **13**, p. 1218 (1979).

3

Entry of chemicals into the environment

Richard A. Conway, P.E.,

Corporate Development Fellow
Union Carbide Corporation
South Charleston, WV

Frank C. Whitmore, Ph.D.,

VERSAR, Inc.
Springfield, VA

Walter J. Hansen

Group Leader, Engineering Department
Union Carbide Corporation
South Charleston, WV

3-1 GENERAL CONSIDERATIONS AND OVERALL SOURCES

An essential aspect of environmental risk analysis for chemicals is to estimate the initial (or entry) concentration of the material in the environmental compartments of concern. The initial concentration is needed for establishing:

1. A dosage level to use in individual determinations of transport and transformation rates and in effect (toxicity) testing.
2. A concentration level in microcosm testing.
3. A starting value to use in mathematical models for establishing expected ambient levels after various time periods.

Both direct entry from a source to a compartment and intercompartmental transfers need to be considered. Any rapid dilution is a key factor. The details of compartmental transport and transformations are discussed in Chapters 4 (aquatic), 5 (atmospheric), and 6 (terrestrial).

Special emphasis is placed in this chapter on manufacturing and product-use sources, although other sources are identified. Because every entry situation is unique, we rely on illustrations for use as checklists and guides.

A model used for estimating rate of injection and the cumulative injected quantity of a pollutant into the general environmental reservoir is described. Polychlorinated biphenyls (PCBs) are discussed as an example of broadly distributed products. References for other approaches are cited. Sources of information on production and use volumes are given.

3-2 ENTRY AND TRANSFER CATEGORIES

Various categories of entry of materials to the environment are cited in Table 3-1. This checklist can aid in avoiding omission of significant, but not always obvious, sources of a material. Figure 3-1 shows entry routes to various compartments and the transfer between compartments. Table 3-2 outlines some additional considerations in determining the likelihood of entry to water and the distribution in area and time. For both transport and transformation mechanisms affecting environmental fate of chemicals, see Table 1-5.

3-3 CHEMICALS MANUFACTURING AND ASSOCIATED WASTE DISPOSAL SOURCES

Chemical manufacturing operations, along with product utilization and energy industries, can be a major source of chemicals in the environment. Controlling the entry of materials to the environment from manufacturing operations is discussed in this section, as well as the means of determining the quantities lost.

3-3.1 Manufacturing Sources and Loss Estimation

The checklist in Table 3-3 identifies potential sources of material loss. For each pertinent source, the quantity lost can be estimated either by a calculated material balance around an operation or by sampling and analysis of the emission. The latter is preferable when possible. For some processes, the U. S. Environmental Protection Agency (Ref. 3-1) has established standard discharges in terms of pound of material per 1000 pounds of product (kg/Mg). Variability must be defined, especially when a close estimate of shorter-term effects is desired (Hamelink, Ref. 3-2) (Table 3-4).

3-3.2 Material Efficiency

A good accounting of all the materials that are involved in the operation is necessary for any manufacture, storage, or distribution of a chemical. Historically, the material balance has been used in the chemical industry for design and production records purposes. With the increasing emphasis on controlling discharges from chemical operations, the material balance is moving towards a complete accounting of all chemicals into and out of an operation. Theoretically, a material

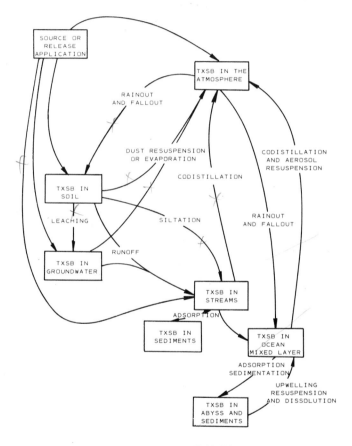

NOTE: DEGRADATION OF TOXIC SUBSTANCE (TXSB)
        CAN OCCUR IN ANY COMPARTMENT.

Fig. 3-1. Pathways for toxic substances (TXSB) in the physicochemical environmental system. *Source*: Lyman, W. J., Harris, J. C., Nelken, L. H., *Research and Development of Methods for Estimating Physiocochemical Properties of Organic Chemicals of Environmental Concern*, Final Report Phase 1, February, 1979, Contract No. DAMD17-78-C-8073, prepared for U. S. Army Medical Research and Development Command by A. D. Little, Inc., Cambridge, MA.

TABLE 3-1. Broad Entry Categories.

Entry Category[a]	Examples
Extraction	Loss of petroleum during production
Processing	Washing agricultural chemicals from foodstuffs, formulating, dividing for use
Manufacturing	Synthesis of chemicals from natural gas
Storage	Storage tank venting during filling
Transportation	Accidental spills
Consumptive use	Charcoal briquets
Confined use	Automatic transmission fluid
Distributed use	Pesticide applications
Disposal	Waste degreasing solvent
Reclamation	Solid waste disposal areas
Natural source	Leaching of fallen leaves

[a]*Source:* Alan Eschenroeder, Arthur D. Little, Inc., 20 Acorn Park, Cambridge, MA 02140.

TABLE 3-2. Forecast of Penetration of a Chemical in Aquatic Ecosystems.

Total Annual Production

    1st year
    2nd year
    Foreseeable production growth after 2nd year

Production Utilizations

    Closed systems only
    Industrial use only
        Less than 200 users in the country
        More than 200 users in the country
    Public use
        Concentrated in some regions
        Homogeneous

Penetration into Aquatic Ecosystems

    Utilization does not lead to usual discharge into wastewater
    Utilization leads to discharge into wastewater
    Chemical enters the environment by other ways and
        may finally enter aquatic ecosystems
        will not finally enter any aquatic ecosystem

Surface Waters

    Chemical received in quantities that are distributed in time and space as urban and common industrial wastewater.
    Different distribution of locations
    Chronic but irregular entry into surface waters

(Continued)

TABLE 3-2. (Continued)

Seasonal or occasional entry into surface waters
Accidental entry into surface waters

Ground Waters

Foreseeable chronic contamination
Accidental contamination only

The questions are answered by "yes" or "no"; then the quantities entering the aquatic environment and their distribution in time and place are determined using the knowledge of such properties as the material's treatability in biological systems and leachability if applied to soils.

Source: Based on Lundahl, P., "Currently Employed Hazard Assessment Schemes for New Chemicals in France," in *Analyzing the Hazard Evaluation Process*, K. Dickson, A. Maki, J. Cairns (eds.), published by Water Quality Section of the American Fisheries Society, Washington, D.C., 1979.

TABLE 3-3. Entry from Manufacturing Operations.

To Surface Water

Treated process wastewaters
Leaks and spills to unrecirculated (once-through) cooling waters
Treated washwaters, condenser waters, scrubber waters
Deep ocean disposal

To Atmosphere

Process stack and vent emissions
Fugitive emissions, e.g., imperfect connections
Storage tank vents
Volatilization from wastewater treatment operations
Volatilization from landfarming, landfilling, and impoundments
Stack emissions from incineration of process wastes

To Soil

Solid wastes, liquid residues, and wastewater treatment sludges to landfarms (soil cultivation) and landfills
Incinerator and boiler ash to landfill
Deepwell injection
Spills and leaks directly to soil
Land treatment of dilute wastewaters

Important Intercompartmental Transport

Stormwater washing chemicals and contaminated silt from soil to surface water
Groundwater flow to surface water, and vice versa
Washout and fallout of chemicals from atmosphere to surface water and soil
Volatilization from surface water and soil

TABLE 3-4. Considerations in Estimating Initial Concentration in River Water from Point Sources.

Quantity Discharged

Basic Data

Assay 24-hr composite samples of wastewater for chemicals of interest; combine result with wastewater flow data to obtain daily quantity discharged. . . . . or calculate quantity by material balance.

Variability

Determine probability of occurrence; data usually fit a log-normal distribution; the maximum daily load approximates the ratio of the 99% to the 50% probability of occurrence or two standard deviations from the measured median . . . . . a frequently encountered variability ratio is 4:2:1 for the quantity discharged on the maximum day, the average day for the maximum month, and the average annual day, respectively.

Hydrodynamics of Receiving Body of Water

Use median daily flow of the river over a typical year for dilution of annual average daily load and use average daily flow for the low month for dilution of average daily load for the maximum month (assume simultaneous occurrence). Compare these concentrations with both short- and long-term no-effect toxicity information for the respective durations of exposure.

For cases where no-effect toxicity data are available and allowable discharges are to be determined, "cross-multiply" the toxicity data for various exposure durations with the average river flows expected for the same time periods to obtain allowable discharge quantities. This assumes no compartmental loss or transformation of the chemical.

Although use of 7-day, 10-yr low flow (rate expected for 7-day period once in 10 yr) for dilution of maximum daily discharge of chemical of interest may be appropriate when considering oxygen-demanding substances, it appears to be an overly conservative approach for toxics. The strategies described above seem to be the better choice for compounds with measured toxicity over various periods of time.

Source: Approach patterned after that of Hamelink, Ref. 3-2.

balance should account for every molecule. Although this is not practical, it must be emphasized that, if losses to the environment are to be minimized, it is necessary that a good balance be established in the design and operation of a chemical enterprise.

Chemicals enter the environment via atmospheric emissions, wastewater effluents, and disposal of solid wastes or liquid residues. All of these means of entry are of concern in the manufacture, storage, distribution, and eventual usage of chemicals.

Atmospheric Emissions. Chemical discharges into community air originate from point sources such as vents, stacks, and other specific, known openings. Emissions also originate from area sources such as surfaces of ponds, areas of land, and areas of equipment that have no specific, known point sources. Chemical concentrations in an emission can range from parts per billion to 100%. Obviously, chemical concentration and total flow, in mass or volume units, together determine the rate of emission from a source.

Emissions or leakage of chemicals from miscellaneous locations in facilities such as packed or gasketed closures and running seals are also of concern. These are so-called fugitive losses, and the aggregate of all such losses can be significant unless minimized by good design and maintenance practices.

Wastewater Effluents. Generally, the design and operation of chemical manufacturing facilities provide for the disposal of wastewaters containing minute to dilute concentrations of chemicals. This includes process wastes in addition to wastewaters from cleaning process equipment, storage vessels, and transportation equipment. These wastewaters are processed by various biological, chemical, and/or physical means for conversion into harmless products, such as carbon dioxide and water, or to remove chemicals from effluent streams before discharge to the environment.

Wastewater treatment is not 100% efficient, and residual chemicals, including intermediates from a treatment process, may be discharged in the effluent to the receiving surface waters.

Liquid and Solid Residues. Most chemical manufacture yields byproduct residues. If recycling is not feasible, then other means of disposal must be used. On occasion, byproduct residue materials can be utilized as raw materials in other manufacturing operations. Many hydrocarbon residues can be burned or incinerated with waste-heat recovery. Combustion or incineration of residue materials may result in emission problems unless steps are taken to avoid them. The combustion of chloride and nitrogen-containing residues are classic examples.

The principal concerns with residues involve disposal via landfill. They involve chemical entry into groundwaters via leaching, entry into surface water runoffs, and human exposures to vapors and liquids at some future date in and around landfill areas. Landfilling toxic or hazardous materials obviously attracts considerable attention and necessitates strict controls for proper containment.

3-3.3 Comparison of Manufacturing Technologies

When comparing two or more processes for the manufacture of a chemical or group of chemicals, the best control technology relative to containment of material losses is often expressed in terms of production ratios. For example, losses to the environment can be expressed in terms of pounds of loss per 1000 pounds of product (kg/Mg) (Ref. 3-1).

The mode of operating a chemical manufacturing facility can affect performance relative to containment of chemicals. Chemical process facilities can be operated in a continuous, discontinuous, or abnormal manner relative to the flow of materials and/or the control of process conditions, principally temperature and pressure.

Most large-volume chemical manufacturing operations utilize around-the-clock, continuous operation. All streams in a continuous process or operation are generally at steady state conditions within the limits of companion control equipment. Although the controls may be manual, analog or digital instrumentation of varying complexity are generally utilized.

When well-designed process equipment is employed within a reasonable range of capacity operation, performance is near optimum and losses to the environment can be minimized. Hence, the degree to which the overall operation releases chemicals into the environment, particularly via atmospheric emissions, becomes primarily a function of the control technology or capacity of the equipment to contain the chemicals within the manufacturing process. For example, the performance of a packed vent scrubber, at steady state operation, is a function of the dimensions of the scrubber relative to the mass flows, the type and efficiency of the packing, and, particularly, the affinity of the scrubbing medium for the chemical(s) to be removed from the vent stream.

Batch operation of chemical manufacturing equipment is typical with small-volume productions. The manufacture of proprietary-type chemicals is an example where materials are often reacted in kettle-type vessels with either an overhead or a kettle product. The flow of materials in batch operations fluctuates, is interrupted, or cycles in discrete steps, usually from flow to no-flow and vice versa. Turndown ratios can be significant and exceed the capacity of equipment. The initial rush of vapors in the overhead from a batch kettle, for example, may initially overload the condenser before the heat input to the kettle is stabilized. Inerts in the kettle system further reduce the capacity of the condenser and must be purged to eliminate the gas film in the condensing and, particularly, the sub-cooling zones of the condenser.

Batch operations very often involve intermittent transfers of chemicals via hose connections, particularly at stations where several transfer lines terminate. Disconnect losses increase the possibility of chemicals entering the environment via atmospheric emissions from spillage and loads on wastewater treatment facilities. Care must be exercised to contain spillage and prevent it from accumulating in earthen, graveled, or similar areas, where leaching or contamination of surface water runoff can occur.

3-4 DILUTION AND TRANSPORT

Important considerations in environmental risk analysis for chemicals include:

Quantity entering each environmental compartment
Dilution and transport within the compartments
Transfer between compartments
Transformations
Duration of exposure of key organisms
Data on concentration versus toxicity for various exposure periods

Analysis may be started either by a manufacturer, with a known or projected quantity of loss, who is seeking an estimate of environmental impact, or by a regulatory person armed with known effects, who is seeking an allowable entry quantity. The question of dilution and transport are important for both. General considerations on these subjects are presented in this section, while transport and transformation within individual compartments are discussed in Section 4-6.

Water. Dilution of compounds in lakes, rivers, and oceans has been described in detail by Metcalf and Eddy (Ref. 3-3). Small, shallow lakes and reservoirs are assumed to be completely mixed, due to wind-induced currents. In small, deep lakes, rapid mixing occurs in the upper layers (epilimnion), as a result of stratification. When submerged outfalls are used in very large lakes and oceans, both the initial dilution rate by turbulent jet mixing and subsequent dispersion as the mixture rises as a plume are important. For disposal in rivers a diffuser system can be used to effect rapid dilution; it would seem possible to estimate mixing, based on stream velocity and depth data, as is reaeration. In estuaries, tidal action increases the dispersion along the axis of the channel. The coefficient of eddy diffusion is used as a measure of mixing; it can be estimated from the roughness coefficient, velocity, and hydraulic radius.

For discharges into rivers, Table 3-4 (Hamelink, Ref. 3-2) covers selection of flow data to be used for calculating dilution. For cases involving analysis based on hypothetical receiving bodies of water, it is useful to refer to the physical dimensions shown in Table 3-5, which were selected by SRI International (Ref-3-4) for studies of this nature. When an analysis involves the widespread production and usage of a commodity material, as opposed to production by a specific manufacturing facility and controlled product usage, the general approach of A. D. Little, Inc., outlined in Ref. 3-5, can be used if factors and coefficients are updated. Table 3-6 gives compartmental data used in the ADL studies.

Air. Atmospheric diffusion models fall into two general categories: (1) short-term models for calculating time-averaged ground-level concentrations for times

TABLE 3-5. Suggested Physical Dimensions and Water Quality
Characteristics for Environmental Analyses.

Parameter	River	Pond	Eutrophic Lake	Oligotrophic Lake
Physical dimension				
Total water volume (m³)	9×10^5	2×10^4	5.5×10^6	5.5×10^6
Inflow (m³/hr)	1.0×10^6	20	9.7×10^2	9.7×10^2
Mean residence time (hr)	8.3×10^{-1}	1.0×10^3	5.7×10^3	5.7×10^3
Pollutant inflow for 1 µg/ml (kg/hr)	1.0×10^3	0.02	0.97	0.97
Water quality				
Total bacteria (cells/ml)	10^6	10^6	10^6	10^2
Active bacteria (cells/ml)	10^5	10^5	10^5	10
pH	7	8	8	6
Sediment loading (µg/ml)	100	300	50	50
Photolysis activity index[a]	0.5	0.2	0.2	1.0
Oxygen reaeration rate (hr⁻¹)	0.04	0.008	0.01	0.01
$[RO_2 \cdot]$ (M)	10^{-9}	10^{-9}	10^{-9}	10^{-9}

[a]Factor to account for differences in light transmission through different types of water.
Distilled water has an index value of 1.

Note: Flowrates between compartments are given in source reference below.

*Source: Environmental Pathways of Selected Chemicals in Freshwater Systems. Part 1
Background and Experimental Procedures*, prepared by SRI International for Environmental Research Laboratory, U.S. Environmental Protection Agency, Athens, GA; available as Document PB-274 548, National Technical Information Service, Springfield, VA
22161.

ranging from several minutes to 24 hr; and (2) long-term models for calculating
monthly, seasonal, and annual ground-level concentrations (Ref. 3-6).

Both the short- and the long-term concentration models are modified versions
of the Gaussian plume model for elevated continuous sources described by Pasquill
(Ref. 3-7). In the short-term model, it is assumed that the plume has Gaussian
vertical and lateral concentration distributions. The long-term model is a sector
model similar in form to the Air Quality Display Model of the U. S. Environmental Protection Agency (Ref. 3-8), in which it is assumed that the vertical
concentration distribution is Gaussian and that the lateral concentration distribution within a sector is rectangular (a smoothing function is used to eliminate
sharp discontinuities at the sector boundaries). The vertical and lateral expansion curves may be determined by using turbulent intensities in simple power law

TABLE 3-6. Compartmental Data for the United States.

Compartment	Subcompartment	Effective Area (m²)	Inventory (10¹⁵ kg)	Annual Flow (10¹⁵ kg)	Time of Residence (yr)
Air (1 mile, 2 mph)		—	16.2	710	0.002
Surface Water	Lakes	1.4×10^{11}	18.8	0.19	100
	Streams	0.25×10^{11}	0.05	1.86	0.3
Ground water	Shallow	—	63.7	0.31	200
	Deep	—	63.7	0.006	—
Ground Moisture	0–1 m	8×10^{11}	0.6	3.1	—
	1–5 m	—	0.4	—	—
	5–10 m	—	0.2	—	—
	10–15 m	—	0.2	—	—
	15–30 m	—	0.2	—	—
	30–50 m	—	0.2	—	—
Atmospheric Moisture		—	0.18	4.8[a]	0.04
Soil	0–1 m	—	15.2	—	—
	1–5 m	—	60.9	—	—
	5–10 m	—	76.1	—	—
	10–15 m	—	76.1	—	—
	15–30 m	—	228.5	—	—
	30–50 m	—	304.4	—	—

[a]Net of short-term reevaporation of approximately 1.2×10^{15} kg per year.

Source: Pre-Screening for Environmental Hazards—A System for Selecting and Prioritizing Chemicals, prepared by A. D. Little, Inc., for Office of Toxic Substances, U.S. Environmental Protection Agency, Washington, D.C.; available as Document PB-267093 from National Technical Information Service, Springfield, VA 22161.

expressions or they may be set equal to the Pasquill–Gifford Curves given by Turner (Ref. 3-9).

These curves may also be modified to include the effects of initial source dimensions. Also, in both the short- and long-term models, buoyant plume rise is calculated by means of the Briggs (Ref. 3-10) plume-rise formulas. An exponent law is used to adjust the surface wind speed to the source height for plume-rise calculations and to the plume stabilization height for concentration calculations. The information required for atmospheric diffusion models is given in Table 3-7 to illustrate the considerations involved.

Some Gaussian models depart a good deal from the above descriptions. Also, increasing in significance are several non-Gaussian approaches to dispersion modeling, particularly in cases where the terrain is complex and for pollutants that participate in atmospheric reactions.

Transport between the troposphere and the stratosphere is discussed in Section 5-1, as is interhemispheric exchange.

Soil. The release, transport, and dilution of chemicals applied to the soil is covered to some extent in Chapter 6. Key parameters include:

• Chemical application rate and type of formulation, e.g., powder, emulsion, water solution.
• Soil properties, e.g., erodability and adsorption capacity.
• Rainfall characteristics.

TABLE 3-7. Input Information Required for Atmospheric Diffusion Models for Continuous Point or Volume Sources.

Source Information

Pollutant emission rate
Volumetric emission rate
Stack exit temperature
Stack height
Length and height of buildings with low level emissions
Pollutant decay coefficient

Meteorological Inputs

Mean wind speed
Mean wind direction
Wind profile exponent
Wind azimuth angle standard deviation
Wind elevation angle standard deviation
Ambient air temperature
Depth of surface mixing layer
Vertical potential temperature gradient

- Slope gradient and length.
- Erosion control measures.
- Vegetative cover.
- Fraction of watershed treated and total watershed area.

Empirical factors have been developed for such parameters for use in soil loss equations (Refs. 3-11, 3-12).

3-5 MODEL FOR ESTIMATING PRODUCT ENTRY RATE AND CUMULATIVE QUANTITY

The model presented herein is designed to use available data to estimate the magnitude of the environmental load at any given time, the partitioning of that total load into the major environmental compartments, and to serve as a guide to additional environmental sampling. Generally, the available information on possible sources provides details on producers and their yearly production rates, the applications in which the chemical has been or is used, and, probably, some information on the useful life of the end use products for that chemical species. In addition, one can usually find some information on the physical and chemical properties of the chemical and data that have been obtained incidental to the manufacture and use of that chemical. The model is designed to utilize these data (and any other information that is available) to form a source function that may then be used in models that deal with the environmental distribution and ultimate fate of the given chemical species.

In order to estimate the rate of injection and the cumulative injected quantity of a particular pollutant into the General Environmental Reservoir (GER), it is necessary to consider a number of processes that ultimately contribute to the injection into the GER. There are losses incident to the original manufacture of the material, losses associated with transport to the sites of end product manufacture, losses incident to end product manufacture, and losses associated with the utilization of the end use product that, after a finite service, is discarded — eventually to serve as a possible source for further injection into the GER by processes associated with the eventual degradation of the discarded end use product.

The following discussion presents a mathematical prescription for estimating the contributions each of these individual sources make to the GER. The prescription (model) will then be applied to the case of the PCBs.

3.5.1 Production and Incidental Losses

Let $r(t)$ be the production rate of the jth (this designation is used in order to simplify the syntax in what follows) chemical species by commercial processes at some time t. The cumulative production from the time of commencement of commercial production up to a time $t = T$ is given as:

$$R(T) = \int_{O}^{T} r(s)\,ds \qquad \text{(Eq. 3-1)}$$

where s has been introduced as an integration variable and $r(s)$ is taken to be a continuous function of s. In practice, there will usually be production data available in the form of yearly production quantities—in which case an attempt should be made to fit a continuous function of production versus time to that data. In the event that no simple function can be discovered, Eq. 3-1 may be rewritten in the following form:

$$R(T) = \sum_{i=1}^{N} r_i(t)\,\Delta(t)_i$$

where Δt_i is taken as 1 year, $r(t)$ is the total yearly production, and N is the number of years from initiation of production to the time T. In any case, in what follows, all the functions used are considered to be continuous functions of time (all the integrals may, alternately, be taken as summations, in which case continuity is not required).

Incident to the manufacturing process, some fraction $\alpha(t)$ of the processed material will be lost to the GER, i.e., the rate of injection of the jth species to the GER by processes incident to the manufacture of the jth species will be given as:

$$q_I(t) = \alpha(t)\,r(t) \qquad \text{(Eq. 3-2)}$$

where $0 < \alpha(t) < 1$.

3-5.2 Losses Associated with End Use Product Manufacture

In general, the jth chemical species will ultimately be utilized in the manufacture of a number, N_j, of end use products. Let $r_i(t)$ be the rate at which the jth chemical species is consigned to the production of the ith end use product. Then,

$$[1 - \alpha(t)]\,r(t) = \sum_{i=1}^{N_i} r_i(t) \qquad \text{(Eq. 3-3)}$$

where the prefactor on the left-hand side represents the available fraction of $r(t)$—available, that is, for useful application.

There are often losses of the jth material in the course of transportation from the source to the utilization site. Let us denote the fraction of $r_i(t)$ lost during transportation as $\alpha_i(t)$. Then the rate of injection of the jth species into the GER from transportation losses, is given as (where all the possible end uses are included):

$$q_{II}(t) = \sum_{i=1}^{N_i} \alpha_i(t) r_i(t) \qquad \text{(Eq. 3-4)}$$

where $0 < \alpha_i(t) < 1$.

Further, the effective rate of admission, $p_i(t)$, of the jth species into the manufacturing site for the ith end use product is given as:

$$p_i(t) = [1 - \alpha_i(t)] \, r_i(t) \qquad \text{(Eq. 3-5)}$$

As before, it is expected that there will be losses of the jth species to the GER by processes incident to the manufacture of the ith end use product. Let the fraction of the jth species lost to the GER by these incidental processes be represented as $\beta_i(t)$, where $0 < \beta_i(t) < 1$. Then the rate of injection of the jth species into the GER from the ith manufacturing process is given by:

$$q_{III_i}(t) = \beta_i(t) \, [1 - \alpha_i(t)] \, r_i(t) \qquad \text{(Eq. 3-6)}$$

and the corresponding rate of injection of the jth species into the GER from all of the i manufacturing processes is given as:

$$q_{III}(t) = \sum_{i=1}^{N_i} \beta_i(t) \, [1 - \alpha_i(t)] \, r_i(t) \qquad \text{(Eq. 3-7)}$$

3-5.3 Losses Associated with End Use Product Service

As suggested by Eq. 3-6, not all the jth species used in the manufacture of the ith end use product result in a useful end product. After accounting for the losses to the GER, the effective rate of incorporation of the jth species into the ith end use product is designated by

$$w_i(t) = [1 - \beta_i(t)] \, [1 - \alpha_i(t)] \, r_i(t) \qquad \text{(Eq. 3-8)}$$

which may be simplified by the observation that, because $\beta_i(t) < 1$ and $\alpha_i(t) < 1$, the product $\alpha_i(t) \beta_i(t) \ll 1$. Making this substitution, Eq. 3-8 becomes:

$$w_i(t) = [1 - \beta_i(t) - \alpha_i(t)] \, r_i(t) \qquad \text{(Eq. 3-8')}$$

There are generally losses of the jth species to the GER by processes incident to the service of the ith end use product. Let it be assumed that these incidental losses amount to a fraction, γ_i, of the material that is in service at time t. It is explicitly assumed that γ_i is not a function of time. Further, it is assumed that the useful service life of the ith end product is λ_i years, after which the product containing the jth species is discarded. Then $W_i(t)$ is defined as the cumulative amount of the jth species contained in the ith end product in service at the time t. There are two cases of interest, as presented below.

Case No. 1; t $\leq \lambda_j$. In this case the service of the ith end use component has been too short to allow a significant contribution to the GER from the discarding of overage ith end use products. The only additional loss to the GER results from processes incidental to the service of the ith end use product. We will let $W_i(t)$ represent the cumulative species j in the in-service ith end use products at time t; similarly, $W_i^o(t)$ represents the total amount of the jth species committed to the manufacture of the ith end use product, up to time t. Then, in terms of the definition of γ_i, the rate of loss to the GER incident to the end use service of the ith product is given as:

$$q_{IV_i}(t) = \gamma_i W_i(t) \qquad \text{(Eq. 3-9)}$$

where γ_i is taken to be constant.
    If we now combine Eq. 3-8′ and 3-9,

$$\frac{dW_i(t)}{dt} = w_i(t) - \gamma_i W_i(t) \qquad \text{(Eq. 3-10)}$$

which has the general solution

$$W_i(T) = e^{-\gamma_i T} \int_0^{T \leq \lambda_i} w_i(t) e^{\gamma_i t} \, dt \qquad \text{(Eq. 3-10)}$$

On the other hand, the gross material committed to the ith end use product is:

$$W_i^o(T) = \int_0^{T \leq \lambda_i} w_i(t) \, dt \qquad \text{(Eq. 3-10)}$$

Combining Eqs. 3-1 and 3-10 ,

$$Q_{IV_i}(T) = W_i^o(t) - W_i(T)$$

$$= \int\limits_0^{T \le \lambda_i} w_i(t)\, dt - e^{-\gamma_i T} \int\limits_0^{T \le \lambda_i} w_i(t)\, e^{\gamma_i t}\, dt \qquad \text{(Eq. 3-11)}$$

Case No. 2; $t > \lambda_i$. When the time under consideration is greater than the useful service life of the ith end use product, Eq. 3-10 must then be modified to consider the residual material that has been discarded due to its being contained in coverage end use products:

$$\frac{dW_i(t)}{dt} = w_i(t) - \lambda_i W_i(t) - (1 - \gamma_i)^{\lambda_i} w_i(t - \lambda_i) \qquad \text{(Eq. 3-12)}$$

for which the general solution is

$$W_i(T) = e^{-\gamma_i T} \int\limits_0^{T} w_i(t)\, e^{\gamma_i t}\, dt - (1 - \gamma_i)^{\gamma_i} e^{-\gamma_i T} \int\limits_0^{T - \lambda_i} w_i(t)\, e^{\gamma_i t}\, dt \qquad \text{(Eq. 3-13)}$$

where the last term on the right-hand side represents the cumulative contribution to the GER arising from the discarding of overage products, $Q_{V_i}(T)$, where

$$Q_{V_i}(T) = (1 - \lambda_i)^{\lambda_i}\, e^{-\lambda_i T} \int\limits_0^{T - \lambda_i} w_i(t)\, e^{\gamma_i t}\, dt \qquad \text{(Eq. 3-14)}$$

Finally, in the epoch where $T > \lambda_i$, the cumulative in-service losses are given by

$$Q_{IV_i}(T) = W_i^o(T) - W_i(T) - Q_{V_i}(T) \qquad \text{(Eq. 3-15)}$$

The cumulative GER may now be computed by combining the (integrated) individual components, that is,

$$GER\,(T \le \lambda_i) = \int\limits_0^{T} (\text{Eqs. } 3\text{-}2 + 3\text{-}4 + 3\text{-}7)\, dt + \text{Eq. } 3\text{-}11 \qquad \text{(Eq. 3-16)}$$

$$GER\ (T > \lambda_i) = \int_0^T (\text{Eqs. 3-2 + 3-4 + 3-7})\,dt + \text{Eq. 3-14} \qquad (\text{Eq. 2-16})$$

As an illustration, the model is applied to the specific case of the PCBs in the following sections.

3-5.4 Estimated Environmental Load of PCBs

Production data on PCBs, as prepared by the Monsanto Company (Refs. 3-13, 3-14) are summarized in Table 3-8 and by the empirical Equation 3-17.

$$R\,(T = 1974) = \int_0^{44} 1.07 \times 10^6\ e^{0.0885t}\,dt \qquad (\text{Eq. 3-17})$$

which indicates that the total production of PCBs (by Monsanto) from 1930 to 1974 was approximately 5.94×10^8 kg (for other estimates see Durfee et al., Reg. 3-14). It will be noted that the data in Table 3-8 extend back to 1957, whereas Eq. 3-17 is extrapolated back to 1930, when PCB production actually began in the United States (Ref. 3-15). From the data (Table 3-8) it is also possible to separate the PCBs used in electrical applications (transformers and capacitors) from those used in nonelectrical applications. The data are summarized in Fig. 3-2 for electrical applications and Fig. 3-3 for nonelectrical applications. The best fit curves for these data are given by Eq. 3-18 and 3-19, respectively.

Electrical applications:

(Eq. $\qquad r_1\,(t) = 2.44 \times 10^6 \exp{(0.051t)}$ kg/yr. $\qquad$ (Eq. 3-18)

Nonelectrical applications:

$$r_2\,(t) = 6.57 \times 10^4 \exp{(0.139t)}\ \text{kg/yr.} \qquad (\text{Eq. 3-19})$$

It should be noted that the application of PCBs to electrical end products is assumed to have begun in 1930 whereas that to nonelectrical applications is assumed to have begun in 1940 and to have been halted in 1972.

   Following Durfee, et al. (Ref. 3-14) we accept the following assignment of parameters for PCBs:

Parameters for CER Estimate for PCBs

$\alpha\,(t) = 0.03$

$\alpha_1\,(t) = \alpha_2\,(t) = 0.01$

TABLE 3-8. Monsanto PCB Manufacturing and Sales (kg $\times$ 10^{-7}) yr^{-1}.

	1957	1958	1959	1960	1961	1962	1963	1964	1965	1966	1967	1968	1969	1970	1971	1972	1973	1974
U. S. Production	(1)	(1)	(1)	1.72	1.66	1.74	2.03	2.30	2.74	2.98	3.40	3.76	3.47	3.86	1.59	1.75	1.91	1.83
Domestic Sales	1.47	1.18	1.42	1.59	1.700	1.72	1.72	2.04	2.34	2.68	2.83	2.95	3.05	3.31	1.54	1.18	1.72	1.54
U. S. Export Sales	(2)	(2)	(2)	(2)	(2)	(2)	0.17	0.19	0.19	0.31	0.37	0.51	0.48	0.64	—	0.29	0.38	0.24
MFG Losses	—	—	—	—	—	—	0.14	0.07	0.21	—	0.20	0.30	(0.06)	(0.09)	—	0.28	(0.19)	0.05
Domestic Sales																		
Heat Transfer	—	—	—	—	—	0.007	0.026	0.042	0.054	0.082	0.104	0.11	0.14	0.18	0.14	—	—	—
Hydraulics	0.073	0.070	0.122	0.113	0.19	0.18	0.18	0.20	0.21	0.19	0.21	0.26	0.36	0.34	0.07	—	—	—
Misc. Industrial	0.032	0.034	0.073	0.073	0.095	0.076	0.069	0.077	0.082	0.082	0.065	0.058	0.049	0.073	0.054	—	—	—
Transformer	0.59	0.26	0.27	0.36	0.29	0.36	0.33	0.36	0.39	0.40	0.50	0.53	0.54	0.63	0.50	1.18	1.72	1.56
Capacitor	0.77	0.64	0.75	0.77	0.72	0.70	0.71	0.88	1.08	1.32	1.35	1.34	1.13	1.21	0.64	—	—	—
Plasticizer	(+)	0.18	0.21	0.28	0.41	0.40	0.42	0.47	0.53	0.61	0.61	0.65	0.75	0.88	0.15	—	—	—
Petroleum Additive	—												0.064					
Domestic Sales Aroclors																		
1242	0.82	0.45	0.62	0.83	0.90	0.94	0.84	1.09	1.43	1.80	1.95	2.04	2.06	2.20	0.99	—	0.28	0.28
1248	0.082	0.12	0.15	0.13	0.18	0.16	0.23	0.24	0.25	0.23	0.21	0.22	0.26	0.19	0.01	—	—	—
1254	0.20	0.30	0.31	0.28	0.29	0.29	0.27	0.29	0.35	0.32	0.32	0.40	0.44	0.56	0.21	0.16	0.36	0.28
1260	0.34	0.27	0.30	0.33	0.29	0.30	0.34	0.39	0.26	0.27	0.29	0.24	0.20	0.22	0.08	—	—	—
All others	0.011	0.017	0.043	0.035	0.039	0.045	0.048	0.057	0.052	0.072	0.072	0.056	0.081	0.14	0.26	0.95	1.07	1.00
	1.45	1.16	1.42	1.73	1.70	1.74	1.73	2.07	2.34	2.69	2.84	2.96	3.04	3.31				

Source: Refs. 3-13, 3-14.

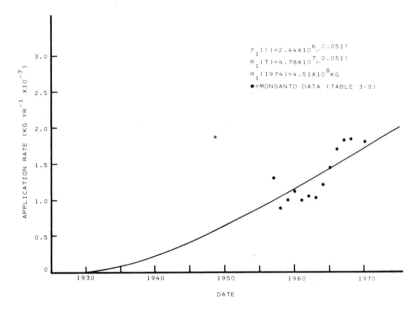

Fig. 3-2. Electrical applications of Arochlors.

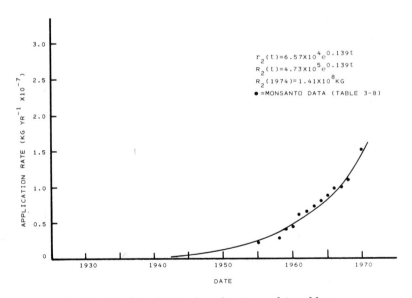

Fig. 3-3. Nonelectrical applications of Arochlors.

$$\beta_1 (t) = \beta_2 (t) = 0.03$$
$$\lambda_1 = 20 \text{ yr}, \lambda_2 = 4 \text{ yr}$$
$$\gamma_1 = 0.01, \gamma_2 = 0.10$$

Note: Subscript $_1$ refers to electrical applications

Subscript $_2$ refers to nonelectrical applications

If these parameters are inserted in the appropriate equations, we may compute the yearly accumulations to the GER resulting from each of the processes that have been treated. The results of these calculations are displayed in Fig. 3-4.

3-6 DATA SOURCES

Production data are useful in first estimates of the environmental impact of a material. Volume of production, manufacturing sites, production methods, uses, wastes generated, and pollution control measures are of interest. Useful references in this regard follow.

Agranoff, J. (ed.), *Modern Plastics Encyclopedia,* McGraw-Hill, Inc.,. New York, 1967-77 (updated annually).

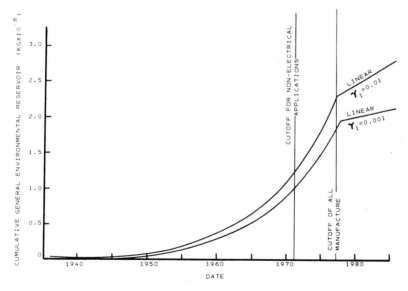

Fig. 3-4.   Estimated general environmental reservoir (GER) from PCBs using parameters from Table 3-8.

A Study of Industrial Data on Candidate Chemicals for Testing, EPA 560/5-77-006, Office of Toxic Substances, U. S. Environmental Protection Agency, Washington, D. C., 1977.

Berg, G. L. (ed.), *Farm Chemicals Handbook, 1976,* Meister Publishing Company, Willoughby, OH, 1976.

Chemical Economics Handbook, SRI International, Menlo Park, CA, 94025, current edition.

Chem Sources – U. S. A., 19th ed., Directories Publishing Co., Inc., Flemington, NJ, 1977.

Crowley, E. (ed.), *Trade Names Dictionary,* Gale Research Co., Detroit, MI, 1974.

Development Document for Effluent Limitations Guidelines and New Source Performance Standards for the Major Organic Products Segment of the Organic Chemicals Manufacturing Point Source Category (similar documents for other categories), U. S. Environmental Protection Agency, Washington, D. C., EPA Document 44011-74-009-9, April 1974 (to be updated 1980).

Directory of Chemical Producers (1975 and 1977), United States of America Chemical Information Services, SRI International, Menlo Park, CA, 1977 (updated annually).

Environmental Protection Agency Compendium of Registered Pesticides, Pesticide Regulation Division, E238, Office of Pesticides Program, U. S. Environmental Protection Agency, Washington, D. C. (updated regularly).

Faith, W. L., Keyes, D. B., Clark, R. L., *Industrial Chemicals,* 3rd ed., John Wiley and Sons, Inc., New York, 1976.

Hawley, Gessner, *The Condensed Chemical Dictionary,* 9th ed., Van Nostrand Reinhold Co., New York, 1977.

Herbicide Handbook of the Weed Science Society of America, 3rd ed., Weed Science Society of America Herbicide Handbook Committee, Champaign, IL, 1974.

IARC Monographs on the Evaluation of the Carcinogenic Risk of Chemicals to Man, International Agency for Research on Cancer, WHO, Lyon, France, Vols. 1-13, 1972-1976.

Jones, M. C., *A Study of Flavors, Fragrances and Essential Oils,* Delphi Marketing Services, Inc., New York, 1976.

Kirk, R. C. and D. F. Othmer (eds.), *Kirk-Othmer Encyclopedia of Chemical Technology,* 2nd ed., John Wiley & Sons, Inc., New York, Vols. 1-22, Supplement, 1963-1972.

McCurdy, P. (ed.), *Chemical Week 1976 Buyers Guide Issue,* October 29, McGraw-Hill, Inc., New York, 1975 (updated annually).

McCutcheon's Detergents and Emulsifiers, North American Edition, McCutcheon Division, MC Publishing Company, Glen Rock, NJ, 07452 (published annually).

Minerals Facts and Problems, Bureau of Mines, U.S. Department of the Interior, U.S. Government Printing Office, Washington, D.C., 1975.

Minerals Yearbook, Bureau of Mines, U.S. Department of the Interior, U.S. Government Printing Office, Washington, D.C. (published annually).

OPD Chemical Buyers Directory 1975-1976, Schnell Publishing Co., Inc., New York, 1975 (updated annually).

Sittig, Marshall, *Pesticide Production Processes*, Chemical Process Review No. 5, Noyes Development Corporation, Park Ridge, NJ, 1967.

Synthetic Organic Chemicals, U.S. Production and Sales, Annual and Preliminary reports, 1971-1975, U.S. International Trade Commission (formerly U.S. Tariff Commission) (updated annually).

The United States Pharmacopeia, 19th Revision, United States Pharmacopeia Convention, Inc., Rockville, Maryland, 1974.

Verschueren, K., *Handbook of Environmental Data on Organic Chemicals*, Van Nostrand Reinhold Co., New York, 1977.

Windholz, M. (ed.), *The Merck Index*, 9th ed., Merck and Co., Inc., Rahway, NJ, 1976.

Wiswesser, W. (ed.), *Pesticide Index*, 5th ed., Entomological Society of America, College Park, MD, 1976.

REFERENCES

3-1. *Development Document for Effluent Limitations Guidelines and New Source Performance Standards for the Major Organic Products Segment of the Organic Chemicals Manufacturing Point Source Category* (similar documents for other categories), U. S. Environmental Protection Agency, Washington, D. C., EPA Document 44011-74-009-9, April 1974 (to be updated 1980).

3-2. Hamelink, J. L., "A Proposed Method for Deriving Effluent Limits from Water Quality Criteria," *Analyzing the Hazard Evaluation Process,* K. Dickson, A. Maki, and J. Cairns (Eds.), published by Water Quality Section of the American Fisheries Society, Washington, D. C., 1979.

3-3. Metcalf and Eddy, Inc., *Wastewater Engineering: Treatment, Disposal, Reuse, 2nd ed.,* revised by G. Tchobanoglous, McGraw-Hill Book Co., New York, pp. 832-859, 1979.

3-4. *Environmental Pathways of Selected Chemicals in Freshwater Systems, Part 1 . . . Background and Experimental Procedures,* prepared by SRI International for Environmental Research Laboratory, U. S. Environmental Protection Agency, Athens, GA; available as Document PB-274 548, National Technical Information Service, Springfield, VA 22161.

3-5. *Pre-Screening for Environmental Hazards – A System for Selecting and Prioritizing Chemicals,* prepared by A. D. Little, Inc., for Office of Toxic Substances, U. S. Environmental Protection Agency, Washington, D. C.; available as Document PB-267 093 from National Technical Information Service, Springfield, VA 22161.

3-6. Seminar on Meteorology and Atmospheric Dispersion, presented by H. E. Cramer Co., Inc., to Union Carbide Corp., South Charleston, WV, March 1974.

3-7. Pasquill, F., *Atmospheric Diffusion,* D. Van Nostrand Co., Ltd., London, 1962.

3-8. *Air Quality Display Model,* U. S. Environmental Protection Agency, Washington, D. C., 1969; available as PB-189 194 from National Technical Information Service, Springfield, Virginia 22161.

3-9. Turner, D. B., *Workbook of Atmospheric Dispersion Estimates,* Office of Air Programs Publ. AP-26, U. S. Environmental Protection Agency, Washington, D. C.; Revised 1970.

3-10. Briggs, G. A., "Some Recent Analyses of Plume Rise Observations," presented at Second International Clean Air Congress, Washington, D. C., December 6-11, 1970.

3-11. McElroy, A. D., *et al., Loading Functions for Assessment of Water Pollution from Nonpoint Sources,* EPA/600/2-76/151, U. S. Environmental Protection Agency, Washington, D. C., 1976; available as PB-253 325 from National Technical Information Service, Springfield, Virgina 22161.

3-12. Smith, C. N., *et al., Transport of Agricultural Chemicals from Small Upland Piedmont Watersheds,* EPA/600/3-78/056, U. S. Environmental Protection Agency, Athens, GA, 1978.

3-13. Monsanto Chemical Co. Tech. Bull. PL-306A, "Aroclor Plasticizers," Monsanto Industrial Chemicals Co., 800 N. Lindbergh Blvd., St. Louis, Missouri 83166.

3-14. Durfee, R. L., *et al., PCBs in the United States, Industrial Use and Environmental Distribution,"* Final Report – EPA Contract 68-01-3259, February 25, 1976, VERSAR, Inc., 6621 Electronic Drive, Springfield, Virginia 22151.;

3-15. Penning, C. H., "Physical Characteristics and Commercial Possibilities for Chlorinated Diphenyl," *Ind. Eng. Chem.,* **22,** pp. 1180, 1930.

4

The risk of chemicals to aquatic environment

**G. R. Southworth, B. R. Parkhurst,
S. E. Herbes, and S. C. Tsai**

Environmental Sciences Division
Oak Ridge National Laboratory
Oak Ridge, Tennessee

4-1 FACTORS DETERMINING THE RISK OF CHEMICALS TO AQUATIC ENVIRONMENTS

The introduction of chemicals to the aquatic environment admits the possibility of affecting the environment in an undesirable manner. The loss of aquatic life or productivity, the creation of toxic hazards for persons drinking or contacting the water, or the creation of unpleasant odors or appearance can reduce the desirability of an aquatic resource. The objective of this chapter is to explore the methods available for identifying materials whose use or manufacture might result in harm to aquatic environments.

Identification of potential problem materials is critical to implementation of the Toxic Substances Control Act of 1976 (TSCA) and similar laws worldwide. One purpose of TSCA is "to assure that innovation and commerce in . . . chemical

This research sponsored by the U. S. Department of Energy under contract W-7405-eng-26 with Union Carbide Corporation. Publication No. 1584, Environmental Sciences Division, Oak Ridge National Laboratory.

substances and mixtures do not present an unreasonable risk of injury to health or the environment." However, a second objective of the act is that "authority over chemical substances and mixtures should be exercised in such a manner as not to impede unduly or create unnecessary barriers to technological innovation." This makes it necessary to accurately assess the potential hazard with a minimal commitment of resources. Given the dual role of protecting public health and environment on the one hand, without unduly restricting innovation and productivity in the chemical industry on the other, a risk analysis program must determine the information essential to making an informed judgement and the most reliable and cost-effective tests for acquiring that information.

4-1.1 Environmental Risk

Before discussing further the hazard analysis specific to aquatic environments, it is necessary to define "Environmental Risk" as it pertains to the aquatic environment. Risk to the aquatic environment is defined for our purposes as the likelihood that the aquatic environment will suffer harm, directly or indirectly, from the introduction of chemicals as a result of human activity. Harm to the aquatic environment falls into two general categories: contamination of the aquatic system or any of its components, that poses a hazard to terrestrial organisms utilizing the system, and production of undesirable changes in structure and/or function of the aquatic ecosystem. This second classification of environmental harm would include effects such as:

- Extinction of species of plants or animals.
- Loss of species from specific aquatic ecosystems.
- Changes in the relative abundance and importance of species comprising the aquatic community.
- Changes in standing crop, size per individual, population age structure, or production within populations of individual species.
- Interference with ecosystem functions of energy conversion and element cycling.
- Changes in physical properties of the system, such as appearance and odor.

The analysis of risk to the aquatic environment involves estimating the probability that harm will result from a specific action (such as the manufacture and marketing of a new chemical), and also the *extent, duration,* and *degree* of damage likely to occur. Such an analysis requires that the concentration of the substance under consideration in aquatic environments be estimated as a function of time and location, and that a prediction of its effects be made. Coupled with this estimate of potential damage, it is necessary to analyze the errors associated with its derivation in order to determine the probability of its accuracy. Since all materials cannot be tested in great depth, a risk analysis system must be developed to screen out substances that are unlikely to become environmental hazards with a small amount of testing, whole also identifying clearly hazardous situations and materials whose potential hazard is ambiguous enough to warrant detailed testing.

4-1.2 Major Aspects of Environmental Effects

The key to aquatic hazard analysis is the determination of environmental concentrations of specific materials and the effects they produce. The major aspects of environmental effects are degree, extent, and duration. Degree refers to the intensity of the effects (degree of harm) or the concentration of contaminants in key components of the ecosystem (degree of contamination). Estimation of the degree of environmental harm can probably never be better than a semiquantitative exercise, due in large part to the variety of components included in the definition of environmental effects. Extent of environmental damage refers to the geographical bounds of harm to the environment. This is simplified for our discussion into three categories — global, regional, and local effects. Duration of effect pertains to the time required for an affected aquatic ecosystem to return to its original condition, or the time required for concentrations in specific components to dissipate, once primary inputs of the toxic agent have been discontinued.

In order to reasonably estimate the extent, duration, and degree of environmental harm that might result from the manufacture, distribution, use, and disposal of a specific chemical, it is necessary to have information on factors that act to determine that risk. These determinants can be classified as:

1. Properties of the substance itself.
2. Properties of its source.
3. Properties of the aquatic environments receiving inputs.

Properties of the substance include its toxicity, its persistence in the environment, and its susceptibility to transport and redistribution in the environment (see Chapter 2). More specific physical, chemical, and toxicological properties combine to determine these general properties. Examples of such specific properties are water solubility, lipid solubility, vapor pressure, photoreactivity, acute toxicity, and carcinogenicity.

Properties of the source of a contaminant in the environment include quantities released, the geographical distribution of sources, the medium (air, water, soil) into which release occurs, and the distribution of releases in time.

Properties of the aquatic environment receiving inputs include physical and chemical characteristics, water renewal rate, trophic status, and species assemblages present.

Thus, the combination of substance's own properties — when, where, how, and how much of it is involved, and which aquatic environments does it enter — determines its impact. In the following sections we discuss which of the above properties create hazards, particularly those of long duration and of global or regional scale, and methodologies for measuring or estimating the hazard. It is hoped such an approach will be useful in determining the most cost-effective information needed for making the initial evaluation of risk to aquatic environments.

4-1.3 Extent of Environmental Effects

The geographical distribution of environmental damage possible from the manufacture, distribution, use, and disposal of a chemical is determined primarily by the distribution and nature of its sources and the way it is transported and redistributed in the environment. The role of these factors in determining the geographical scale of inputs to aquatic ecosystems is covered in Chapter 3.

Once a material is released to the environment, physical, chemical, and biological processes act to redistribute it and may transform it into forms that are hazardous or less hazardous. A schematic of processes acting upon substances released directly to aquatic systems is depicted in Figure 4-1. Initially, the material is partitioned between aqueous and particulate phases by *sorption*, in which some of the substance is adsorbed or absorbed by particulate matter suspended in the water. Sorption also occurs at the interface between water and sediments. Sorption to the sediments and sedimentation of suspended particles acts to accumulate substances with high sorption affinity in sediments. Sorption is generally viewed as an equilibrium process, resulting from a dynamic exchange of the sorbed substance between water and particles. Thus, removal of dissolved material from the water will be counterbalanced in some cases by desorption of the material.

ORNL-DWG 79-19164

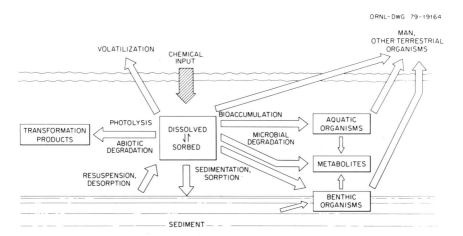

Fig. 4-1. Generalized pathways and processes of environmental transport of contaminants.

Thus sediments may act as both a source and a sink for foreign substances. In riverine systems, sediments are also transported downstream and ultimately deposited in lakes, estuaries, or the ocean. Deposition of uncontaminated sediment on top of a contaminated layer may prevent its serving as a source to the overlying water.

That fraction of the substance remaining in solution may be acted on by various removal processes, including volatilization, photolysis, oxidation, reduction, hydrolysis, radioactive decay, and microbial degradation. Material within the sediments may also be removed by these processes, with the exceptions of volatilization and photolysis. The rates of these processes vary with the chemical properties of the substance under consideration and characteristics of the environment.

The transport of chemicals in aquatic ecosystems is also mediated by the biota, which take up such materials from the water and food and may accumulate and/or metabolize them. High levels of accumulation are possible, resulting in organisms attaining concentrations of materials potentially hazardous to animals preying on them. Determination of the potential for a given substance to accumulate in food organisms thus becomes an important factor in hazard analysis.

Environmental transport of substances released to the land or atmosphere may be important in delivering such materials to aquatic ecosystems. Via runoff or groundwater, rainwater and snow melt can carry substances from land to aquatic environments. Substances in the atmosphere can enter aquatic environments by precipitation (wetfall), by settling of dust particles containing the sorbed material (dryfall), and by direct dissolution from air to water.

4-1.3.1 Global Scale Hazards

Perhaps the most significant hazard attributable to pollutants derived from human activity is global scale environmental damage or contamination. At this time examples of actual damage are not yet apparent, although several examples of global contamination are evident. Radionuclides from atomic weapons testing (Ref. 4-1), fluorocarbons from aerosol propellants and refrigeration systems (Ref. 4-2), carbon dioxide from the combustion of organic fuels (Ref. 4-3), and DDT (Ref. 4-4) are examples of global contaminants.

Identification of possible global hazards is made more complicated by the complex, indirect way in which global effects may occur. For example, the potential threat of increased atmospheric CO_2 to aquatic ecosystems is not toxicological, but stems instead from the changes in hydrologic cycle patterns and water temperatures likely to occur if elevated atmospheric CO_2 alters global climate. Although toxicological screening tests would undoubtedly overlook such indirect effects, there clearly are properties of both substances and their sources that are required to create a potential global hazard. If substances possessing such properties can be identified, we can greatly reduce the list of substances requiring evaluation as possible indirect threats to global aquatic environments.

Four general properties necessary for the creation of a global hazard are listed in Table 4-1. These are the operation of a global dispersal mechanism, extremely large releases of a substance to the environment, long persistence of the material in the environment, and the capability of causing harm at low ambient concentrations. Materials that do not possess all four properties are unlikely to present global scale environmental hazards to aquatic ecosystems. More specific properties (Table 4-2) act to enhance the potential for global hazard to aquatic environment. A substance need not possess all these properties in order to be a global hazard, but each may contribute to creating a problem.

In order to pose a hazard on a global scale, a substance obviously needs to be distributed to aquatic ecosystems all over the world, or throughout those aquatic ecosystems (the oceans) whose extent is global. The most straightforward means of achieving worldwide dispersal of a chemical is through atmospheric transport. Materials volatile enough to vaporize can be transported long distances and be introduced to aquatic systems through precipitation, aerosol deposition, and direct interfacial transfer. Physical and chemical properties of a substance, as well as properties of its source, determine its potential for global dispersal via atmospheric transport.

In order for the atmosphere to act as a source of worldwide contamination to aquatic systems, a contaminant must be volatile enough to exist as a gas in appreciable quantities. It must also dissolve into water in contact with the gas to a degree that will result in appreciable aqueous concentrations at ambient atmospheric concentrations. However, this partitioning cannot be so strong that the substance is readily scavenged by precipitation, as this will act to decrease its atmospheric residence time and prevent global transport. These simple physical properties (volatility and gas–water partitioning) greatly reduce the number of chemicals possessing global atmospheric transport properties. Resistance to photochemical transformation in the atmosphere is an additional requirement for global transport, although reaction products must be considered as possible hazards.

TABLE 4-1 Properties Necessary for Risk of Global Hazard.

Global dispersal mechanism
    Atmosphere
    Oceans
    Global source distribution

Very large quantities released to the environment

Long persistence in environment, aquatic and atmospheric

Capability of causing harm at low ambient concentrations

TABLE 4-2. Specific Properties of Source and Substance Acting to Enhance the Potential for Global Hazard to Aquatic Environments.

Source properties
    Global distribution
    Very large quantities released
    Direct release to atmosphere
    Direct release to ocean

Substance properties
    Propensity for transport in atmosphere
    Capability of transfer from atmosphere to water
    Long persistence in environment
    High mobility in soil
    High toxicity
    High accumulation potential in aquatic biota
    Long turnover time in aquatic biota
    Potential for indirect effects

The volatility of a chemical is indicated by its *equilibrium vapor pressure*, a measure of the concentration of the substance in the gas above an amount of the material in its liquid or solid phase at a given temperature. Equilibrium vapor pressure is a thermodynamic property, not a kinetic parameter, and although it sets an upper limit on attainable atmospheric concentration, it is not directly indicative of a substance's *rate* of volatilization. A second property pertinent to atmospheric transportability is the Henry's Law coefficient. This is an equilibrium coefficient that describes the partitioning of substances between gas and water at a given temperature. In this chapter it is expressed as:

$$H = [x]_{air} / [x]_{water} \qquad \text{(Eq. 4-1)}$$

where

H = Henry's Law coefficient

$[x]_{air}$ = the molar concentrations of x in air at 1×10^5 Pa and $25°C$

$[x]_{water}$ = the molar concentration of x in water in equilibrium with $[x]_{air}$

H is also commonly defined as the ratio of concentration per unit mass or the ratio of mole fractions, or, as used here, molar ratios. Care must be taken to be certain of the underlying definition of H used when comparing values, as the magnitude of the dimensionless H varies considerably with the units used in defining it.

Table 4-3 provides equilibrium vapor pressures and Henry's Law coefficients for several chemicals whose environmental distribution is widespread. Several substances whose atmospheric distribution is widespread, DDT and PCBs, (Refs. 4-4, 4-5, 4-6) have low equilibrium vapor pressures, indicating that high vapor

TABLE 4-3. Vapor Pressures and Henry's Law Coefficients for
Several Environmental Contaminants at 25°C.

Substance	Vapor pressure (Pa)	Reference	Henry's Law Coefficient, H^a	Reference
Dimethyl Phthalate	1.3	4-7	2×10^{-2}	4-7
Mercury	1.3×10^{-1}	4-8	5×10^{-1}	4-8
DDT	1.3×10^{-5}	4-8	2×10^{-3}	4-8
Atrazine	4×10^{-5}	4-9	1×10^{-7}	4-9
SO$_2$	Gas		2×10^{-4} [b]	4-10
Benz(a)pyrene	9×10^{-7}	4-11	1×10^{-5}	4-12
PCB (Aroclor 1254)	10×10^{-3}	4-8	2×10^{-1}	4-8
CO$_2$	Gas		1×10^{0}	4-13

[a]Molar concentration ratio, gas–water at 1×10^5 Pa and 25°C.
[b]Includes hydration, $SO_2 + H_2O \rightleftharpoons H_2SO_3$.

pressure is not a requisite for atmospheric transport. There is considerable variation noted in H, ranging from 10^{-7} for atrazine, which partitions predominantly into water, to about 1.1 for CO_2, which distributes evenly between gas and water. The relationship between H, the atmospheric concentration of a substance (expressed as vapor pressure), and its concentration in water equilibrated with that air is shown in Fig. 4-2 for a 1-cm^3 volume of water equilibrated with 22 liters of water-saturated air. A 100-ng/liter concentration of a material in water corresponds to 1×10^{-9} to 2×10^{-10} M for substances with molecular weight 100 to 500. Thus, a material with an equilibrium vapor pressure of $< 1 \times 10^{-6}$ Pa would not produce aqueous concentrations of ~100 ng/liter unless the entire atmosphere was saturated with it and H was $< 1 \times 10^{-3}$. Substances with equilibrium vapor pressures $< 1 \times 10^{-9}$ Pa could not ever realistically be expected to produce aqueous concentrations as high as 100 ng/liter in rainfall due to vapor scavenging. As H increases, higher atmospheric concentrations are required to produce a given aqueous concentration.

The relative ease with which a substance is scavenged by rain can determine its persistence in the atmosphere. Substances with low Henry's Law coefficients ($H < 1 \times 10^{-4}$) are readily scavenged by rainfall, suggesting relatively short atmospheric residence. From Fig. 4-2, it can be seen that those compounds for which atmospheric transport has been shown to provide an important input to aquatic ecosystems (DDT, PCBs), H falls between 10^{-1} and 10^{-4}

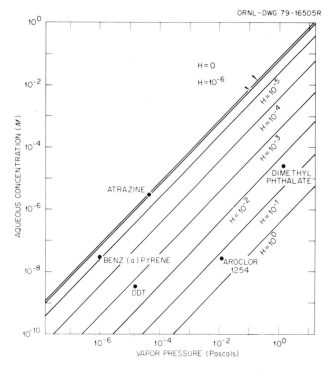

Fig. 4-2. Calculated concentration of a chemical in 1 cm^3 of water equilibrated with 22.4 ℓ of water saturated air vs the vapor pressure of the chemical in the air prior to equilibrium. Data points for specific chemicals are for equilibrium vapor pressure at 25°C.

Source magnitude is a third simple property that helps to further define hazard potential. Large amounts of material must be released to the atmosphere in order to produce significant levels of global contamination, particularly if the atmospheric residence time is short. Direct atmospheric inputs of hundreds or thousands of metric tons per year are thus likely to be requirements for global contamination of aquatic ecosystems. Such releases are similar to the total annual production volume of many of the most widely used pesticides (Table 4-4).

Other properties that act to determine global aquatic hazard due to atmospheric transport and deposition are persistence in the atmosphere, mobility in soil, and persistence in the aquatic environment. The atmosphere is a highly reactive environment, with intense solar radiation directly degrading many substances and producing reactive radicals that act to degrade others (Ref. 4-20). Many common gaseous pollutants are subject to rapid chemical transformation in the atmosphere (Ref. 4-17). Rapid transformation reduces the steady state atmos-

TABLE 4-4. Annual Production and/or Environmental
Release of Specific Chemicals in the United States.

Chemical	Production Volume (metric tons/yr)	Year	Reference
DDT	7.2×10^4	1961	4-14
PCB	3.3×10^4	1970	4-15
Mercury	1.9×10^3	1973	4-16
2, 4-D	1.9×10^4	1970	4-9
Benzene[a]	4.5×10^5	1971	4-16
Atrazine	4.1×10^4	1971	4-9
SO_2^b	1.3×10^8	1970	4-17
Benz(a)pyrene	8×10^2	1971–1973	4-16
CO_2	5×10^9	1970	4-3
Phthalates	4.5×10^5	1969	4-18
Mirex[c]	2.7×10^2	1965	4-19

[a]Auto emissions only.
[b]Global production.
[c]Peak production at Niagara Falls, New York only.

pheric concentration attainable, and thus reduces the potential for direct inputs to aquatic ecosystems. However, this also may act as a source of hazardous reaction products. Atmospheric contaminants in rainfall will be transported to lakes and streams only if they are not retained by soil within the watershed. Substances readily retained by soil pose a threat primarily to aquatic ecosystems where direct precipitation of the water surface plays a larger role in the water budget than terrestrial runoff or groundwater flow. In such an instance, global atmospheric contamination may produce local or regional scale impacts on aquatic environments.

Persistence within the aquatic system also enhances the global hazard potential, as rapid removal, coupled with the low concentrations delivered via atmospheric transport, would act to keep aqueous concentrations low.

Substances lacking physical–chemical properties that would make them candidates for long-range atmospheric transport may still be dispersed globally if the sources to the atmosphere themselves are globally distributed. An example would be emissions from combustion of organic fuels, such as jet aircraft emissions. Stratospheric release is another means by which materials with properties not consistent with long-range atmospheric transport can achieve global distribution. Radioactive particulates from nuclear weapons tests were globally distributed as a result of being injected into the stratosphere, where low rates of precipitation scavenging and vertical mixing prevented rapid removal (Ref. 4-1).

In addition to atmospheric dispersal, the oceans may provide a means by which contaminants may achieve near-global distribution. Oceanic mixing times are long and volumes are quite large, indicating that substances capable of posing a global contamination hazard due to ocean dispersal would have to be quite persistent in the water column and released in great quantities.

Other contributing factors in the creation of a global hazard involve the biological behavior and effects of materials. These factors, including toxicity, bioconcentration potential, and residence time in biota, are covered in the sections dealing with duration and degree of effect.

A final and very important factor in global contamination is the potential for indirect effects. A good example of an indirect effect is the toxic effect of aluminum in lakes in the Adirondack region of New York State. Acidic precipitation, due primarily to oxides of sulfur and nitrogen, mobilizes aluminum from the soil and acidifies lakes to levels at which aluminum remains in solution at toxic concentrations (Ref. 4-21). Identifying possible indirect effects of pollutants is probably best approached by identifying global and regional contaminants and relying on the scientific community to investigate the possible indirect mechanisms of ecological effects. A few, such as chemical transformation to more toxic forms, and mobilization of other toxic substances, may be checked in advance, but setting up an *a priori* list of indirect effects to investigate may result in less obvious possibilities being overlooked.

4-1.3.2 Regional-Scale Hazards

The problem of regional scale environmental harm is in many ways more difficult to address than global contamination. With only one planet to consider, the simplest global models can contain only a few compartments (such as land, water, biota, and atmosphere), and the constraints posed by such things as finite atmospheric volume, etc., are obvious. For regional problems, there exist an unlimited number and variety of regions, with almost any chosen definition of "region." Good models for predicting environmental contamination patterns are fundamental to addressing the problem.

The possibility of environmental harm on a regional basis is a function of properties of the regional environment as well as properties of the source and substance. The properties necessary for existence of a regional hazard (defined for our discussion as an impact occurring ⩾25 km from a discrete point or area source, but not worldwide in scope) are similar to those required for global hazard (Table 4-5).

The specific properties that act to enhance the potential regional hazard are quite similar to those enhancing global hazard differing primarily in magnitudes

TABLE 4-5. Properties Required for Regional Hazard.

Regional dispersal mechanism

Release and persistence of quantities sufficient to contaminate a large area

Capability of causing harm at low concentrations

Presence of aquatic ecosystems in the region

of rates and fluxes (Table 4-6). Thus, important source properties would be wide distribution within the region (such as numerous discrete sources or broad distribution of a material, as in pesticide application), large quantities of material released to the environment, direct aquatic inputs, and direct atmospheric inputs. These properties are not required (thus, material not directly released to water or atmosphere may also be regional hazards), but they increase the likelihood of regional contamination.

Properties of the substance are very important in determining regional hazard, in part because of the role environmental transport plays in regional scale contamination. Processes that prevent a substance from being a global hazard, such as precipitation scavenging, may cause regional hazards. In several cases it has been demonstrated that atmospheric transport of toxic substances causes regional damage to aquatic ecosystems. In Scandanavia, many lakes and ponds (Ref. 4-22) have been adversely affected by acid rain resulting from discharges to the

TABLE 4-6. Specific Properties of Source, Substance, and Environment Acting to Enhance the Potential for Regional Hazard to Aquatic Environments.

Source properties
  Regional distribution of sources
  Large quantities released
  Direct release to aquatic systems
  Direct atmospheric release

Substance properties
  Atmospheric transportability
  Propensity for transfer from atmosphere to aquatic ecosystem
  High mobility in soil and water
  Resistance to degradation in atmosphere, soil, and water
  High toxicity
  High bioaccumulation potential

Environment properties
  Regional scale aquatic ecosystems
  Regional assemblages of unique or highly sensitive biota
  Presence of critical habitats where local impacts could produce regional effects
  Specific physical, chemical, and biological properties that enhance persistence
  Regional variations in climate, geology, and land use that enhance input or persistence

atmosphere hundreds of miles away.   In Lake Michigan and Lake Superior, atmospheric inputs of PCBs seem to be maintaining relatively high levels of PCBs in water and biota (Refs. 4-5, 4-6).

Two key properties associated with regional atmospheric transport are equilibrium vapor pressure and Henry's Law coefficient, as in global transport.   However, the existence of a gaseous phase is not required for regional atmospheric dispersal, as particulates are readily transported over 100 km in the air (Refs. 4-23, 4-24).   Substances introduced to the atmosphere as aerosols or transformed into aerosols are readily input to aquatic ecosystems.   Materials with appreciable vapor pressure $(VP \geqslant 1 \times 10^{-5}$ Pa) and low values of H $(H \leqslant 1 \times 10^{-3})$ are readily scavenged by precipitation and input to aquatic systems.   If H is high, inputs will be spread over a larger area and are less likely to attain hazardous concentrations in precipitation.   As was the case in global transport, equilibrium vapor pressure places a constraint on atmospheric transportability of substances requiring volatilization to be input to the atmosphere.   Kinetic factors, primarily the rate of volatilization, play a larger role in determining the atmospheric concentration on a regional scale, where short-term processes dominate.

The mobility of a substance in soil in contact with water is a key property for substances that may be distributed on terrestrial systems, either by atmospheric input or direct application.   Substances not tightly bound by soil particles will be leached into runoff and ground water and input to surface waters (Ref. 4-25).   Soil mobility also affects the likelihood of substances disposed of in landfills.   Persistence in the terrestrial environment also plays a key role in determining the extent to which terrestrial ecosystems may serve as reservoirs and sources of contamination for aquatic systems (Chapter 6).

The persistence and redistribution of a substance upon reaching the aquatic environment is critical in determining the extent of contamination.   Persistence is determined by a substance's susceptibility to various removal processes and is a function of properties of both the substance and the environment.

For large lakes and estuaries, where the water renewal rate is much slower than in rivers, removal processes are probably more significant in restricting the extent of contamination.   That is, relatively slower removal processes may remove a considerable fraction of a contaminant before it is widely dispersed, due to the slower bulk movement of water.   Unless the rates of such removal processes are quite rapid, continuous releases of contaminants to aquatic ecosystems are likely to result in regional scale contamination, particularly in riverine ecosystems.   Regional distribution and transport of contaminants generally entails considerable dilution of the materials in environmental media.   Therefore, substances that are toxic at very low concentrations or that bioaccumulate to a very high degree are more likely to pose hazards on a regional scale than less toxic, less bioaccumulable materials, because, even after substantial dilution, such materials may still exist at hazardous levels.

Properties of the environment play a critical role in determining the regional contamination potential of a substance. The nature of the aquatic environment itself can determine whether a substance represents a local or regional hazard or no hazard at all. Because the aquatic environment itself can act as the major distribution mechanism for contaminants, a fundamental property enhancing regional contamination potential is the presence of regional scale ecosystems. Examples of such systems in the United States are the individual Great Lakes, large rivers, and bays and estuaries. A single source of contamination located on such a system can produce large-scale regional contamination. Examples of such instances include the contamination of Lake Ontario with Mirex (Ref. 4-19), the James River Estuary with Kepone (Ref. 4-26), and the upper Hudson River with PCBs (Ref. 4-27).

The presence of regional assemblages of unique, endangered, or highly sensitive biota increases the susceptability of a region to environmental harm. Organisms tend to be regional in distribution, often limited to specific drainage basins. In a similar vein, the presence of regional assemblages of unique or pristine ecosystems must be considered. While such systems may or may not be more susceptible to environmental harm than systems already impacted by man, levels of contamination that are acceptable in the impacted systems are likely to result in the loss of the attributes of the pristine system that make it unique. Thus, standards set to protect areas subject to human impact might be less stringent than those where wilderness ecosystems are part of the region under consideration.

The presence of critical habitats, where localized impacts could cause regional scale environmental damage acts to increase the potential for regional hazards. Examples of such a situation would be a toxic discharge that blocks access of fish to upstream spawning areas, or the spill of a highly toxic material in an area where populations that are normally distributed throughout a region are concentrated at certain times of the year.

The physical, chemical, and biological properties of regional scale ecosystems play a key role in determining potential for regional scale harm. Physical properties, such as water residence time, depth, temperature, wind and current velocities, surface area and dimensions, flow characteristics, and the relative importance of precipitation, runoff, and groundwater as water inputs, act to influence the removal rate of contaminants in the system. Similarly, chemical properties, such as pH, ionic strength and composition, light transparency, suspended solids loading, composition, and settling rate, and nutrient concentrations, influence processes acting to compartmentalize or transform contaminants. Finally, the biotic community itself can be a major transport system, either by translocating contaminants through the movements of contaminated organisms, or by biological degradation of contaminants. The interaction of these environmental properties with properties of the source and substance is best evaluated by means of a model that predicts substance's behavior as a function of such properties. However, past experience with environmental contaminants suggests that certain properties of regional scale ecosystems make them highly susceptible to contamination with persistent contaminants. Among these are long water residence time, low rates of sedimentation, and considerable depth. This combination of properties minimizes the role of sedimentation, water outflow, and interfacial removal processes (direct sorption, volatilization) in preventing the buildup of contamination.

Regional variations in climate, geology, and land use may affect the potential for regional contamination, particularly in the case of airborne contamination. The ability of the soil to neutralize or retain contaminants borne in air and precipitation can vary greatly with geological and land use factors. In areas of very thin soil cover and high precipitation, such as Scandanavia, airborne acids are readily transported to aquatic systems. The storage of contaminants in snowpack and subsequent release in spring can produce sudden, large fluxes of contaminants.

4-1.3.3 Local Scale Contamination

The problem of local-scale contamination is made tractable by the limits placed on the systems considered. Local scale contamination is a site-specific problem, with point or small area sources. In order to estimate the ambient environmental concentration it is necessary to know the nature and magnitude of a contaminant's inputs to an aquatic system, the magnitude of the receiving system, and the rate of degradation, removal, and redistribution of the substance in the system. To determine contamination on a local scale, intermediate and long-range transport models and good information on rates of removal processes are not as critical as they are for regional scale contaminants, although detailed small-scale models, such as plume models, may be vital. Distinguishing regional versus local scale contaminants, however, is not a clearcut process. Since the evaluation of materials on a local scale in inherently simpler than a regional approach, it is perhaps advisable to screen materials first for their potential hazard on a local basis (where sources are discrete enough to allow this), using a set of conservative simplifying assumptions and hypothetically retaining all releases within the defined locale. If a substance presents little or no hazard under such an analysis, it undoubtedly presents even less on a regional scale.

4-1.4 Duration of Environmental Effects

There are two facets to the duration of environmental effects — one relating to the duration of the contamination itself once the primary source has been removed, and the other to the duration of the impact of the contamination once contamination is removed. Duration of the contamination itself after the primary source has been removed is determined by the persistence of the substance in the various compartments of the aquatic environment, and the possibility of continuing inputs from secondary sources. The properties of the source, substance, and environment that act to enhance the duration of contamination are listed in Table 4-7.

Source properties that increase the duration of contamination include the presence of external secondary sources, such as soil contamination, waste disposal sites, atmospheric inputs, and a large reservoir of material distributed throughout routes of commerce that will ultimately be disposed of. An example of a material having such secondary sources are PCBs. Despite the major cutback in PCB manufacture in the United States in 1971, atmospheric inputs of PCBs to the upper Great Lakes continue at high levels (Refs. 4-5, 4-6), presumably as a result of PCB volatilization from materials currently in use or in disposal sites. Another example of a secondary source creating long-term contamination problems is the mercury leaching from abandoned waste disposal sites on the North Holston River in Virginia (Ref. 4-28), which has contaminated over 100 km of stream.

Persistent chemicals can remain in aquatic systems long after primary inputs have been eliminated. Sediments can act to accumulate and store contaminants, which can then be released back into the water through desorption. The biota itself can act as a reservoir of contamination for substances having high accumulation potential and slow elimination kinetics, with organisms retaining high levels of contamination throughout their lifetimes even after aqueous sources are removed.

Environmental characteristics affect the duration of chemical contamination. The residence time of water in the system, or flushing rate, has a direct impact

TABLE 4-7. Properties of Source, Substance, and Environment Acting to Extend the Duration of Contamination of Aquatic Systems After the Primary Source is Discontinued.

Source properties
  Diffuse secondary sources which act as external reservoirs

Substance properties
  Long persistence in the aquatic environment
  Accumulation and storage in sediments
  Accumulation and storage in biota

Environment properties
  Long residence time of water in the system
  Readily reversible exchange of material between water and sediment
  Presence of long-lived biota that accumulate contaminants
  Other biological, chemical, and physical properties of the system that favor slow degradation rates

on the removal rate of contaminants. Many environmental properties influence degradation rates. Light transparency of the water and suspended solid loadings influence photolysis rate and temperature influences abiotic and biotic processes such as volatilization and microbial degradation. The relative importance of desorption from bottom sediments as a secondary source is affected by the depth of the water and the rate at which new sediments are deposited above contaminated sediments. Thus, a given desorption flux may have more impact on a shallow body of water than a deep one, where the contaminant is more diluted.

The characteristics of the biota present also influence the duration of contamination. Long-lived species may retain high levels of some contaminants for the remainder of their lifetimes, making systems containing such organisms take longer to reflect changes in contaminant loading than systems in which long-lived biota are less dominant.

The *duration of the effect* of prior contamination is determined primarily by properties of the environment and the degree and extent of environmental damage (Table 4-8). Properties of the environment include the presence of unique species, whose loss would result in extinction, and the rate of immigration or colonization by organisms, which provides a supply of organisms for recovery of decimated populations. The presence of populations with low intrinsic rates of population growth would tend to retard the return of changes in population size or age structure to their original state. Immigration rate, which is a property of individual systems, would also be a variable responsive to the extent of contamination. Thus, regional scale damage to aquatic ecosystems could destroy much of the base from which recovery of local aquatic systems could originate, resulting in effects of long duration. Degree of damage obviously plays a key role in determining the duration of effects. Complete mortality of all representatives of a given species within a system means recovery must rely on colonization from other sources. Return to conditions existing before environmental changes occurred is not necessarily assured even if adequate colonization is present. Environmental changes often promote the expansion of populations of organisms that were not formerly as abundant. Once established, such popula-

TABLE 4-8. Properties of Environment Associated with Duration of Effect.

Presence of unique species

Rates of immigration of organisms into impacted systems

Extent of environmental harm

Degree of environmental harm

Intrinsic rate of growth of affected populations

Changes that are difficult to reverse

tions may be able to resist expansion of previously important species whose relative competitive status was reduced under contaminated conditions. Newly important species may be able to modify the environment in such a way as to prevent or delay restoration of preimpact conditions.

4-1.5 Degree of Environmental Effect

Degree of contamination. As previously defined, harm to the environment was divided into environmental contamination and ecological change. Degree of contamination is a straightforward concept, represented by the instantaneous concentrations of a given contaminant in the components of an aquatic ecosystem. As such, it is a one-dimensional concept, dependent on a consideration of duration and extent of the contamination. Properties acting to determine the degree of contamination are listed in Table 4-9. This list appears incomplete because many of the properties influencing degree of contamination were covered under the discussions of duration and extent.

Source properties determining the degree of contamination include the magnitude and time course of contaminant input, and the physical and chemical form of the input material. In addition to the environmental factors involved, the maximum concentration of a contaminant attained in an aquatic system at a given point is, thus, a function of how much and how fast it is added. It is therefore necessary to be able to represent sources as time-dependent rate functions. Such functions may range from virtually instantaneous inputs (small spills) to cyclical pulse inputs to continuous, relatively stable inputs. The physical state of the material (liquid, solid, or gas) when it enters the water, while perhaps more appropriately a property of the substance itself, acts to determine the time

TABLE 4-9. Properties of Source, Substance, and Environment Influencing Degree of Contamination.

Source properties
    Magnitude and time course of input
    Physical and chemical form of input material

Substance properties
    Partitioning behavior among components of the aquatic system
    Disposition of the substance within organisms
    Solubility limits, dissolution rates

Environment properties
    Volume of the receiving water
    Water renewal rate
    Type of biota present
    Physical and chemical properties of the system
    Duration and Extent of Contamination

course of input to the water, due to possible thermodynamic and kinetic limitations of finite solubility units and dissolution rates.

Properties of the substance that influence degree of contamination include its partitioning behavior, deposition within organisms, solubility limits, and dissolution rates. The most important of these factors are probably partitioning behavior and metabolic disposition, which are instrumental in determining the degree to which a substance will be accumulated in fish, mollusks, and other aquatic life. These same two factors also determine the amount of time the contaminant will remain within an organism.

Properties of chemicals that determine bioaccumulation potential differ for organic and inorganic contaminants. Uptake of organic chemicals by aquatic organisms is primarily a passive process, driven by diffusion to and passive transport through membranes and redistribution within the organism according to physicochemical properties, foremost of which is the affinity for lipids. Continued exposure to organic chemicals leads to attainment of a equilibrium concentration within the organism, in which rates of uptake and removal are equivalent. The potential for accumulation can be simply depicted by a *concentration factor*, the ratio of concentration of a contaminant in an aquatic organism to its concentration in the water to which the organism is exposed. The *concentration factor* is a valid concept when the uptake rate is directly proportional to the concentration of the contaminant in the water, and the mode of uptake is direct uptake of dissolved material from solution. Fortunately, these two assumptions are generally valid for organic chemicals (Ref. 4-29). Because bioconcentration of organics is primarily a redistribution of dissolved organics between aqueous and lipid phases, measures of lipophilicity have been shown to correlate well with the bioaccumulation of organics (Refs. 4-30, 4-31, 4-32). In particular, the octanol–water partition coefficient has achieved considerable popularity (Refs. 4-30, 4-32).

In addition to water–lipid partitioning, however, disposition of the organic material within the organism may play a critical role in determining the degree of bioconcentration. Many fish and other aquatic organisms have enzyme systems capable of transforming lipid–soluble compounds into more polar, water-soluble substances that can be more readily excreted than the parent compound (Ref. 4-23). Thus, the susceptibility of a substance to enzymatic oxidation or conjugation can determine its bioaccumulation potential, particularly for highly lipophilic substances that, in the absence of metabolic elimination mechanisms, would be likely to accumulate to high levels.

The range of bioaccumulation exhibited by organic chemicals is very great, ranging from concentration factors of 10^5–10^6 for several persistent lipophilic substances to less than 1 for highly polar materials. Some examples of experimentally determined concentration factors are listed in Table 4-10. Inorganic compounds can also exhibit very high concentration above ambient levels in biota.

TABLE 4-10. Laboratory-Derived Concentration Factors for Chemicals in Fish.

Substance	Concentration Factor	Organism	Reference
DDT	1×10^5	Fathead minnow	4-34
PCB (Aroclor 1254)	1×10^5	Fathead minnow	4-35
Kepone	1×10^4	Sheepshead minnow	4-36
Hexachlorobenzene	8×10^3 [a]	Rainbow trout	4-30
Di-2-Ethylhexylphthalate	5×10^2	Fathead minnow	4-37
Anthracene	5×10^2	Fathead minnow	4-38
p-dichlorobenzene	2×10^2 [a]	Rainbow trout	4-30
Acridine	1×10^2	Fathead minnow	4-39
Carbon tetrachloride	2×10^1 [a]	Rainbow trout	4-30

[a]Estimated from kinetic parameters.

The accumulation of inorganic substances by aquatic organisms is a more complex phenomenon than accumulation of organics. This is in part due to the active nature of the uptake process, with inorganic ions being taken up against a concentration gradient, and the less rapid kinetics of the process. Because many inorganics are essential trace nutrients or mimic trace nutrient behavior in organisms, their levels within organisms are affected by complex physiological homeostatic mechanisms, and thus uptake and elimination rates tend to not be proportional to either external or internal contaminant concentrations. The slower membrane permeability rates of inorganics lengthens the kinetic approach to equilibrium, and thus tends to increase the significance of ingesting contaminated food (food chain accumulation). Estimations of likely bioaccumulation behavior of new inorganic chemicals based on specific physicochemical properties is thus difficult, due in large part to the lack of a clearly defined relationship between exposure concentrations and internal concentrations in organisms.

Properties of the environment that influence the degree of contamination include the volume and renewal rate of the receiving water body, the kinds of biota present, and physicochemical properties of the aquatic system, such as temperature, suspended solids' characteristics, and loading and ionic composition. The role of water volume and renewal rate is obvious, as they determine the maximum aqueous concentration a given input flux can produce. The level of contamination present in the biota is influenced to a large degree by characteristics of individual species and organisms. The degree of bioconcentration of lipophilic substances is influenced by the lipid content of organisms. Thus, oily, high lipid fish tend to contain higher residues of organic contaminants than leaner fish from the same environment (Refs. 4-40, 4-41). Materials whose bioaccumulation potential is affected significantly by metabolic elimination accumulate to the highest levels in organisms that possess little such capability. Thus, polycyclic aromatic hydrocarbons appear to attain higher concentrations in bivalve mollusks than in fish, which have higher levels of mixed function oxidase activity (Refs. 4-42, 4-43).

Materials whose uptake and elimination kinetics are quite slow, such as methyl mercury, tend to build up continuously in aquatic organisms, and thus the age and growth rate of individual organisms influence the contamination level (Refs. 4-44, 4-45). For instances in which food chain magnification may occur, the composition and length of various food webs may be significant.

Physical and chemical properties of the system also influence accumulation of contaminants in biota. Uptake and elimination rates are influenced by temperature and oxygen concentration. The amount and nature of suspended solids in the water are important in determining biotic contamination, as sorption to this material reduces the dissolved concentration of contaminant. The ionic composition of the water is influential in the accumulation of inorganic substances (Ref. 4-46).

In addition to these factors, the extent and duration of contamination must be considered in order to develop a complete picture of contamination within an aquatic ecosystem.

4-1.5.2 Degree of Environmental Damage. The degree of environmental damage resulting from chemical contamination may be manifest in a wide spectrum of ecological changes. Examples of such ecological changes were enumerated previously, and are restated in Table 4-11. The degree of damage resulting from environmental contamination is closely tied to the extent, duration, and degree of contamination. Its prediction is therefore a complex exercise incorporating both the uncertainties associated with those previous estimations and those associated

TABLE 4-11. Ecological Changes Produced by Chemical Contamination in Aqueous Environments.

Extinction of species of plants or animals

Loss of species from specific aquatic ecosystems

Changes in the relative abundance and importance of species comprising the aquatic community

Changes in standing crop, size per individual, population age structure, or production within populations of individual species

Interference with ecosystem functions of energy conversion and element cycling

Changes in physical properties of the system, such as appearance and odor.

with prediction of effect, given a known description of contaminant levels as a function of time and space.

The first four examples of environmental harm (Table 4-11) involve changes in structural aspects of the biotic component of the ecosystem. One mechanism behind such changes is direct toxic action of the contaminant at the individual level, affecting an organism's ability to survive, reproduce, or grow. Such individual effects may be manifested at the population level as any of the changes noted previously. In natural communities, where populations of individual species serve as competitors, predators, or prey of other species, changes in one population may produce changes in others. This may be particularly the case when predators are impacted, as they may tend to control community composition by attenuating interspecific competition (Ref. 4-47). On the other hand, toxicological effects at the individual level may be counterbalanced by compensatory mechanisms that prevent a population level effect from occurring. A hypothetical example would be a toxicant-induced decrease in reproductive success in organisms having compensatory mechanisms that would offset such additional mortality via decreased natural mortality.

Despite the fact that predicting population and suprapopulation level effects is the goal of analysis of the risk of environmental contaminants, little toxicological work has focused on these levels of organization. Thus, most information on the toxicological properties of specific chemicals is at the individual level.

Properties that are instrumental in determining the degree of environmental damage are listed in Table 4-12. The most commonly determined property of chemicals is acute toxicity, which is used to estimate the concentrations at which environmental damage will or will not occur. Acute toxicity refers to the toxicological response of an organism following short-term exposure to a toxic

TABLE 4-12. Properties of Substance and Environment that Influence the Degree of Environmental Damage.

Substance properties
    Acute toxicity to aquatic and terrestrial organisms
    Chronic toxicity to aquatic and terrestrial organisms
    Mutagenicity, carcinogenicity, and teratogenicity to terrestrial organisms
    Indirect effects
    Duration of contamination
    Extent of contamination
    Degree of contamination

Environment properties
    Nature of the species assemblage present
    Presence of pristine, unique, or highly desirable aquatic resources
    Presence of sport or commercial fisheries
    Interrelationships among various aquatic environments

substance. In aquatic systems this arbitrary exposure time is usually 24 to 96 hr, and the response monitored is death or immobilization. Knowledge of the acute toxicity of a substance to several aquatic species provides a basis for predicting effects when aqueous concentrations exceed toxic levels, but provides little information that can be used to establish the effects of concentrations somewhat below the acutely toxic level. The primary utility of acute toxicity tests has been to use the acute toxicity as a reference point to estimate an environmental concentration that will produce no observable effects. Thus, the acute toxicity, or LC_{50}, is multiplied by an application factor, usually 0.01, to estimate a safe concentration.

Acutely toxic concentrations of chemicals in aquatic environments produce unmistakable evidence of their existence and thus are readily avoided; it next becomes necessary to avoid producing undesirable changes in aquatic environments when toxicant levels are low enough that dead organisms are not obvious. A property of the chemical substance useful for addressing this problem is its *chronic toxicity* (while listed as properties of a substance, acute and chronic toxicity are also clearly properties of individual species and environmental conditions). Chronic toxicity refers to those toxicological manifestations occurring after more than 96 hr exposure, and includes factors other than mortality alone. Thus, alterations in growth rate, reproductive ability, ability to survive, and maturation are factors commonly monitored in chronic toxicity determinations. Chronic evaluations are more sensitive than acute tests, and thus provide a reference point closer to the actual no-effect level. However, the interactive forces present in natural communities (competition, predation, etc.) are excluded from such estimations, and thus chronic toxicity as well as acute toxicity is of limited utility in attempting to predict changes may occur in aquatic environments if chronic toxicity levels are approached or exceeded.

4-2 ENVIRONMENTAL FATE STUDIES

A significant body of information has been published in the field of aquatic environmental risk analysis, most notably under the guidance of the American Society of Testing and Materials (Refs. 4-48, 4-49, 4-50). The approach used has consisted of (1) estimating the concentration of a substance anticipated in the components of aquatic environment, (2) estimating the toxicity of the substance to aquatic and terrestrial biota, and (3) comparing expected environmental concentrations with the minimum concentrations estimated to adversely impact aquatic communities or produce unacceptable concentrations in the biota, water, or sediment. This approach entails measuring or estimating the properties previously outlined associated with the source, substance, and environment that act to determine environmental risk. The information needed falls into two broad categories — that needed to determine *environmental fate* (degree, extent, and duration of contamination) and that needed to determine *environmental effects* — (degree, extent, and duration of environmental harm).

In order to estimate the concentrations of foreign substances in aquatic environments it is necessary to know how much has been input to the aquatic environment and its fate within that system. Knowledge of the persistence and distribution of chemicals in aquatic ecosystems is essential to predicting the extent and duration of environmental impact associated with a chemical discharge. Efforts to date have most successfully focused on determining environmental fate (Refs. 4-51, 4-52, 4-53, 4-54). Less effort has been directed toward the difficult problem of predicting inputs of specific substances to aquatic systems based on their physical and chemical properties and the volume and pattern of their manufacture, distribution, use, and disposal.

Three approaches have been used for evaluating the fate of chemicals in the aquatic environments. These are:

1. Laboratory estimation of the rates of individual transport and transformation processes, from laboratory experiments, and incorporation of those rates into a predictive model.
2. Observation of the fate of specific substances in model ecosystems, or microcosms, and extrapolation to natural environments.
3. Observations of the fate of specific substances under field conditions, in which a contaminated ecosystem provides the study site.

Each approach has its advantages and disadvantages, and all have shortcomings that cause imprecision in predicting the fate of a given toxicant in a given system.

4-2.1 Process Rate Studies

The first approach, based on the rates of individual processes, is most readily applicable to the evaluation of new chemicals, and is the approach proposed by EPA for TSCA testing protocols (Ref. 4-55). The fate of chemicals in the aquatic environment is determined by the abiotic processes of sorption, photolysis, hydrolysis, oxidation, reduction, complexation, dissolution, precipitation, volatilization, sedimentation, resuspension, and radioactive decay, and the biological processes of microbial degradation, and uptake and metabolism by plants and animals. Laboratory studies of the rates of these processes under varying conditions of important environmental parameters enables researchers to develop functions depicting relationships between process rates and environmental factors. Such functions can then be used in estimating persistence under specific environmental conditions (Refs. 4-10, 4-56). The following is a brief discussion of the individual processes and their relationship to specific properties of chemicals.

When a chemical enters an aquatic ecosystem, it is subjected to natural processes that result in its partition and dispersion among various components of the system (transport processes) and its chemical alteration into other substances (transformation processes). These processes are studied in the laboratory in order to estimate chemical concentrations in water, sediment, and biota as a function of time and location, and to help identify its transformation products and estimate their concentrations.

Two basic concepts are useful in this approach to investigating the fate of chemicals: the *equilibrium coefficient* and the *rate coefficient*. The equilibrium coefficient defines the thermodynamic equilibrium relationship between concentrations of a substance in two compartments. The equilibrium coefficient (dimensionless) for a given substance (A) is described:

$$K = [A]_a/[A]_b \qquad\qquad \text{(Eq. 4-2)}$$

where the numerator is the concentration of A in compartment a and the denominator is the concentration of A in compartment b. In aquatic studies, b is usually the aqueous phase.

While the equilibrium coefficient describes how a substance will be distributed in the environment at equilibrium, the rate coefficient describes the speed of the process. Depicting such processes as first-order rate expressions (Eq. 4-2) has proved a useful simplification (Ref. 4-51):

$$d[A]_a/dt = k[A]_a \qquad\qquad \text{(Eq. 4-3)}$$

k = the rate coefficient of a process;
$[A]_a$ = the concentration of A in compartment a, and
$d[A]_a/dt$ = the instantaneous rate of change of $[A]_a$ with time.

Both rate coefficients (k) and equilibrium coefficients (K) may vary as functions of environmental characteristics such as temperature, pH, dissolved oxygen, wind and water motion, light intensity, and properties of the components of individual compartments. In general, transport processes are driven by equilibrium considerations (although equilibrium may never be attained) while transformational processes are essentially nonequilibrium (Ref. 4-57).

4-2.1.1 Sorption

Immediately upon entering surface waters, a chemical partitions into dissolved and sorbed (particle bound) forms. The extent of this partitioning is described by the equilibrium coefficient (usually referred to as distribution coefficient, K_d) and amount of particulates in suspension (Ref. 4-58). The distribution coefficient is defined as:

$$K_d = [A]_{\text{particulates}}/[A]_{H_2O} \qquad\qquad \text{(Eq. 4-4)}$$

where $[A]_{particulates}$ is the concentration of A sorbed ($\mu g\ A/g$ particulate) and $[A]_{H_2O}$ is concentration of A in solution ($\mu g\ A/g\ H_2O$).

This expression is actually a simplification of the empirical Freundlich Adsorption Isotherm,

$$K = [A]_{particulates} / [A]_{H_2O}^{1/n} \qquad \text{(Eq. 4-5)}$$

in which $n = 1$. Such a simplification is generally valid at the low suspended solids loadings typical of natural waters. The reduction of the dissolved concentration of a substance due to sorption by suspended particulates is described by the expression:

$$F = \frac{1}{K_d[s] + 1} \qquad \text{(Eq 4-6)}$$

where F is the fraction of the total aqueous concentration present in dissolved form and $[s]$ is the concentration of suspended particulates (g particulate/g H_2O).

Typical distribution coefficients for the sorption of several chemicals onto sediments or soils having widespread inputs or distribution in the environment are listed in Table 4-13. Those organic chemicals having low aqueous solubility (DDT, PCB, Mirex) display considerable affinity for particulates, while more water-soluble materials, such as Atrazine and 2, 4-D, do not. At the suspended particulate loadings of 10–50 mg/liter (typical of many natural waters; (Ref. 4-65) a substance with $K_d \leqslant 1 \times 10^4$ would be less than 33% sorbed.

The sorbent characteristics most important in determining K_d are particle size and organic carbon content (Refs. 4-51, 4-59, 4-66, 4-67). The sorption coefficient (K_d) increases with decreasing particle size and increasing organic carbon content. Environmental parameters affecting K_d most are temperature, pH, and

TABLE 4-13. Distribution Coefficients (K_d) for Several Widespread Pollutants onto Sediments and Soils.

Substance	K_d	Sorbent	References
Lindane	3×10^2	Sediment	4-59
PCB (Aroclor 1254)	7×10^4	Sediment	4-60
Benz(a)pyrene	8×10^4	Sediment	4-10
2, 4-D	3×10^0	Soil	4-61
Cs-137	$3.5–10 \times 10^3$	Sediment	4-62
Sr-90	$4–12 \times 10^1$	Sediment	4-62
DDT	1×10^5	Soil	4-61
Mirex	5×10^5	Sediment	4-10
Atrazine	$1–5 \times 10^0$	Soil	4-63
2,4,5-T	$0.3–3 \times 10^0$	Soil	4-64

salt content. Since sorption is generally exothermic (heat evolving), K_d usually increases with decreasing temperature. The pH of water is critical for compounds that ionize within the range of pH 5-10, and acts to a lesser degree to influence sorbent character. Dissolved salts are most important in estuarine and oceanic environments, where competing ions act to influence K_d.

Description of the partitioning of substances between dissolved and sorbed phases is vital to the evaluation of environmental fate via process rate studies, particularly if it is assumed that certain processes, such as volatilization, act only on the dissolved fraction and others, such as sedimentation, only on the sorbed fraction.

Measurement of sorption coefficients is usually made by dispersing a known amount of sorbent (soil or sediment) in water containing the substance being investigated. After allowing time for equilibration (usually a rapid process, <2 hr), the sorbent is removed by centrifugation or filtration and the amount of material sorbed determined by the change in aqueous concentration or by direct measurement of the amount of substance associated with sorbent. Some important experimental considerations follow.

1. Keep sorbent concentrations low, or at levels typical of natural environments (<100 ppm).
2. Aqueous concentrations of the substance under consideration should be as low as feasible, unless higher concentrations are likely in the environment.
3. Caution must be exercised in using filtration techniques, as hydrophobic materials display a high affinity for nearly all fine pore filters, and will be sorbed by the filter itself. Combusted glass fiber filters are often suitable in such instances. Volatile materials may be lost in vacuum filtration.
4. Determining the variation in K_d with changes in sorbent: sorbate ratio is desirable, but usually not necessary at low concentrations of both sorbent and sorbate.
5. Determining variation in K_d with temperature provides useful information.
6. Important data to collect on sorbent characteristics are organic carbon content, ion exchange capacity (when relevant), and particle size distribution.

See Refs. 4-25, 4-55, 4-56, and 4-58 for techniques for determining sorption coefficients.

Material sorbed to particulates is removed from the water column by sedimentation of the particulate material. Sedimentation rates are determined by the size, density, and concentration of the particulate material and the density, viscosity, and turbulence of the water (Refs. 4-38, 51, 68). A rate coefficient for removal of a contaminant from the water column is obtained by multiplying the first-order rate coefficient for removal of particulates times the fraction of the contaminant sorbed (Ref. 4-38). Removal of contaminants by sorption-sedimentation is thus most significant for those substances with high sorption distribution coefficients (K_d), particularly in systems having high suspended particulate loadings and low levels of turbulence.

4-2.1.2 Volatilization

Many chemical substances possess appreciable volatility, and thus are subject to movement from water to air. Even substances with relatively low vapor pressures may readily pass from water to air if they are quite hydrophobic (Ref. 4-8). The movement of chemicals across the air-water interface is governed by (Ref. 4-69):

- Transport from the bulk water to the interface
- Equilibrium distribution of the substance between air and water
- Transport away from the interface into the bulk air mass.

Environmental parameters affecting volatilization include (1) flow velocity and turbulence in the water, (2) temperature of the air and water, (3) wind velocity and turbulence, (4) ambient concentration of the chemical in the air, and (5) waves and surface active substances in the water (Refs. 4-70, 4-71).

Attempts to predict removal of chemical substances from water have focused on empirical measurement of volatilization in the laboratory (Refs. 4-72, 4-73), application of mass transfer models (Refs. 4-8, 4-69, 4-70), and correlation with reaeration rates (Refs. 4-57, 4-74). The mass transfer model approach is perhaps the most versatile of these techniques, in that the rate of volatilization is expressed as a function of variables that are properties of the chemical and its environment. These variables are the following: (1) the liquid-phase mass transfer coefficient, which varies as a function of current and wind velocity and, to a lesser extent, as a function of temperature and molecular structure of the chemical; (2) the Henry's Law constant (H), a distribution coefficient describing the equilibrium partitioning of a specific chemical between air and water, which varies as a function of temperature in each phase; and (3) the gas-phase mass transfer coefficient, which varies as a function of wind velocity and, to a lesser extent, with current, temperature, and molecular properties. While simplistic, the mass transfer model makes it possible to estimate volatilization rates in a given system based on only a single, easily determined parameter of the chemical, H (Ref. 4-75).

The estimation of volatilization from empirical comparison with oxygen uptake (Ref. 4-74) provides a direct means of estimating volatilization in a field situation in which reaeration rates are known for the system. Because reaeration provides only a measure of the role of liquid phase transfer in controlling overall mass transfer, it is therefore inappropriate for estimating volatilization of substances when mass transfer is gas-phase controlled. Because volatilization is gas-phase controlled when materials volatilize at a slow rate (low H), the reaeration ratio method is likely to underestimate the volatilization rate for low volatility substances. Because volatilization of such substances is probably not likely to be a major pathway of environmental transport, this shortcoming is not of great significance.

Empirical measurement of volatilization (Refs. 4-72, 4-73) from small containers provides rapid identification of volatile and nonvolatile solutes, and can be roughly extrapolated to environmental situations after adjusting for differences in depth and mixing in air and water.

A more complex situation involves spills of immiscible liquids on water. Evaporation from such systems can be estimated by mass transfer equations (Refs. 4-76, 4-77) for single component mixtures, but accurate prediction of evaporation rates in complex mixtures is far more difficult. In general, evaporation of spills is controlled by temperature and heat transfer, wind velocity, spill thickness, and liquid (chemical) phase properties.

4-2.1.3 Sorption–Desorption in Bedded Sediments

Contaminants in aquatic systems may be removed from the water column by sorption into the sediment bed of the system, or they may be added to the water column by desorption from the sediments. Thus, sediments can act to ameliorate a short-term problem by reducing aqueous concentrations but may increase the duration of aqueous contamination through the slow release of contaminants back into the water.

Sorption–desorption processes appear to be controlled by the affinity of the contaminant for the sediment (distribution coefficient, K_d), which varies as a function of the parameters listed previously, and by the rate of delivery of the contaminant to and away from the water sediment interface, which should vary as a function of water turbulence, velocity, contaminant concentration, and diffusion of the contaminant through the sediment bed. Diffusion into the sediment bed varies with sediment composition and density, oxygen concentration, and K_d. The direction of contaminant movement (into or out of sediments) is determined by the net magnitude of the coupled sorption–desorption rates.

Sorption rates of materials from water can be expressed using a one-dimensional mass transfer coefficient similar to the approach used to conceptualize gas exchange rates (Refs. 4-69, 4-78). Uptake coefficients for movement from water to sediments of 0.1–0.2 cm/hr were observed in still water, and increased up to 32 cm/hr in flowing water under laboratory conditions (Refs. 4-78, 4-79). This suggests that uptake of materials with high sorptive affinity for sediments by previously uncontaminated sediments is initially limited by aqueous phase processes (delivery to the interface). Uptake slows under chronic exposure conditions and is no doubt limited by diffusion into the sediments (removal from the interface) (Ref. 4-78). In systems where such chronic exposures occur, net movement from water into sediments probably has an insignificant effect on aqueous concentration because of the establishment of equilibrium conditions. Under acute exposure conditions, rapid movement of contaminants into sediments would occur initially, but the overall mass of contaminant sequestered by bedded sediments would probably be quite small, due to the involvement of only the uppermost surface of sediments (<1 mm) in the process (Ref. 4-78).

Desorption of contaminants from bedded sediments may provide a significant input of toxicants to the water column (Refs. 4-79, 4-80). Transfer coefficients calculated from Halter and Johnson (Ref. 4-60) of about 3×10^{-6} cm/hr (flux = transfer coefficient times concentration of contaminant in the sediment) for PCB are large enough to produce aqueous concentrations in some systems that are near equilibrium. The major factors affecting the relative importance of desorption in affecting aqueous contaminant concentrations in a completely mixed system are water depth, water turnover time, and the rates of removal processes such as microbial degradation, photolysis, etc. (Ref. 4-80).

Methods for measuring sediment-water fluxes of contaminants can be found in Refs. 4-60, 4-63, 4-78, 4-79, 4-80.

4-2.1.4 Bioaccumulation

The accumulation of chemical substances from the environment by aquatic organisms can pose a hazard to other organisms ingesting them. Bioaccumulation behavior of organic and inorganic substances differs in that direct uptake of organic substances by aquatic organisms is directly related to aqueous concentrations, which is not so in the uptake of inorganic substances, due to homeostatic mechanisms and competitive uptake of other ions.

The simplicity of bioaccumulation of organic compounds, and its dependence on the partition chemistry of the chemical substances, has facilitated the conceptualization of bioaccumulation of organics as an equilibrium process in which elimination of the substance by metabolic and excretory processes counterbalances uptake from food, water, and sediment (Ref. 4-82). Rates of uptake and elimination can be estimated in laboratory studies (Refs. 4-30, 4-83) and equilibrium concentrations of substances in the test biota can be observed or calculated (Ref. 4-35). Concentration factors ([substance] $_{fish}$/[substance] $_{H_2O}$) so determined are well correlated with the physical properties of the substances, especially the H_2O-n-octanol partition coefficient, which can be used to predict bioaccumulation potential of a substance via direct uptake (Refs. 4-30, 4-35). The special process of direct uptake and accumulation of a contaminant from water is called bioconcentration.

The importance of accumulation of organic chemicals by ingesting contaminated food in relation to direct uptake from water should increase with increasing direct bioaccumulation potential of a substance, since direct uptake will be limited by the amount of water reaching the gills whereas uptake from food depends on the substance's concentration in the food, which increases with increasing bioconcentration potential. Such an increase in toxicant accumulation up the food chain, or biomagnification, would be expected for compounds such as DDT, PCB, Mirex, and methyl mercury, which have extremely high direct bioaccumulation potential (Ref. 4-29). Evidence of biomagnification of organics within aquatic systems is ambiguous, with some researchers detecting no evidence (Ref. 4-84), while others have attributed the higher concentrations of organics in terminal aquatic predators to the intake of contaminated food (Refs. 4-85, 4-86). As a general principle, it appears as though substances with very high bioconcentration potential (direct aqueous uptake concentration factors $\geqslant 10^4$) such as PCB, DDT, and methyl mercury, are taken up at substantial rates from both food and water (Refs. 4-29, 4-40, 4-85, 4-87), while for substances with less bioaccumulation potential, direct aqueous uptake (Refs. 4-29, 4-39) dominates the bioaccumulation behavior. Even for substances with a large predicted food chain component, some species of predator fish exhibit concentrations of organics less than or equal to that in prey species in systems where other species exhibit biomagnification (Refs. 4-19, 4-88).

Environmental factors affecting bioaccumulation of organics are primarily those affecting metabolic rate of organisms — temperature, dissolved oxygen concentration, and food availability (Ref. 4-44). The organism's own properties also play a dominant role, with factors such as size, surface area/volume ratio, lipid content, growth rate, food ration, and age affecting the concentrations of organics found in contaminated fish (Refs. 4-41, 4-44, 4-89, 4-90). Physiological responses, such as enzyme induction related to the metabolism of contaminants, act to increase the elimination rate and thus reduce bioaccumulation (Refs. 4-33, 4-43).

Properties of chemical substances that determine bioaccumulation behavior are relative affinity for lipids versus water, which is estimated by octanol–water partitioning (Refs. 4-30, 4-35), and susceptibility to metabolic degradation (Refs. 4-37, 4-43). Measurement of bioconcentration of organics is most ideally carried out by exposing test organisms to at least two widely disparate ($> 10X$) unchanging concentrations of the organic substance at a constant temperature in a flowthrough system that flushes out degradation products and metabolites. Aqueous concentrations and concentrations in the test organisms are monitored frequently until a near equilibrium condition is attained (Ref. 4-82). In practice, such a system is difficult to attain, due to the long time required to bring substances having high bioconcentration factors to near equilibrium. For the latter materials, an estimate of bioconcentration is obtained using short-term exposures to the test substance to determine an uptake rate and then placing organisms in uncontaminated water to measure the elimination rate (Refs. 4-30, 4-91). The equilibrium concentration factor is estimated from the ratio of uptake and elimination rates. When metabolic degradation is a significant component of the elimination rate, the results of experimental determination of the elimination rate usually deviate considerably from the first-order model they are assumed to fit. Under such conditions, the ratio of uptake to elimination tends to be an unreliable estimate of bioaccumulation potential.

Even in the absence of data derived from organism tests, the strong correlation of bioaccumulation to water solubility and octanol–water partitioning enables prediction of bioaccumulation potential at a level of precision adequate for a preliminary assessment (Ref. 4-35).

For more precise estimates of movement of contaminants through specific pathways, realistic long-term bioaccumulation studies may be required. These will incorporate models that will account for growth and metabolism, uptake from both water and food, and nonuniform tissue distribution of contaminants.

Many species have been used for bioaccumulation studies in fresh and salt water. At present, the fathead minnow (*Pimephales promelas*) and rainbow trout (*Salmo gairdneri*) are probably the most widely used. Bioconcentration potential varies among fish species (Ref. 4-35), creating some problems for extrapolating from one species to another. The problem of extrapolation is probably more one of imprecision than of possible gross misestimation of bioaccumulation. Thus, concentration factors obtained in one species of fish are probably within a factor of 2 or 3 of concentration factors for other fish, if metabolism is not the major determinant of bioconcentration potential. The role of metabolism in determining bioconcentration potential has not received sufficient study to permit interspecies comparison of its importance.

Invertebrates used in bioaccumulation studies include various species of *Daphnia*, crayfish, shrimp, crabs, clams, snails, and mussels. Studies on accumulation of contaminants by specific species are warranted when the organism is an important human food source, causing a likelihood of contamination to exist. Comparison and evaluation of bioconcentration potential of various chemicals is perhaps best served by standardization of test species, but the prediction of the biological compartmentalization of contaminants in specific ecosystems is better served by the diversification of test species. The resolution of this dichotomy may lie in more extensive comparison of the results of standardized laboratory tests with field studies, as has been done for several substances (Ref. 4-90).

Excellent examples of methodologies used in evaluating the bioconcentration of contaminants include Refs. 4-30, 4-35, 4-82, 4-91.

4-2.1.5 Photolysis

Sunlight acts to degrade many organic chemicals in water. Photolysis is likely to be the dominant degradative pathway for many foreign substances in the aquatic environment, with rates of photodegradation being quite rapid for some substances. Benz(a)pyrene, for example, exhibited a half life of <3 hr in a calculation based on quantum yield (Ref. 4-56). Many pesticides are readily photodegraded (Refs. 4-92, 4-93). The rate at which a substance is degraded is determined by properties of the substance (absorption spectrum and quantum yield) and those of the environment — (intensity and spectral composition of incident light, and transparency and depth of the water). Photolysis may proceed as a direct process, in which the substance absorbing light energy is transformed, or by indirect processes, in which other substances absorb the light energy and the transformation of the primary substance follows the transfer of energy or reaction with reactive species produced by photoexcitation of the secondary substance. The presence of such secondary substances (photosensitizers) in natural waters can influence photolysis rates.

Suspended particulates in the water influence photolysis by absorbing light, scattering light, and adsorbing dissolved contaminants. Light scattering effects are complex, but result in enhanced photolysis rates in the region of scattering (Ref. 4-94). The susceptibility of sorbed substances to photolysis is less than that of dissolved material, but not necessarily zero (Ref. 4-95). Rates and reaction pathways of photolysis can vary with pH, primarily for substances that ionize within the range of pH typical of natural waters (Ref. 4-53).

Rates of photolysis of chemicals in aquatic environments are usually estimated by one of two techniques. The technique most widely used involves exposing aqueous solutions of the substance being investigated to a narrow spectrum of light of measured intensity, and determining the quantum yield (fraction of photoexcited molecules that undergo chemical transformation). The quantum yield is coupled with an estimation of the amount of light absorbed per unit time (determined by the incident light intensity and spectrum, the absorption spectrum of the substance, and the absorption spectrum and depth of the water mass being considered). This complex calculation is carried out using a computer model (Ref. 4-96). The following are some advantages of this approach:

1. The use of artificial light sources permits continuous exposure to very intense illumination, aiding in determining low rates of photolysis that would take weeks to determine in sunlight exposures.
2. The experimental system can be used at any time of day or year.
3. Quantum yields and absorption spectra, once obtained, can be used to estimate photolysis in various environmental situations.

Some disadvantages of this method are
1. The need for specialized apparatus.
2. The requirement of complex calculations.
3. The inability to account for variation in nautral light intensity due to clouds and haze.

The second approach consists of exposing aqueous solutions of the substance to natural sunlight for a period of time and then measuring the change in concentration (Ref. 4-38). Advantages of this procedure include the following:

1. The procedure is simple, straightforward, and requires little specialized apparatus.

2. Rates can be estimated without use of a computer.
3. Direct estimation of photolysis rate is possible at a specific time and place.

Disadvantages are the following:
1. Source intensity variations with time and weather cause imprecision in determinations.
2. Source intensity is limited.

The current ASTM proposed protocols for photolysis testing suggest the application of both sunlight exposures and quantum yield determinations in a testing program.   References giving further details of photolysis testing procedures include (Refs. 4-38, 4-53, 4-55, 4-56, 4-96). Reference 4-96 provides an excellent account of the methodology and theory of aqueous photolysis. A "proposed standard practice for conducting photolysis tests," developed by the American Society for Testing and Materials is currently being completed and provides excellent procedural guidelines.

A few key points for determination of photolysis rates include the following:

1. The use of quartz-faced vessels for sunlight exposures of substances absorbing at <310 nm or if indirect photolysis is being investigated.
2. The substance should be dissolved in water at as low a concentration as practical, with a minimum amount of organic solvent present. Light transmission through the full depth of the exposure solution should exceed 90% at all wavelengths for sunlight rate determinations.
3. Appropriate dark controls are important in establishing that changes in the exposed solutions are due to photolysis.
4. Sorption of highly hydrophobic substances to container surfaces may remove a substantial portion of the substance from solution. Appropriate choice of exposure vessels and minimizing length of exposure may lessen this problem.

4-2.1.6 Other Abiotic Pathways

Two other major abiotic transformation pathways for organic substances in water are oxidation and hydrolysis (Refs. 4-11, 4-56). Hydrolysis is a major factor in the environmental degradation of some pesticides (Refs. 4-92, 4-93). Rates of hydrolysis are generally second-order reactions, with the rate highly dependent upon pH. Estimations of hydrolysis rates are performed by monitoring the disappearance of the substance as a function of time at various controlled pH. Oxidation reactions occur through the interaction of substances with free radicals naturally occurring in water. Reaction kinetics for free-radical oxidation can be readily measured in laboratory studies (Refs. 4-11, 4-56). However, predicting oxidation rates in natural waters requires a knowledge of the minute steady-state concentration of free radicals in the water. The prediction of oxidation rates in natural waters thus must rely at the present time on relatively untested assumptions of natural free-radical concentrations.

4-2.1.7 Microbial Transformations

Heterotrophic microorganisms, found in virtually all aquatic environments, derive most of their growth and energetic requirements by degrading dissolved and particulate organic matter to simpler molecules. They are capable of degrading many xenobiotic compounds as well as naturally occurring substances and thus constitute a major mechanism for removal or transformation of contaminants. The chemical pathways of degradation are similar to abiotic routes — oxidation, hydrolysis, reduction, and substitution reactions.

Environmental conditions influence the number of microorganisms present, rates of microbial degradation, and metabolic pathways. The major environmental parameters affecting microbial transformations are temperature and dissolved oxygen concentration. Temperature simply alters the rate of transformation (Ref. 4-97) while the dissolved oxygen concentration may affect both the rate (Ref. 4-98) and the pathway, reductive or oxidative, that is followed (Ref. 4-99).

Microbial populations can undergo alteration in response to either short-term or chronic influxes of contaminants. Microbial species capable of utilizing petroleum hydrocarbons, which had been numerically and functionally unimportant in an estuarine beach ecosystem, rapidly multiplied and dominated the community within several days after an oil spill (Ref. 4-100). Similarly, rates of transformation of polycyclic aromatic hydrocarbons were far more rapid in a stream sediment receiving a continuous input of the contaminants than in a pristine stream sediment (Ref. 4-101). Thus the anticipated rate of microbial transformation of a contaminant in an aquatic system is dependent upon the magnitude (high versus low concentration) and continuity (acute versus chronic) of the contaminant source.

Two experimental approaches have been used to estimate rates of microbial degradation of substances in natural waters. The first involves adding the xenobiotic to natural water samples at levels comparable to those anticipated and then monitoring the disappearance of substrate or appearance of transformation products (Ref. 4-101) in order to determine the rate of degradation. This degradation rate is recognized as a second-order process, i.e., dependent upon both substrate concentration and the abundance of microorganisms capable of degrading the substrate. At sufficiently low substrate concentrations, the rate expression is thus:

$$\frac{d[x]}{dt} = k\,[x]\,[m]$$ (Eq. 4-7)

where

  $[x]$ = the substrate concentration (moles/liter)
  $[m]$ = the concentration of microorganisms (number of organisms/liter)
  k = second-order rate coefficient (liter organism^{-1} hr^{-1})

To experimentally determine the second-order rate coefficient it is necessary to measure the concentration of living microorganisms and the rate of change of substrate concentration in an experimental system in which, ideally, the concentration of microorganisms does not change during the course of the investigation. Isolation and quantification of specific microorganisms populations at levels found in natural waters is a difficult process, and thus efforts to date have focused on using plate counts of colony-forming units to estimate the size of the total heterotrophic microbial community (Ref. 4-99). Generally the specific substrate-degrading organisms are assumed to constitute a constant proportion of the total heterotrophic community for purposes of predicting rates of microbial degradation in the environment. This assumption may be valid for compounds that are degraded by many strains of microorganisms at roughly comparable rates. However, for compounds that may be degraded by specialized microorganisms that constitute a relatively small and widely variable fraction of the microbial community (e.g. polycyclic aromatic hydrocarbons), the assumption is probably not valid. To circumvent this difficulty, some workers have considered the rate of degradation as a "pseudo" first-order process, in which k and $[m]$ are combined as a first-order rate coefficient, which thus varies with time and location (Ref. 4-101). An implicit assumption of this approach is that microbial degradation rate is a system-specific property; the objective is to define the rate as a range of observed first-order degradation coefficients that are set by temperature, exposure history, and other parameters.

  The second experimental approach to estimating degradation rates in the environment entails measuring the degradation rate under enriched conditions, in which a large amount of substrate is added to a culture or water sample and incubated until there develops a large population of microorganisms capable of degrading the substrate. The kinetics of degradation are then analyzed as a function of microbial biomass using Monod kinetics (Ref. 4-56). The rate coefficients thus obtained are coupled with an estimate of the biomass of substrate-degrading microbes present in the natural water to provide an estimate of environmental degradation rates. This latter step may introduce uncertainties of several orders of magnitude into the estimation of rates of microbial degradation in natural waters and sediment if the population of substrate-degrading microorganisms is a quantitatively small fraction of the total heterotroph community. More precise estimations of microbial degradation rates in a particular aquatic system can be obtained with very low substrate concentrations and short incubation times, but analytical problems associated with the very low substrate concentrations often present difficulties.

Techniques employed for measuring rates of substrate degradation include direct measurement of remaining substrate, oxygen utilization, evolution of CO_2 or $^{14}CO_2$ from isotopically labeled substrates, and quantitation of multiple transformation products. The use of high substrate concentrations is often necessary for analytical measurement but may far exceed levels anticipated for chronic contamination of aquatic systems (although levels may be comparable to acute "spill" situations). High substrate levels may be toxic to microorganisms, and may induce or depress degradative enzyme systems that are nonfunctional at low concentrations of xenobiotics. Alternatively, microbial populations that are numerically and functionally unimportant in the initial inoculum may rise rapidly to dominance during the assay. Both changes in enzyme and population levels may result in "lag" periods followed by rapid substrate utilization; such dramatic changes in degradation rates make kinetic interpretations difficult.

Each method of degradation measurement possesses inherent disadvantages. Oxygen uptake is only useful for examining aerobic transformations and will only detect oxidative metabolism. Measurement of CO_2 evolution underestimates transformation rates of metabolic pathways that produce incomplete transformation to CO_2. Spectrophotometric techniques employed alone without use of chromatographic separation procedures may not differentiate between residual substrate and transformation products with similar spectral characteristics. Many of these difficulties can be avoided by using tracer techniques in which all transformation products are quantified, but these are very time consuming and require substrates of high specific activity, and the availability of isotopically labeled xenobiotic compounds, which may be prohibitively expensive. In each technique, careful experimental design, and particularly incorporation of appropriate controls, is vital.

The most critical step in choosing a microbial degradation assay procedure is a clear statement of the objective: whether the rate measured is to reflect a maximal rate under high substrate concentrations (i.e., an acute spill) or a steady-state rate in a chronically contaminated stream (i.e., a continuous low-level discharge). An assay procedure may then be selected to obtain the required rate information. The investigator must bear in mind the environmental and experimental constraints of the rate information obtained, and utilize appropriate judgement in data interpretation.

4-2.2 Environmental Fate Modeling

An environmental fate model is essentially a formalized collection of hypotheses of a contaminant's transport and transformation through a defined aquatic system. The necessity for such models in hazard assessment is apparent: contaminant fate must be predicted prior to contaminant release (whether intentional or inadvertent) so that decisions can be made concerning whether such a release is environmentally acceptable.

All models of contaminant behavior in aquatic systems attempt to combine mathematical descriptions of (1) the hydrodynamics of the receiving water, and (2) the dynamics of contaminant addition and loss from the system. The use for which the model is designed will determine the degree of accuracy with which each expression must be defined. A basic axiom is "simple is good, simpler is better" (Ref. 4-102), thus the best model is the least complex one that can adequately define the system.

Hydrodynamic aspects of contaminant movement may usually be modeled with greater accuracy than contaminant transformations and source/sink behavior. Processes of advection and turbulent diffusion dominate most water movement and are defined in a more predictable manner than are contaminant-specific terms (Ref. 4-102). Hydrodynamic models are also readily testable at specific sites through use of dye, brine, or other conservative tracers.

In general, hydrodynamic complexity increases in the following sequence:

1. Steady state, unidirectional flow (e.g., a river during constant flow).
2. Nonsteady state unidirectional flow (e.g., a river during and following a storm).
3. Steady state, multidimensional flow (e.g., a lake).
4. Nonsteady state multidimensional flow (e.g., a tidal estuary).

Complexity of the contaminant transport/transformation aspect (and generally, uncertainty of the fate prediction as well) varies inversely with both the degree of definition of the source and with degree of conservatism of the contaminant. A similar generalized sequence of increasing complexity may thus be outlined:

1. Steady state point discharge of a conservative contaminant, or one whose transformations are relatively simple and well defined (e.g., discharge of heated water, sewage, or water-soluble radionuclides).
2. Steady state point discharge of contaminants that undergo relatively complex transformations (e.g, trace organics or sediment-binding radionuclides)
3. Nonsteady state point discharges (e.g., oil spills).
4. Diffuse, nonsteady state inputs of contaminants whose transformations are relatively simple or well defined (e.g., rainfall washout or surface runoff of sulfate).

5. Diffuse, nonsteady state inputs of contaminants whose transformations are complex or are not well defined (e.g., atmospheric deposition of organic contaminants).

Geographic scale is not directly related to model complexity; some local-scale (near-field) models, such as thermal plume models, may incorporate a high degree of complexity to provide predictions with sufficient resolution. In contrast, regional (far-field) or even global models may be less complex if the hazard assessment does not require a high degree of resolution and precision in fate prediction.

Two general applications of contaminant transport models have been employed in hazard assessment. The first is a *site-specific* attempt to predict contaminant concentrations in the vicinity of a known source (i.e., either an intentional or unintentional discharge). The physical and hydrodynamic characteristics of the actual water body must therefore be defined sufficiently to produce a model prediction with whatever level of accuracy is needed. The second application is *non-site specific* — rather than predicting contaminant behavior in one location, the objective is to describe, in general terms, contaminant behavior in representative water bodies so that comparisons can be made of transport behavior among several potential contaminants.

Historically, most applications of contaminant transport models have been site specific. Perhaps the earliest and simplest (and most widely tested) aquatic contaminant transport models are those describing the oxygen level in rivers as a function of sewage input. Such models are one dimensional, and assumptions of steady state source, constant river flowrate, and constant rate of oxygen consumption have been used for over 50 years since the earliest work (Ref. 4-103). Thus, most modifications of these simple models have refined the relationship between hydrodynamics and rate of oxygen movement into water (Ref. 4-103). Because volatilization rates of highly volatile compounds are directly related to oxygen transfer rates, reaeration models may be used to estimate rates of contaminant volatilization from rivers (Ref. 4-74). Reaeration models are generally phenomenological, or are based on observed relationships between stream characteristics and process rates, rather than on fundamental principles. Extrapolation of a model from one river section to a dissimilar one, or between different rivers, without calibration of the model by measurement of oxygen levels, may produce unsatisfactory results (Ref. 4-103).

More complex two- or three-dimensional models have been developed to describe behavior of an effluent jet in the vicinity of the discharge (i.e., near-field models). Dunn *et al.* (Ref. 4-104) reviewed several dozen of the more widely used models and concluded that, for the most accurate ones, predicted and measured isotherms from heated water discharges differ by no more than a lateral distance factor of 2. Espey *et al.* (Ref. 4-102) concurred that a factor of 2 in predicted contaminants concentrations should be attainable in most situations, and accuracy within 25% is possible under favorable conditions. Because near-field models have been widely used in power plant siting studies, they may be considered to be well validated for most applications.

A second class of near-field models are those that have been developed, in part, to simulate two-dimensional oil spill transport behavior. Such spills are subject to combined effects of wind, waves, and currents. Both phenomenological models based on observed dispersion of tracers (Ref. 4-102) and probabilistic (Ref. 4-105) models employing computerized statistical techniques have been used to predict spill movement.

Most models are developed for point source contaminant discharges, whether continuous or transient (i.e., spills). Several models containing program elements that have been developed and verified to describe nonpoint sources include the Storm Water Management Model (Ref. 4-106) and the Stanford Watershed Model IV (SWM) (Ref. 4-107), although apparently neither has been applied to potentially hazardous contaminants to date.

Several other nonpoint source models, based at least in part on the SWM, have have been used and at least partially validated in modeling of suspended solids (Ref. 4-108) and pesticides (Ref. 4-109). A Unified Transport Model (UTM) developed at Oak Ridge National Laboratory (Ref. 4-110) combines runoff and groundwater movement submodels to predict contaminant movement. The UTM was satisfactorily applied to movement of Pb, Zn, Cd, and Cu introduced by airborne deposition in a Missouri watershed (Ref. 4-111), and thus may be considered valid at least for heavy metals. Application of the UTM to transport of organic contaminants is currently underway.

A number of far-field (i.e., relatively far from the source) river, lake, and estuary models have been developed during the past 15 years, primarily for use in water quality management. Several have been adopted for use in radionuclide fate modeling and presumably could be modified for site-specific modelings of transport of other contaminants. Espey et al. (Ref. 4-102) have reviewed the utility of existing models different systems, and conclude that the capability is well developed for one-dimensional modeling of transport of contaminants that undergo little interaction with sediments. As system complexity increases (i.e., requiring two-dimensional riverine and estuarine modeling), difficulties in model verification create more uncertainty about applicability. Several such models, however, have been verified for water quality parameters (Refs. 4-112, 4-113) and may be directly applicable to contaminant transport in complex surface waters.

Although most far-field riverine models can be used for discontinuous inputs (i.e., spills), few models have been specifically designed to monitor dissolution and subsequent transport of spilled contaminants. One model (Ref. 4-114) considers a river to be a series of linked well-mixed compartments. By empirically manipulating parameter values, data on spilled chloroform movement in the Mississippi River were adequately fit by the model. The predictive capability of the model, however, has not been evaluated.

All the models thus far mentioned have been applied in site-specific studies of contaminant transport. An alternative model application is non-site specific comparison of contaminant transport in a representative water body. The former application requires description of the system hydrodynamics, which, depending on the system modeled, may be relatively simple or extremely complex. In the latter application, development of a detailed and comprehensive submodel of the chemical and biological transport and transformations of the contaminant is emphasized. A considerably simplified hydrological model component may thus be utilized, particularly for model application to elucidation of contaminant behavior in stylized, "representative" ecosystems. This approach of "evaluative modeling" has been developed and refined at the Athens (GA) Environmental Research Laboratory, and has recently been summarized (Ref. 4-51). Simplified, stylized ecosystems have been employed to predict polycyclic aromatic hydrocarbon transport in rivers and lakes under both steady state (Refs. 4-38, 4-57) and discontinuous source (Ref. 4-11) conditions. Two justifications for the approach are (1) it permits semiquantitative prediction of contaminant behavior under widely varying environments, and (2) it can pinpoint what source/environment combinations are of sufficient interest to require further site-specific modeling (Ref. 4-51).

Increasing interest in contaminant fate modeling has stimulated development of user-oriented models that may be employed with minimal preparation. One such model is the Exposure Analysis Modeling System (EXAMS), which is presently under development for the EPA by the Athens Environmental Research Laboratory (Ref. 4-115). EXAMS provides a time and distance distribution of contaminant concentrations in water, sediment, and biota resulting from either dispersed or point source inputs. Either stylized or site-specific hydrodynamics may be specified; additional input data are physicochemical characteristics of both contaminant and receiving water. At present, only steady state concentrations are predicted, although it is proposed that dynamic capabilities be developed in the future (L.H. Burns, personal communication).

There are two general shortcomings that hinder the usage of currently available contaminant transport models: (1) insufficient treatment of sediment-water interactions, and (2) insufficient model verification. Because many hydrophobic organic contaminants are concentrated to high levels in sediments, the sediment–water contaminant flow may largely determine aquatic environmental behavior. Of 24 radionuclide transport models reviewed by Espey (Ref. 4-102) only 6 provided for sediment interactions. Of these only two — both two-dimensional models developed by Onishi (Refs. 4-116, 4-117) — attempted to model physical transport of sediments, which may be a dominant mechanism of contaminant dispersion under certain conditions. Another sediment transport model, CHNSED (Ref. 4-118) can also simulate transport of sediment-bound contaminants.

Perhaps a more serious shortcoming of contaminant transport modeling is the present inadequacy of model verification. Many models have been *calibrated,* i.e., model parameters have been adjusted to give a close fit between predicted and observed concentrations over a particular time interval. With the exception of several water quality models, however, none has been sufficiently *verified* (i.e., model predictions tested over a different time period and, optimally, one in which major environmental parameters vary substantially). Verification is essential to confirm that the model incorporates major transport mechanisms in both a qualitatively and quantitatively correct manner. One difficulty is that inherent environmental variability may lead to inconclusive model verification, as has been observed in transport of polycyclic aromatic hydrocarbons (Ref. 4-119). Compilation of large data bases on contaminant levels and movement, such as the work in which Huggett and coworkers at Virginia Institute of Marine Sciences are currently engaged for Kepone in the James River, may prove sufficient for verification of aquatic contaminant fate models. Because hazard assessment screening of contaminants will almost certainly require use of model predictions, model verification must remain a high priority in contaminant fate research.

4-2.3 Microcosms, Monitoring, and Field Studies

Another method for experimentally evaluating the environmental fate of chemicals is the model ecosystem, or microcosm. A key assumption of this approach is that a small-scale replica of an ecosystem can sufficiently mimic the function of the actual ecosystem to provide an accurate estimation of the fate of contaminants. If this assumption is met, the microcosm approach makes it possible to determine environmental fate with far less expenditure of time and resources than process rate studies. Numerous materials have been investigated in microcosms (Refs. 4-119, 4-120, 4-121, 4-122), and there has been developed a wealth of information for comparing relative persistence and compartmentalization of materials in aquarium-sized systems.

Methods are needed to meaningfully extrapolate from microcosm behavior of a contaminant to its behavior in far larger natural systems. Because processes that are affected by factors of environmental scale, such as interfacial processes (photolysis, volatilization, uptake, and degradation in sediments), are not distinguished from factors less affected by scale (abiotic and biotic degradation, which occur throughout the water column), processes can occur in microcosms at rates far in excess of what they would be in deeper waters. Similarly, the attempt to estimate bioaccumulation potential in a system in which the contaminant is rapidly decreasing in concentration obfuscates the relationship between exposure concentration and concentrations in biota. This can result in an estimate of equilibrium concentration factor that is considerably in error (Ref. 4-52). Another disadvantage of using microcosms for fate studies is the inability to distinguish between equilibrium and nonequilibrium processes.

The microcosm approach as used by some can be criticized for trying to extract too much information from a single experiment. This is not to say that it is without value, as it is an excellent system in which to evaluate assumptions of the process rate method and evaluate the rates of individual processes under more realistic conditions. When anthracene was evaluated in 80-liter aquatic microcosms under white fluorescent lighting, it was found to rapidly disappear from the water column and to accumulate in sediments and biota (Ref. 4-119). An evaluation of the environmental behavior of anthracene using the process rate approach predicted rapid photolysis, sorption by sediments and organic detritus, and moderate accumulation by zooplankton and fish (Ref. 4-38). The rate of disappearance noted in the microcosm was not inconsistent with that predicted by individual process studies. It is likely that experimental approaches using process rate studies and well-controlled microcosm studies will provide a data base that makes each approach more readily extrapolated to natural environments.

The objective of monitoring programs and field studies in an environmental risk analysis of a chemical is to provide field verification of the laboratory-based analyses and to act as a final hazard assessment of the chemical. Such studies thus provide the last chance of identifying hazards before excessive damage or contamination results.

Some of the "experiments" that have been most successful in determining the extent and duration of environmental damage resulting from chemical pollution have been the case histories of the use of materials such as DDT and PCBs. Case histories of such contamination provide an excellent opportunity for observing the combinations of chemical properties and environmental conditions that act together to create a hazard.

The Laurentian Great Lakes have proved to be an enormous aquatic laboratory for evaluating chemicals on an after-the-fact basis. The high bioaccumulation potential of hydrophobic substances that are not subject to rapid metabolic degradation was demonstrated in the Great Lakes for methyl mercury, DDT, PCBs, Mirex, and other chlorinated pesticides (Refs. 4-19, 4-45, 4-123). These systems have also demonstrated the extent of the area that can be contaminated by persistent, lipophilic substances. Curtailing inputs of these materials has turned these same systems into an experiment of the recovery rate of contaminated ecosystems. The accelerated eutrophication of Lake Erie provides an example of widespread ecosystem scale changes resulting, at least partially, from the transformation of man-made chemicals (polyphosphates) in the environment.

The value of case histories of environmental contamination is to make us aware of the possible scale of environmental damage possible from naively marketing and using synthetic chemicals, and to make us aware of what properties should be taken as warnings of potential hazards. Lessons learned from past mistakes will hopefully ensure that they are not repeated.

The above-mentioned examples illustrate that environmental monitoring, even when conducted in a relatively unsystematic fashion, has in many instances been an effective "safety valve" in environmental risk analysis. However, the routine use of field monitoring to evaluate the fate and effects of chemicals in the aquatic environment is subject to several limitations. Field monitoring experiments tend to be very expensive in terms of manpower, time, and money. Because they are performed in real ecosystems, lack of environmental control renders them subject to deficiencies in experimental design that often limits the value of the data gathered. Given proper site selection criteria and careful statistical analysis prior to designing a sampling regimen, field monitoring studies can play a vital role in substantiating the initial risk analysis or in detecting problems not illuminated in the initial laboratory analysis. The use of a "shotgun" approach to monitoring, in which poorly designed studies are carried out with inadequate resources on a large number of sites, is no substitute for a comprehensive generic risk analysis carried out at an earlier stage. Such widespread application of mini-monitoring programs is not likely to generate useful information and utilizes resources that could be better used elsewhere.

A key decision in environmental monitoring is the determination of where problems are likely to develop if assumptions or methodologies used in the prior assessment were inappropriate. Then, it is possible to design a monitoring effort, with the formulation of testable hypotheses and statistically sound sampling schemes.

4-3 ANALYZING THE ECOLOGICAL EFFECTS OF CHEMICALS IN THE AQUATIC ENVIRONMENT

The goal of effects testing is to predict the impact of chemical contaminants on aquatic ecosystems at various concentration levels and exposure durations. As discussed previously, limitations in aquatic testing procedures presently preclude accurate predictions of effects above the single species population level. Therefore, to assess environmental risk the results of toxicity tests with selected species components of the ecosystem are generally used to predict chronic toxicity threshold concentrations of chemicals. These results are then used to predict the chronic toxicity threshold concentration for the ecosystem as a whole. A limitation to this approach is that chronic toxicity thresholds can generally only be experimentally determined for a limited number of aquatic species. Many more species are amenable to acute toxicity testing and, as a result, application factors have been developed for estimating chronic toxicity threshold concentrations from acute toxicity data (Section 4-3.2). The following sections will discuss the methods used with potential aqueous chemical contaminants for determining acutely toxic concentrations, chronic toxicity threshold concentrations, and application factors.

4-3.1 Acute Toxicity Testing

Acute toxicity can be defined as the severe effects suffered by organisms from short-term exposure to toxic chemicals.

The objective of acute toxicity testing is to determine the dose or concentration of a particular chemical or chemicals that will elicit a specific response or measurable endpoint from a test organism or population in a relatively short period of time, usually 1 to 10 days. This is in contrast to chronic toxicity tests, which investigate the effects of exposure to a chemical over prolonged periods, often over entire life cycles.

The endpoints measured in acute toxicity studies can include any response that an organism or population may exhibit as a result of a chemical stimulus; however, the endpoint most commonly used in acute toxicity studies with aquatic biota is the death or immobilization of the test organism. These criteria are used as end points because they are easily determined and have obvious biological and ecological significance.

Acute toxicity tests have two general applications in environmental risk analyses. One application is for toxicological screening. The purpose of screening tests is to determine whether the chemical or solution being tested is biologically active with respect to the end point being measured. Essentially, screening tests provide "yes or no" answers, i.e., a chemical is toxic or nontoxic, mutagenic or nonmutagenic, carcinogenic or noncarcinogenic, and so on, at the dosages tested. Generally, screening tests do not attempt to determine the level of response that a particular chemical stimulus will produce, but only whether it will elicit the desired response. The second type of application is for acute toxicity determinations. The objective here is to measure the degree of biological response produced by a particular level of chemical stimulus. Most commonly, this is calculated in the form of the median lethal concentration, or LC_{50}, which is the estimated concentration of the test material which will kill or immobilize 50% of the test organisms in a predetermined length of time. The 50% effect level is used because it is the most reproducible response and can be estimated with the highest confidence. The LC_{50} is expressed in terms of the duration of the test which produced the desired response, e.g., 96-hr LC_{50}.

Test species. Any aquatic species can potentially be used in an acute toxicity test. However, except in cases where a particular species is selected because of special interest or concern, the group of test species selected should (1) include representatives of all major trophic levels, (2) include a range of sensitivities to the type of chemical being tested, (3) include at least some species for which a large toxicological data base exists for comparative purposes, and (4) include species that are maintainable in the laboratory in a healthy condition for a minimum of one week.

Most routine acute toxicity tests with aquatic animals are conducted using the species listed in Table 4-14. The advantages to using these species are that they meet most of the criteria listed above. A considerable data base exists on the acute toxicity of a large number of chemicals to these species under a wide variety of conditions. These data facilitate comparisons of the acute toxicities of different chemicals to the same species and comparisons of the toxicities of the same chemical to different species.

It is often desirable to use the most sensitive life stages of the organisms such as the eggs and young of fish, larvae, and early instars of invertebrates, as tests with these life stages generally provide the most conservative estimates of potential toxic effects.

Species of aquatic plants are not included in Table 4-15 because the methods used to test the toxicity of chemicals to plants differ considerably from the tests used with animals. Toxicity tests with plants are discussed below.

Toxicity Tests with Aquatic Plants. The development of testing procedures for studying the toxic effects of chemicals on plants has centered on algae and duckweed (*Lemna* spp.). Because of their short generation times, effects can be measured over several generations of both algae and duckweed in a relatively short period. The parameters generally measured in toxicity studies with algae are photosynthesis and growth, while only growth is measured with duckweed. Effects on photosynthesis can be measured by a number of well-established methods including oxygen production, $^{14}CO_2$ uptake, photosynthetic pigment concentration, ATP production, and cell counts. No generally accepted standardized method has been established for conducting toxicity studies with algae, although the U.S. EPA has developed the bottle test, after which many other methods are patterned. Examples of several commonly used toxicity test methods for algae and duckweed are provided in Table 4-15.

Exposure Systems. Various methods have been developed for exposing aquatic organisms in the laboratory to chemicals to test their toxic effects. The mode of delivery of the chemical as well as various physical, chemical, and biological factors will considerably affect the toxicity of chemicals to aquatic biota. Fairly standardized methods have been developed to facilitate comparisons of results obtained for different chemicals and aquatic organisms. Three general types of toxicant delivery systems are used in toxicity testing: flowthrough, static, and renewal.

TABLE 4-14. Aquatic Species Used in Acute Toxicity Tests.

Common Name	Species	Habitat
Freshwater Vertebrates		
Bluegill	*Lepomis macrochirus*	Warm lakes and ponds
Brook trout	*Salvelinus fontinalis*	Cold, high-quality streams and lakes
Channel catfish	*Ictalurus punctatus*	Warm rivers and lakes
Coho salmon	*Onchorhynchus kisutch*	Cold, high-quality streams
Fathead minnow	*Pimephales promelas*	Warm ponds
Flagfish	*Jordenalla floridae*	Semitropical swamps and lagoons
Frog	*Rana* sp.	Warm ponds, marshes
Goldfish	*Carassius auratus*	Warm lakes
Green sunfish	*Lepomis cyanellus*	Warm lakes
Rainbow trout	*Salmo gairdneri*	Cold, high-quality streams and lakes
Toad	*Bufo* sp.	Warm ponds
Freshwater Invertebrates		
Amphipods	*Gammarus lacustris, G. fasciatus, G. pseudolimnaeus*	Small streams
Crayfish	*Cambarrus* sp., *Orconectes* sp., *Procambarus* sp. *Pacifastacus leniusculus*	Streams, lakes, ponds
Daphnids	*Daphnia magna, D. pulex, D. pulicaria*	Pelagic zones of lakes, ponds
Mayflies	*Baetis* sp., *Ephemerella* sp., *Hexagenia limbata, H. bilinata*	High-quality streams and lakes
Midges	*Chironomus* sp.	Lakes, ponds, streams
Planaria	*Dugesia tigrina*	Slow streams
Snails	*Physa integra, P. heterostropha, Amnicola limosa*	Streams
Stoneflies	*Pteronarcys* sp.	Fast, clear streams
Marine Vertebrates		
English sole	*Parophrys vetulus*	Benthic

TABLE 4-14, (continued)

Common Name	Species	Habitat
Flounder	*Paralichthys dentatus* *P. lethostigma*	Benthic
Herring	*Clupea harengus*	Pelagic, North Atlantic
Longnose killifish	*Fundulus similis*	Estuarine
Mummichog	*Fundulus heteroclitus*	Estuarine
Pinfish	*Lagodon rhomboides*	Estuarine
Sanddab	*Citharichthys stigmaeus*	Benthic
Sheepshead minnow	*Cyprinodon variegatus*	Estuarine
Silverside	*Menidia* sp.	Estuarine
Spot	*Leiostomus xanthurus*	Estuarine
Starry flounder	*Platichthys stellatus*	Benthic
Threespine stickleback	*Gasterosteus aculeatus*	Estuarine
Tidepool sculpin	*Oligocottus maculosus*	Estuarine
Marine Invertebrates		
Copepods	*Acartia clausi, A. tonsa*	Pelagic
Crabs		
Blue crab	*Callinectes sapidus*	Benthic, estuarine
Fiddler crab	*Uca* sp.	Benthic, littoral
Green crab	*Carcinus maenas*	Benthic
Shore crab	*Hemigraspus* sp., *Pachygraspus* sp.	Benthic, littoral
Oyster	*Crassostrea virginica*	Benthic, estuarine
Polychaetes	*Capitella capitata*	Benthic
Shrimp		
Bay shrimp	*Crangon nigricauda*	Estuarine
Grass shrimp	*Palaemonetes pugio,* *P. intermedius, P. vulgaris*	Benthic, estuarine, marine
Mysid shrimp	*Mysidopsis bahia*	Benthic, estuarine
Sand shrimp	*Crangon septemspinosa*	Benthic, marine, and estuarine
Penaed shrimp	*Panaeus aztecus, P. duorarum, P. setiferus*	Benthic, marine, and estuarine
Others	*Pandalus danae, P. jordani*	Benthic, marine, and estuarine

Sources: Refs. 4-124, 4-125, 4-126

TABLE 4-15. Aquatic Plant Toxicity Testing Methods.

Test	References for Method
Algal	4-127, 4-128
Algal photosynthesis	4-129, 4-130, 4-131, 4-132, 4-133, 4-134
Duckweed growth	4-135

Flowthrough methods. Flowthrough exposure systems subject the test organisms to relatively fresh solutions of the toxic material flowing into and out of the exposure chambers. The flow can be continuous or intermittent; however, it must be able to maintain a constant concentration of the test material, a constant water temperature, and the dissolved oxygen concentration of the water at 90% saturation while flushing waste products out of the system. It is generally accepted that, flowrates must provide at least five complete water volume changes per 24 hr.

Many types of toxicant delivery systems have been designed for use in flowthrough exposure systems. The most preferred and recommended system is the proportional diluter. First developed by Mount and Brungs (Ref. 4-136), the proportional diluter has been modified and improved for a wide variety of applications. Table 4-16 lists these applications and provides references to detailed descriptions of the system used. Peristaltic pumps have also found wide application for the delivery of a series of concentrations of a toxic chemical to aquatic organisms in flowthrough exposure systems (Refs. 4-139, 4-145).

Flowthrough systems are the preferred methods for acute toxicity studies. They are especially recommended for highly volatile materials and with materials that are not persistent in water, as the concentration of the test material and the water temperature can be maintained at fairly constant levels and the dissolved oxygen concentration can be kept near saturation for prolonged periods. The major disadvantage of flowthrough systems and proportional diluters are their

TABLE 4-16. Applications of Proportional Diluters for Use with Several Classes of Toxic Chemicals.

Type of Chemical	References for Recommended Methods
Soluble chemicals	4-137
Insoluble chemicals	4-138, 4-139, 4-140
Volatile chemicals	4-141
Oil	4-142
Effluents	4-143, 4-144

complexity, which requires considerable attention and maintenance if they are to function properly.

Static systems. Static exposure systems are much simpler in design and operation than flowthrough ones. They generally consist of exposure vessels in which the test organisms are subjected to the same test solution for the duration of the test. The solution is not changed or renewed. The advantages of this type of exposure system are its simplicity and low cost; however, problems that commonly arise in static systems are (1) decreases in the concentration of the test material through loss from volatilization, microbial degradation or transformation, sorption, or organism uptake and (2) low dissolved oxygen concentration occurring if the test material has a high biochemical oxygen demand (BOD) or as a result of the accumulation and microbial degradation of fecal material. Due to these limitations, static exposure systems are generally applied in short-term tests ($\leqslant 96$ hr), with nonvolatile or slowly degradable materials and with a low loading (biomass/volume of water) of test organisms. Static systems are also generally the only ones that can be used if a very limited quantity of the chemical to be tested is available or if disposal of wastewaters is critical. Some examples of the use of static exposure systems in acute toxicity testing are provided in Refs. 4-124, 4-125, 4-126.

Renewal systems. Renewal exposure systems are a compromise between flowthrough and static exposure systems. The apparatus used is essentially the same as in a static system; however, instead of exposing the test organisms to the same solution throughout the test, the test organisms are either periodically transferred to fresh solutions or new test material is added to the exposure chamber to maintain the initial concentration. This circumvents some of the disadvantages of the static system, such as depletion of oxygen and decrease in test material concentration. The disadvantages of this approach are the increased handling of the test organisms, which increases stress and the possibility of injury, and the fact that problems can still arise from dissolved oxygen depletion if the test material has a high BOD. Additionally, the concentration of the test material may fluctuate between changes of solutions. It is also a more labor-intensive method.

Renewal exposure systems are of primary use with small organisms (e.g., *Daphnia*) that may be flushed out of flowthrough systems or are very sensitve to currents (e.g., copepods). They are also useful when only a limited amount of test material is available but a prolonged test is required. A detailed description of the renewal system for use in toxicity studies is given by Ref. 4-146.

Several articles and reports have been published describing detailed methods and procedures for use in acute toxicity tests with aquatic biota. Recommended procedures are provided by the U.S. Environmental Protection Agency (Refs. 4-139, 4-147), the American Public Health Association (Ref. 4-125), and the American Society for Testing and Materials (Ref. 4-126).

Data Analysis. A variety of methods can be used to calculate the LC_{50} and its confidence limits, including those of Litchfield and Wilcoxon (Ref. 4-148), probit analysis (Ref. 4-149), and the moving average (Ref. 4-150). Computerized programs for doing the calculations are also available (Ref. 4-151, 4-152). For a detailed discussion of the statistical methods used in toxicity testing see Ref. 4-150.

Considerations with Mixtures. The analysis of the potential environmental risks of releases of complex mixtures of chemicals, such as industrial or municipal effluents or leachates from solid wastes, requires special consideration. Several experimental approaches are available for analyzing the toxicity (or other effects) of mixtures to aquatic biota. The choice of approach depends on the objective of the research. The objective may be to study the mixture as a whole or to determine the toxicity of each component or fraction of the mixture. Each of these approaches will be discussed in more detail below.

Whole effluent testing. In an analysis of the toxicity of a complex mixture of chemicals, such as an industrial effluent, the effluent is essentially treated as a single chemical compound. The LC_{50} is calculated as a percent concentration of the effluent in the receiving water. Detailed methods for measuring the toxicity of effluents to aquatic organisms are provided by the U.S. Environmental Protection Agency (Ref. 4-144).

Testing the components of effluents. If the objective of the environmental risk analysis is to identify and quantify the hazards of the biologically active components of complex effluents, several analytical approaches can be used. The most obvious and direct approach is to perform a complete chemical analysis of the effluent and then, by screening, identify the biologically active components and individually analyze them for their potential environmental fate and effects. The limitation to this approach is the large number of individual chemicals (in the hundreds or thousands) in many industrial and municipal effluents. Therefore, undertaking comprehensive environmental risk analyses on each individual chemical rapidly becomes prohibitively expensive. In addition, tests done solely on single chemical components of effluents will not provide information on such chemical interactions as synergism or antagonism. Among the analytical approaches developed to handle the large number of chemicals found in some effluents are chemical fractionation, representative class compounds, and selected component testing. Chapter 11 provides a description of a study of a complex mixture using wastewater from synthetic fuel production.

Chemical fractionation can be used to separate complex aqueous effluents or mixtures into several less complex fractions, each with components of a more similar chemical nature. The toxicity to aquatic biota of each of the fractions can then be determined, thus identifying those that are biologically active (Ref. 4-153). Up to 14 fractions have been separated from synthetic fuel process effluents for use in mutagenesis testing (Ref. 4-154). The toxicities of the individual components of the fractions can also be determined if the fractions are chemically analyzed.

Another useful approach to studying the environmental effects of complex mixtures of chemicals is to classify the components based on chemical structure, behavior, molecular weight, and so on, and then select representative compounds from each class. These compounds are then intensively studied and the results extrapolated to the class as a whole. This approach, called the representative class compound approach, has been used in the study of the environmental behavior of aqueous effluents from synthetic fuel processes (Ref. 4-148).

If the mixture being studied is composed of relatively few chemicals or if the biologically active components are few, it is often practical to determine the toxicity of each of these components. First, the chemical composition of the effluent is determined. Second, acute or chronic toxicity determinations are made for each of the components. Third, the contribution of each of the components (as a percent) to the toxicity of the whole effluent is calculated. The completeness of the analysis can be determined by comparing the toxicity of a synthetic effluent to the toxicity of the original effluent. Interactions between the components of the effluent can also be tested (Ref. 4-155). Detailed descriptions of the methods used to analyze the toxic components of mixtures are provided by Brown (Ref. 4-156) and Parkhurst et al. (Ref. 4-157).

4-3.2 Chronic Toxicity Testing

Chronic toxicity tests measure the effects of prolonged exposure of organisms to a toxic chemical. Due to the increase in duration, the significance of chronic toxicity lies not in the immediate survival of individual organisms, as in acute toxicity tests, but in sublethal effects on reproductive and growth processes and the long-term maintenance of populations, communities, and ecosystems. Chronic toxicity testing implies an analysis at a higher level of ecological complexity than acute toxicity testing and should ultimately include analyses at population, community, or ecosystem levels.

The objective of chronic toxicity testing is to determine if prolonged exposure to the concentrations of a chemical expected to be present in the aquatic environment will have significant adverse effects on aquatic ecosystems. This is accomplished by estimating chronic toxicity threshold concentrations for a number of selected species that inhabit the aquatic ecosystem to receive the chemical being tested. From these data, the chronic toxicity threshold concentration for the aquatic ecosystem as a whole is predicted. The chronic toxicity threshold concentration for the ecosystem is taken as that of the most sensitive species tested. Three assumptions must be made here: first, that the species selected for testing are representative of the range of sensitivities to the chemical found in the species comprising the ecosystem; second, that the chronic toxicity threshold determined for the most sensitive species is the chronic toxicity threshold for the ecosystem; and third, that species and species-level properties are the most sensitive properties of ecosystems to toxic chemicals. Next, the ecosystem chronic toxicity threshold concentration is compared to the expected water concentration of the chemical. If the threshold concentration is exceeded, it is concluded that significant chronic toxicity effects will result. The three assumptions listed above are critical to this analysis. Their validity has not been substantiated.

The following discussion will briefly describe (1) the methods used to measure the chronic toxicity effects of chemicals on selected species of aquatic ecosystems and (2) the methods used to calculate chronic toxicity threshold concentrations.

Chronic Toxicity Tests. Three categories of tests are commonly used to predict the chronic effects of toxic chemicals or aquatic organisms. The first category, life-cycle toxicity tests, measures the effects chronic exposure to a chemical has on reproduction, growth, survival, and other parameters over one or more generations of a population of test organisms. The second category, tests on most sensitive life stages, measures the effects chronic exposure to a chemical has on the survival and growth of the toxicologically most sensitive life stages of a species, for example, eggs and larvae of fishes. The third category, functional tests, measures the effects of chemicals on various physiological functions of individual aquatic organisms. The data from all three categories of tests are used to estimate chronic toxicity threshold concentrations.

The methods of toxicant delivery and test organism exposure utilized in chronic toxicity tests are basically the same in design and function as those used in acute toxicity tests. The primary distinction between the dosing systems for acute and chronic toxicity tests is that, in chronic toxicity tests, the apparatus must be capable of being maintained in constant use for periods ranging from one month to longer than a year.

Life-cycle toxicity tests with aquatic animals. In life-cycle toxicity tests, groups of test organisms are exposed to a series of concentrations of the test chemical. The tests are initiated with either eggs, larvae, or juveniles and are continued until the test organisms have (or should have) reproduced. The tests can continue through several generations, if desired. Chemical concentrations range from those having significant adverse effects on survival, growth, and reproduction to at least one not having any significant effect on these parameters, compared to controls.

The species of aquatic animals that can be used in life-cycle toxicity tests are limited to those which can complete their life cycles under laboratory conditions. Table 4-17 lists those species most commonly used for life-cycle toxicity tests and lists references providing detailed instructions for performing the tests.

TABLE 4-17. Aquatic Species Used for Life-Cycle Toxicity Tests.

Common Name	Species	Habitat	Reference
Brook trout	*Salvelinus fontinalis*	Cold streams and lakes	4-158
Daphnid	*Daphnia magna, D. pulex*	Warm, freshwater lakes and ponds	4-146
Fathead minnow	*Pimephales promelas*	Warm, freshwater lakes and ponds	4-159
Grass shrimp	*Palaemonetes* sp.	Estuaries	4-160
Midge	*Chironomus* sp.	Warm, freshwater lakes, ponds, and streams	4-161
Mysid shrimp	*Mysidopsis bahnia*	Estuaries	4-162
Sheepshead minnows	*Cyprinodon variegatus*	Estuaries	4-163

Tests with most sensitive life stages. Because of the considerable time and expense involved in conducting life-cycle toxicity tests, especially with vertebrates, methods have been developed for utilizing tests with the most sensitive life stages of aquatic organisms, primarily fishes, to predict chronic toxicity threshold concentrations. The life stages of fishes most sensitive to chemical toxicants are generally embryos and/or larvae (Ref. 4-164). It has been found that estimates of chronic toxicity threshold concentrations calculated from embryo–larval toxicity tests do not differ significantly from those calculated from entire life-cycle toxicity tests (Ref. 4-164).

Fish embryo–larval toxicity tests are initiated by exposing groups of embryos or larvae, generally in a flowthrough toxicant delivery system, to a series of concentrations of the test chemical. The range of concentrations used should span the expected effects and include concentrations producing no significant effects. The organisms are generally exposed for 60 to 90 days. Parameters to be measured include survival, growth, and teratogenesis.

Aside from the time and expense saved in using embryo–larval tests to estimate chronic toxicity thresholds, an additional advantage is that they can study a much larger number of species than life-cycle toxicity tests. Thus, estimates of chronic toxicity thresholds can be made for a much wider variety of species from a much wider range of habitats and trophic levels than are possible with life-cycle toxicity tests.

A list of species that have been used in embryo–larval toxicity tests and reference to the methods used is provided in Table 4-18. It is also possible to use, in these tests, any additional species for which fertilized eggs can be secured, and then hatched and raised in the laboratory for a minimum of 30 days within the confines of a chronic toxicity exposure system.

Use of functional tests in chronic toxicity testing. Fish and other aquatic organisms are known to respond physiologically and behaviorally to exposure to

TABLE 4-18. Aquatic Species Used in Embryo-Larval Toxicity Tests[a].

Common Name	Species
Bluegill	*Lepomis macrochirus*
Brook trout	*Salvelinus fontinalis*
Brown trout	*Salmo trutta*
Channel catfish	*Ictalurus punctatus*
Coho salmon	*Onchorhyncus kisutch*
Fathead minnow	*Pimephales promelas*
Flagfish	*Jordanella floridae*
Lake trout	*Salvelinus namaycush*
Northern pike	*Esox lucius*
Rainbow trout	*Salmo gairdneri*
Smallmouth bass	*Micropterus dolomieui*
White sucker	*Catostonus commersoni*

[a]From McKim, (Ref. 4-164).

sublethal concentrations of toxic chemicals. Some of these functional responses are a result of the mode of action of the toxicant while others are adaptive responses by the organisms to the toxicant. Among the types of functional responses that have been measured in fish and in invertebrates exposed to toxic chemicals are changes in blood biochemistry, histology, swimming performance, avoidance, respiration, enzyme activities, sensory perception and disease resistance. Good reviews on the use of function responses of fish in environmental toxicology are provided by Refs. 4-165, 4-166, 4-167, 4-168, 4-169, 4-170, 4-171, 4-172. Many of these functional responses are sensitive indicators of sublethal toxic effects; however, their usefulness in environmental risk analyses for chemicals is limited by several factors. First, many of the effects are transitory and gradually disappear as the animal acclimates to the stress. Second, the variability in sampling tends to be large, reducing the precision of the analyses. Last, there have been few, if any, studies undertaken to determine the relationship between functional responses measured in individual organisms and the effects of the toxic chemical on survival, growth, and reproduction at the population level. Interpretation of these response at the population level or above is thus precluded.

Calculating chronic toxicity threshold concentrations. The statistic that is generally used to estimate chronic toxicity threshold concentrations is the MATC, or maximum acceptable toxicant concentration (Ref. 4-173). The MATC is

defined as the highest test concentration at which no significant effects are observed on survival, growth, or reproduction during a full life cycle of exposure to
the toxic chemical. The precision with which the MATC is estimated will depend
on the dilution series used in chronic toxicity tests. The actual MATC will fall
between the highest concentration causing no effect and the lowest concentration causing significant effects. As a result, the MATC is generally given as the
range between these two values. The closeness of these values to each other,
determines the precision of the estimate. Since the MATC is really an estimate
of the highest concentration that will produce no significant observed effects
and has nothing to do with the acceptability of the effects, Maki (Ref. 4-174)
has suggested that the term MATC be replaced by NOEC (no observed effects
concentration).

Application factors. In aquatic toxicology, application factors are mathematical formulas that are used to estimate the MATC from acute toxicity data (LC_{50}).
The rationale behind using application factors is that (1) it is often prohibitively
expensive and/or time comsuming to determine NOECs using life cycle or embryo-
larval toxicity tests for all the species desired, (2) methods are available only for
performing these tests on a limited number of species, and (3) acute toxicity tests
can be run on a much larger selection of species in a much shorter time.

Several application factors for calculating MATCs from acute toxicity data
have been proposed. Some are simply arbitrary fractions of acute toxicity concentrations, such as 0.1 or 0.01 of the 96 hr LC_{50} (Ref. 4-165). However, these
arbitrary fractions have not been found to be generally applicable among different
species and are often over- or under-conservative. The most widely accepted
application factor has been that of Mount and Stephán (Ref. 4-173). This application factor is used, under the assumption that the ratio of MATC to LC_{50} for
a given toxicant varies little from species to species. An application factor can
thus be determined experimentally using this ratio MATCs.

$$\text{Application Factor} = \frac{\text{MATC (mg/liter) (Species A)}}{48\text{- or }96\text{-hr } LC_{50} \text{ (mg/liter) (Species A)}}$$

The MATC for other species (Species B) is then estimated from the relationship,
MATC (Species B) = LC_{50} (Species B) $\times$ application factor (Species A). Application factors should, however, be used with caution until the validity of the
underlying assumptions has been established.

4-3.3 System-Level Tests

The concept of testing the effects of possible contaminants on communities of
aquatic organisms rather than only on single species is appealing. The extrapolation from observed effects on laboratory communities to predicted effects on

natural communities seems to be less than that from a single species LC_{50} to predicted effects on natural communities.

Microcosms, ponds, and artificial and natural streams were used previously in toxicity tests on entire communities (Refs. 4-175, 4-176). While such methods hold considerable promise, the ecological significance of the observed changes in test systems and how those observations can yield predictions of effects in full-scale communities have not yet been unambiguously determined. In general, community-level tests are too expensive and time consuming to use as screening tools.

Monitoring for ecological damage, however, can be a highly effective tool for evaluating chemicals, albeit after a decision has been made to manufacture and use a material. The composition of communities of aquatic plants and animals is very sensitive to stressors such as toxic chemicals. Changes in measures of species diversity, the loss or gain of particular "indicator species," or changes in the standing crop of organisms are likely results of toxicants and can be readily detected. Such measures have been applied routinely to water pollution investigations (Refs. 4-177–4-183). If contaminant levels anticipated to be safe as a result of single species tests prove to be harmful in natural aquatic communities, carefully designed monitoring studies can detect environmental damage and can alert users and manufacturers to a problem early enough to minimize environmental damage. However, some of the same limitations and criticisms of system-level toxicity tests can be applied to the use of monitoring programs for evaluating contaminants. First, there is the same difficulty in interpreting the ecological significance of changes observed in the structure and/or function of the aquatic ecosystem being studied. Unless the changes are severe, there is no general concensus as to what signifies adverse changes. The types of monitoring programs that are needed to detect less than catastrophic changes in aquatic ecosystems are generally very time consuming and expensive. Finally, monitoring programs are most feasible for analyzing the local effects of point-source contaminants and are not practical for analyzing regional effects of either point or non-point contaminants.

REFERENCES

4-1. Machta, L. Status of global radioactive-fallout predictions. pp. 369–404. IN *Radioactive Fallout from Nuclear Weapons Tests,* A. W. Klement, Jr. (ed.). U. S. Atomic Energy Commission, Washington, D. C. 1965.

4-2. Gutowsky, H. S. *Halocarbons: Effects on Stratospheric Ozone.* National Academy of Sciences, Washington, D. C. 1976. 352 pp.

4-3. Olson, J. S., H. A. Pfuderer, and Y. H. Chan. Changes in the global carbon cycle and the biosphere. ORNL/EIS-109. Oak Ridge National Laboratory, Oak Ridge, Tennessee. p. 4. 1978.

4-4. Woodwell, G. N., P. P. Craig, and H. A. Johnson. DDT in the biosphere: Where does it go? *Science,* **174**:1101-1107 (1971).

4-5. Eisenreich, S. J., and G. J. Hollod. Accumulation of polychlorinated biphenyls (PCBs) in surficial Lake Superior sediments, atmospheric deposition. *Environ. Sci. Technol.,* **13**(5):569-573 (1979).

4-6. Murphy, T. J., and C. P. Rzeszutko. Precipitation inputs of PCBs to Lake Michigan. *J. Great Lakes Res.,* **3**(3-4):305-312 (1977).

4-7. Mellan, Ibert. *Industrial Plasticizers.* The Macmillan Co., New York. 1963.

4-8. Mackay, D., and P. J. Leinonen. Rate of evaporation of low solubility contaminants from water bodies to atmosphere. *Environ. Sci. Technol.,* **9**(1975):1178-1180 (1979).

4-9. Hazardous Materials Advisory Committee, Herbicide Report, EPA-SAB-74-001. U. S. Environmental Protection Agency, Washington, D. C. 1974.

4-10. Hales, J. M., and S. L. Sutter. Solubility of sulfur dioxide in water at low concentrations. *Atmos. Environ.,* **7**:997-1001 (1973).

4-11. Smith, J. H., W. R. Mabey, N. Bohonos, B. R. Holt, S. S. Lee, T. W. Chou, D. C. Bomberger, and T. Mill. Environmental pathways of selected chemicals in freshwater systems. Part II. Laboratory studies. EPA-600/7-78-074. U. S. Environmental Protection Agency, Washington, D. C. 1978.

4-12. Southworth, G. R. The role of volatilization in removing PAH from natural waters. *Bull. Envir. Contam. Toxicol.,* **21**:507-514 (1979).

4-13. Ruttner, F. *Fundamentals of Limnology,* University of Toronto Press, Toronto, Canada. 1953. 295 pp.

4-14. Metcalf, R. L. A century of DDT. *J. Agric. Food Chem.,* **21**(4):511 (1972).

4-15. Kornreich, M., B. Fuller, J. Dorigan, P. Walker, and L. Thomas. Environmental impact of polychlorinated biphenyls. MTR-7006. Mitre Corp., McLean, Virginia. 1976.

4-16. McRae, A., L. Whelchel, and H. Rowland (eds.). *Toxic Substances Control Source Book.* Aspen Systems Publication, Germantown, Maryland. p. 98. 1978.

4-17. Rasmussen, K. H., M. Taheri, and R. L. Kabel. Global emissions and natural processes for removal of gaseous pollutants. *Water Air Soil Pollut.,* **4**:33-64 (1975).

4-18. Mathur, J. P. Phthalate esters in the environment: Pollutants or natural products. *J. Environ. Qual.,* **3**(3):3(3):189-197 (1974).

4-19. Kaiser, K. L. E. The rise and fall of mirex. *Environ. Sci. Technol.,* **12**(5):520-528 (1978).

4-20. Nelson, N. (Chairman, Committee for the Working Conference on Principles of Protocols for Evaluating Chemicals in the Environment). *Principles for Evaluating Chemicals in the Environment,* National Academy of Sciences, Washington, D. C.

4-21. Cronan, C. S., and C. L. Schofield. Aluminum leaching response to acid precipitation: Effects on high-elevation watersheds in the northeast. *Science,* **204**(4390):304-306 (1979).

4-22. Wright, R. F., D. Torstein, E. T. Gjessing, G. R. Hendry, A. Henriksen, N. Johannessen, and I. P. Muniz. Impact of acid precipitation on freshwater ecosystems in Norway. *Water Air Soil Pollut.,* 6:483-499 (1976).

4-23. Scinn, W. G. N. Some approximations for the wet and dry removal of particles and gases from the atmosphere. *Water Air Soil Pollut.,* 7:513-543 (1977).

4-24. Bjorseth, A., and G. Lunde. Long-range transport of polycyclic aromatic hydrocarbons. *Atmos. Environ.,* 13:45-53 (1979).

4-25. Browman, M. G., and G. Chesters. The solid-water interface: Transfer of organic pollutants across the solid water interface. *Fate of Pollutants in the Air and Water Environments,* Part 1, Vol. 8, F. H. Suffet (ed.). John Wiley & Sons, Inc., New York. 1977.

4-26. Shupe, S. J., and G. W. Dawson. Current disposition of kepone in residuals in the Hopewell, Virginia, area. BNWL-SA-6488 Battelle Pacific Northwest Laboratories, Richland, Washington. 1977.

4-27. Nadeau, R. J., and R. A. Davis. Polychlorinated biphenyls in the Hudson River (Hudons Falls – Fort Edward, New York State). *Bull. Environ. Contam. Toxicol.* 16:436-444 (1976).

4-28. Turner, R. R., and S. E. Lindberg. Behavior and transport of mercury in river-reservoir system downstream of inactive chloralkali plant. *Environ. Sci. Technol.,* 12(8): 918-923 (1978).

4-29. Macek, K. J., S. R. Petrocelli, and B. H. Sleight, III. Considerations in assessing the potential for, and significance of, biomagnification of chemical residues in aquatic food chains. pp. 251-268 IN Aquatic Toxicology, L. L. Marking and R. A. Kimerle (eds.). ASTM STP 667. American Society for Testing and Materials, Philadelphia, Pennsylvania. 1979.

4-30. Neely, W. B., D. R. Branson, and G. E. Blau. Partition coefficient to measure bioconcentration potential of organic chemicals in fish. *Environ. Sci. Technol.,* 8(13): 1113-1115 (1974).

4-31. Southworth, G. R., J. J. Beauchamp, and P. K. Schneider. Bioaccumulation potential and acute toxicity of synthetic fuels effluents in freshwater biota: Azaarenes. *Environ. Sci. Technol.,* 12(9):1062-1066 (1978).

4-32. Lu, P. Y., R. L. Metcalf, and E. M. Carlson. Environmental fate of five radiolabeled coal conversion by-products evaluated in a laboratory model ecosystem. *Environ. Health Perspect.,* 24:201-208 (1979).

4-33. Lee, R. F., R. Sauerheber, and G. H. Dobbs. Uptake, metabolism, and discharge of polycyclic aromatic hydrocarbons by marine fish. *Mar. Biol.,* 17(3):201-208 (1972).

4-34. Jarvinen, A. W., M. J. Hoffman, and T. W. Thorslund. Long term toxic effects of DDT food and water exposure on fathead minnows (*Pimephales promelas*). *J. Fish. Res. Board Can.,* 34(11):2049-2103 (1977).

4-35. Veith, G. D., D. L. DeFoe, and B. V. Bergstedt. Measuring and estimating the bioconcentration factor of chemicals in fish. *J. Fish. Res. Board Can.,* 36(9):1040-1048 (1979).

4-36. Bahner, L. H., A. J. Wilson, Jr., J. M. Sheppard, J. M. Patrick, Jr., L. R. Goodman, and G. E. Walsh. Kepone bioconcentration, accumulation, loss, and transfer through estuarine food chains. *Chesapeake Sci.,* **14**(3):299-308 (1977).

4-37. Mayer, F. L. Residue Dynamics of Di(2-ethylhexyl) phthalate in fathead minnows (*Pimephales promelas*). *J. Fish. Res. Board Can.,* **33**(11):2610-2613 (1976).

4-38. Southworth, G. R. Transport and transformations of anthracene in natural waters. pp. 359-380. IN *Aquatic Toxicology.* L. L. Marking, and R. A. Kimerle (eds.), ASTM STP 667. American Society for Testing and Materials, Philadelphia, Pennsylvania. 1979.

4-39. Southworth, G. R., B. R. Parkhurst, and J. J. Beauchamp. Accumulation of acridine from water, food, and sediment by the fathead minnow, *Pimephales promelas.* *Water Air Soil Pollut.,* **12**:331–341 (1979).

4-40. Reinert, R. E., and H. L. Bergman. Residues of DDT in lake trout (*Salvelinus namaycush*) and coho salmon (*Oncorhynchus kisutch*) from the Great Lakes. *J. Fish. Res. Board Can.,* **31**(2):191-199 (1974).

4-41. Roberts, J. R., A. S. W. deFrietas, and M. A. Gidney. Influence of lipid pool size on bioaccumulation of the insecticide chlordane by northern redhorse suckers (*Moxostoma macrolepidotum*). *J. Fish. Res. Board Can.,* **34**(1):89-97 (1977).

4-42. Lee, R. F., R. Sauerheber, and A. A. Benson. Petroleum hydrocarbons: Uptake and discharge by the marine mussel, *Mytilus edulis. Science,* **177**:344-346 (1972).

4-43. Payne, J. F. Mixed function oxidases in marine organisms in relation to petroleum hydrocarbon metabolism and detection. *Mar. Pollut. Bull.,* **8**(5):112-116 (1977).

4-44. Norstrom, R. J., A. E. McKinnon, and A. S. W. deFreitas. A bioenergetics-based model of pollutant accumulation by fish. Simulation of PCB and methylmercury residue levels in Ottawa River yellow perch (*Perca flavescens*). *J. Fish. Res. Board Can.,* **33**(2):248-267 (1976).

4-45. Kelly, T. M., J. D. Jones, and G. R. Smith. Historical changes in mercury contamination in Michigan walleyes (*Stizostedion vitreum vitreum*). *J. Fish. Res. Board Can.,* **32**(10):1745-1754 (1975).

4-46. Vanderploeg, H. A., D. C. Parzyck, W. H. Wilcox, I. R. Kercher, and S. V. Kaye. Bioaccumulation factors for radionuclides in freshwater biota. ORNL-5002. Oak Ridge National Laboratory, Oak Ridge, Tennessee. 1975.

4-47. Paine, R. T. Foodweb complexity and species diversity. *Am. Nat.,* **100**:65-75 (1966).

4-48. Mayer, F. L., and J. L. Hamelink (eds.). *Aquatic Toxicology and Hazard Evaluation.* ASTM STP 634. American Society for Testing and Materials, Philadelphia, Pennsylvania. 1977.

4-49. Cairns, J., Jr., K. L. Dickson, and A. W. Maki, eds. *Estimating the Hazard of Chemical Substances to Aquatic Life.* ASTM STP 657. American Society for Testing and Materials, Philadelphia, Pennsylvania. 1978.

4-50. Marking, L. L., and R. A. Kimerle. *Aquatic toxicology.* ASTM STP 667. American Society for Testing and Materials, Philadelphia, Pennsylvania. 1979. 387 pp.

4-51. Baughman, G. L., and R. R. Lassiter. Prediction of environmental pollutant concentration. pp. 35-54. IN *Estimating the Hazard of Chemical Substances to Aquatic*

Life, John Cairns, Jr., K. L. Dickson, and A. W. Maki (eds.). ASTM STP 657. American Society for Testing and Materials, Philadelphia, Pennsylvania. 1978.

4-52. Branson, D. L. Predicting the fate of chemicals in the aquatic environment from laboratory data. pp. 55-70. IN *Estimating the Hazard of Chemical Substances to Aquatic Life,* John Cairns, Jr., K. L. Dickson, and A. W. Maki (eds.), ASTM STP 657. American Society for Testing and Materials, Philadelphia, Pennsylvania. 1976.

4-53. Stern, A. M., and C. R. Walker. Hazard assessment of toxic substances: environmental fate testing of organic chemicals and ecological effects testing. pp. 81-131. IN *Estimating the Hazard of Chemical Substances to Aquatic Life,* John Cairns, Jr., K. L. Dickson, and A. W. Maki (eds.), ASTM STP 657. American Society for Testing and Materials, Philadelphia, Pennsylvania. 1978.

4-54. Branson, D. R. A new capacitor fluid – A case study in product stewardship. pp. 44-61. IN *Aquatic Toxicology and Hazard Evaluation,* F. L. Mayer and J. L. Hamelink (eds.), ASTM STP-634. American Society for Testing and Materials, Philadelphia, Pennsylvania. 1977.

4-55. Environmental Protection Agency, Office of Toxic Substances, Toxic Substances Control Act, Premanufacture Testing of New Chemical Substances. Fed. Reg. 44(53):16240-16292. 1979.

4-56. Smith, J. H., W. R. Mabey, N. Bohonos, B. R. Holt, S. S. Lee, T. W. Chou, D. C. Bomberger, and T. Mill. Environmental pathways of selected chemicals in freshwater systems. Part I. Background and experimental procedures. EPA-600/7-77-113. U. S. Environmental Protection Agency, Washington, D. C. 1977.

4-57. Herbes, S. E., G. R. Southworth, D. L. Shaffer, W. H. Griest, and M. P. Maskarinec. Critical pathways of polycyclic aromatic hydrocarbons in aquatic environments. *The Scientific Basis for Toxicity Assessment.* H. Witschi, Ed. Elsevier/North-Holland Biomedical Press, Amsterdam, The Netherlands, pp. 113-128.

4-58. Herbes, S. E. Partitioning of polycyclic aromatic hydrocarbons between dissolved and particulate phases in natural waters. *Water Res.,* 11:493-496 (1975).

4-59. Lotse, E. G., P. A. Graetz, G. Chesters, G. B. Lee, and L. W. Newland. Lindane adsorption by lake sediments. *Environ. Sci. Technol.,* 2(5):353-357 (1968).

4-60. Halter, M. T., and H. E. Johnson. A model system to study the desorption and biological availability of PCB in hydrosoils. pp. 178-195. IN *Aquatic Toxicology and Hazard Evaluation,* F. L. Mayer and J. L. Hamelink (eds.). ASTM STP 634. American Society for Testing and Materials. 1977.

4-61. Guring, C. A. I., and J. W. Hamaker. *Organic Chemicals in the Soil Environment.* Marcel Decker, Inc., New York. p. 87, 89. 1972.

4-62. Harris, C. I. Adsorption, movement, and phytotoxicity of monuron and S-triazine herbicides in soil. *Weeds,* 14(1):6-10 (1966).

4-63. Lerman, A., and T. A. Ligtzke. Uptake and migration of tracers in lake sediments. *Limnol. Oceanogr.,* 20(4):497-509 (1975).

4-64. O'Connor, G. A., and J. U. Anderson. Soil factors affecting the adsorption of 2,4, 5-T. *Soil Sci. Soc. Am. Proc.,* 38:433-436 (1974).

4-65. Briggs, J. C., and J. F. Ficke. Quality of rivers of the United States, 1975 water year – Based on the National Stream Quality Accounting Network (NASQAN). U. S. Geological Survey Open-File Report 78-200, Reston, Virginia. May 1977.

4-66. Pionke, H. B., and G. Chesters. Pesticide-sediment-water interactions. *J. Environ. Qual.,* **2**(1):29-45 (1973).

4-67. Karickhoff, S. W., D. S. Brown, and T. A. Scott, T. A. Sorption of hydrophobic pollutants on natural sediments. *Water Res.,* **13**:241-248 (1979).

4-68. Fields, D. E. CHNSED: Simulation of sediment and trace contaminant transport with sediment/contaminant interaction. ORNL/NSF/EATC-19. Oak Ridge National Laboratory, Oak Ridge, Tennessee. 1976.

4-69. Liss, P. S., and P. G. Slater. Flux of gases across the air-sea interface. *Nature,* **247**: 181-184 (1974).

4-70. Liss, P. S. Processes of gas exchange across an air-water interface. *Deep-Sea Res.,* **20**:224-238 (1973).

4-71. Cohen, Y., W. Cocchio, and D. Mackay. Laboratory study of liquid-phase controlled volatilization rates in presence of wind waves. *Environ. Sci. Technol.,* **12**(5):553-558 (1978).

4-72. Dilling, W. L., N. B. Teferatiller, and G. J. Kallos. Evaporation rates and reactivities of methylene chloride, chloroform, 1, 1, 1-trichloromethane, trichloroethylene, tetrachloroethylene, and other chlorinated compounds in dilute aqueous solutions. *Environ. Sci. Technol.,* **9**(9):833-838 (1975).

4-73. Dilling, W. L. Interphase transfer processes. II. Evaporation rates of chloro methanes, ethanes, ethylenes, propanes, and propylenes from dilute aqueous solutions: Comparisons with theoretical predictions. *Environ. Sci. Technol.,* **11**(4):405-409 (1977).

4-74. Hill, J., H. P. Kolling, D. F. Paris, N. L. Wolfe, and R. G. Zepp. Dynamic behavior of vinyl chloride in aquatic ecosystems. EPA-600/3-76-001. U. S. Environmental Protection Agency, Washington, D. C. 1976.

4-75. Mackay, D., W. Y. Shiu, and R. P. Sutherland. Determination of air-water Henry's Law constants for hydrophobic pollutants. *Environ. Sci. Technol.,* **13**(3):333-337 (1979).

4-76. Harrison, W., M. A. Winnik, P. T. Y. Kwong, and D. Mackay. Crude oil spills, disappearance of aromatic and aliphatic components from small sea-surface slicks. *Environ. Sci. Technol.,* **9**(3):231-234 (1975).

4-77. Mackay, D., and R. S. Matsugu. Evaporation rates of liquid hydrocarbon spills on land and water. *Can. J. Chem. Eng.,* **51**:434-439 (1973).

4-78. Kudo, A., and J. S. Hart. Uptake of inorganic mercury by bed sediments. *J. Environ. Qual.,* **3**(3):273-278 (1974).

4-79. Kudo, A., and E. F. Gloyna. Transport of ^{137} Cs -II. Interaction with bed sediments. *Water Res.,* **5**:71-79 (1971).

4-80. Theis, T. L., and P. J. McCabe. Retardation of sediment phosphorus release by fly ash application. *J. Water Pollut. Control Fed.,* **50**(12):2666-2676 (1978).

4-81. Kudo, A., D. C. Mortimer, and J. S. Hart. Factors influencing desorption of mercury from bed sediments. *Can. J. Earth Sci.,* **12**:1036-1040 (1975).

4-82. Hamelink, J. T. Current bioconcentration test methods and theory. pp. 149-161. IN *Aquatic Toxicology and Hazard Evaluation,* F. L. Mayer, and J. L. Hamelink (eds.). American Society for Testing and Materials, Philadelphia, Pennsylvania. 1977.

4-83. Neely, W. B. Estimating rate constants for the uptake and clearance of chemicals by fish. *Environ. Sci. Technol.*, **13**(12):1506-1510 (1979).

4-84. Hamelink, J. L., R. C. Waybrant, and R. C. Ball. A proposal: Exchange equilibria control the degree chlorinated hydrocarbons are biologically magnified in lentic environments. *Trans. Am. Fish. Soc.*, **100**(2):207-214 (1971).

4-85. Macek, J. M., and S. Korn. Significance of food chain in DDT accumulation by fish. *J. Fish. Res. Board Can.*, **27**(8):1496-1498 (1970).

4-86. Thomann, R. V. Size dependent model of hazardous substances in aquatic food chain. EPA-600/3-78-036. U. S. Environmental Protection Agency, Washington, D. C. 1978.

4-87. Huckabee, J. W., R. A. Goldstein, S. A. Janzen, and S. E. Woock. Methylmercury in a freshwater food chain. pp. 199-216. IN *Preceedings: International Conference on Heavy Metals in the Environment.* International Joint Commission, October 27-31, 1975, Toronto, Ontario. 1975.

4-88. Haile, C. L., G. D. Veith, G. F. Lee, and W. C. Boyle. Chlorinated hydrocarbons in the Lake Ontario ecosystem (IFYGL). EPA-660/3-75-022. U. S. Environmental Protection Agency, Washington, D. C. 1975.

4-89. Eberhardt, L. L. Some methodology for appraising contaminants in aquatic systems. *J. Fish. Res. Board Can.*, **32**(10):1852-1859 (1975).

4-90. Spacie, A., and J. L. Hamelink. Dynamics of trifluralin accumulation in river fishes. *Environ. Sci. Technol.*, **13**(7):817-822 (1979).

4-91. Blanchard, F. A., I. T. Takahashi, H. C. Alexander, and E. A. Bartlett. Uptake, clearance, and bioconcentration of ^{14}C-sec-butyl-4-chlorodiphenyl oxide in rainbow trout. pp. 162-177 IN *Aquatic Toxicology and Hazard Evaluation,* F. L. Mayer and J. L. Hamelink (eds.). ASTM STP 634. American Society for Testing and Materials, Philadelphia, Pennsylvania. 1977.

4-92. Wolfe, N. L., R. G. Zepp, and R. C. Hollis. Carbaryl, propham, and chloropropham: A comparison of the rates of hydrolysis and photolysis with the rate of biolysis. *Water Res.*, **12**:565-571 (1978).

4-93. Wolfe, N. L., R. G. Zepp, G. L. Baughman, R. C. Fincher, and J. A. Gordon. Chemical and photochemical transformation of selected pesticides in aquatic systems. EPA-600/3-76-067. 1976.

4-94. Miller, G. C., and R. G. Zepp. Effects of suspended sediments on photolysis rates of dissolved pollutants. *Water Res.*, **13**:453-459 (1979).

4-95. Miller, G. C., and R. G. Zepp. Photoreactivity of aquatic pollutants sorbed on suspended sediments. *Environ. Sci. Technol.*, **13**(7):860-863 (1979).

4-96. Zepp, R. G., and D. M. Cline. Rates of direct photolysis in the aquatic environment. *Environ. Sci. Technol.*, **11**:359-366 (1977).

4-97. Ward, D. M., and T. D. Brock. Environmental factors influencing the rate of hydrocarbon oxidation in temperate lakes. *Appl. Environ. Microbiol.*, **31**(5):764-772 (1976).

4-98. Raghu, K., and I. C. MacRae. Biodegradation of the gamma isomer of benzene hexachloride in submerged soils. *Science*, **154**:263-264 (1966).

4-99. Alexander, M. Microbiological aspects of soil pollution. pp. 78-80. IN *Proceedings: International Symposium on Identification and Measurement of Environmental Pollutants.* National Research Council of Canada, Ottawa. 1971.

4-100. Pierce, R. J., Jr., A. M. Cundell, and R. W. Traxler. Persistence and biodegradation of spilled residual fuel oil on an estuarine beach. *Appl. Microbiol.,* 24(5):646-652 (1975).

4-101. Herbes, S. E., and L. R. Schwall. Microbial transformation of polycyclic aromatic hydrocarbons in pristine and petroleum contaminated sediments. *Appl. Environ. Microbiol.,* 35(2):306-316 (1978).

4-102. Espey, W. H., Jr., E. F. Gloyna, and R. D. Taylor. Appendix: Hydrologic transport of radionuclides. pp. 55-72. IN *Proceedings: Workshop on the Evaluation of Models Used for the Environmental Assessment of Radionuclide Releases,* F. O. Hoffman, D. L. Shaeffer, C. W. Miller, and C. T. Garten (eds.). CONF-770901. Technical Information Center, Oak Ridge, Tennessee. 1978. 131 pp.

4-103. Nemerow, N. C. *Scientific Stream Pollution Analysis.* McGraw-Hill, New York. 1974.

4-104. Dunn, W. E., A. J. Policastro, and R. A. Paddock. Surface thermal plumes: Evaluation of mathematical models for the near and complete field. ANL/WR-75-3, Part 1. Argonne National Laboratory, Argonne, Illinois. 1975.

4-105. Tayfun, M. A., and H. Wang. Monte Carlo simulation of oil slick movements. ASCE J. Waterw. Harbors Coastal Eng. Div. 99(WW3):309-324 (1973).

4-106. Metcalf and Eddy, Inc., University of Florida and Water Res. Eng., Inc. Storm Water Management Model, Vol. I-IV. Water Quality Office, Water Pollution Control Research Series 11024, DOC 07/71. U. S. EPA, Washington, D. C. 1971.

4-107. Crawford, N. H., and R. K. Linsley. Digital simulation in hydrology: Stanford Watershed Model IV. Stanford University Technology Report No. 39, Stanford, California. 1966.

4-108. Donigan, H. S., Jr., and N. H. Crawford. Modeling nonpoint pollution from the land surface. 600/3-76-083. U. S. EPA, Athens, Georgia. 1976.

4-109. McElroy, A. D., S. Y. Chiu, J. W. Nebgen, A. Aleti, and F. W. Bennett. Loading functions for assessment of water pollution from nonpoint sources. 600/2-76-151. U. S. EPA, Washington, D. C. 1976.

4-110. Van Hook, R. I., and W. D. Shults. Ecology and analysis of trace contaminants: Progress report, October 1974-December 1975. ORNL/NSF/EATC-22. Oak Ridge National Laboratory, Oak Ridge, Tennessee. 1976.

4-111. Munro, J. K., Jr., R. J. Luxmore, C. L. Begovich, K. R. Dixon, A. P. Watson, M. R. Patterson, and D. R. Jackson. Application of the Unified Transport Model to the movement of Pb, Cd, Zn, Cu, and S through the Crooked Creek Watershed. ORNL/NSF/EATC-28. Oak Ridge National Laboratory, Oak Ridge, Tennessee. 1976.

4-112. Chen, C. W., and G. T. Orlob. Ecologic simulation for aquatic environments. pp. 475-587. IN *Systems Analysis and Simulation in Ecology,* Vol. III, B. C. Patten (ed.). Academic Press, New York. 1975.

4-113. Leendertse, J. J. A water quality simulation model for well-mixed estuaries and coastal seas, Vol. I. Principles of computation. The Rand Corp., Report No. RM-6230-RC. 1970.

4-114. Neely, W. B., G. E. Blau, and T. Alfrey, Jr. Mathematical models predict concentration-time profiles resulting from chemical spill in a river. *Environ. Sci. Technol.,* **10**(1): 72-76 (1976).

4-115. Baughman, G. L., and L. A. Burns. Transport and transformations of chemicals in the environment: A perspective. IN *Handbook of Environmental Chemistry,* O. Hutzinger (ed.). Springer-Verlag, Heidelburg (in press).

4-116. Onishi, Y. Mathematical simulation of sediment and radionuclide transport in the Columbia River. Battelle Pacific Northwest Laboratory Report No. 228, Richland, Washington. 1977.

4-117. Onishi, Y. Finite element models for sediment and contaminant transport in surface waters — Transport of sediments and radionuclides in the Clinch River. Battelle Pacific Northwest Laboratory Report No. 2227, Richland, Washington. 1977.

4-118. Herbes, S. E., G. R. Southworth, and W. H. Griest. Field site evaluation of aquatic transport of polycyclic aromatic hydrocarbons. pp. 221-230. IN *Potential Health and Environmental Effects of Synthetic Fossil Fuel Technologies.* CONF-780903. Technical Information Center, Oak Ridge, Tennessee. 1979.

4-119. Giddings, J. M., B. T. Walton, G. K. Eddlemon, and K. G. Olson. Transport and fate of anthracene in aquatic microcosms. pp. 312-320. IN *Proceedings of the Workshop: Microbial Degradation of Pollutants in Marine Environments.* EPA-600/9-79-012. U. S. Environmental Protection Agency, Washington, D. C. 1979.

4-120. Lu, P. Y., R. L. Metcalf, N. Plummer, and D. Mandel. The environmental fate of three carcinogens: benzo(a)pyrene, benzidine, and vinyl chloride evaluated in laboratory model ecosystems. *Arch. Environ. Contam. Toxicol.,* **6**:129-142 (1977).

4-121. Metcalf, R. L., I. P. Kapoor, P. Y. Lu, C. K. Schuth, and P. Sherman. Model ecosystem studies of the environmental fate of six organochlorine pesticides. *Environ. Health Perspect.,* **4**:35-44 (1973).

4-122. Pupp, C., R. C. Lao, J. J. Murray, and F. P. Roswell. Equilibrium vapor concentrations of some polycyclic aromatic hydrocarbons, As_4O_6, and SeO_2 and the collection efficiencies of these air pollutants. *Atmos. Environ.,* 8 15:925 (1974).

4-123. Waller, W. T., and G. F. Lee. Evaluation of observations of hazardous chemicals in Lake Ontario during the international field year for the Great Lakes. *Environ. Sci. Technol.,* **13**(1):79-85 (1979).

4-124. U. S. Environmental Protection Agency. Methods for acute toxicity tests with fish, macroinvertebrates, and amphibians. EPA-660/3-75-009, Washington, D. C. 1975. 67 pp.

4-125. American Public Health Association, American Water Works Association, and Water Pollution Control Federation. Standard Methods for the Examination of Water and Wastewater, 14th ed. Washington, D. C. 1976. 1193 pp.

4-126. American Society for Testing and Materials, Standard Practice for Conducting Basic Acute Toxicity Tests with Fishes, Macroinvertebrates, and Amphibians. American Society for Testing and Materials, Philadelphia, Pennsylvania. 1979. 32 pp.

4-127. U. S. Environmental Protection Agency. Algal assay procedure: Bottle test. National Environmental Research Laboratory — Corvallis, Oregon. 1971. 82 pp.

4-128. Hannan, P. J., and C. Patouillet. Effect of mercury on algal growth rates. pp. 93-101. IN *Biotechnology and Bioengineering* 14. 1972.

4-129. Jensen, A., B. Rystad, and S. Nelson. Heavy metal tolerance of marine phytoplankton. I. The tolerance of three algal species to zinc in coastal seawater. *J. Exp. Mar. Bio. Ecol.*, **15**:145-157 (1974).

4-130. Erickson, S. I., W. Lackie, and T. E. Maloney. A screening technique for estimating copper toxicity to estuarine phytoplankton. *J. Water Pollut. Control Fed.*, **42**: 270-278 (1970).

4-131. Woolery, M., and R. A. Lewin. The effects of lead on algae. IV. Effects of Pb on respiration and photosynthesis of *Phaeodactylum tricornutum* (Baccillario phyceae). *Water Air Soil Pollution*, **6**:25-31 (1976).

4-132. Herang, J. C., and E. F. Gloyna. Effect of organic compounds on photosynthetic oxygenation — I. Chlorophyll destruction and suppression of photosynthetic oxygen production. *Water Res.*, **2**:347-366 (1968).

4-133. Overnell, J. The effects of heavy metals on photosynthesis and loss of cell potassium in two species of marine algae, *Dunatiella terliolecta* and *Phaeodactylum tricornutung. Mar. Biol.*, **29**:99-103 (1975).

4-134. Winters, K., J. C. Batterton, and C. Van Boalen. Phenalen-1-one: Occurrence in a fuel oil and toxicity to microalgae. *Environ. Sci. Technol.*, **11**:270-272 (1977).

4-135. Walbridge, C. T. A flow-through testing procedure with duckweed (*Lemna minor*). EPA 600/3-77-108, U. S. EPA. Washington, D. C. 1977.

4-136. Mount, D. I., and W. A. Brungs. A simplified dosing apparatus for fish toxicological studies. *Water Res.*, **1**:21-29 (1967).

4-137. Lempke, A. E., W. A. Brungs, and B. J. Halligan. Manual for construction and operation of toxicity-testing proportional diluters. EPA-600/3-78-072. U. S. EPA, Washington, D. C. 70 pp. 1978.

4-138. DeFoe, D. L. Multichannel toxicant injection system for flow-through bioassays. *J. Fish. Res. Board Can.*, **32**:544-546 (1975).

4-139. Birge, W. J., J. A. Black, J. E. Hudson, and D. M. Bruser. Embryo-larval toxicity tests with organic compounds. pp. 131-147. IN *Aquatic Toxicology*, L. L. Marking, and R. A. Kimerle (eds.). ASTM STP 667. American Society for Testing and Materials, Philadelphia, Pennsylvania. 1979.

4-140. Veith, G. D., and V. M. Comstock. Apparatus for continuously saturating water with hydrophobic organic chemicals. *J. Fish. Res. Board Can.*, **32**:1849-1851 (1975).

4-141. Benville, P. E., Jr., and S. Korn. A simple apparatus for metering volatile liquids into water. *J. Fish. Res. Board Can.*, **31**:367-368 (1973).

4-142. Lichatowich, J. A., P. W. O'Keefe, J. A. Strand, and W. L. Templeton. Development of methodology and apparatus for the bioassay of oil. pp. 659-666. IN *Proceedings: Joint Conference on Prevention and Control of Oil Spills.* American Petroleum Institute, Environmental Protection Agency, and U. S. Coast Guard, Washington, D. C. 1973.

4-143. Riley, C. W. Proportional diluter for effluent bioassays. *J. Water Pollut. Control Fed.*, **47**:2620-2626 (1975).

4-144. Peltier, W. Methods for measuring the acute toxicity of effluents to aquatic organisms. EPA-600/4-78-012. U. S. EPA, Washington, D. C. 1978. 52 pp.

4-145. Maki, A. W.   Modifications of continuous-flow toxicity test methods for small aquatic organisms. *Prog. Fish Cult.,* **39**:172-174 (1977).

4-146. American Society for Testing and Materials.   Proposed Standard Practice for Conducting Renewal Life Cycle Toxicity Tests with Daphnids. American Society for Testing and Materials, Subcommittee E 3523, Philadelphia, Pennsylvania. 1979.

4-147. U. S. Environmental Protection Agency.   Directory of short-term tests for health and ecological effects.   EPA-600/1-78-052.   U. S. EPA, Research Triangle Park, North Carolina. 1978. 206 pp.

4-148. Litchfield, J. T., Jr., and F. Wilcoxon.   A simplified method of evaluating dose-effect experiments. *J. Pharmacol. Exp. Ther.,* **96**:99-113 (1949).

4-149. Finney, D. J.   Probit Analysis, 3rd ed.   Cambridge University Press, London. 1971. 333 pp.

4-150. Stephan, C. E.   Methods for calculating an LC_{50}.   pp. 65-84.   IN Am Soc. Test. Mats. Spec. Tech. Pub. 634, Philadelphia, Pennsylvania. 1977.

4-151. Barr, A. J., J. H. Goodnight, J. P. Sall, and J. T. Helwig. A User's Guide to SAS 76. SAS Institute, Inc., Raleigh, North Carolina. 1976.

4-152. Stephan, C. E., K. A. Busch, R. Smith, J. Burke, and R. W. Andrew.   A computer program for calculating an LC_{50}.   U. S. Environmental Protection Agency, ERL-Duluth, Duluth, Minnesota (in preparation).

4-153. Parkhurst, B. R., C. W. Gehrs, and I. B. Rubin.   Value of chemical fractionation for identifying the toxic components of complex aqueous effluents. pp. 122-130. IN *Aquatic Toxicology,* L. L. Marking and R. A. Kimerle (eds.). ASTM STP 667. American Society for Testing and Materials. 1979.

4-154. Rubin, I. B., M. R. Guerin. S. A. Hardigree, and J. L. Epler. Fractionation of synthetic crude oils from coal for biological testing. *Environ. Res.,* **12**:358-365 (1976).

4-155. Marking, L. L., and V. K. Dawson.   Method for assessment of toxicity or efficacy of mixtures of chemicals. IN   Investigation in Fish Control, No. 67. U. S. Department of Interior, Fish and Wildlife Service, Washington, D. C. 1975.

4-156. Brown, V. M.   The calculation of the acute toxicity of mixtures of poisons to rainbow trout. *Water Res.,* **2**:723-733 (1968).

4-157. Parkhurst, B. R., A. S. Bradshaw, J. L. Forte, J. L., and G. P. Wright. An evaluation of the acute toxicity to aquatic biota of a coal conversion effluent and its major components. *Bull. Environ. Contam. Toxicol.* **23**:349-356 (1979).

4-158. U. S. Environmental Protection Agency.   Recommended procedure for brook trout, *Salvelinus fontinalis* (Mitchell), partial chronic tests.   Environmental Research Laboratory − Duluth Mimeo. 12 pp. 1971.

4-159. U. S. Environmental Protection Agency.   Recommended bioassay procedure for fathead minnows, *Pimephales promelas* (Rafinesque), chronic tests.   Environmental Research Laboratory − Duluth Mimeo. 13 pp. 1972.

4-160. Tyler-Schroeder, D. B.   Use of the grass shrimp (*Palaemonetes pugio*) in a life-cycle toxicity test. pp. 159-170. IN *Aquatic Toxicology,* L. L. Making and R. A. Kimerle (eds.), ASTM STP 667. American Society for Testing and Materials, Philadelphia, Pennsylvania.

4-161. Maki, A. W. The life history, culture and toxicity testing of the midge, Paratanytarsus parthenogenica. Special Symposium of the Entomological Society of America, Washington, D. C. 1977.

4-162. Mimmo, D. W., L. H. Bhaner, R. A. Rigby, J. M. Sheppard, and A. J. Wilson. *Mysidopsis bahia:* An estuarine species suitable for life-cycle toxicity tests to determine the effects of a pollutant. pp. 109-116. IN *Aquatic Toxicology and Hazard Evaluation,* ASTM STP 634. American Society for Testing and Materials, Philadelphia, Pennsylvania. 1977.

4-163. Schimmel, S. C., and D. J. Hansen. Sheepshead minnow (*Cyprinodon variegatus*): An estuarine fish suitable for chronic (entire life-cycle) bioassays. pp. 392-398. IN *Proc., 28th Annual Conference of the Southeastern Association of Game and Fish Commissioners.* 1974.

4-164. McKim, M. Evaluation of tests with early life stages of fish for predicting long term toxicity. *J. Fish. Res. Board Can.,* 34:1148-1154 (1977).

4-165. Sprague, J. B. Measurement of pollutant toxicity to fish. III. Sublethal effects and "safe" concentrations. *Water Res.,* 5:245-266 (1971).

4-166. Sprague, J. B. Current status of sublethal tests of pollutants on aquatic organisms. *J. Fish. Res. Board Can.,* 33:1988-1992 (1976).

4-167. Koeman, J. H., and Strik, J. J. *Sublethal Effects of Toxic Chemicals on Aquatic Animals.* Elsevier Scientific Publishing Co., New York. 1975. 234 pp.

4-168. McKim, J. M., G. M. Christensen, J. H. Tucker, D. A. Benoit, and M. J. Lewis. Effects of pollution on freshwater fish – A literature review. *J. Water Pollut. Control Fed.,* 45:1370-1407 (1973).

4-169. McKim, J. M., G. M. Christensen, J. H. Tucker, D. A. Benoit, and M. J. Lewis. Effects of pollution on freshwater fish. *J. Water Pollut. Control Fed.,* 46:1540-1591 (1974).

4-170. McKim, J. M., D. A. Benoit, K. E. Biesinger, W. A. Brungs, and R. E. Siefert. Effects of pollution on freshwater fish. *J. Water Pollut. Control Fed.,* 47:1711-1820 (1975).

4-171. Brungs, W. A., J. H. McCormick, T. W. Neiheisel, R. L. Spehar, C. E. Stephan, and G. N. Stokes, Effects of pollution on freshwater fish. *J. Water Pollut. Control Fed.,* 49:1425-1492 (1977).

4-172. Brungs, W. A., R. W. Carlson, W. B. Horning, II, J. H. McCormick, R. L. Spehar, and J. D. Yount. Effects of pollution on freshwater fish. *J. Water Pollut. Control Fed.,* 50:1582-1636 (1978).

4-173. Mount, D. I., and C. E. Stephan. A method for establishing acceptable toxicant limits for fish – Malathion and the butoxyethanol ester of 2, 4-D. *Trans. Am. Fish. Soc.,* 96:185-193 (1967).

4-174. Maki, A. W. Correlations between *Daphnia magna* and fathead minnow (*Pimephales promelas*) chronic toxicity values for several classes of test substances. *J. Fish. Res. Board Can.,* 36:411-421 (1979).

4-175. Giddings, J. M., and G. K. Eddlemon. Some ecological and experimental properties of complex aquatic microcosms. *Int. J. Environ. Stud.,* 13(2):119-123 (1979).

4-176. U. S. Environmental Protection Agency. Design and evaluation of laboratory ecological system studies. EPA-600/3-77-022. U. S. EPA, Washington, D. C. 1977.

4-177. Hynes, H. B. N. *The Biology of Polluted Waters.* University of Toronto Press, Toronto, Canada. 1974.

4-178. Weber, C. I. Biological field and laboratory methods for measuring the quality of surface waters and effluents. EPA-670/4-73-001. U. S. EPA, Washington, D. C. 1973.

4-179. Wihlm, J. L. Range of diversity index in benthic macroinvertebrate populations. *J. Water Pollut. Control Fed.,* **42**(5):R221-R224 (1970).

4-180. Hilsenhoff, W. L. Use of arthropods to evaluate water quality of streams. Technical Bulletin No. 100. Wisconsin Department of Natural Resources, Madison, Wisconsin. 1977.

4-181. Cole, R. A. Stream community response to nutrient enrichment. *J. Water Pollut. Control Fed.,* **45**(9):1874-1888 (1973).

4-182. MacKenthun, K. M. *The Practice of Water Pollution Biology,* U. S. Department of the Interior, Federal Water Pollution Control Association, Washington, D. C., 1969.

4-183. Patrick, R. Biological measures of stream conditions. *Sewage Ind. Wastes,* **22**:926-938 (1950).

5

Atmospheric
environment

Wendell L. Dilling, Ph. D.

Research Specialist
Environmental Sciences Research Laboratory
The Dow Chemical Company
Midland, Michigan

5-1 INTRODUCTION, TRANSPORT, TRANSFORMATION, AND LOSS MECHANISMS

5-1.1 Introduction

When released into the environment, the major portion of a compound having a high vapor pressure and a low solubility in water will be found in the atmosphere. A discussion of the quantitative aspects of a compound's partitioning among the various phases of the environment has already been presented in Chapter 2.

Chemicals may be introduced into the atmosphere by direct release or by evaporation from the aquatic or terrestrial environments.

5-1.2 Transport

After entering the atmosphere, a compound is subjected to various meteorological forces, such as wind and rain, that move it into various regions of the atmosphere and, in some instances, back to the aquatic or terrestrial environments.

The atmosphere is a complex region of the environment and has been divided into several sections, primarily on the basis of temperature (Fig. 5-1). This chapter deals only with the troposphere, the lowest region of the atmosphere, because most compounds that enter the atmosphere are released into and destroyed in the troposphere. If a compound persists in the troposphere, then it may be necessary to study its degradation under stratospheric conditions. Few organic compounds are likely to persist in the stratosphere because of the high-energy ultraviolet radiation ($\sim$200 nm) present there (Refs. 5-1, 5-2, 5-3).

The movement of compounds into different regions of the atmosphere occurs at different rates. A schematic representation of the movement of compounds between the northern and southern troposphere and the stratosphere is shown in Fig. 5-2. The approximate global average lifetimes (τ) and half-lives ($t_{1/2}$) as defined by Eq. 5-1 are given in Table 5-1 (Refs. 5-4, 5-5, 5-6).

$$\tau = \frac{t_{1/2}}{0.693} = \frac{1}{k} \qquad \text{(Eq. 5-1)}$$

The mixing or dispersion of chemicals within either the northern or southern troposphere is faster than the mixing between these two regions, and both of these mixing processes are much faster than the transfer of the materials to the stratosphere. Rainout and washout provide lifetimes of one week or more for aerosols and soluble gases in the lower 5 km of the troposphere (Ref. 5-4).

5-1.3 Transformation and Loss Mechanisms

Compounds that are emitted or transferred to the atmosphere may disappear by chemical reaction there or by transport back to the aquatic or terrestrial environments.

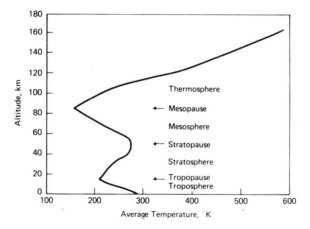

Fig. 5-1. Regions of the atmosphere and average temperature profile. *Source:* Heicklen, J., *Atmospheric Chemistry*, New York: Academic Press, 1976.

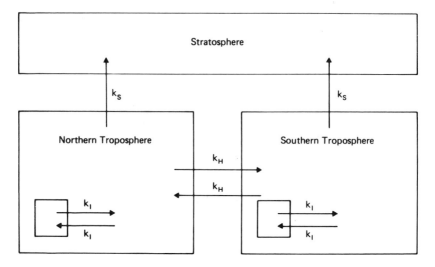

Fig. 5-2. Movement of chemicals among regions of the atmosphere. The small squares within the northern and southern tropospheres represent volume elements within those regions.

Compounds in the troposphere may react by at least three types of processes: (1) direct photolysis by sunlight, (2) electronic energy transfer, and (3) reaction with variuos reactive species normally present in the troposphere.

To undergo direct photolysis in sunlight, a compound (C) must absorb at least a portion of the ultraviolet (UV) or visible radiation from the sun (Scheme 5-1). The absorption of sunlight, however, does not guarantee the decomposition

Scheme 5-1 Direct Photolysis

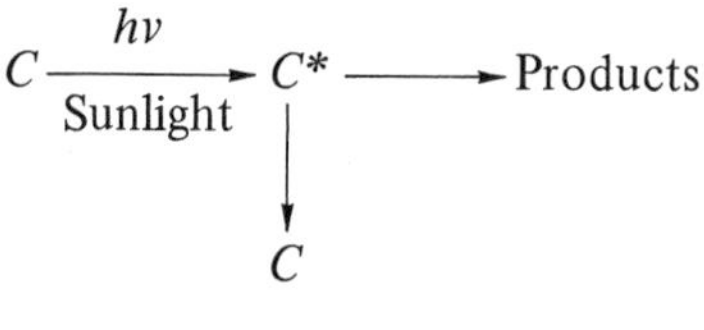

$$C \xrightarrow[\text{Sunlight}]{h\nu} C^* \longrightarrow \text{Products}$$
$$\downarrow$$
$$C$$

of the compound. Various energy-dissipating routes, which leave the absorbing species unchanged chemically, are available (Refs. 5-7, 5-8). A photolysis experiment is necessary to determine if direct absorption of light does indeed lead to a chemical reaction.

Compounds that do not absorb sunlight directly may still undergo reactions via their excited states by an energy transfer process (Scheme 5-2). A different compound that absorbs sunlight may act as a sensitizer (S), i.e., an electronic

TABLE 5-1. Approximate Average Lifetimes and Half-Lives for Exchange Among Various Regions of the Atmosphere.

Process	Lifetime (τ)	Half-life ($t_{1/2}$)
Intrahemispheric exchange (k_I)	1 month	0.7 month
Interhemispheric exchange (k_H)	1 year	0.7 year
Tropospheric $\longrightarrow$ stratospheric transfer (k_S)	30 years	20 years

Scheme 5-2 Energy Transfer

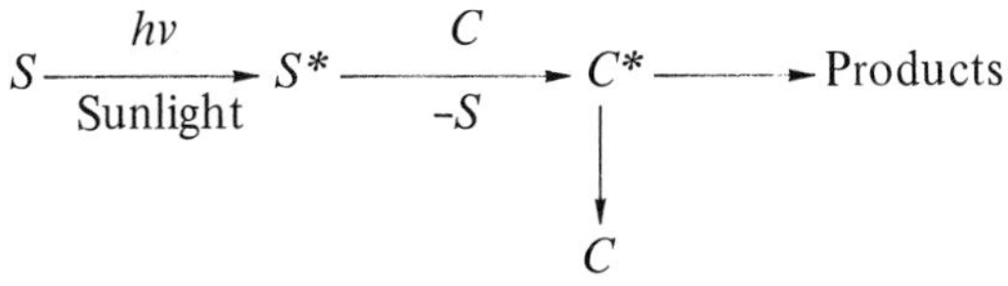

$$S \xrightarrow[\text{Sunlight}]{h\nu} S^* \xrightarrow[-S]{C} C^* \longrightarrow \text{Products}$$

$$C^* \downarrow$$

$$C$$

energy transfer agent. This transfer of energy to a compound (C) raises it to an excited state (C^*) from which reaction may or may not occur. Again, experimentation is the only method for determining whether these processes result in the decomposition of the compound in question. Carbonyl or aromatic compounds are typical sensitizers. The importance of electronic energy transfer for the decomposition of compounds in the troposphere is not known.

Probably the most important mechanism for the destruction of organic compounds (C) in the troposphere is their reactions with reactive species (X) normally present in the troposphere (Scheme 5-3). These reactive species are generally formed by reactions, some of which are initiated by sunlight, of

Scheme 5-3 Reaction With Reactive Species

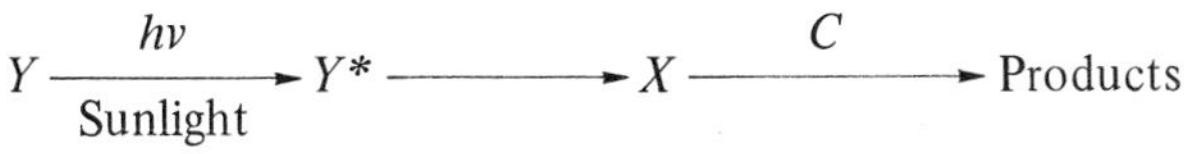

$$Y \xrightarrow[\text{Sunlight}]{h\nu} Y^* \longrightarrow X \xrightarrow{C} \text{Products}$$

naturally occurring constituents (Y) of the troposphere. According to present theory, important examples of reactive species, X, are hydroxyl radicals ($\cdot$OH) and ozone (O_3).

These three types of destruction mechanisms can occur in both the gas phase and in condensed media. Homogeneous gas-phase reactions are thought to predominate in the decomposition of compounds in the troposphere. Under some conditions, however, heterogeneous reactions on the surface of particulate matter or solution-phase reactions in aerosols may be significant processes. Little is known about the importance of these surface reactions in the decomposition of organic compounds in the troposphere.

5-2 MATERIAL PROPERTIES

A compound's physical properties control the location and rate at which it moves in the environment, and its chemical properties control its rate of decomposition and the nature of the products.

5-2.1 Physical Properties

It is possible to estimate the partitioning of a compound between the atmosphere and the hydrosphere if its vapor pressure and solubility in water are known. A dimensionless partition coefficient, H, can be calculated by using Eq. 5-2, where C_{air} and C_{water} are the equilibrium concentrations in the air and

$$H = \frac{C_{air}}{C_{water}} = \frac{16.04\, PM}{TS} \qquad \text{(Eq. 5-2)}$$

water phases, respectively, P is the vapor pressure of the pure solute in mm of Hg, M is the gram molecular weight of the solute, T is the temperature in degrees Kelvin, and S is the solubility of the solute in water in milligrams per liter ($\sim$ppm) (Ref. 5-9). A high vapor pressure and a low solubility in water promote movement of a chemical to the atmosphere.

The volume of the troposphere to a height of 10 km (5.1×10^{18} m^3) is $\sim$140 times as large as that of the hydrosphere (3.6×10^{16} m^3) to a depth of 100 m, the average mixing depth (Ref. 5-10). If a compound with a partition coefficient of 1.0, e.g., CCl_4 (Ref. 5-9), were stable long enough to allow complete mixing in these portions of the troposphere and hydrosphere, more than 99% would move to the trosposphere. Even a compound having $H \approx 0.01$, e.g., $CHCl_2 CHCl_2$ (Ref. 5-9), would be present in the troposphere at equilibrium to an extent of $> 50\%$.

An estimate of the rate of transfer from the hydrosphere to the troposphere for a compound that is slightly soluble in water can be made by using Eq. 5-3,

$$t_{1/2} = 2.18 \times 10^{-4} \left(\frac{1.04}{H} + 100 \right) dM^{1/2} \qquad \text{(Eq. 5-3)}$$

where $t_{1/2}$ is the half-life for solute evaporation in days and d is the water depth in meters (Ref. 5-9). For example, the half-life for the transfer of CCl_4 from the ocean, in the upper 100 m, to the troposphere is 27 days. Equation 5-3 assumes that the original solute concentration in the troposphere is negligible. These transfer rates and equilibria and their dependence on the physical properties of compounds are discussed in greater detail in Chapter 2.

The electronic absorption spectrum (UV or visible) is a physical property that is important with respect to a compound's stability in the troposphere. As mentioned in the previous section, these spectra indicate if direct photolysis by sunlight is possible.

5-2.2 Chemical Properties

The chemical properties that are important in the determination of a compound's atmospheric behavior include the reactivity of its excited state(s) and the reactivity of its ground state with such species as $\cdot OH$, O_3, singlet molecular oxygen ($^1O_2$), atomic oxygen (O), and possibly a few other reactive species.

Compounds that absorb sunlight and then dissociate in the gas phase will decompose. For example, acetone absorbs sunlight from the short-wavelength limit of tropospheric sunlight ($\sim$290 nm) to $\sim$330 nm (Ref. 5-7). Photodissociation gives methyl radicals, acetyl radicals, and carbon monoxide (Eq. 5-4) (Ref. 5-7).

$$\underset{CH_3CCH_3}{\overset{O}{\overset{\|}{}}} \xrightarrow{h\nu} \underset{CH_3C\cdot}{\overset{O}{\overset{\|}{}}} + \cdot CH_3 + CO \qquad \text{(Eq. 5-4)}$$

The radicals react further, primarily with molecular oxygen (O_2) to give oxidation products.

A few compounds, such as o-hydroxybenzoyl derivatives, have energy-dissipating routes available, at least in solution (Eq. 5-5) (Ref. 5-7). The compound is left unchanged after irradiation.

$$\text{(Eq. 5-5)}$$

Other energy-dissipating mechanisms, such as internal conversion and luminescence, are also possible. Most compounds that absorb sunlight will eventually decompose by direct photolysis, but the rate may be too slow to be competitive with other routes. The reactivity of compounds with free radicals is often an approximate guide to their reactivities or decomposition rates under tropospheric conditions, because $\cdot OH$ is one of the principal reactants in the troposphere (Refs. 5-11, 5-12). Most organic compounds containing carbon–carbon multiple bonds or carbon–hydrogen bonds react with $\cdot OH$ (Eqs. 5-6, 5-7). The rates of these processes vary according to the nature of the substituents and are

$$\underset{/}{\overset{\backslash}{}}C{=}C\underset{\backslash}{\overset{/}{}} + \cdot OH \longrightarrow HO{-}\overset{|}{C}{-}\overset{/}{\underset{\backslash}{C}}\cdot \qquad \text{(Eq. 5-6)}$$

$$-\overset{|}{\underset{|}{C}}{-}H + \cdot OH \longrightarrow -\overset{|}{\underset{|}{C}}\cdot + H_2O \qquad \text{(Eq. 5-7)}$$

governed by the same principles that apply to other free radical reactions (Refs. 5-13, 5-14). Olefins usually react faster than saturated hydrocarbons. Only the perhaloalkanes are inert to hydroxyl radicals (Ref. 5-15).

Under conditions of low $[\cdot OH]$, reactions of O_3 with some compounds, particularly polyalkylated olefins, may be significant destruction routes. Other reactions, such as those with $^1O_2$ (Ref. 5-16), may also occur. Sometimes several different reactions, such as direct photolysis and reactions with $\cdot OH$ and O_3, may all be significant with respect to destruction of the compound in the troposphere. For most compounds under most conditions, however, the reaction with $\cdot OH$ appears to be the major decomposition route.

Ionic reactions may be important in some tropospheric reactions, particularly in condensed phases, such as aerosols. For example, the hydrolysis of acyl halides (Eq. 5-8) in aqueous aerosols or raindrops may be the destruction mechanism for these compounds, which could arise from the photooxidation of halogenated ethylenes (Refs. 5-17, 5-18).

$$
\begin{array}{c}
\quad\quad O \quad\quad\quad\quad\quad\quad\quad\quad O \\
\quad\quad \| \quad\quad\quad\quad\quad\quad\quad\quad \| \\
RCX \ + \ H_2O \longrightarrow RCOH \ + \ HX
\end{array} \quad\quad (Eq. 5-8)
$$

5-3 ATMOSPHERIC PROPERTIES

5-3.1 Composition

The troposphere contains a large number of compounds (Table 5-2), mostly at trace levels. Undoubtedly many other compounds not shown in the table are also present. Where possible, the globally, seasonally, diurnally averaged composition of the troposphere is given in Table 5-2. Exceptions are noted in the footnotes. The original references should be consulted for the ranges and variations. The concentrations of some of the trace components are highly variable. Analyses from urban areas, e.g., Los Angeles (Ref. 5-28) and New York (Ref. 5-44), have been excluded from Table 5-2.

N_2, O_2, and Ar make up >99.9% (gas volume or molar basis) of the dry troposphere. Only O_2 enters into chemical reactions of interest to this discussion. Most of the chemical reactions occurring in the troposphere involve the trace constituents. The majority of the compounds in Table 5-2 are of natural origin. A few are probably from anthropogenic sources.

Aerosols or particulate matter may significantly affect the decomposition of some compounds in the troposphere, although little is known about these processes. A representative aerosol number density profile is shown in Fig.

TABLE 5-2. Average Composition of Troposphere.

Species	Concentration[a]	References
N_2	78.1%	5-4
$O_2(^3\Sigma \bar{g})$[b]	20.9%	5-4
Ar	0.9%	5-4
H_2O	0.75%[c]	5-24
CO_2	320 ppm	5-4
Ne	18 ppm	5-25
He	5.2 ppm	5-25
CH_4	1.5 ppm	5-4
Kr	1.1 ppm[d]	5-26
$C_{10}H_{16}$[e]	$\geqslant$1 ppm[f]	5-4
H_2	500 ppb	5-3, 5-4
N_2O	330 ppb	5-27
CO	100 ppb[g]	5-4
Xe	87 ppb[h]	5-24
O_3	40 ppb[h, i, j]	5-24
$CH_3(CH_2)_3OH$	10 ppb[h, k]	5-4, 5-28, 5-29
HF	10 ppb[h]	5-30
NH_3	6 ppb	5-4
(cyclopentane structure)[1]	6 ppb[m]	5-31
CH_3CH_3	1 ppb	5-6, 5-32
$CH_3(CH_2)_3CH_3$	1 ppb[h]	5-28
Cl_2	1 ppb[h, n]	5-30
NO_2	1 ppb[h, o, p]	5-24
$CH_3(CH_2)_4CH_3$	800 ppt[h]	5-28
CH_3Cl	750 ppt	5-3
CH_3COCH_3	700 ppt[h]	5-28
HCHO	600 ppt[q]	5-4
CH_3CHO	600 ppt[h, r]	5-28
$CH_3(CH_2)_2CH_3$	500 ppt[h]	5-28
HCl	500 ppt	5-3
NO	500 ppt[h, s]	5-24
$CH_2=CH_2$	400 ppt[h]	5-28
$HC\equiv CH$	250 ppt	5-32
CCl_2F_2	220 ppt	5-3, 5-6
H_2S	200 ppt	5-4
SO_2	200 ppt[t]	5-4
$CH_3OH + CH_3CH_2OH$	200 ppt[h]	5-28
CCl_3F	130 ppt	5-3, 5-6
CCl_4	130 ppt	5-3
C_6H_6[u]	100 ppt[h]	5-28
CH_3CCl_3	70 ppt	5-3, 5-6, 5-10
HNO_3	60 ppt[h]	5-24

(Continued)

TABLE 5-2. (Continued).

Species	Concentration[a]	References
CH_2Cl_2	30 ppt	5-3, 5-6, 5-32
$CHCl_3$	20 ppt	5-3
$COCl_2$	20 ppt[h]	5-33
I_2	20 ppt[h]	5-30
$CF_2ClCFCl_2$	18 ppt[h]	5-6
$HO_2\cdot$	16 ppt[v, w]	5-3
CF_2ClCF_2Cl	11 ppt[h]	5-6
CCl_3CCl_3	7 ppt[h, x]	5-6
$CHCl=CCl_2$	5 ppt	5-3
$CCl_2=CCl_2$	5 ppt	5-3
$CHFCl_2$	5 ppt[h]	5-6
CH_3Br	5 ppt[h]	5-3, 5-6
CH_2BrCH_2Br	5 ppt[h]	5-6
$^1O_2\,(^1\Delta_g)^y$	3 ppt[z]	5-34
CH_3I	1 ppt[d, a']	5-6, 5-35
SF_6	0.2 ppt	5-6, 5-32
$\cdot OH$	0.07 ppt[b', c']	5-3, 5-6, 5-10 5-36–5-40
$Cl\cdot$	<0.01 ppt[c', d']	5-41, 5-42
$O(^3P)^b$	4×10^{-5} ppt[v, w]	5-3
Rn	6×10^{-8} ppt[d]	5-25
$O(^1D)^x$	$<10^{-9}$ ppt[v, w]	5-3, 5-34

(a) Molar or gas volume basis. Values are based on dry composition of troposphere except for H_2O value. The tropospheric average is given unless specified otherwise. For other values see also Refs. 5-19–5-23. (b) Ground state. (c) Midlatitude mean at the earth's surface; mean at 10 km is 70 ppm. Extremes for entire troposphere is 10 ppm–4% (Ref. 5-4). (d) At sea level. (e) Terpenes. (f) In local areas. No tropospheric average known. (g) National primary air quality standard, 9000 ppb/8 hr. (h) Near earth's surface. (i) Range of tropospheric concentrations 10-100 ppb (Ref. 5-4). (j) National primary air quality standard, 120 ppb/1 hr. (k) The original publication (Ref. 5-43) stated a concentration of 190 ppb. (l) α-Pinene. (m) Maximum observed in pine forest in NC. (n) Probably too high based on current understanding of tropospheric chemistry. (o) Noontime surface concentrations range from 0.5 to 4 ppb (Ref. 5-4). (p) National primary air quality standard, 50 ppb annual average. (q) Range 400-700 ppt. (r) Range 0.1-1.1 ppb. (s) Noontime surface concentrations range from 200 to 2000 ppt (Ref. 5-4). (t) National primary air quality standard, 140,000 ppt/24 hr. (u) Benzene. (v) Noontime, midlatitude surface. (w) Assuming 2.5×10^{19} molecules of air cm^{-3} at the earth's surface. (x) Northern hemisphere. (y) Excited state. (z) At 10 km. (a') Over ocean. (b') 10^6 radicals cm^{-3}. The agreement among various workers is ± factor of 2 or 3 from value given. (c') Assuming 1.4×10^{19} molecules of air cm^{-3} for tropospheric average. (d') $<10^5$ atoms cm^{-3}; based on recent model calculations (Ref. 5-41).

5-3 as a function of altitude for a clear standard atmosphere. Average logarithmic size distribution functions for tropospheric aerosols at three altitudes over the mid-continental United States and over the open ocean are shown in Fig. 5-4. There is information available on the composition of tropospheric aerosols (Ref. 5-24). Vegetation, soil dust, and sea spray are the major constituents. The average tropospheric concentration of sulfate aerosols is $2\mu g\ m^{-3}$ (Ref. 5-4).

5-3.2 Conditions

The average temperature, pressure, and molecular density as functions of altitude in the troposphere and stratosphere are shown in Table 5-3 (Ref. 5-28).

Solar radiation intensity and spectral distribution in the troposphere (Fig. 5-5) play a major role in determining the types and rates of chemical reactions occurring there. The short-wavelength limit of ~290–300 nm (Fig. 5-6) is critical in determining which compounds can decompose by direct photolysis and which cannot. This short-wavelength, or high-energy per photon, limit is critical for lighting systems used for simulating tropospheric sunlight in labora-

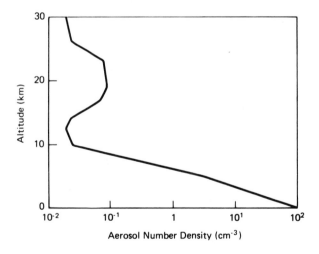

Fig. 5-3. Representative aerosol number density profile as a function of altitude. *Source*: Reference Technical Report AFCRL–64–740, Sept. 1964, L. Elterman, Atmospheric Attenuation Model, 1964, in the Ultraviolet, Visible, and Infrared Regions for Altitudes to 50 km.

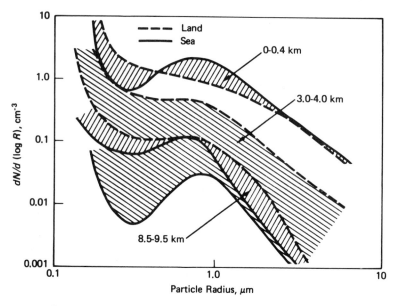

Fig. 5-4. Average tropospheric aerosol particle size distributions. *Source*: National Oceanic and Atmospheric Administration, National Aeronautics and Space Administration, United States Air Force, *U. S. Standard Atmosphere, 1976*, U. S. Government Printing Office, Washington, D. C., 1976.

TABLE 5-3. Average Properties of Troposphere and Stratosphere at Various Altitudes.

Altitude, km	Temperature, °K (°C)	Pressure, mm Hg	Molecular density, molecules cm⁻³
0	291 (18)	760	2.5×10^{19}
5	266 (-7)	370	1.3×10^{19}
10	231 (-42)	180	7.7×10^{18}
15	211 (-62)	85	3.9×10^{18}
20	219 (-54)	39	1.7×10^{18}
25	227 (-46)	18	7.7×10^{17}
30	235 (-38)	8.6	3.6×10^{17}
35	252 (-21)	4.3	1.7×10^{17}
40	268 (-5)	2.2	8.1×10^{16}
45	274 (1)	1.2	4.3×10^{16}
50	274 (1)	0.66	2.3×10^{16}

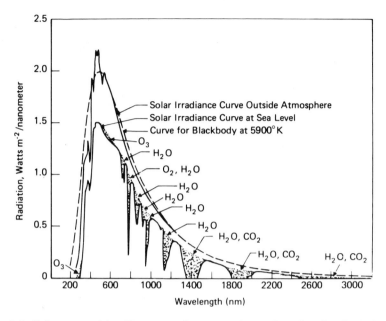

Fig. 5-5. Solar spectral irradiance as a function of wave-length. The shaded areas show light absorbed in the atmosphere by the molecules indicated. Reproduced from *Halocarbons: Effects on Stratospheric Ozone*, National Academy of Sciences, Washington, D. C., p. 183, 1976, with permission.

tory experiments in which the types and rates of tropospheric decomposition reactions are determined. The use of lighting systems with radiation of < 290 nm may result in reactions that do not occur in the real troposphere. There are available more extensive discussions and tabulations of tropospheric sunlight (Refs. 5-26, 5-28, 5-45–5-50).

5-4 ATMOSPHERIC CHEMISTRY

Many homogeneous gas-phase reactions and some heterogeneous reactions on or in aerosols or particles occur in the troposphere. The heterogeneous reactions are not as well understood as the homogeneous ones. An abbreviated listing of the most significant reactions will be presented here; only a qualitative discussion will be given. Other references should be consulted for the rate constants and a more extensive discussion of these reactions (Refs. 5-4, 5-26, 5-28, 5-51, 5-52).

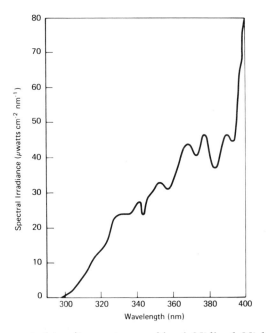

Fig. 5-6. Solar spectral irradiance at ground level, Midland, Michigan, clear sky, noon, mid-June; 3.0 mW cm^{-2} in 290–400 nm range. Measured with International Light 700, 760, 780 Spectroradiometer system.

5-4.1 Inorganic Reactions

The most significant inorganic tropospheric reactions, regarding chemical fate in the troposphere, are shown in Scheme 5-4. These reactions illustrate the major interconversions that inorganic compounds of oxygen, nitrogen, hydrogen, sulfur, and chlorine undergo. The rates of these processes and other factors lead to the average tropospheric concentrations shown in Table 5-2.

5-4.2 Organic Reactions

The principal reactions of organic compounds in the troposphere are shown in Scheme 5-5 (Refs. 5-4, 5-26, 5-28, 5-53). The reactions that account for the oxidation of saturated hydrocarbons (RH) are similar to those that convert CH_4, the natural hydrocarbon that occurs at the highest concentration in the troposphere, to CO_2. The reactions of olefins with ·OH and, in some circumstances, O_3, are important decomposition routes for the olefins.

The chemistry of photochemical smog is not discussed in detail in this chapter. However, many of the reactions involved in this phenomenon are shown in Schemes 5-4 and 5-5. Other references should be consulted for a more detailed discussion of smog chemistry (Refs. 5-28, 5-54–5-56).

Scheme 5-4 Inorganic Tropospheric Reactions

$$O_3 \xrightleftharpoons[M]{hv,\ 290-350\ nm} O(^1D) + O_2 \tag{Eq. 5-9}$$

$$O_3 \xrightleftharpoons[M]{hv,\ 300-1180\ nm} O(^3P) + O_2 \tag{Eq. 5-10}$$

$$O(^1D) + H_2O \longrightarrow 2 \cdot OH \tag{Eq. 5-11}$$

$$O_3 + NO \longrightarrow O_2 + NO_2 \tag{Eq. 5-12}$$

$$O_3 + NO_2 \longrightarrow O_2 + NO_3 \tag{Eq. 5-13}$$

$$NO_2 \xrightarrow{hv,\ 290-430\ nm} O(^3P) + NO \tag{Eq. 5-14}$$

$$NO_2 + \cdot OH \xrightleftharpoons[hv]{M} HNO_3 \tag{Eq. 5-15}$$

$$NO_2 + NO_3 \xrightleftharpoons[\text{Dark or } hv]{M} N_2O_5 \tag{Eq. 5-16}$$

$$NO_2 + NO + H_2O \xrightleftharpoons{} 2 HNO_2 \tag{Eq. 5-17}$$

$$NO + \cdot OH \xrightleftharpoons[hv]{M} HNO_2 \tag{Eq. 5-18}$$

$$NO + HO_2 \cdot \longrightarrow NO_2 + \cdot OH \tag{Eq. 5-19}$$

$$NO + NO_3 \longrightarrow 2 NO_2 \tag{Eq. 5-20}$$

$$NO_3 \xrightarrow{hv} NO + O_2 \tag{Eq. 5-21}$$

$$HNO_3 + \cdot OH \longrightarrow H_2O + NO_3 \tag{Eq. 5-22}$$

$$H_2 + \cdot OH \longrightarrow H \cdot + H_2O \tag{Eq. 5-23}$$

$$H_2O_2 + \cdot OH \longrightarrow HO_2 \cdot + H_2O \tag{Eq. 5-24}$$

$$H_2O_2 \xrightarrow{hv} 2 \cdot OH \tag{Eq. 5-25}$$

$$\cdot OH + HO_2 \cdot \longrightarrow H_2O + O_2 \qquad \text{(Eq. 5-26)}$$

$$2\,HO_2 \cdot \longrightarrow H_2O_2 + O_2 \qquad \text{(Eq. 5-27)}$$

$$H \cdot + O_2 \xrightarrow{\;M\;} HO_2 \cdot \qquad \text{(Eq. 5-28)}$$

$$H_2S + \cdot OH \longrightarrow HS \cdot + H_2O \qquad \text{(Eq. 5-29)}$$

$$H_2S + O_3 \xrightarrow{\;\text{On surface}\;} SO_2 + H_2O \qquad \text{(Eq. 5-30)}$$

$$HS \cdot + O_2 \longrightarrow \longrightarrow SO_2 \qquad \text{(Eq. 5-31)}$$

$$SO_2 + HO_2 \cdot \longrightarrow SO_3 + \cdot OH \qquad \text{(Eq. 5-32)}$$

$$SO_2 + \cdot OH \xrightarrow{\;M\;} HSO_3 \cdot \qquad \text{(Eq. 5-33)}$$

$$HSO_3 \cdot \longrightarrow \longrightarrow H_2SO_4 \qquad \text{(Eq. 5-34)}$$

$$SO_3 + H_2O \xrightarrow{\;\text{Heterogeneous}\;} H_2SO_4 \qquad \text{(Eq. 5-35)}$$

$$Cl \cdot + O_3 \longrightarrow ClO \cdot + O_2 \qquad \text{(Eq. 5-36)}$$

$$ClO \cdot + NO \longrightarrow Cl \cdot + NO_2 \qquad \text{(Eq. 5-37)}$$

$$HCl + \cdot OH \longrightarrow Cl \cdot + H_2O \qquad \text{(Eq. 5-38)}$$

$$Cl_2 \xrightarrow{\;h\nu\;} 2\,Cl \cdot \qquad \text{(Eq. 5-39)}$$

M = third body

Scheme 5-5 Organic Tropospheric Reactions

$$\underset{(RH)}{CH_4} + \cdot OH \longrightarrow \underset{(R\cdot)}{\cdot CH_3} + H_2O \qquad \text{(Eq. 5-40)}$$

$$\underset{(R\cdot)}{\cdot CH_3} + O_2 \xrightarrow{\;M\;} \underset{(RO_2\cdot)}{CH_3O_2 \cdot} \qquad \text{(Eq. 5-41)}$$

$$\underset{(RO_2\cdot)}{CH_3O_2 \cdot} + NO \longrightarrow \underset{(RO\cdot)}{CH_3O \cdot} + NO_2 \qquad \text{(Eq. 5-42)}$$

$$2\,\underset{(RO_2\cdot)}{CH_3O_2 \cdot} \longrightarrow 2\,\underset{(RO\cdot)}{CH_3O \cdot} + O_2 \qquad \text{(Eq. 5-43)}$$

$$CH_3O_2\cdot + HO_2\cdot \longrightarrow CH_3O_2H + O_2 \qquad \text{(Eq. 5-44)}$$
$$(RO_2\cdot) \qquad\qquad\qquad (RO_2H)$$

$$CH_3O\cdot + O_2 \longrightarrow HCHO + HO_2\cdot \qquad \text{(Eq. 5-45)}$$
$$(R_2CHO\cdot) \qquad\qquad (R_2CO)$$

$$HCHO \xrightarrow{\;hv\;} H\overset{\centerdot}{C}O + H\cdot \qquad \text{(Eq. 5-46)}$$
$$(R_2CO) \qquad\quad (R\overset{\centerdot}{C}O) \;\; (R\cdot)$$

$$HCHO \xrightarrow{\;hv\;} H_2 + CO \qquad \text{(Eq. 5-47)}$$

$$HCHO + \cdot OH \longrightarrow H\overset{\centerdot}{C}O + H_2O \qquad \text{(Eq. 5-48)}$$
$$(RCHO) \qquad\qquad\quad (R\overset{\centerdot}{C}O)$$

$$CH_3O_2H \xrightarrow{\;hv\;} CH_3O\cdot + \cdot OH \qquad \text{(Eq. 5-49)}$$
$$(RO_2H) \qquad\qquad (RO\cdot)$$

$$CH_3O_2H + \cdot OH \longrightarrow CH_3O_2\cdot + H_2O \qquad \text{(Eq. 5-50)}$$
$$(RO_2H) \qquad\qquad\qquad (RO_2\cdot)$$

$$H\overset{\centerdot}{C}O + O_2 \longrightarrow HO_2\cdot + CO \qquad \text{(Eq. 5-51)}$$

$$CO + \cdot OH \longrightarrow CO_2 + H\cdot \qquad \text{(Eq. 5-52)}$$

$$R\overset{\centerdot}{C}O \longrightarrow R\cdot + CO \qquad \text{(Eq. 5-53)}$$

$$R\overset{\centerdot}{C}O + O_2 \xrightarrow{\;M\;} RCO_3\cdot \qquad \text{(Eq. 5-54)}$$

$$RCO_3\cdot + NO \longrightarrow RCO_2\cdot + NO_2 \qquad \text{(Eq. 5-55)}$$

$$RCO_3\cdot + NO_2 \longrightarrow RCO_3NO_2 \qquad \text{(Eq. 5-56)}$$

$$RCO_2\cdot \longrightarrow R\cdot + CO_2 \qquad \text{(Eq. 5-57)}$$

$$RH + Cl\cdot \longrightarrow R\cdot + HCl \qquad \text{(Eq. 5-58)}$$

$$\overset{\textstyle OH}{\underset{\textstyle |}{}}$$
$$R_2C{=}CR_2 + \cdot OH \xrightarrow{\;M\;} R_2C{-}\overset{\centerdot}{C}R_2 \qquad \text{(Eq. 5-59)}$$

$$\overset{\textstyle OH}{\underset{\textstyle |}{}} \qquad\qquad \overset{\textstyle OH}{\underset{\textstyle |}{}}$$
$$R_2C{-}\overset{\centerdot}{C}R_2 \longrightarrow \longrightarrow R_2C{-}CR_2O\cdot \qquad \text{(Eq. 5-60)}$$

$$\overset{\textstyle OH}{\underset{\textstyle |}{}}$$
$$R_2C{-}CR_2O\cdot \longrightarrow R_2\overset{\centerdot}{C}OH + R_2CO \qquad \text{(Eq. 5-61)}$$

$$R_2\overset{\cdot}{C}OH \longrightarrow \quad \longrightarrow HOCR_2O\cdot \qquad \text{(Eq. 5-62)}$$

$$HOCR_2O\cdot \longrightarrow \cdot OH + R_2CO \qquad \text{(Eq. 5-63)}$$

$$R_2C{=}CR_2 + O_3 \longrightarrow \overset{\displaystyle R_2C{-}CR_2}{\underset{\displaystyle O}{O\qquad O}} \qquad \text{(Eq. 5-64)}$$

$$\overset{\displaystyle R_2C{-}CR_2}{\underset{\displaystyle O}{O\qquad O}} \longrightarrow R_2CO + R_2\overset{\cdot}{C}O_2\cdot \qquad \text{(Eq. 5-65)}$$

$$R\overset{\cdot}{C}HO_2\cdot \longrightarrow ROH + CO \qquad \text{(Eq. 5-66)}$$

$$R\overset{\cdot}{C}HO_2\cdot \longrightarrow RH + CO_2 \qquad \text{(Eq. 5-67)}$$

$$R_2\overset{\cdot}{C}O_2\cdot + O_3 \longrightarrow R_2CO + 2\,O_2 \qquad \text{(Eq. 5-68)}$$

R = alkyl, substituted alkyl, or H (aryl or halogen in some cases)
M = third body

5-5 ATMOSPHERIC PERSISTENCE PREDICTION—TEST METHODS

Predicting a compound's persistence in the troposphere requires determining or estimating its rate of decomposition under conditions found in the troposphere. Because these measurements can rarely be undertaken in the real atmosphere, various laboratory methods have been developed to aid in their determination or estimation.

5-5.1 Direct Photolysis

The vapor-phase ultraviolet and visible absorption spectra of a compound provide an indication of whether its decomposition by direct photolysis is possible. If the compound absorbs radiation at wavelengths longer than $\sim$290 nm, it might decompose by direct absorption of tropospheric solar radiation. As noted previously, however, direct absorption of tropospheric solar radiation is no guarantee of eventual decomposition. A maximum rate of decomposition by direct photolysis can be calculated from the solar spectral irradiance (Figs. 5-5, 5-6) and the extinction coefficients of the compound by assuming that each photon absorbed leads to the decomposition of one molecule of the compound, i.e. $\phi_{decomp} = 1.0$ (quantum yield). An experimental measurement (Ref. 5-7) is necessary to determine the decomposition rate by direct photolysis, however. If a compound does not absorb in the UV or visible region at wavelengths longer than $\sim$290 nm, then it will not decompose by direct photolysis.

5-5.2 Rate of Reaction with Hydroxyl Radicals

Two general laboratory methods have been used for measuring decomposition rates of compounds in the troposphere. One involves the determination of the rate constant (k) for the gas-phase reaction of ·OH with the compound (Eq. 5-69), and the other involves subjecting the compound to simulated tropospheric

$$\text{Compound} + \cdot\text{OH} \xrightarrow{\quad k \quad} \text{Products} \qquad \text{(Eq. 5-69)}$$

conditions. A correlation of these two methods for a series of hydrocarbons is shown in Fig. 5-7.

Rate constants for the reactions of ·OH with compounds have been measured in both static and flow systems. Hydroxyl radical concentrations have been determined by several techniques, including resonance fluorescence, magnetic resonance, and mass spectroscopy. A detailed description of these techniques is beyond the scope of this chapter; the references cited in Table 5-4 and Ref. 5-12 should be consulted for details of these techniques.

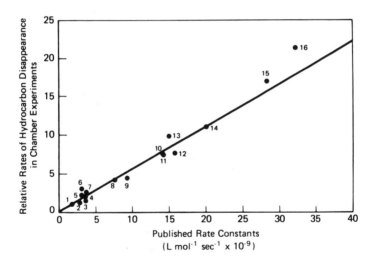

Fig. 5-7. Comparison of relative rates of hydrocarbon disappearance determined by environmental chamber method with selected published rate constants for reaction of those hydrocarbons with ·OH. The following compounds are shown: 1, $CH_3(CH_2)_2CH_3$; 2, $(CH_3)_2CHCH_2CH_3$; 3, $C_6H_5CH_3$ (toluene); 4, $(CH_3)_2$-$CH(CH_2)_2CH_3$;   5, $CH_3(CH_2)_4CH_3$;   6, $CH_2=CH_2$;   7, $(CH_3CH_2)_2CHCH_3$; 8, p-$C_6H_4(CH_3)_2$;   9, o-$C_6H_4(CH_3)_2$;   10-11, m-$C_6H_4(CH_3)_2$;   12, 1,2,3-$C_6H_3(CH_3)_3$; 13, $CH_2=CHCH_3$; 14, 1,2,4-$C_6H_3(CH_3)_3$; 15, 1,3,5-$C_6H_3(CH_3)_3$; 16, cis-$CH_3CH=CHCH_3$. Reprinted with permission from Darnall, K. R., Lloyd, A. C., Winer, A. M., and Pitts, J. N., Jr., "Reactivity Scale for Atmospheric Hydrocarbons Based on Reaction with Hydroxyl Radical," *Environ. Sci. Technol.*, 10, pp. 692–696, July, 1976. Copyright by the American Chemical Society.

TABLE 5-4. Rate Constants for Gas Phase Reactions of Selected Hydrocarbons, Halohydrocarbons, and Other Compounds with ·OH — Calculated Tropospheric Half-Lives and Lifetimes.

Compound	$k_{Compound + \cdot OH}$, cm³ molecule⁻¹ sec⁻¹ [a] −8°C	25°C [b]	References	Tropospheric $t_{1/2}$ (τ) of Compound, yr [c] — 5 × 10⁵ ·OH cm⁻³ −8°C	25°C	1 × 10⁶ ·OH cm⁻³ −8°C	25°C	2 × 10⁶ ·OH cm⁻³ −8°C	25°C
CH₄	3.7×10^{-15}	7.6×10^{-15}	5-52	12. (17.)	5.8 (8.3)	5.9 (8.6)	2.9 (4.2)	3.0 (4.3)	1.4 (2.1)
CH₃CH₃	1.8×10^{-13}	3.0×10^{-13}	5-52	0.24 (0.35)	0.15 (0.21)	0.12 (0.18)	0.073 (0.11)	0.061 (0.088)	0.037 (0.053)
CH₃CH₂CH₃	9.3×10^{-13}	1.2×10^{-12}	5-57	0.047 (0.068)	0.036 (0.051)	0.024 (0.031)	0.018 (0.026)	0.012 (0.017)	0.0089 (0.013)
CH₃(CH₂)₂CH₃	2.1×10^{-12}	2.7×10^{-12}	5-52	0.021 (0.030)	0.016 (0.024)	0.010 (0.015)	0.0082 (0.012)	0.0052 (0.0075)	0.0041 (0.0059)
(CH₃)₂CHCH₃	2.0×10^{-12}	2.4×10^{-12}	5-52	0.022 (0.031)	0.019 (0.027)	0.011 (0.016)	0.0093 (0.013)	0.0054 (0.0078)	0.0046 (0.0067)
(CH₃)₂CHCH₂CH₃	—	3.3×10^{-12}	5-58	—	0.013 (0.019)[d]	—	0.0066 (0.0095)[d]	—	0.0033 (0.0048)[d]
C(CH₃)₄	5.8×10^{-13}	8.3×10^{-13}	5-57	0.075 (0.11)	0.053 (0.076)	0.038 (0.054)	0.026 (0.038)	0.019 (0.027)	0.013 (0.019)
CH₃(CH₂)₄CH₃	—	6.3×10^{-12}	5-58	—	0.0070 (0.010)[d]	—	0.0035 (0.0050)[d]	—	0.0017 (0.0025)[d]
(CH₃)₂CH(CH₂)₂CH₃	—	5.3×10^{-12}	5-58	—	0.0083 (0.012)[d]	—	0.0041 (0.0060)[d]	—	0.0021 (0.0030)[d]
(CH₃CH₂)₂CHCH₃	—	7.1×10^{-12}	5-58	—	0.0062 (0.0089)[d]	—	0.0031 (0.0044)[d]	—	0.0015 (0.0022)[d]
(CH₃)₂CHCH(CH₃)₂	7.8×10^{-12}	7.4×10^{-12}	5-57	0.0056 (0.0081)	0.0059 (0.0086)	0.0028 (0.0041)	0.0030 (0.0043)	0.0014 (0.0020)	0.0015 (0.0021)
(CH₃)₃CCH(CH₃)₂	5.2×10^{-12}	5.4×10^{-12}	5-57	0.0085 (0.012)	0.0081 (0.012)	0.0043 (0.0062)	0.0041 (0.0059)	0.0021 (0.0031)	0.0020 (0.0029)
CH₃(CH₂)₆CH₃	7.5×10^{-12}	8.7×10^{-12}	5-57	0.0059 (0.0085)	0.0051 (0.0073)	0.0029 (0.0042)	0.0025 (0.0036)	0.0015 (0.0021)	0.0013 (0.0018)
(CH₃)₃CCH₂CH(CH₃)₂	3.1×10^{-12}	3.7×10^{-12}	5-57	0.014 (0.020)	0.012 (0.017)	0.0071 (0.010)	0.0059 (0.0085)	0.0035 (0.0051)	0.0030 (0.0043)
(CH₃)₃CC(CH₃)₃	7.9×10^{-13}	1.1×10^{-12}	5-57	0.056 (0.080)	0.040 (0.058)	0.028 (0.040)	0.020 (0.029)	0.014 (0.020)	0.010 (0.014)
c-C₆H₁₂ [e]	7.0×10^{-12}	8.0×10^{-12}	5-57	0.0062 (0.0090)	0.0055 (0.0079)	0.0031 (0.0045)	0.0027 (0.0039)	0.0016 (0.0023)	0.0014 (0.0020)
CH₂=CH₂	9.4×10^{-12}	8.0×10^{-12}	5-59	0.0047 (0.0067)	0.0055 (0.0079)	0.0023 (0.0034)	0.0027 (0.0040)	0.0012 (0.0017)	0.0014 (0.0020)
CH₂=CHCH₃	3.2×10^{-11}	2.5×10^{-11}	5-59	0.0014 (0.0020)	0.0017 (0.0025)	0.00069 (0.00099)	0.00086 (0.0013)	0.00034 (0.00050)	0.00043 (0.00062)
CH₂=CHCH=CH₂	—	7.7×10^{-11}	5-58	—	0.00057 (0.00082)[d]	—	0.00028 (0.00041)[d]	—	0.00014 (0.00021)[d]
C₆H₆	9.3×10^{-13}	1.1×10^{-12}	5-60	0.047 (0.068)	0.038 (0.055)	0.024 (0.034)	0.019 (0.028)	0.012 (0.017)	0.0096 (0.014)
C₆H₅CH₃	7.5×10^{-12}	5.8×10^{-12}	5-60	0.0058 (0.0084)	0.0076 (0.011)	0.0029 (0.0042)	0.0038 (0.0055)	0.0015 (0.0021)	0.0019 (0.0027)
o-C₆H₄(CH₃)₂	1.8×10^{-11}	1.6×10^{-11}	5-60	0.0025 (0.0035)	0.0027 (0.0040)	0.0012 (0.0018)	0.0014 (0.0020)	0.00061 (0.00089)	0.00069 (0.00099)
m-C₆H₄(CH₃)₂	2.5×10^{-11}	2.5×10^{-11}	5-60	0.0018 (0.0026)	0.0018 (0.0026)	0.00089 (0.0013)	0.00089 (0.0013)	0.00044 (0.00064)	0.00045 (0.00064)
p-C₆H₄(CH₃)₂	2.0×10^{-11}	1.8×10^{-11}	5-60	0.0022 (0.0031)	0.0024 (0.0034)	0.0011 (0.0016)	0.0012 (0.0017)	0.00054 (0.00078)	0.00059 (0.00086)
C₆H₅CH₂CH₃	—	8.0×10^{-12}	5-58	—	0.0055 (0.0079)[d]	—	0.0028 (0.0040)[d]	—	0.0014 (0.0020)[d]
C₆H₅CH(CH₃)₂	—	6.1×10^{-12}	5-58	—	0.0071 (0.010)[d]	—	0.0036 (0.0052)[d]	—	0.0018 (0.0026)[d]

Atmospheric lifetime / reaction-rate data table (units of rate constants in cm³ molecule⁻¹ s⁻¹; lifetime-type entries dimensionless, with values in parentheses):

Compound	k	k	Ref.	(1)	(2)	(3)	(4)	(5)	(6)
CH_3Cl	3.0×10^{-14}	4.8×10^{-14}	5-52	1.5 (2.1)	0.92 (1.3)	0.74 (1.1)	0.46 (0.66)	0.37 (0.54)	0.23 (0.33)
CH_2Cl_2	8.4×10^{-14}	1.3×10^{-13}	5-52	0.52 (0.76)	0.33 (0.48)	0.26 (0.38)	0.17 (0.24)	0.13 (0.19)	0.083 (0.12)
$CHCl_3$	6.5×10^{-14}	1.0×10^{-13}	5-52	0.68 (0.98)	0.42 (0.61)	0.34 (0.49)	0.21 (0.30)	0.17 (0.24)	0.11 (0.15)
CCl_4	—	$< 4. \times 10^{-14}$	5-15	>11. (>16.)	>11. (>16.)	>5. (>8.)	>5. (>8.)	—	>3. (>4.)
CH_2FCl	2.5×10^{-14}	4.0×10^{-14}	5-15, 5-61	1.8 (2.6)	1.1 (1.6)	0.89 (1.3)	0.56 (0.80)	0.45 (0.64)	0.28 (0.40)
CHF_2Cl	2.5×10^{-15}	4.3×10^{-15}	5-15, 5-61, 5-62	18. (26.)	10. (15.)	8.8 (13.)	5.1 (7.3)	4.4 (6.4)	2.5 (3.7)
$CHFCl_2$	1.6×10^{-14}	2.7×10^{-14}	5-15, 5-38, 5-61	2.7 (3.9)	1.6 (2.3)	1.3 (1.9)	0.81 (1.2)	0.67 (0.97)	0.40 (0.58)
CF_2Cl_2	—	$< 4 \times 10^{-16}$	5-61	>110. (>160.)	>110. (>160.)	>55. (>79.)	>55. (>79.)	—	>27. (>40.)
$CFCl_3$	—	$< 5. \times 10^{-16}$	5-15	>88. (>130.)	>88. (>130.)	>44. (>63.)	>44. (>63.)	—	>22. (>32.)
CH_3Br	2.8×10^{-14}	3.8×10^{-14}	5-15, 5-37	1.6 (2.3)	1.1 (1.7)	0.79 (1.1)	0.57 (0.83)	0.40 (0.57)	0.29 (0.41)
CH_3CH_2Cl	—	3.9×10^{-13}	5-63	—	—	0.11 (0.16)	0.056 (0.081)	—	0.028 (0.041)
CH_3CHCl_2	—	2.6×10^{-13}	5-63	—	—	0.17 (0.24)	0.084 (0.12)	—	0.042 (0.061)
CH_2ClCH_2Cl	—	2.2×10^{-13}	5-63	—	—	0.20 (0.29)	0.10 (0.14)	—	0.050 (0.072)
CH_3CCl_3	9.6×10^{-15}	1.9×10^{-14}	5-52	4.6 (6.6)	2.4 (3.4)	2.3 (3.3)	1.2 (1.7)	1.1 (1.6)	0.59 (0.86)
CH_3CF_2Cl	1.6×10^{-15}	3.0×10^{-15}	5-61, 5-63	28. (40.)	15. (21.)	14. (20.)	7.3 (10.)	7.0 (10.)	3.6 (5.2)
CH_2ClCF_3	—	1.1×10^{-14}	5-63	—	4.2 (6.0)	—	2.1 (3.0)	—	1.0 (1.5)
$CHFClCF_3$	—	1.2×10^{-14}	5-63	—	3.5 (5.1)	—	1.8 (2.6)	—	0.89 (1.3)
$CHCl_2CF_3$	—	2.8×10^{-14}	5-63	—	1.5 (2.2)	—	0.77 (1.1)	—	0.39 (0.56)
CF_2ClCF_2Cl	—	$< 5. \times 10^{-16}$	5-63	—	>88. (>130.)	—	>44. (>63.)	—	>22. (>32.)
$CF_2ClCFCl_2$	—	$< 3. \times 10^{-16}$	5-61, 5-63	—	>150. (>210.)	—	>73. (>110.)	—	>37. (>53.)
CH_2BrCH_2Br	—	2.5×10^{-13}	5-63	—	0.18 (0.25)	—	0.088 (0.13)	—	0.044 (0.063)
$CH_2{=}CHCl$	—	6.0×10^{-12}	5-64	—	0.007 (0.011)	0.0077 (0.011)	0.0037 (0.0053)	—	0.0018 (0.0026)
$CHCl{=}CCl_2$	2.9×10^{-12}	2.4×10^{-12}	5-65	0.015 (0.022)	0.019 (0.027)	0.21 (0.31)	0.0093 (0.013)	0.0038 (0.0056)	0.0046 (0.0067)
$CCl_2{=}CCl_2$	1.0×10^{-13}	1.7×10^{-13}	5-52	0.43 (0.62)	0.26 (0.38)	—	0.13 (0.19)	0.11 (0.15)	0.065 (0.094)
$CF_2{=}CFCl$	—	$7. \times 10^{-12}$	5-64	—	0.0063 (0.009)	—	0.0031 (0.0045)	—	0.0016 (0.0023)
CH_3CHO	1.8×10^{-11}	1.6×10^{-11}	5-66, 5-67	0.0024 (0.0035)	0.0027 (0.0039)	0.0012 (0.0018)	0.0014 (0.0019)	0.00061 (0.00088)	0.00068 (0.00097)
NH_3	1.1×10^{-13}	1.6×10^{-13}	5-52	0.39 (0.56)	0.28 (0.40)	0.20 (0.28)	0.14 (0.20)	0.098 (0.14)	0.070 (0.10)

(a) The value given is the preferred (Ref. 5-52) or average of those values reported or was calculated from data given in the reference cited. (b) The rate constants given in Ref. 5-57 were measured at 32°C. The rate constants given in References 5-63 and 5-64 were measured at 23°C; these latter rate constants are probably within 5% of the values at 25°C. (c) 0.1 year = 37 days, 0.01 year = 3.7 days, 0.001 year = 8.8 hours, 0.0001 year = 53 minutes. (d) At 32°C. (e) Cyclohexane.

Calculations of average lifetime or half-life for a compound in the troposphere requires knowledge of the average steady-state concentration, $[\cdot OH]_{trop.}$, of $\cdot OH$ in the troposphere. Although the globally, seasonally, diurnally averaged tropospheric $\cdot OH$ concentration is not well defined, an average value is $\sim 1 \times 10^6$ cm^{-3} (Table 5-2). There is an uncertainty in this value of at least a factor of 2 or 3.

The lifetime (τ) or half-life $(t_{1/2})$ in years can be calculated by using the pseudo-first-order rate equation (5-70), where the units for k and $[\cdot OH]_{trop.}$

$$\tau = \frac{t_{1/2}}{0.693} = \frac{6.17 \times 10^{-8}}{k[\cdot OH]_{trop.}} \qquad \text{(Eq. 5-70)}$$

are cm^3 molecule^{-1} sec^{-1} and radicals cm^{-3}, respectively. The conversion factor for the second-order rate constants in different units, Fig. 5-7 and Eq. 5-70, is 1.661×10^{-21} cm^3 molecule^{-1} sec^{-1} (liters mole^{-1} sec^{-1})$^{-1}$.

Table 5-4 lists the average tropospheric half-lives and life-times for a series of hydrocarbons, halohydrocarbons, and other compounds. These values were obtained using Eq. 5-70 with $[\cdot OH]_{trop.}$ values of 5×10^5 cm^{-3}, 1×10^6 cm^{-3}, and 2×10^6 cm^{-3}, to give an indication of the uncertainty in the half-lives and lifetimes. Results are given, where possible, for reactions at both $25°C$ and $-8°C$, the latter being the weighted average temperature of the troposphere (Refs. 5-37, 5-61). The rate constants at $-8°C$ were not available for all compounds. Comparisons among most of the compounds can be made from the data at $25°C$. An indication of the effect of structure on the tropospheric lifetime can be observed by examining the data in Table 5-4. Further compilations of $\cdot OH$ reaction rate data are available (Refs. 5-12, 5-28, 5-52, 5-57).

More refined calculations of decomposition rates than those achieved with Eq. 5-70 can be performed (e.g., Ref. 5-40), but these are not discussed here.

Reactions of a compound with O_3, O, or other reactive species may contribute to the tropospheric decomposition of the compound. Rate data for some of these reactions have been compiled elsewhere (Ref. 5-28, 5-52). Compared to the reactions with $\cdot OH$, these reactions with O_3, O, or other species appear to contribute only slightly to destructing a compound in the troposphere (Ref. 5-57, 5-68).

5-5.3 Reactions Under Simulated Tropospheric Conditions

The second general method for determining decomposition rates of compounds in the troposphere involves subjecting the compound to simulated tropospheric conditions and measuring the decrease in concentration of the compound as a function of time. These experiments have been performed in a variety of reaction chambers (Refs. 5-17, 5-18, 5-44, 5-68–5-91), ranging from a simple 2 liter bulb (Fig. 5-8), constructed of quartz or borosilicate glass, e.g., Pyrex®,

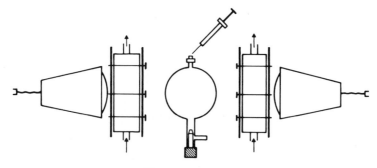

Fig. 5-8. Simple quartz or Pyrex® bulb reactor with vacuum stopcock and septum. *Note:* **Light from two 275-W sunlamps (medium pressure Hg arc lamps) passes through water filters to control the temperature of the bulb. Reactants are introduced into the bulb and samples are removed for analysis via a gas-tight syringe.**

to an elaborate 5500-liter "smog chamber" (Fig. 5-9). Even larger chambers have been described (Refs. 5-76, 5-80, 5-88).

The portions of the reaction chambers that are in actual contact with the reactants have been constructed of various types of glass, such as Pyrex® and quartz, and such metals as aluminum or stainless steel, often coated with Teflon FEP®, or plastic film such as Teflon FEP®. At least a portion of the container must be transparent to UV radiation down to ~300 nm; the glass and plastic film fulfill this requirement. The most satisfactory container materials are those that are most inert to the reactants, intermediates, and products. Teflon FEP® film appears to be the most inert (Ref. 5-86), but many compounds either diffuse through this film or dissolve in it (Ref. 5-92). In general, Teflon FEP® is suitable for use only with compounds that decompose rapidly. The length of time for which a run can be performed is a function of the compound's diffusivity (Ref. 5-93); runs lasting more than a few days often cannot be easily performed.

Reaction chambers are often made large so that the surface/volume ratio is low. This low surface area minimizes contact between gas-phase reactants and walls, thus minimizing reactions or other processes that occur on the walls. Large chambers, in contrast to small ones, however, have the disadvantages of more leakage problems, nonevacuability without special construction, more difficult temperature control, and the requirement of a large air purification system, all of which add to the cost. Even low surface/volume ratios do not assure that reactions or other processes on the walls have been reduced to acceptable levels.

The absence of significant surface effects may be demonstrated by showing that the reactions, and their rates, are the same in containers constructed of different materials or are the same in a single container in which the surface/volume ratio is varied by adding small pieces of container material such as glass chips to the chamber. Whereas the former method only shows that the surface effect is

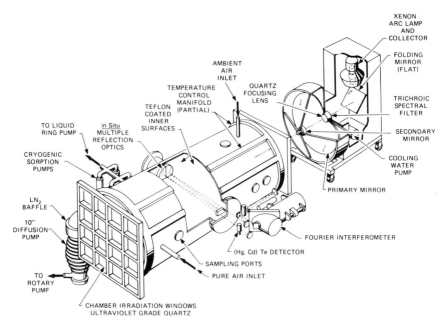

Fig. 5-9.   Schematic of evacuable smog chamber solar simulator facility at the Statewide Air Pollution Research Center, University of California, Riverside. *Note*: The radiation source is a 25-kW xenon arc lamp. *Source*: Finlayson, B. J., and Pitts, J. N., Jr., "Photochemistry of the Polluted Troposphere," *Science*, 192, 4235, pp. 111–119, April, 1976. Copyright 1976 by the American Association for the Advancement of Science.

the same in the various containers, the probability of identical surface effects on different surfaces is low.

Illuminating a surface such as Pyrex® glass has resulted in reactions that do not occur on these surfaces in the dark (Ref. 5-94). Keeping light from striking the chamber surface, however, does not assure that surface processes have been eliminated.

Both artificial sunlight sources and natural sunlight have been used for photo-decomposition studies under simulated tropospheric conditions. The former include Hg arc lamps, Xe arc lamps, and fluorescent lamps, all filtered to remove radiation of wavelengths shorter than $\sim$290 nm.

Ambient air and/or synthetic purified air, to which known concentrations of tropospheric trace constituents, such as NO, NO_2, O_3, have been added, have beed used for decomposition studies under simulated tropospheric conditions. Use of the latter type of air can be criticized for not representing real tropospheric air. On the other hand, the composition of ambient air varies from place to place and from time to time so that experiments often are not reproducible.

Further, the composition of ambient air is usually unknown unless extensive analyses are performed. Using purified air is recommended under most circumstances although arguments can be made for performing decomposition studies with both types of air.

A variety of initiating species, such as NO, NO_2, O_3, and Cl_2, have been employed for these decomposition studies. The first three initiators eventually produce ·OH in the presence of water vapor, as outlined in Scheme 5-4. Chlorine atoms from the solar photolysis of Cl_2 react with many organic compounds in a manner similar to the reactions of ·OH, particularly in H· abstraction processes. The concentrations of these initiators and of water have varied greatly in different studies. Concentrations close to those shown in Table 5-2 are recommended for determining average tropospheric decomposition rates.

The concentration of the organic compounds under study has often varied over several orders of magnitude. A recommended procedure uses the lowest concentration possible. This concentration will ordinarily be several orders of magnitude higher than the expected real tropospheric concentration of that compound. It is useful to determine the dependence of the decomposition rate on the concentration of the compound. At sufficiently low concentrations, the decomposition is expected to be kinetically first order in the compound under study. Such a demonstration lends confidence to the assertion that the decomposition rate at real tropospheric concentrations will be the same as that observed in a laboratory study.

Table 5-5 gives the results for some half-lives obtained from studies of the decomposition rates of several hydrocarbons and halohydrocarbons under simulated tropospheric conditions. These examples show the variability of results obtained by this method. For comparison, the half-lives of the same compounds calculated by the ·OH reaction rate method and by the tropospheric concentration–release rate method discussed below are also shown in Table 5-5. Some of the variability among the half-lives determined by the simulation method for the same compounds is due to the different reaction conditions used, particularly the concentration of some of the reactants, such as NO_2. Many of these experiments were designed primarily for determining the reactivity of the organic compounds under urban smog conditions in which the concentrations of NO_2, O_3, ·OH (Ref. 5-38), etc., are often higher than in rural "nonpolluted" air. Determining the decomposition rates of compounds under average tropospheric conditions by the simulation method should be performed under conditions as similar as possible to those of the average troposphere. Where this is not possible, experiments should be designed to allow extrapolation to average tropospheric conditions, as discussed previously.

A comparison of the ·OH reaction rate method (·OH method) and the simulation method for determining average decomposition rates of compounds in the troposphere shows the advantages and disadvantages of each. The precision of the ·OH method depends on the accuracy with which $[·OH]_{trop.}$

is known. Since this concentration is not known precisely, the decomposition rate will have the same uncertainty as this concentration. The relative decomposition rates of various compounds, however, can be precisely determined. The validity of the $\cdot$OH method depends on the validity of the theory that the major decomposition route of compounds in the troposphere is reaction with $\cdot$OH.

TABLE 5.5. Rates of Decomposition of Hydrocarbons and Halohydrocarbons Under Tropospheric Conditions.

Compound	$\cdot$OH Reaction Rate Method[a]	Simulation Method[b]	Tropospheric Concentration[c]– Anthropogenic Release Rate Method[d]
C_6H_6	8.8 days	4.4 days (Ref. 5-96)[e]	4 yr[f]
CH_3Cl	270 days		230 yr[g]
CH_2Cl_2	96 days	1.5 days (Ref. 5-95) $>$ 30 days (Ref. 5-90) $\geqslant$ 70 days (Ref. 5-97) $>$ 90 days (Ref. 5-90) 230 ± 115 days (Ref. 5-17)	130 days[h]
$CHCl_3$	120 days	0.6 day (Ref. 5-98) 160 ± 80 days (Ref. 5-17)	9 yr[i]
CH_3CCl_3	2.3 yr	$>\sim$ 100 days[j] (Ref. 5-44) 180 ± 90 days (Ref. 5-17)	2.8 yr[k]
$CCl_2=CCl_2$	78 days	0.26 day (Ref. 5-98) 0.63 day (Ref. 5-95) 0.84 day (Ref. 5-44) 1.5 days (Ref. 5-99) 2.9 days (Ref. 5-90) 4.3 days (Ref. 5-100) 4.9 days (Ref. 5-90) 4.9 days (Ref. 5-18) 80 ± 40 days (Ref. 5-17)	24 days[l]

(a) Data from Table 5-4 for average $[\cdot OH]_{trop.}$ of 10^6 cm^{-3}. (b) In diurnal days. Calculated by assuming that the reactions were first order in the organic compound and that the amount of sunlight for a dirunal day equaled one-third that of continuous average sunlight for 24 hr. For average sunlight, the photolysis rate of NO_2 $[k_1(NO_2)]$ was assumed to be 0.34 min^{-1} (Ref. 5-95). The decomposition rate was assumed to be proportional to the light intensity. (c) Assuming average tropospheric concentrations shown in Table 5-2. (d) Assuming all of compound was released in the northern troposphere except as noted. (e) See the original references for experimental details. (f) Assuming worldwide release rate (9.1×10^{10} g yr^{-1}) is twice the U. S. release rate for solvent use (10^8 lb yr^{-1}) in 1968 (Ref. 5-28). (g) Assuming a release rate of 7.9×10^9 g yr^{-1} (Ref. 5-3). (h) Assuming a release rate of 3.5×10^{11} g yr^{-1} (Ref. 5-3). (i) Assuming a release rate of 1.24×10^{10} g yr^{-1} (Ref. 5-3). (j) Lower limit of 95% confidence interval calculated from data shown in Figure 4 of Ref. 5-44. (k) Assuming a release rate of 3.242×10^{11} g yr^{-1} (Ref. 5-3) and distribution over both northern and southern troposphere. (l) Assuming a release rate of 6.09×10^{11} g yr^{-1} (Ref. 5-3).

As indicated in Table 5-5 and the preceding discussion, the simulation method gives a range of decomposition rates or half-lives that are dependent on the reaction conditions. Using appropriate conditions and extrapolations as necessary, it is possible to obtain reliable decomposition rates of compounds under average tropospheric conditions.

The ·OH method is better suited than the simulation method to the study of slowly decomposing compounds. The ·OH method is applicable to compounds with average tropospheric half-lives up to ~5-15 years, but the simulation method is only applicable to compounds with half-lives of less than about one year.

With the techniques presently available, the determination of ·OH reaction rates requires more complicated equipment and procedures than does the simulation method. However, the simulation method may require several experiments in order to exclude extraneous reactions, such as those on the chamber surface, or to perform extrapolations to real tropospheric conditions.

Finally, the ·OH method by itself gives no information about the decomposition products, but the simulation method can give this information if the analytical techniques are sufficiently sensitive.

5-5.4 Validation

An approximate validation of these methods for persistence determination is available for those compounds that have been released to the atmosphere over a period of years and for which tropospheric concentrations and release rates are in an approximate steady state condition. The compounds for which release rates are known most reliably are those of anthropogenic origin, although the rates are usually not known with a high degree of accuracy. An average lifetime (τ) or half-life ($t_{1/2}$) of a compound in the atmosphere can be calculated from Eq. 5-71 (Ref. 5-28):

$$\tau = \frac{t_{1/2}}{0.693} = \frac{\text{Mass of compound in atmosphere}}{\text{Release rate}} \qquad \text{(Eq. 5-71)}$$

The mass of a compound in the troposphere can be calculated if the average tropospheric concentration, C_t, is known:

Mass of compound in troposphere (g) =

$$\frac{C_t \times 10^{-12} \text{ mole compound (mole air)}^{-1} \times M \text{ g compound (mole compound)}^{-1} \times 4 \times 10^{21} \text{ g air (troposphere)}^{-1}}{29 \text{ g air (mole air)}^{-1}}$$

$$= 1.4 \times 10^8 \, C_t M \qquad \text{(Eq. 5-72)}$$

where C_t is in ppt (molar or gas volume basis) and M is the gram molecular weight.

The tropospheric half-lives calculated by Eq. 5-71 for several compounds are shown in the last column of Table 5-5. For these approximate calculations, the compounds with half-lives of less than one year, all except CH_3CCl_3, were assumed to be released into and reside in only the northern troposphere. The major portions of these compounds are released into the northern hemisphere. The exchange rates (Table 5-1) between the northern and southern tropospheres and the stratosphere are slow enough that only a small fraction of each compound was able to leave the northern troposphere. The mass of the troposphere (Ref. 5-10) given in Eq. 5-72 must be divided by two when a compound is assumed to reside in only one hemisphere of the troposphere.

The agreement between the half-lives determined by the ·OH method, Eq. 5-70, and those calculated by the tropospheric concentration–release rate method, Eq. 5-71, is reasonable for CH_2Cl_2, CH_3CCl_3, and $CCl_2=CCl_2$ (Table 5-5) when the uncertainties in the input data are considered. The obvious disparity in the half-lives of C_6H_6, CH_3Cl, and $CHCl_3$ calculated by the two methods is presumably caused by large errors in either the average tropospheric concentrations or the anthropogenic release rates, or more likely by large natural sources of the compounds.

More exact calculations with more detailed mathematical models based on release rates and atmospheric concentrations have been performed (Refs. 5-10, 5-40).

5-6 ATMOSPHERIC FATE PREDICTION–TEST METHODS

5-6.1 Methods

Determining the fate or decomposition products of a compound in the troposphere generally requires identification of the reaction products from laboratory experiments in which the compound is subjected to simulated tropospheric conditions. The laboratory may be outdoors if natural sunlight is used as the light source, but the reaction mixture must be confined within a chamber.

Identification and quantification of the reaction products is performed by the same types of techniques used in any other chemical study. Gas chromatography, mass spectrometry, and infrared spectroscopy are well suited and often used for qualitative and quantitative analyses of gas-phase samples. Gas-phase products are sometimes condensed and analyzed by these and numerous other techniques. Derivatization is sometimes applied to products that cannot be easily analyzed directly. A detailed discussion of the analytical techniques used in these studies is outside the scope of this chapter.

Other studies, often of a mechanistic nature, related to reactions occurring in the troposphere, such as photo-oxidations, are often valuable in delineating the reactions and products resulting from the release of compounds into the atmosphere.

Primary interest usually focuses on the ultimate decomposition products, but intermediate products are also important, particularly if they themselves have significant lifetimes. The importance of identifying an intermediate product will depend on its properties, e.g., its toxicity, the amount formed, and its lifetime and concentration in the environment. Some of the techniques employed to determine the decomposition products from compounds under tropospheric conditions and some problems associated with interpreting these determinations will be outlined in the following discussion for four specific compounds: CH_2Cl_2, $CCl_2=CCl_2$, parathion, and trifluralin.

5-6.2 Examples

Methylene Chloride. Irradiation of CH_2Cl_2 in ambient air with a filtered Xe arc lamp gave nearly quantitative yields of CO_2 and HCl (Eq. 5-73) (Ref. 5-17).

$$CH_2Cl_2 + \text{ambient air} \xrightarrow{h\nu,\ >290\ nm} CO_2 + HCl \quad \text{(Eq. 5-73)}$$
$$\phantom{CH_2Cl_2 + \text{ambient air} \xrightarrow{h\nu,\ >290\ nm}} (99\%)\ (95\%)$$

Photolysis of a mixture of CH_2Cl_2 and Cl_2, the latter as an initiator, in dry air gave these products and CO and $COCl_2$ (Eq. 5-74) (Ref. 5-101). The Cl_2 photolyzed to give $Cl\cdot$, which initiated the decomposition of CH_2Cl_2 in a manner similar to what would have been expected had $\cdot OH$ been the initiator.

$$CH_2Cl_2 + \text{dry air} + Cl_2 \xrightarrow[\text{5 minutes}]{h\nu} CO_2 + CO + HCl + COCl_2$$

$$\begin{array}{lll} \text{20 ppm} & \text{5 ppm} & (12\ ppm)(5\ ppm)\ (38\ ppm)(2\ ppm) \\ \text{(19 ppm reacted)} & & \text{(Eq. 5-74)} \end{array}$$

A mechanistic study of the $Cl\cdot$ sensitized photooxidation of CH_2Cl_2 showed the formation of HCOCl, an unstable intermediate that decomposed thermally to HCl and CO (Eq. 5-75) (Ref. 5-102).

$$CH_2Cl_2 + O_2 + Cl_2 \xrightarrow[32 \pm 2°C]{h\nu,\ 365.5\ nm} HCOCl + HCl + COCl_2 + CO$$

$$\begin{array}{lll} \text{4-19} & \text{6-80} & \text{2-9} \\ \text{torr} & \text{torr} & \text{torr} \end{array} \qquad \underbrace{\Phi = 49 \pm 3 \qquad \Phi = 4.4}_{\text{Major products}}$$

$$\Phi_{-CH_2Cl_2} = 62 \qquad \qquad \text{(Eq. 5-75)}$$

The mechanism proposed on the basis of this study is shown in Scheme 5-6. CO (Eqs. 5-74, 5-75) is an intermediate oxidation product and would be further oxidized to CO_2 (Eq. 5-52) under real tropospheric conditions.

$COCl_2$ (Eqs. 5-74, 5-75) is probably not a product under real tropospheric conditions because the suggested mechanism (Scheme 5-6) indicates that it is formed from an intermediate, $(CHCl_2 O)_2$, that arises via a reaction that is bimolecular in the peroxy radical, $CHCl_2 O_2 \cdot$. The concentration of this peroxy radical would ordinarily be so low ($\ll$30 ppt) that the reaction forming the peroxide, $(CHCl_2 O)_2$, would be insignificant compared with other reactions

Scheme 5-6 Cl· Sensitized Photooxidation of $CH_2 Cl_2$

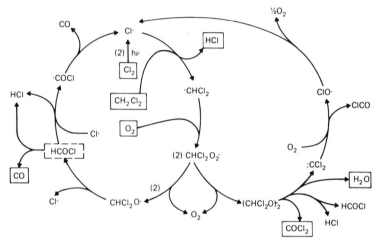

in the troposphere. It is more likely, analogous to (Eq. 5-42), that $CHCl_2 O_2 \cdot$ will react with NO to give $CHCl_2 O\cdot$ and NO_2.

Tetrachloroethylene. Irradiation of $CCl_2 = CCl_2$ in ambient air with a filtered Xe arc lamp, and subsequent reaction of the decomposition products with $H_2 O$, gave mainly $CCl_3 CO_2 H$ and minor amounts of CO_2 and HCl (Scheme

Scheme 5-7 Photolysis of $CCl_2 = CCl_2$ in Ambient Air

$$CCl_2 = CCl_2 + \text{ambient air} \xrightarrow{hv, > 290 \text{ nm}} CCl_3 COCl + COCl_2$$

$$\Big\downarrow H_2 O$$

$$\underset{(95\%)}{CCl_3 CO_2 H} + \underset{(5\%)}{CO_2 + HCl}$$

5-7) (Ref. 5-17). A related reaction gave a different product distribution

$$CCl_2 = CCl_2 + \text{purified air} + NO_2 + H_2O \xrightarrow[30°C, 2.3 \text{ hr}]{hv}$$

5.0 ppm   1.8 ppm ~100 ppm
(0.37 ppm reacted)

$$CCl_3COCl + COCl_2 + HCO_2H + CO + HCl + O_3$$
(~0.1 ppm) (0.12 ppm) (0.55 ppm) (0.27 ppm) (0.42 ppm) (0.07 ppm)

(Eq. 5-76)

(Eq. 5-76) (Ref. 5-18).   A third study identified several additional products: CCl_4, $CHCl_2COCl$, and $CHCl_3$ (Scheme 5-8) (Ref. 5-103).   The CCl_3COCl, which was identified as its isopropyl ester, was suggested as the precursor of CCl_4 (Ref. 5-103).

Scheme 5-8 Photolysis of $CCl_2=CCl_2$ in Purified Air

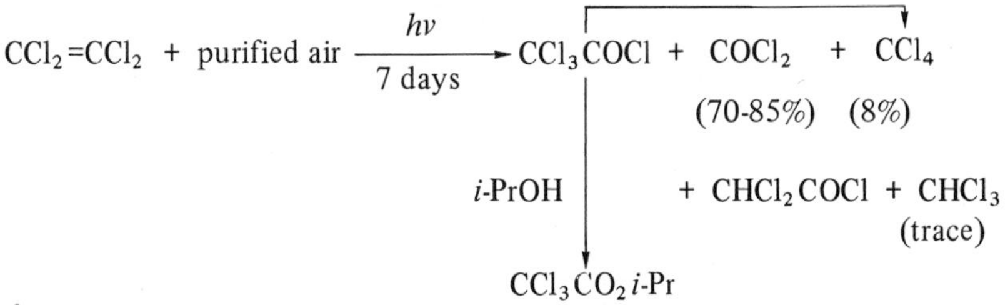

$$CCl_2 = CCl_2 + \text{purified air} \xrightarrow[7 \text{ days}]{hv} CCl_3COCl + COCl_2 + CCl_4$$

(70-85%)   (8%)

i-PrOH   + $CHCl_2COCl$ + $CHCl_3$
                                                        (trace)

$$CCl_3CO_2\text{-}i\text{-Pr}$$

Mechanistic studies of the Cl· sensitized photo-oxidation of $CCl_2=CCl_2$ showed the formation of the same two products (Eq. 5-77) (Ref. 5-104, 5-105) that were observed in each of the above three studies (Schemes 5-7 and 5-8 and Eq. 5-76).   A chain mechanism involving addition of Cl· to $CCl_2=CCl_2$ was proposed (Scheme 5-9) (Ref. 5-104, 5-105).   It was suggested that the peroxide $(CCl_3CCl_2O)_2$ results from the termination reactions (Ref. 5-105).

$$CCl_2 = CCl_2 + O_2 + Cl_2 \xrightarrow[32°C]{hv, 366 \text{ nm}} CCl_3COCl + COCl_2$$

1-3   23-440   2 Φ = 80-260 Φ = 30-80
torr   torr   torr

$\Phi_{-CCl_2=CCl_2}$ = 90-280   (Eq. 5-77)

The similarity of the products in the simulation experiments (Schemes 5-7 and 5-8 and Eq. 5-76) and the sensitized oxidation (Eq. 5-77) suggests that a Cl· chain reaction may also have occurred in the simulation experiments (Ref. 5-64).   Under real tropospheric conditions, any Cl· formed by decomposition of $CCl_3CCl_2O\cdot$ or $CCl_3O\cdot$ (Scheme 5-9) or similar species would have reacted preferentially with other tropospheric species, e.g., O_3, CH_4, and CH_3CH_3, relative rates 20,000, 4400, and 2800, respectively, rather than with $CCl_2=CCl_2$, relative rate 1.0. These relative rates (rate = k[reactant] [Cl$\cdot$]) are derived from the average tropospheric concentrations (Table 5-2) and the rate constants, 1.0×10^{-11} (Ref. 5-52), 6.1×10^{-14} (Ref. 5-52), 5.8×10^{-11} (Ref. 5-52), and 4.2×10^{-12} cm^3 molecule^{-1} sec^{-1} (Ref. 5-106), respectively, at -8°C. Because Cl$\cdot$ appears only in very low concentrations in the troposphere (Table 5-2), it is not expected to react significantly with $CCl_2=CCl_2$. The more likely initiation reaction of $CCl_2=CCl_2$ with $\cdot OH$ may (Ref. 5-107, 5-108) or may not (Ref. 5-64) produce CCl_3COCl (Scheme 5-10).

Additional insight into the identity of the products from $CCl_2=CCl_2$ in the troposphere can be gained by performing several additional experiments. The possibility that CCl_3COCl or CCl_4 was formed by surface-catalyzed reactions

Scheme 5-9 Cl$\cdot$ Sensitized Photooxidation of $CCl_2=CCl_2$

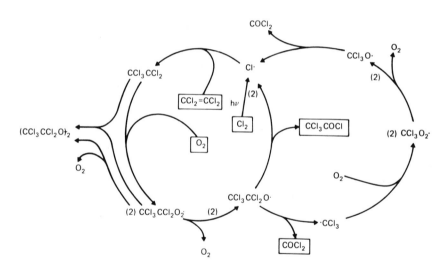

should be examined. Reactions in the presence of Cl$\cdot$ scavengers show whether Cl$\cdot$ is required for the formation of CCl_3COCl. The effect of changing the initial concentration of $CCl_2=CCl_2$ on the product distribution in reactions in which $\cdot OH$ was formed might help indicate whether the Cl$\cdot$ sensitized photooxidation of $CCl_2=CCl_2$ was responsible for forming CCl_3COCl. In the simulation experiments, $CCl_2=CCl_2$ was presumably the only source of Cl$\cdot$.

Scheme 5-10 Possible ·OH Initiated Oxidation of CCl₂=CCl₂

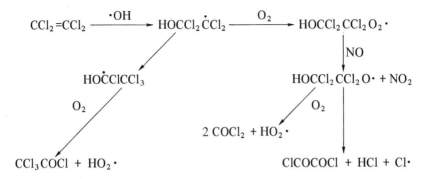

If CCl_3COCl is indeed a product of the tropospheric decomposition of $CCl_2=CCl_2$, other tests should be applied to determine whether CCl_4 can really be expected to be a product in the troposphere or is only an artifact resulting from the manner in which the simulation was performed. There are at least two reactions of CCl_3COCl (Scheme 5-11) that could be responsible for the CCl_4 formation in the simulation experiments. Reaction of $\cdot CCl_3$ with Cl_2 or $Cl\cdot$ could also account for the formation of CCl_4 (Ref 5-104). It would be informative to determine the validity of the photochemical process (Eq. 5-79) under tropospheric conditions. CCl_3COCl may absorb sunlight slightly in the troposphere (Ref. 5-109).

Scheme 5-11 Possible Routes of Formation of CCl₄ from CCl₃COCl

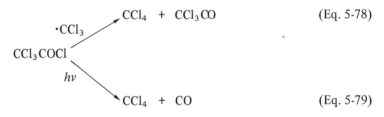

Parathion. Irradiating the insecticide, parathion (*PT*), in a flow apparatus under simulated tropospheric conditions with a 275-W sunlamp (Ref. 5-47) (medium pressure Hg arc) gave paraoxon (*PO*) and several other products (Eq. 5-80) (Ref. 5-110).

$$O_2N-\langle\!\!\!\!\!\!\rangle-\overset{\overset{S}{\|}}{O}P(OEt)_2 \quad + \quad \text{clean dry air} \quad \xrightarrow[\sim 6\ \text{hr}]{hv,\ > 300\ \text{nm}}$$

PT

0.4 ppm

$$O_2N-\langle\!\!\!\!\!\!\rangle-\overset{\overset{O}{\|}}{O}P(OEt)_2 \quad + \quad O_2N-\langle\!\!\!\!\!\!\rangle-OH$$

PO

(60%)   (10%)

$$+ \quad \overset{\overset{S}{\|}}{HOP(OEt)_2} \quad + \quad \overset{\overset{O}{\|}}{HOP(OEt)_2} \qquad \text{(Eq. 5-80)}$$

(3%)   (7%)

It is possible that some oxidant (e.g., O_3) was present in the air as previous static experiments had indicated that *PT* was stable to irradiation in air (Ref. 5-110). *PT* on glass surfaces and dust particles in the presence of UV radiation decomposed to *PO* (Ref. 5-110); decomposition of *PT* on the glass reactor walls may also have occurred in the flow experiment.

In a similar flow experiment with <10 ppb O_3 in the air and a 150-W Xe arc lamp (λ >300 nm), *PT* decomposed with a half-life of 41 min (Ref. 5-111). In the presence of 1-3 ppm O_3, *PT* had a half-life of 23 min in the light and 14.6 days in the dark (Ref. 5-111). No reaction was observed in the dark in the absence of O_3.

Trifluralin. Photolysis of the herbicide, trifluralin (*TF*), in a static reactor under simulated tropospheric conditions with a 275-W sunlamp ($\sim$2/3 of sunlight irradiance in the 300–400 nm range) gave eight identified products (Eq. 5-81) (Ref. 5-112). The results were the same whether or not the light was allowed

$$\begin{array}{c} N(n\text{-}Pr)_2 \\ O_2N\diagdown\underset{\diagdown}{\langle}\!\!\!\!\!\!\!\!\underset{|}{\bigcirc}\!\!\!\!\!\!\!\!\diagup NO_2 \\ CF_3 \end{array} \quad + \quad \text{ambient air} \quad \xrightarrow[25\text{-}30^\circ C,\ 12\ \text{days}]{hv,\ > 300\ \text{nm}}$$

TF

5 ppm
(4.3 ppm reacted)

(Eq. 5-81)

to strike the glass reactor walls. The N-propyldinitrotoluidine (*PD*) was the major product at short reaction times. When photolyzed, the N-propylbenzimidazole (*PB*) was easily dealkylated to the benzimidazole (*BI*), while *BI* resisted further irradiation (8 days). *TF* was converted to *PB* when irradiated in the absence of O_2.

In a similar reaction chamber operated under flow conditions with a 150-W Xe arc lamp ($\lambda > 300$ nm), dry air, and <10 ppb O_3, *TF* decomposed with a half-life of 117 min (Ref. 5-111). The half-life in the presence of 1-3 ppm O_3 was 47 min in the light and 12.5 days in the dark. No reaction occurred in the dark in the absence of O_3.

5-6.3 Validation

Validating the identity of the decomposition products involves releasing the compound into the atmosphere and subsequently collecting and identifying the products. This type of experiment is inherently difficult because of the dispersion (Ref. 5-56) of the compound and its decomposition products. Increasing the time interval between release and collection further dilutes the compounds and makes product identification more difficult. The intrusion of compounds unconnected with the study at hand many complicate the analysis by making it necessary to determine exactly which products resulted from the compound released.

Only compounds that decompose reasonably rapidly in the troposphere are amenable to this type of validation. Usually the time interval between the release of a compound and collection of its decomposition products cannot exceed a few hours, or a few days at the most. The time interval depends on meteorological conditions, especially the wind velocity. Further complications may arise if the compound transfers by adsorption or dissolution from the atmosphere to the aquatic or terrestrial environment, undergoes chemical transformations there, and then is again released to the atmosphere. Such transformations are irrelevant to processes that occur in the atmosphere, but may be significant from an overall environmental viewpoint. The only validations of which we are aware for identifying tropospheric decomposition products are field studies of the pesticides, *PT* and *TF*, the laboratory studies of which were discussed in the preceding two sections.

Parathion. *PT* (700 g) in 480 liters of H_2O was sprayed in a 0.32 ha plum orchard at 4-5 p.m. in sunlight (Ref. 5-110). Air was sampled with high-volume air samplers 100 m downwind (4.8 km hr^{-1}) during the spraying. The concentration of *PT* at the sampling site was 1620 ± 160 ng m^{-3} (~140 ppt). The only product observed above the background concentration was *PO* (403 ± 10 ng m^{-3}, ~37 ppt), [*PO*]/[*PT*] = 0.25 ± 0.02 (wt. basis). At the same time, 400 m downwind, the concentrations were 123 ± 2 ng m^{-3} *PT* and 60 ± 4 ng m^{-3} *PO*, [*PO*]/[*PT*] = 0.49 ± 0.04. *PT* was also converted to *PO* in the air sampling equipment, [*PO*]/[*PT*] = 0.16 ± 0.07. Five hours later (9-10 p.m.), after dark, when the wind speed was 1.6 km hr^{-1}, the concentrations at 100 m were 760 ± 140 ng m^{-3} *PT*, 52 ± 2 ng m^{-3} *PO*, [*PO*]/[*PT*] = 0.07 ± 0.02; at 400 m the concentrations were 270 ± 14 ng m^{-3} *PT*, 36 ± 7 ng m^{-3} *PO*, [*PO*]/[*PT*] = 0.14 ± 0.03 (Ref. 5-110). These results suggest that *PT* is converted to *PO* in air in the presence of sunlight but possibly not in the dark.

Continued air sampling on 21 succeeding days, both in the orchard and downwind, showed further increases in the [*PO*]/[*PT*] ratio, up to 1.6 (Ref. 5-110). Under these conditions, in which the average wind speed was 1.6 km hr^{-1}, some of the *PT*→*PO* conversion may have occurred on surfaces such as leaves or the ground, or in water or natural plant oil solutions. Desorption or evaporation of *PT* and *PO* would then return them to the gas phase. The *PT* was assumed to be present in the air in the vapor phase, but could have been in solution or the liquid state as a fine aerosol (Ref. 5-110).

Further field releases of *PT* were performed; *PO* was confirmed as a decomposition product and information on the rate of the transformation was obtained (Ref. 5-111). Aircraft released *PT* 3 m above the ground, where the O_3 concentration was 100-320 ppb. Air sampling 161 m downwind showed that the concentrations were 1296 ng m^{-3} *PT* and 226 ng m^{-3} *PO;* 402 m downwind the concentrations were 411 ng m^{-3} *PT* and 176 ng m^{-3} *PO.* The wind speed was such that the transit times were 0.95 and 2.38 min to the two sampling sites, respectively. A half-life of 5 min was calculated for this decomposition, which was approximately first order in *PT,* assuming that no products other than *PO* were formed. Similar experiments in the absence of sunlight resulted in a half-life of 131 min.

Trifluralin. Five grams of *TF* were volatilized into the air within 25 m of a high-volume air sampler located 1 m above the ground, and the air was collected for 15 min; analysis of the trapped materials showed that the conversion of *TF* to *PD* was 0.55% [94 μg *TF*/15 m^3 air ($\sim$460 ppt), 0.52 μg *PD*/15 m^3 air ($\sim$3 ppt)] (Ref. 5-112). The major product, *PD,* formed in the laboratory study (Eq. 5-81) at short reaction times was thus confirmed to be a product under real tropospheric conditions. The products that were formed more slowly, e.g., *PB* and *BI,* were not detected.

In further field experiments, *TF* was released into the air and sampled downwind (Ref. 5-111). At 50 m, 15 sec transit time, the concentrations were 50 μg m^{-3} *TF* and 395 ng m^{-3} *PD*; at 100 m, 30 sec transit time, the concentrations were 43.4 μg m^{-3} *TF* and 654 ng m^{-3} *PD*; and at 150 m, 45 sec transit time, they were 16.7 μg m^{-3} *TF* and 374 ng m^{-3} *PD*. These results are consistent with a decomposition that is first order in *TF* and with a half-life for decomposition of 21 min if no other products were formed. Similar releases on other dates resulted in half-lives of 63 and 193 min, which were presumably caused by lower sunlight intensities. In one experiment, lindane, which decomposed to a negligible extent on the time scale of these experiments, was released as a tracer or internal standard along with *TF.* The decrease in concentration of *TF* was calculated from its ratio to the lindane rather than from its ratio to the sum of [*TF*] and [*PD*]. The results were comparable to those obtained previously, indicating that no products other than *PD* were formed in major amounts.

5-7 EFFECTS AND RELATIVE RISK IN THE ATMOSPHERIC ENVIRONMENT

Effects

Only a brief discussion of the effects of atmospheric constituents will be presented here. The references cited should be consulted for more detailed discussions.

From a human's standpoint, the most important effect of compounds or materials in the atmosphere involves human health (Refs. 5-21, 5-55, 5-56, 5-113-5-116). Other important effects on humans and animals are eye, nose, and throat irritation (Ref. 5-55), odors (Ref. 5-55), and visibility limitation

(Refs. 5-55, 5-56). Plant damage caused by air pollutants has been discussed (Refs. 5-21, 5-55, 5-56, 5-117). Damage to various materials, including metal corrosion, rubber cracking, paint deterioration, and soiling has been reviewed (Refs. 5-21, 5-55, 5-56, 5-115).

Several of these adverse effects on living and nonliving systems are thought to be caused by photochemical oxidants. These are not compounds emitted as such but are formed in the atmosphere from the interaction of organic compounds, nitrogen oxides, O_2, and sunlight (Ref. 5-28). The main photochemical oxidant is O_3; also present are smaller quantities of peroxyacyl nitrates (RCO_3-NO_2) and other oxidizing species that react with KI to give I_2.

Smoke (Ref. 5-55), dusts, fumes, and mists (Refs. 5-55, 5-56) may affect several of the areas mentioned previously. Weather modification may result from the emission of compounds to the atmosphere (Ref. 5-55).

Measurement of Effects

Measurement of health effects of air pollutants has been discussed (Ref. 5-55). Effects such as plant damage are often determined by observation (Ref. 5-117) and comparison with observations made in the same area at an earlier date (baseline survey) or by comparison with other similar areas (Ref. 5-118–5-121).

Relative Risk Rating

A standardized sequence of tests for determining the fate and effect of compounds in the atmosphere has not been developed, although several groups, e.g., the American Society for Testing and Materials, are working to develop such tests. Some suggested tests are discussed in Section 5-2.

An estimation of the ambient concentration of a compound in air may be made from data on the amount released over a given period of time or the amount formed from another material released, the partitioning among the various regions of the environment, and the loss rate. The predicted ambient concentration is compared with the effect level to determine the risk.

REFERENCES

5-1. Crutzen, P.J., "Estimates of Possible Variations in Total Ozone Due to Natural Causes and Human Activities," *Ambio*, 3, pp. 201–10, 1974.

5-2. Rowland, F. S., and Molina, M. J., "Chlorofluoromethanes in the Environment," *Rev. Geophys. Space Phys.*, 13, pp. 1–35, February, 1975.

5-3. Panel on Atmospheric Chemistry, *Halocarbons: Effects on Stratospheric Ozone*, National Academy of Sciences, Washington, 1976.

5-4. Levy, H., II., "Photochemistry of the Troposphere," *Adv. Photochem.*, 9, pp. 369–524, 1974.

5-5. Neely, W. B., "Material Balance Analysis of Trichlorofluoromethane and Carbon Tetrachloride in the Atmosphere," *Sci. Total Environ.*, 8, pp. 267–74, 1977.

5-6. Singh, H. B., Salas, L. J., Shigeishi, H., and Scribner, E., "Atmospheric Halocarbons, Hydrocarbons, and Sulfur Hexafluoride: Global Distributions, Sources, and Sinks," *Science*, 203, pp. 899–903, 1979.

5-7. Calvert, J. G., and Pitts, J. N., Jr., *Photochemistry*, John Wiley and Sons, Inc., New York, 1966.

5-8. Turro, N. J., *Modern Molecular Photochemistry*, The Benjamin/Cummings Publishing Co., Inc., Menlo Park, CA, 1978.

5-9. Dilling, W. L., "Interphase Transfer Processes. II. Evaporation Rates of Chloro Methanes, Ethanes, Ethylenes, Propanes, and Propylenes from Dilute Aqueous Solutions. Comparisons with Theoretical Predictions," *Environ. Sci. Technol.*, 11, pp. 405–9, April, 1977.

5-10. Neely, W. B., and Plonka, J. H., "Estimates of Time-Averaged Hydroxyl Radical Concentration in the Troposphere," *Environ. Sci. Technol.*, 12, pp. 317–21, March, 1978.

5-11. Darnall, K. R., Lloyd, A. C., Winer, A. M., and Pitts, J. N., Jr., "Reactivity Scale for Atmospheric Hydrocarbons Based on Reaction with Hydroxyl Radical," *Environ. Sci. Technol.*, 10, pp. 692–6, July, 1976.

5-12. Atkinson, R., Darnall, K. R., Lloyd, A. C., Winer, A. M., and Pitts, J. N., Jr., "Kinetics and Mechanisms of the Reactions of the Hydroxyl Radical with Organic Compounds in the Gas Phase," *Adv. Photochem.*, 11, pp. 375–488, 1979.

5-13. Walling, C., *Free Radicals in Solution*, John Wiley and Sons, Inc., New York, 1957.

5-14. Kochi, J. K., Ed., *Free Radicals*, John Wiley and Sons, New York, Vols. I and II, 1973.

5-15. Howard, C. J., and Evenson, K. M., "Rate Constants for the Reactions of OH with CH_4 and Fluorine, Chlorine, and Bromine Substituted Methanes at 296 K," *J. Chem. Phys.*, 64, pp. 197–202, January, 1976.

5-16. Pitts, J. N., Jr., "Photochemical Air Pollution: Singlet Molecular Oxygen as an Environmental Oxidant," *Adv. Environ. Sci.*, 1, pp. 289–337, 1969.

5-17. Pearson, C. R., and McConnell, G., "Chlorinated C_1 and C_2 Hydrocarbons in the Marine Environment," *Proc. Roy. Soc. London, B*, 189, pp. 305–32, 1975.

5-18. Gay, B. W., Jr., Hanst, P. L., Bufalini, J. J., and Noonan, R. C., "Atmospheric Oxidation of Chlorinated Ethylenes," *Environ. Sci. Technol.*, 10, pp. 58–67, January, 1976.

5-19. Sanders, H. J., "Chemistry and the Atmosphere," *Chem. Eng. News*, 44, pp. 1A–54A, March 28, 1966.

5-20. National Science Board, National Science Foundation, *Patterns and Perspectives in Environmental Science*, U. S. Government Printing Office, Washington, 1972.

5-21. Committee on Environmental Improvement, *Cleaning Our Environment, A Chemical Perspective*, 2nd ed., American Chemical Society, Washington, 1978.

5-22. Bailey, R. A., Clarke, H. M., Ferris, J. P., Krause, S., and Strong, R. L., *Chemistry of the Environment*, Academic Press, New York, 1978.

5-23. Graedel, T. E., *Chemical Compounds in the Atmosphere*, Academic Press, New York, 1978.

5-24. National Oceanic and Atmospheric Administration, National Aeronautics and Space Administration, United States Air Force, *U. S. Standard Atmosphere, 1976*, U. S. Government Printing Office, Washington, 1976.

5-25. Weast, R. C., Ed., *Handbook of Chemistry and Physics*, 57th ed., CRC Press, Cleveland, 1976.

5-26. McEwan, M. J., and Phillips, L. F., *Chemistry of the Atmosphere*, John Wiley and Sons, New York, 1975.

5-27. Rasmussen, R. A., and Pierotti, D., "Global and Regional N_2O Measurements," *Pageoph*, **116**, pp. 405–13, 1978.

5-28. Heicklen, J., *Atmospheric Chemistry*, Academic Press, New York, 1976.

5-29. Robinson, E., and Robbins, R. C., *Sources, Abundance, and Fate of Gaseous Atmospheric Pollutants*, Stanford Research Institute, Project PR-6755, Menlo Park, CA, 1968.

5-30. Israel, H., and Israel, G. W., *Trace Elements in the Atmosphere*, Translated by STS, Inc., Ann Arbor Science Publishers Inc., Ann Arbor, MI, 1973.

5-31. Bufalini, J. J., "Factors in Summer Ozone Production in the San Francisco Air Basin," *Science,* **203**, p. 81, January, 1979.

5-32. Cronn, D. R., Harsch, D. E., and Robinson, E., "Tropospheric and Lower Stratospheric Profiles of Halocarbons and Related Chemical Species," *Preprints of Papers Presented at the 176th National Meeting, ACS, Div. Environ. Chem.*, **18**, 2, pp. 360–2, September, 1978.

5-33. Singh, H. B., "Phosgene in the Ambient Air," *Nature,* **264**, pp. 428–9, December, 1976.

5-34. Penzhorn, R.-D., Güsten, H., Schurath, V., and Becker, K. H., "Halogenierte Kohlenwasserstoffe in der Atmosphäre: Geschwindigkeitskonstanten für die Löschung von metastabilen Sauerstoffmolekülen," *Staub–Reinhalt. Luft,* **35**, pp. 95–8, March, 1975.

5-35. Lovelock, J. E., Maggs, R. J., and Wade, R. J., "Halogenated Hydrocarbons in and over the Atlantic," *Nature,* **241**, pp. 194–6, January, 1973.

5-36. Cox, R. A., Derwent, R. G., Eggleton, A. E. J., and Lovelock, J. E., "Photochemical Oxidation of Halocarbons in the Troposphere," *Atmos. Environ.,* **10**, pp. 305–8, 1976.

5-37. Davis, D. D., Machado, G., Conaway, B., Oh, Y., and Watson, R., "A Temperature Dependent Kinetics Study of the Reaction of OH with CH_3Cl, CH_2Cl_2, $CHCl_3$, and CH_3Br," *J. Chem. Phys.,* **65**, pp. 1268–74, August, 1976.

5-38. Perry, R. A., Atkinson, R., and Pitts, J. N., Jr., "Rate Constants for the Reaction of OH Radicals with $CHFCl_2$ and CH_3Cl over the Temperature Range $298–423°K$, and with CH_2Cl_2 at $298°K$," *J. Chem. Phys.,* **64**, pp. 1618–20, February, 1976.

5-39. Crutzen, P. J., and Fishman, J., "Average Concentrations of OH in the Troposphere, and the Budgets of CH_4, CO, H_2, and CH_3CCl_3," *Geophys. Res. Lett.,* **4**, pp. 321–4, August, 1977.

5-40. Crutzen, P. J., Isaksen, I. S. A., and McAfee, J. R., "The Impact of the Chlorocarbon Industry on the Ozone Layer," *J. Geophys. Res.,* 83, pp. 345-63, January, 1978.

5-41. Logan, J., Harvard University, The Center for Earth and Planetary Physics, Cambridge, Massachusetts, private communication, June, 1979.

5-42. Wofsy, S. C., and McElroy, M. B., "HO_x, NO_x, and $C1O_x$: Their Role in Atmospheric Photochemistry," *Can. J. Chem.,* 52, pp. 1582-91, 1974.

5-43. Cavanagh, L. A., Schadt, C. F., and Robinson, E., "Atmospheric Hydrocarbon and Carbon Monoxide Measurements at Point Barrow, Alaska," *Environ. Sci. Technol.,* 3, pp. 251-7, March, 1969.

5-44. Lillian, D., Singh, H. B., Appleby, A., Lobban, L., Arnts, R., Gumpert, R., Hague, R., Toomey, J., Kazazis, J., Antell, M., Hansen, D., and Scott, B., "Atmospheric Fates of Halogenated Compounds," *Environ. Sci. Technol.,* 9, pp. 1042-8, November, 1975.

5-45. Pettit, E., "Measurements of Ultra-violet Solar Radiation," *Astrophys. J.,* 75, pp. 185-221, 1932.

5-46. Stair, R., and Johnston, R. G., "Preliminary Spectroradiometric Measurements of the Solar Constant," *J. Res. Natl. Bur. Stand.,* 57, pp. 205-11, October, 1956.

5-47. Koller, L. R., *Ultraviolet Radiation*, 2nd ed., John Wiley and Sons, Inc., New York, 1965.

5-48. Gates, D. M., "Spectral Distribution of Solar Radiation at the Earth's Surface," *Science,* 151, pp. 523-9, February, 1966.

5-49. Green, A. E. S., Ed., *The Middle Ultraviolet: Its Science and Technology*, John Wiley and Sons, Inc., New York, 1966.

5-50. Böer, K. W., "The Solar Spectrum at Typical Clear Weather Days," *Sol. Energy,* 19, pp. 525-38, 1977.

5-51. Demerjian, K. L., Kerr, J. A., and Calvert J. G., "The Mechanism of Photochemical Smog Formation," *Adv. Environ. Sci. Technol.,* 4, pp. 1-262, 1974.

5-52. Hampson, R. F., Jr., and Garvin, D., *Reaction Rate and Photochemical Data for Atmospheric Chemistry–1977*, National Bureau of Standards Spec. Publ. 513, U. S. Government Printing Office, Washington, 1978.

5-53. Falls, A. H., and Seinfeld, J. H., "Continued Development of a Kinetic Mechanism for Photochemical Smog," *Environ. Sci. Technol.,* 12, pp. 1398-1406, December, 1978.

5-54. Leighton, P. A., *Photochemistry of Air Pollution*, Academic Press, New York, 1961.

5-55. Faith, W. L., Atkisson, A. A., Jr., *Air Pollution*, 2nd ed., Wiley-Interscience, New York, 1972.

5-56. Seinfeld, J. H., *Air Pollution, Physical and Chemical Fundamentals*, McGraw-Hill Book Co., New York, 1975.

5-57. Lloyd, A. C., Darnall, K. R., Winer, A. M., and Pitts, J. N., Jr., "Relative Rate Constants for Reaction of Hydroxyl Radical with a Series of Alkanes, Alkenes, and Aromatic Hydrocarbons," *J. Phys. Chem.,* 80, pp. 789-94, 1976.

5-58. Greiner, N. R., "Hydroxyl Radical Kinetics by Kinetic Spectroscopy. VI. Reactions with Alkanes in the Range 300-500°K," *J. Chem. Phys.,* 53, pp. 1070-6, August, 1970.

5-59. Atkinson, R., Perry, R. A., and Pitts, J. N., Jr., "Rate Constants for the Reaction of OH Radicals with Ethylene over the Temperature Range 299-425°K," *J. Chem. Phys.*, **66**, pp. 1197–1201, February, 1977.

5-60. Perry, R. A., Atkinson, R., and Pitts, J. N., Jr., "Kinetics and Mechanism of the Gas Phase Reaction of OH Radicals with Aromatic Hydrocarbons over the Temperature Range 296-473°K," *J. Phys. Chem.*, **81**, pp. 296–304, 1977.

5-61. Watson, R. T., Machado, G., Conaway, B., Wagner, S., and Davis, D. D., "A Temperature Dependent Kinetics Study of the Reaction of OH with CH_2ClF, $CHCl_2F$, $CHClF_2$, CH_3CCl_3, CH_3CF_2Cl, and $CF_2ClCFCl_2$," *J. Phys. Chem.*, **81**, pp. 256–62, 1977.

5-62. Atkinson, R., Hansen, D. A., and Pitts, J. N., Jr., "Rate Constants for the Reaction of OH Radicals with CHF_2Cl, CF_2Cl_2, $CFCl_3$, and H_2 over the Temperature Range 297-434°K," *J. Chem. Phys.*, **63**, pp. 1703–6, September, 1975.

5-63. Howard C. J., and Evenson, K. M., "Rate Constants for the Reactions of OH with Ethane and Some Halogen Substituted Ethanes at 296°K," *J. Chem. Phys.*, **64**, pp. 4303–6, June, 1976.

5-64. Howard, C. J., "Rate Constants for the Gas-phase Reactions of OH Radicals with Ethylene and Halogenated Ethylene Compounds," *J. Chem. Phys.*, **65**, pp. 4771–7, December, 1976.

5-65. Chang, J. S., and Kaufman, F., "Kinetics of the Reactions of Hydroxyl Radicals with Some Halocarbons: $CHFCl_2$, CHF_2Cl, CH_3CCl_3, C_2HCl_3, and C_2Cl_4," *J. Chem. Phys.*, **66**, pp. 4989–94, June, 1977.

5-66. Atkinson, R., and Pitts, J. N., Jr., "Kinetics of the Reactions of the OH Radical with HCHO and CH_3CHO over the Temperature Range 299-426°K," *J. Chem. Phys.*, **68**, pp. 3581–4, April, 1978.

5-67. Niki, H., Maker, P. D., Savage, C. M., and Breitenbach, L. P., "Relative Rate Constants for the Reaction of Hydroxyl Radical with Aldehydes," *J. Phys. Chem.*, **82**, pp. 132–4, 1978.

5-68. Finlayson, B. J., and Pitts, J. N., Jr., "Photochemistry of the Polluted Troposphere," *Science*, **192**, pp. 111–9, April, 1976.

5-69. Schuck, E. A., and Doyle, G. J., *Photooxidation of Hydrocarbons in Mixtures Containing Oxides of Nitrogen and Sulfur Dioxide*, Air Pollution Foundation, San Marino, CA, 1959.

5-70. Rose, A. H., Jr., and Brandt, C. S., "Environmental Irradiation Test Facility," *J. Air Pollut. Contr. Assoc.*, **10**, pp. 331–5, August, 1960.

5-71. Tuesday, C. S., "The Atmospheric Photooxidation of *trans*-Butene-2 and Nitric Oxide," *Chemical Reactions in the Lower and Upper Atmosphere*, Proceedings of an International Symposium, Interscience Publishers, New York, pp. 15–49, 1961.

5-72. Rose, A. H., Jr., Stahman, R. C., and Korth, M. W., "Dynamic Irradiation Chamber Tests of Automatic Exhaust, Part I," *J. Air Pollut. Contr. Assoc.*, **12**, pp. 468–73, 478, October, 1962.

5-73. Altshuller, A. P., and Cohen, I. R., "Structural Effects on the Rate of Nitrogen Dioxide Formation in the Photo-Oxidation of Organic Compound-Nitric Oxide Mixtures in Air," *Int. J. Air Wat. Pollut.*, **7**, pp. 787–97, 1963.

5-74. Tuesday, C. S., "Atmospheric Photo-Oxidation of Olefins," *Arch. Environ. Health*, **7**, pp. 188–201, 1963.

5-75. Tuesday, C. S., D'Alleva, B. A., Heuss, J. M., and Nebel, G. J., *The General Motors Smog Chamber*, Research Publication GMR-490, General Motors Corp., Warren, MI, 1965.

5-76. Brunelle, M. F., Dickinson, J. E., and Hamming, W. J., *Effectiveness of Organic Solvents in Photochemical Smog Formation*," Los Angeles: Air Pollution Control District, 1966.

5-77. Altshuller, A. P., Kopczynski, S. L., Lonneman, W. A., Becker, T. L., and Slater, R., "Chemical Aspects of the Photooxidation of the Propylene-Nitrogen Oxide System," *Environ. Sci. Technol.*, 1, pp. 899–914, November, 1967.

5-78. Dimitriades, B., "Methodology in Air Pollution Studies Using Irradiation Chambers," *J. Air Pollut. Contr. Assoc.*, 17, pp. 460–6, July, 1967.

5-79. Stevens, E. R., and Burleson, F. R., "Analysis of the Atmosphere for Light Hydrocarbons," *J. Air Pollut. Contr. Assoc.*, 17, pp. 147–53, March, 1967.

5-80. Scofield, F., Levy, A., and Miller, S. E., *Paint Industry Smog Chamber. I. Design and Validation of a Smog Chamber*, Scientific Circular No. 797, National Paint, Varnish and Lacquer Association, Inc., Washington, 1969.

5-81. Dimitriades, B., *On the Function of Hydrocarbon and Nitrogen Oxides in Photochemical-Smog Formation*, U. S. Bur. Mines Rep. Invest. 7433, U. S. Department of the Interior, Washington, 1970.

5-82. Doyle, G. J., "Design of a Facility (Smog Chamber) for Studying Photochemical Reactions under Simulated Tropospheric Conditions," *Environ. Sci. Technol.*, 4, pp. 907–16, November, 1970.

5-83. Ripperton, L. A., and Lillian, D., "The Effect of Water Vapor on Ozone Synthesis in the Photo-oxidation of Alpha-pinene," *J. Air Pollut. Contr. Assoc.*, 21, pp. 629–35, October, 1971.

5-84. Kocmond, W. C., Kittelson, D. B., Yang, J. Y., and Demerjian, K. L., *Determination of the Formation Mechanisms and Composition of Photochemical Aerosols*, First Annual Summary Report, Contract No. 68-02-0557, CAPA-8-71, Calspan Corp., Buffalo, 1973.

5-85. Crosby, D. G., and Moilanen, K. W., "Vapor-Phase Photodecomposition of Aldrin and Dieldrin," *Arch. Environ. Contam. Toxicol.*, 2, pp. 62–74, 1974.

5-86. Jaffe, R. J., Smith, F. C., Jr., and Last, K. W., *Study of Factors Affecting Reactions in Environmental Chambers Final Report on Phase II*, EPA-650/3-74-004-a, U. S. Environmental Protection Agency, Washington, 1974.

5-87. Smith, G. N., "Photo-Chemical Reaction Apparatus,"U. S.Patent 3,858,051, December 31, 1974.

5-88. Jeffries, H., Fox, D., and Kamens, R., *Outdoor Smog Chamber Studies, Effect of Hydrocarbon Reduction on Nitrogen Dioxide*, EPA-650/3-75-011, U. S. Environmental Protection Agency, Washington, 1975.

5-89. Pitts, J. N., Jr., and Finlayson, B. J., "Mechanisms of Photochemical Air Pollution," *Angew. Chem. Intern. Ed.*, 14, pp. 1–15, January, 1975; *Angew. Chem.*, 87, pp. 18–33, 1975.

5-90. Dilling, W. L., Bredeweg, C. J., and Tefertiller, N. B., "Organic Photochemistry. Simulated Atmospheric Photodecomposition Rates of Methylene Chloride, 1, 1,-1-Trichloroethane, Trichloroethylene, Tetrachloroethylene, and Other Compounds," *Environ. Sci. Technol.*, 10, pp. 351–6, April, 1976.

5-91.   Akimoto, H., Hoshino, M., Inoue, G., Sakamaki, F., Washida, N., and Okuda, M., "Design and Characterization of the Evacuable and Bakable Photochemical Smog Chamber," *Environ. Sci. Technol.,* **13**, pp. 471-5, April, 1979.

5-92.   Dilling, W. L., and Goersch, H. K., "Organic Photochemistry. XVI. Tropospheric Photodecomposition of Methylene Chloride," *Preprints of Papers Presented at the 176th National Meeting, ACS, Div. Environ. Chem.,* **18**, 2, pp. 144-7, September, 1978.

5-93.   Singh, H. B., Salas, L., Lillian, D., Arnts, R. R., and Appleby, A., "Generation of Accurate Halocarbon Primary Standards with Permeation Tubes," *Environ. Sci. Technol.,* **11**, pp. 511-3, May, 1977.

5-94.   Crosby, D. G., and Moilanen, K. W., "Vapor-Phase Photodecomposition of DDT," *Chemosphere,* **4**, pp. 167-72, 1977.

5-95.   Joshi, S. B., and Dimitriades, B., "Reactivities of Organics under Pollutant Transport Conditions," *Preprints of Papers Presented at the 173rd National Meeting, ACS, Div. Environ. Chem.,* **17**, 1, pp. 20-2, March, 1977.

5-96.   Heuss, J. M., and Glasson, W. A., "Hydrocarbon Reactivity and Eye Irritation," *Environ. Sci. Technol.,* **2**, pp. 1109-16, December, 1968.

5-97.   Wilson, K. W., and Doyle, G. J., *Investigation of Photochemical Reactivities of Organic Solvents*, SRI Project PSU-8029, Stanford Research Institute, Menlo Park, CA, September, 1970.

5-98.   Joshi, S. B., and Bufalini, J. J., "Halocarbon Interferences in Chemiluminescent Measurements of NO_x," *Environ. Sci. Technol.,* **12**, pp. 597-9, May, 1978.

5-99.   Lillian, D., Singh, H. B., Appleby, A., Lobban, L., Arnts, R., Gumpert, R., Hague, R., Toomey, J., Kazazis, J., Antell, M., Hansen, D., and Scott, B., "Fates and Levels of Ambient Halocarbons," *Removal of Trace Contaminates from the Air*, Deitz, V. R., Ed., ACS Symposium Series 17, American Chemical Society, Washington, pp. 152-8, 1975.

5-100. Winer, A. M., Lloyd, A. C., Darnall, K. R., and Pitts, J. N., Jr., "Relative Rate Constants for the Reaction of the Hydroxyl Radical with Selected Ketones, Chloroethanes, and Monoterpene Hydrocarbons," *J. Phys. Chem.,* **80**, pp. 1635-9, 1976.

5-101. Spence, J. W., Hanst, P. L., and Gay, B. W., Jr., "Atmospheric Oxidation of Methyl Chloride, Methylene Chloride, and Chloroform," *J. Air Pollut. Contr. Assoc.,* **26**, pp. 994-6, October, 1976.

5-102. Sanhueza, E., and Heicklen, J., "Chlorine-Atom Sensitized Oxidation of Dichloromethane and Chloromethane," *J. Phys. Chem.,* **79**, pp. 7-11, 1975.

5-103. Singh, H. B., Lillian, D., Appleby, A., and Lobban, L., "Atmospheric Formation of Carbon Tetrachloride from Tetrachloroethylene," *Environ. Lett.,* **10**, pp. 253-6, 1975.

5-104. Huybrechts, G., Olbregts, J., and Thomas, K., "Gas-phase Chlorine-photosensitized Oxidation and Oxygen-inhibited Photochlorination of Tetrachloroethylene and Pentachloroethane," *Trans. Faraday Soc.,* **63**, pp. 1647-55, 1967.

5-105. Mathias, E., Sanhueza, E., Hisatsune, I. C., and Heicklen, J., "The Chlorine Atom Sensitized Oxidation and the Ozonolysis of C_2Cl_4," *Can. J. Chem.,* **52**, pp. 3852-62, 1974.

5-106. Chiltz, G., Goldfinger, P., Huybrechts, G., Martens, G., and Verbeke, G., "Atomic Chlorination of Simple Hydrocarbon Derivatives in the Gas Phase," *Chem. Rev.*, **63**, pp. 355–72, 1963.

5-107. Butler, R., Snelson, A., and Solomon, I. J., "The Homogenous Gas Phase Hydrolysis of Some Acid Chlorides," *Preprints of Papers Presented at the 174th National Meeting, ACS, Div. Environ. Chem.*, **17**, 2, pp. 59–60, August, 1977.

5-108. Snelson, A., Butler, R., and Jarke, F., *Study of Removal Processes for Halogenated Air Pollutants*, EPA-600/3-78-058, U. S. Environmental Protection Agency, Research Triangle Park, NC, 1978.

5-109. Rehman, R. A., Samuel, R., and Sharf-ud-Din, "On the Absorption Spectra of Some Organic Molecules in the Vapour State," *Indian J. Phys.*, **8**, pp. 537–45, 1933.

5-110. Woodrow, J. E., Seiber, J. N., Crosby, D. G., Moilanen, K. W., Soderquist, C. J., and Mourer, C., "Airborne and Surface Residues of Parathion and Its Conversion Products in a Treated Plum Orchard Environment," *Arch. Environ. Contam. Toxicol.*, **6**, pp. 175–91, 1977.

5-111. Woodrow, J. E., Crosby, D. G., Mast, T., Moilanen, K. W., and Seiber, J. N., "Rates of Transformation of Trifluralin and Parathion Vapors in Air," *J. Agr. Food Chem.*, **26**, pp. 1312–6, November, 1978.

5-112. Soderquist, C. J., Crosby, D. G., Moilanen, K. W., Seiber, J. N., and Woodrow, J. E., "Occurrence of Trifluralin and Its Photoproducts in Air," *J. Agr. Food Chem.*, **23**, pp. 304–9, 1975.

5-113. Carnow, B. W., and Carnow, V., "Air Pollution, Morbidity, and Mortality and the Concept of No Threshold," *Adv. Environ. Sci. Technol.*, **3**, pp. 127–56, 1974.

5-114. Snodderly, D. M., Jr., "Biomedical and Social Aspects of Air Pollution," *Adv. Environ. Sci. Technol.*, **3**, pp. 157–281, 1974.

5-115. Committee for the Working Conference on Principles of Protocols for Evaluating Chemicals in the Environment, *Principles for Evaluating Chemicals in the Environment*, National Academy of Sciences, Washington, 1975.

5-116. Finklea, J. F., Shy, C. M., Moran, J. B., Nelson, W. C., Larsen, R. I., and Akland, G. G., "The Role of Environmental Health Assessment in the Control of Air Pollution," *Adv. Environ. Sci. Technol.*, **7**, pp. 315–89, 1977.

5-117. Jacobson, J. S., and Hill, A. C., *Recognition of Air Pollution Injury to Vegetation: A Pictorial Atlas*, Air Pollution Control Association, Pittsburgh, 1970.

5-118. Curran, T. C., and Hunt, W. F., "Interpretation of Air Quality Data with Respect to the National Ambient Air Quality Standards," *J. Air Pollut. Contr. Assoc.*, **25**, pp. 711–4, July, 1975.

5-119. Johnson, W. B., and Ruff, R. E., "Observational Systems and Techniques in Air Pollution Meteorology," Air–AMS Workshop on Meteorology and Environmental Assessment, American Meteorological Society, Boston, 1975.

5-120. Willis, B. H., and Henebry, W. M., "What You Need to Know about Environmental Impact Statements," *Hydrocarbon Processing,* **55**, pp. 87–93, October, 1976.

5-121. Ott, W. R., and Hunt, W. F., "A Quantitative Evaluation of the Pollutant Standard Index," *J. Air Pollut. Contr., Assoc.*, **26**, pp. 1050–4, November, 1976.

6

Terrestrial environment

D. A. Laskowski
C. A. I. Goring
P. J. McCall '
R. L. Swann
Agricultural Products Department
The Dow Chemical Company
Midland, Michigan

6.1 GENERAL CONSIDERATIONS

Terrestrial environments are different from those of air and water. This is the supposition made by most people concerned with environmental risk analyses, but is this true? Are these compartments so different that information collected from one is not applicable to another? Real differences do exist between the compartments, but close examination shows that these differences may be more quantitative than qualitative. This is particularly true for terrestrial and aquatic environments. It turns out that things happening to chemicals in water can also happen in soil. They may differ only in the degree.

Information gleaned from the atmosphere may not be so readily related to aquatic and terrestrial compartments. Many of the reactions occurring in air are not likely to be found in the other two systems. Nor are the biochemical and surface phenomena characteristic of water and soils necessarily a part of air chemistry. Although similarities in the general reactivity of a chemical may be observed for the three compartments, reaction rates and products may be totally different in the air compartment.

Nevertheless, regardless of the chemical, there is usually a thread of consistency interlinking a chemical's reactions in all of the environments. We will talk more about this consistency as we discuss terrestrial systems and their similarities to and differences from aquatic systems. We will purposely avoid repetition of the findings of others in terrestrial environmental research because space does not permit a comprehensive review. Instead we will focus on objectives, data requirements, and methodology for environmental fate analysis. After briefly describing the physical aspects of terrestrial systems, we will discuss objectives and data requirements for environmental chemistry. We will examine the methodology necessary for some of the key environmental measurements. Finally we will discuss interpretation of the kind of data obtained so that readers will appreciate the ways it can be used in risk analysis.

For more information on terrestrial environments, readers are encouraged to turn to excellent general reviews on soil by Buckman and Brady (Ref. 6-7), soil analytical methods by Black *et al.* (Ref. 6-5), soil microbiology by Alexander (Ref. 6-1), fate of chemicals in soil by Goring and Hamaker (Ref. 6-29) and Guenzi *et al.* (Ref. 6-31), degradation of herbicides by Kearney and Kaufman (Ref. 6-43), and general soil biochemistry by McLaren *et al.* (Refs. 6-59, 6-60, 6-62).

6-1.1 Characteristics of Terrestrial Environments

The terrestrial environment, like an aquatic one, can be viewed as a four-compartment system: air, water, solids, and biota. However, whereas the terrestrial environment consists primarily of solid particles covered by thin films of water, the aquatic environment consists mainly of water in contact with small amounts of suspended solids and a large contiguous envelope of modified soil or bottom sediments. Water buffers the system against desiccation and temperature fluctuation. Because terrestrial environments lack this buffer, they experience large variations in temperature and moisture. Terrestrial systems contain much more reactive surface area and biotic loadings, and less variability of these factors than aquatic systems. Other obvious differences between the two environments are outlined in Table 6-1.

Because of the high solids-to-water ratio, the environmental chemistry of a terrestrial system takes place mainly in thin water films sandwiched between reactive colloidal surfaces and layers of air that vary in thickness, tortuosity, and continuity. The colloidal surfaces are usually coated with a layer of organic matter. Because they contain a net negative charge, they are also surrounded by clusters of positively charged ions that decline rapidly in concentration with increasing distance from the colloidal surface. This results in cationic materials being held tightly by the exchange complex of the soil solids unless the concentrations of other cations are sufficiently high to compete for sites on the exchange

TABLE 6-1. Comparison of Terrestrial with Aquatic Environments.

Item	Terrestrial	Aquatic
General characterization	Heterogeneous	Homogeneous
Gradients	Numerous, large	Less numerous, smaller
Temperature	Highly variable, diurnal pattern ($< 0°C$ to $50°C$)	Less variable, Yearly pattern
Moisture	Highly variable	Constant
Reactive surfaces	High level, less variable	Low level, highly variable
Biotic levels	High level, less variable populations	Low level, highly variable populations
Respiration	Mainly aerobic, transient anaerobic zones	Aerobic with discrete anaerobic zone

complex.   Reverse concentration gradients can exist for negatively charged compounds. Anions may be repulsed by the colloid surfaces so that concentrations in the water actually increase with increasing distance from the colloidal surface.

Another consequence of the cation exchange complex is possible enhancement of acid hydrolysis. Colloids in soil on the acidic side of soil pH tend to be surrounded by an acid medium approximately two pH units lower than what exists or is measured in the free water away from the colloids. This effect provides acidic surfaces that can promote the acid hydrolysis of compounds in the vicinity of such surfaces.

Expressions of such cationic and anionic gradients in soil systems depend mainly on the thickness of the water films surrounding the soil particles, and upon overall concentrations of dissolved ionic species. Saline solutions reduce the gradients by neutralizing the residual charges and by competing for exchange sites on the clay complex.

Colloidal surfaces in soil are in large measure coated with organic matter. The interactions between this organic coating and organic chemicals is of interest to the environmental chemist. The coated surfaces can contain stabilized enzymes that are biochemically active and capable of degrading the chemicals. They can also contain free radicals capable of reacting with susceptible functional groups of molecules, and metal complexes that can participate in catalyzed reactions. Microbes attach themselves to the surfaces and provide potentially billions of intact cells that can become involved in the degradation of molecules.

The complex structural arrangement of the colloidal surfaces must be carefully considered. The outer surface of large stable aggregates is composed of groupings of smaller aggregates and discrete particles. The inner surfaces are comprised of smaller aggregates and individual particles buried deep within the mother aggregate. Watery pathways leading to the interior are smaller and more tortuous than those connecting the exterior aggregate surfaces. There is, therefore, a critical time requirement for equilibration and true expression of a chemical's properties with the totality of an aggregate's surface. For the environmental chemist, such a system has the following implications: extraction methods are not valid unless proven against aged samples; mobility is not totally what it appears if judged only by experiments with fresh, unaged samples; and studies that involve leaching require knowledge of the rate of water movement in the soil column or soil profile. This last implication results from the dual system of pores in soil, i.e., large pores between aggregates and small pores within aggregates.

Plant roots permeate the soil structure intimately; organisms move about freely within the soil environment. Larger forms of animals burrow continuously through it, and microorganisms occur in vast quantitites in every crumb or speck of dust. It has been estimated (Ref. 6-1) that bacteria alone can account for 0.015 to 0.15% of the total weight of the top 6 in. of the soil profile. The total weight of living matter including plant roots in the same 6 in. can be from 0.3% to as much as 1% of the total weight (Ref. 6-7). This population represents a highly concentrated, very active biochemical force to interact with chemicals that reach soil. As a rule, competition for food is very keen. Any substrate susceptible to attack is rapidly consumed by the "zymogenous" microorganisms, which are normally at a near-starvation level. From a resting state they burst into frenzied activity when new food becomes available and return again to somnolence when the food is exhausted. The "autochthonous" microorganisms, on the other hand, appear to scavenge among the remnants of food distasteful to the zymogens, and are responsible for the slow but steady breakdown of the more refractory materials that comprise the soil organic matter. Generally, this group does not show the spectacular responses to organic amendments so characteristic of the zymogens.

In toto, the terrestrial system is heterogeneous. Gradients of all types are the rule, and the patterns of these gradients change constantly with time. Studies performed in the outdoors under the terms dictated by nature limit investigators of terrestrial environments to passive observation. At best, considerable effort must be devoted to monitoring the surroundings during an experiment in order to decipher environmental fate and extrapolate findings to other locations or situations.

6-1.2 The Continuity of Environmental Chemistry

There is an alternative to the carrying out of studies in the outdoors. Knowledge of the laws of nature can be marshalled and used to create laboratory experiments designed to measure key environmental properties. These properties can then be employed in predicting the fate and impact of chemicals in the environment. The prime requisite of this approach is that careful delineation of the problem and proper definition of objectives be undertaken before plunging into experimentation.

However, once the delineation has taken place one recognizes a distinct continuity in the environmental chemistry within and between environmental compartments. It begins with the overall goal of fate characterization as related to environmental toxicology, continues on into mental conceptualization of environmental compartments, and winds up at the same property measurements for fate analysis regardless of the system being considered. Regardless of the compartment, the goal is the determination of a chemical's exposure or its breakdown products to the inhabitants of all environmental compartments. This translates directly into assessment of *migration* between and *residuality* within each compartment, be it air, water, soil, or biota. The compartment makes no difference; the objectives are identical and are answerable by the same discrete set of facts dealing with physical and chemical properties.

Conceptually, fate characterization can be viewed as shown in Fig. 6-1. In this model water acts as the central source of distribution of the chemical to other compartments. The fate analysis itself focuses on the degradation and

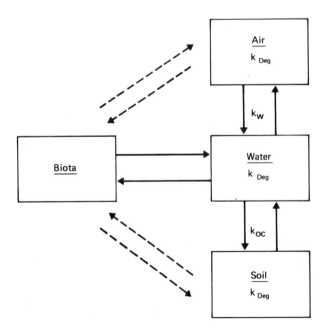

Fig. 6-1. Model for environmental fate analysis.

partitioning constants (K_{ow}, K_w, and K_{oc}) shown in the figure. The degradation constants generally describe rates of reaction, but the partition constants can represent rate or equilibrium values, depending upon the purposes of the experimenter.

Thus the data base for environmental chemistry narrows rapidly to the set of constants shown in Fig. 6-1. Water serves as the central compartment in this model, from which interaction with the other compartments takes place.

6-2 ASSESSMENT OF EXPOSURE

The model presented in Fig. 6-1 may seem too simple to be useful when one considers the vast complexities of the environment. However, the model's usefulness relates directly to the scope of an experimenter's objectives. Narrow objectives will call for measurements under very specific environmental conditions and a focus on the smaller variations and idiosyncracies of the system. However, broad objectives will allow a great deal of simplification and a focusing on only the general trends of a system rather than on its detailed composition.

Similarly, questions regarding environmental fate tend to be divisible: there are general questions dealing with environmental trends over wide areas, and there are specific questions focusing on exact fate at specific locations. Preliminary analysis of fate in the environment always tends to be general in nature. It is quite appropriate in this instance to employ the simplified system shown in Fig. 6-1. This makes it possible to recognize trends without becoming anesthetized by the morass of details that engulf the experimenter at narrower levels of conception. Out of mental necessity the investigator must begin thought processes on a broad plane, and only after perceiving the general behavioral pattern of the chemical will it be possible to efficiently comprehend and extrapolate findings from one pond, field, or waste stream, to other ponds, fields, or streams. This is especially true for the terrestrial environment as it is such a complex, heterogeneous system.

6-2.1 Environmental Chemistry Objectives

To perform risk analysis, environmental chemistry data must ultimately be related to the toxicity of materials. This is accomplished by interpreting toxicological data in relation to estimates of environmental exposure. Without information on exposure it is impossible to place a chemical's toxicity into proper perspective. Specifically, it is necessary to estimate the combined effects of length of contact and level of concentration of chemicals to which biota are exposed. Experiments designed to study exposures should be constructed in order to establish concentration patterns with respect to time, and should focus on the chances for general contamination through migration of the chemical from the point of application.

Terrestrial environmental studies have frequently dealt only with the collection of concentration data under actual use conditions. This was accomplished through monitoring programs that measured concentrations of chemicals in the environment as a result of practical field applications. Most of the existing concentration data have been collected in this manner, and monitoring is still a very popular tool. The technique does provide a direct measure of the fate of the chemical in the environment, and if a sufficient number of data points are obtained, one can extrapolate somewhat to other similar situations.

Unfortunately it is difficult to establish mass balances under actual field situations, and thus one never knows whether the measured declines in concentrations result from destruction of the material or simply dilution because of transfer to other portions of the environment. A second disadvantage of field monitoring studies is the cost of measuring all the parameters that influence the fate of a compound. Such characterization must be done if one wishes to use the data effectively for prediction and extrapolation to other field environments.

For these reasons, objectives in recent years have focused more on the measurement of physical and chemical properties under closely controlled, well-defined laboratory conditions. This places a bigger burden on the experimenter to be much more knowledgeable about what the data mean and how they can be used to describe environmental trends of chemicals, i.e., the tendency to migrate, to bioconcentrate, and to degrade. Investigators can no longer afford to be passive observers and reporters of observed phenomena. They must be able to recognize and place in perspective artifacts caused by the artificiality of the laboratory system, and they must make sure that the measurements made have real scientific substance and are not superficial demonstrations of environmental phenomena that have little predictive value.

6-2.2 Data Requirements

It is not enough to "observe" the leaching of a chemical through a soil profile; instead the property that controls this movement, i.e., the sorption of the chemical to soil solids, is what must actually be measured. The dissipation pattern of a chemical from soil is a useful piece of information, but the chemical's Henry's Law constant, its soil adsorption constant, and its volatilization and decomposition rates must also be known in order to meaningfully assess the dissipation. These properties and others shown in Table 6-2 provide the scientific basis for analyzing the fate of chemicals in the environment.

Water Solubility and Vapor Pressure. These two properties are becoming increasingly important in fate analysis because they can be used to estimate many of the other properties shown in Table 6-2. Water solubility in particular is

TABLE 6-2. Data Requirements for Fate Analysis.

Water solubility (25°C)[a]
Vapor pressure (25°C)
Octanol–water ratio
Bioconcentration factor
Soil adsorption constant
Water–air ratio (25°C)
Degradation rate
constant (25°C) in
Water
Air
Soil
Biota

[a]Use of 25°C is somewhat arbitrary but serves to standardize data at a temperature not unreasonable for terrestrial environments.

assuming a key role in the preliminary analysis of a chemical's fate since methods are now available for estimating bioconcentration factor (Refs. 6-45, 6-14), octanol–water partition coefficient (Ref. 6-14), and soil sorption constant (Ref. 6-45) from a knowledge of water solubility. Water solubilities in conjunction with vapor pressures can also be used to evaluate the volatility potential of chemicals from soil and water. All of these uses are in addition to the practical importance of being armed with information on the vapor pressures and water solubilities of chemicals in order to conduct meaningful laboratory experiments.

Methods for measuring water solubilities and vapor pressures are not discussed in this chapter because they are covered in Chapter 2. However, their importance and their central position in fate analysis for terrestrial and aquatic environments does need to be stressed. Later, in Section 6-2.3, some methodology will be given to show how vapor pressures and water solubilities can be used in fate analysis.

Octanol–Water Ratio and Bioconcentration Factor. These properties are extremely important environmentally because they are indicators of migration into biota. Measurement of the octanol–water ratio, in effect, provides a measure of the more complex process of bioconcentration, as the two have been shown to be correlated (Refs. 6-14, 6-45). They also correlate to soil sorption, making it possible to estimate soil sorption constants from either value (Ref. 6-45). Octanol–water ratios or bioconcentration factors are an integral component of the overall expression that can be used to describe exposure of living organisms to chemicals, which is why they are listed in Table 6-2.

Chapter 2 discusses the measurement of these properties so this topic is not discussed further in this chapter. However, the use of octanol–water ratios for risk analysis will be illustrated in section 6-23.

Soil Adsorption Constant. The soil adsorption constant (K_{oc}) indicates the tendency for a chemical to partition between water and soil solids. It is needed to assess leaching through soil, volatilization from soil, and volatilization from aquatic systems. When combined with an estimate of stability in soil or water it is possible to evaluate whether leaching and volatilization are apt to be significant migration pathways for chemicals in aquatic and terrestrial environments.

K_{oc} is defined as the amount of chemical adsorbed per unit of soil organic carbon in $\mu g/g$, divided by the $\mu g/ml$ of chemical in the water. It remains relatively constant from soil to soil. It can be used to calculate the soil adsorption coefficient (K_d) for any soil provided that the organic carbon content of that soil is known. Definition of K_d is the concentration of chemical adsorbed by soil in $\mu g/g$ on an oven-dry soil weight basis divided by the concentration of chemical in the water in $\mu g/ml$. The relationship between K_{oc} and K_d is shown in Eq. 6-1:

$$K_{oc} = [(K_d)(100)]/\% \text{ Organic Carbon} \qquad \text{(Eq. 6-1)}$$

The usefulness of K_{oc} for fate analysis lies in its ability to express adsorption independently of soil type. This is possible because adsorption of many organic chemicals relates directly to content of soil organic carbon. Many experimenters have observed this correlation (Ref. 6-34); Goring (Ref. 6-26) took advantage of it and developed the concept of a soil sorption constant that was independent of the solid adsorbent.

Adsorption of some chemicals such as metals, perhaps polymers, and positively charged molecules is not necessarily correlated with organic carbon content, and in this case some other soil property such as cation exchange capacity may serve better than organic carbon. It is recommended that the experimenter check the correlation with organic carbon before proceeding with its use. This can be done simply by measuring adsorption of a compound in several soils or sediments differing in organic carbon content.

The measurements are not difficult if the compound does not degrade rapidly in the soil-water systems. All that is required is a sample of chemical, a method of analysis for the chemical in the water of the soil-water mixture, a centrifuge capable of 10,000 to 20,000 rpm, sealed metal or glass centrifuge tubes that will withstand the centrifugal force, some soil of known organic carbon content, and some water. The common procedure is to prepare water solutions of the material and then add enough soil to cause significant reduction of the chemical's concentration in the water of the mixture. Between 4 and 24 hr are allowed for equilibration, the solids are centrifuged to the bottom of the tube, and the concentration of chemical remaining in the supernatant is measured. The difference between this value and that of a water control containing chemical but no soil equals the chemical sorbed to soil. The sorption coefficient K_d is calculated directly from these data, and then the sorption constant K_{oc} is derived from the K_d and soil organic carbon values.

The process is more difficult if the chemical does not prove to be stable during the incubation period. In this case specific assay for the chemical must be carried out on the soil solids as well as the soil solution.

K_{oc} constants can be obtained indirectly by estimation from known values for water solubility, octanol-water ratio, or bioconcentration factor. Several researchers (Refs. 6-6, 6-45) have used existing data to develop regression equations expressing the relationships of these properties with K_{oc}. Although Briggs (Ref. 6-6) was the first to make this comparison, the most extensive treatment is by Kenaga and Goring (Ref. 6-45). Their regression equations are reproduced in Table 6-3.

This procedure is acceptable if the experimenters' objectives are compatible with the errors involved in such estimates. As Kenaga and Goring point out, expected error is 1 to 1.5 orders of magnitude. This compares to expected standard deviations of roughly $\pm$ 100% (95% C. L.) of mean K_{oc} values when they are actually measured in several soils. In the future, estimation errors from regressions might be less than the present 1.5 orders of magnitude because of an improved data base for K_{oc} values. At present, K_{oc} values have not been extensively determined, and many found in the literature were developed from experiments not intended for this kind of measurement. Thus, better correlations of K_{oc} with other properties will probably be achieved eventually. The situation should improve as better values become available for a larger number of chemicals.

Once a value for K_{oc} becomes available it can be used in conjunction with the calculated water-air ratio to assess leaching, volatilization, and runoff characteristics of chemicals from soil. It also is necessary for evaluation of volatilization and partitioning into sediments in aquatic systems. More could be said about this key environmental property, but excellent reviews on sorption in soil by Weber (Ref. 6-83), Hance (Ref. 6-36), Osgerby (Ref. 6-61), and Hamaker and Thompson (Ref. 6-34) make it unnecessary. Those wishing to delve deeper into this topic are encouraged to seek out these references.

Water-Air Ratio (K_w). This constant is the same as Henry's Law constant, only it is its inverse for the sake of convenience and is expressed in different units than the normal Henry's Law value. It is defined here as the ratio of chemical concentration in water (ppm) over the concentration of chemical in air ($\mu g/cm^3$) at equilibrium. Temperature is standardized at 25°C.

K_w is useful because it provides a measure of the tendency of a chemical to partition between water and air. It can be used in estimating the volatilization of a chemical from water or soil systems. When it is combined with the soil sorption constant and an estimate of the chemical's decomposition rate in each system, it is possible to evaluate the overall likelihood of volatilization being a significant mechanism of migration from water and soil into air. Hartley (Ref. 6-37), Call (Ref. 6-10), and Goring (Ref. 6-26) discussed water-air ratios or Henry's Law constants more than 20 years ago, and used them to evaluate move-

TABLE 6-3. Estimation of K_{oc} from Water Solubility, Octanol–Water Partition Coefficient, or Bioconcentration Factor.

Property	Regression Equation	± Order of Magnitude (95% C. L.) from Calculated Value	Correlation Coefficient r	Number of Data Points n
Water solubility (S), ppm	$\log K_{oc} = 3.64 - 0.55 \,(\log S)$	1.23	−0.84	106
Octanol–water ratio (K_{ow})	$\log K_{oc} = 1.377 + 0.544 \,(\log K_{ow})$	1.37	0.86	45
Bioconcentration factor (BCF) (flowing water system)	$\log K_{oc} = 1.963 + 0.681 \,(\log BCF)$	1.52	0.87	13
Bioconcentration factor (BCF) (terrestrial aquatic static system)	$\log K_{oc} = 1.886 + 0.681 \,(\log BCF)$	1.14	0.91	22

Source: Kenaga, E. E., and Goring, C. A. I., "Relationship Between Water Solubility, Soil Sorption, Octanol–Water Partitioning, and Bioconcentration of Chemicals in Biota," ASTM Third Aquatic Toxicology Symposium, New Orleans, LA, Oct. 15–20, 1978.

ment of chemicals through and from soil. More recently, Mackay and Leinonen (Ref. 6-56) and Mackay and Cohen (Ref. 6-55) revived the environmental chemist's awareness of Henry's Law constants and demonstrated their use in predicting volatilization rates of chemicals from bodies of water. Mackay discusses Henry's Law constants in Chapter 2 of this book.

Water-air ratios can be measured by a procedure similar to that described by Mackay and Cohen (Ref. 6-55). It consists of bubbling air through a column of water containing the dissolved chemical at a slow enough rate to ensure equilibration with the solution. The distribution ratio of the chemical between water and the air can then be determined by direct measurement of the concentrations of chemical in the solution and in the air.

More commonly, K_w constants are estimated by calculation from water solubility and vapor pressure. Hamaker (Dow Chemical, personal communication) developed a simple expression for this calculation by assuming that the partial vapor pressure of a chemical over a saturated water solution is the same as the vapor pressure of the pure compound. Then it was assumed that the change in partial pressure is linearly related to the change in solution concentration, so that:

$$K_w = \text{conc. liquid/conc. vapor} \qquad \text{(Eq. 6-2)}$$

$$K_w \cong \text{water solubility/saturated vapor conc.} \qquad \text{(Eq. 6-3)}$$

From the perfect gas law:

$$PV = (g/M)(RT) \qquad \text{(Eq. 6-4)}$$

where

g = weight of vapor
M = molecular weight
R = molar gas constant
T = absolute temperature
P = saturated vapor pressure
V = volume of gas

Rearranging Eq. 6-4,

$$\text{Saturated Vapor Conc.} = g/V = PM/RT \qquad \text{(Eq. 6-5)}$$

Combination and rearrangement of Eq. 6-5 with Eq. 6-3 yields

$$K_w = (\text{Water Solubility})(RT/PM) \qquad \text{(Eq. 6-6)}$$

If the units for K_w are the following:

$$K_w = (\mu g/cm^3)/(\mu g/cm^3)$$

then

water solubility (S) = ppm

vapor pressure (P) = mm Hg°

M = g/mole
R = 82.06 cm³ atm deg^{-1} mole^{-1}
g = 10^6 μg
V = gas volume in cm³
Atm. = 760 mm Hg°

and the final expression becomes:

$$K_w = [(S)(82.06)(760)(273.15 + °C)] / [(P)(M)(10^6)] \quad \text{(Eq. 6-7)}$$
$$= [(S)(0.062366)(273.15 + °C)] / [(P)(M)] \quad \text{(Eq. 6-8)}$$

This expression is very useful because few K_w values for environmental chemicals have actually been measured. The only means of obtaining them at present is from water solubility and vapor pressure calculations.

As pointed out by Mackay and Wolkoff (Ref. 6-57) and again by Mackay and Cohen (Ref. 6-55), the calculations may be in error for compounds existing as solids at the temperature of consideration. These authors suggest that it is the vapor pressure of the liquid state that is needed for such calculations. If the material exists as a solid at the temperature under consideration, then an error of unknown magnitude can arise from use of the solid state vapor pressures. This error might be as much as fourfold, but its seriousness will not be known until after more measurements of K_w have been made. In cases where the objective is to classify volatilization potential in a very general manner, as for example in the first estimate of the environmental fate of chemicals, it is entirely possible that the differences between calculated and measured values will be quite tolerable.

K_w is a valid measure of volatility potential from water alone but not from a mixture of sediments or soil and water. Here the volatility is modified by sorption to the solids, and it is the combined expression of adsorption and water-air partitioning that must be used in the evaluation. Goring (Ref. 6-27) provided a tool for expression of potential soil volatility when he combined the two partitioning constants K_{oc} and K_w in the following manner:

$$\% \text{ in soil air} = (100x)/[x + y(K_w) + (zdw/100)(K_w K_{oc})] \quad \text{(Eq. 6-9)}$$

where

x = volume % of soil air = 25
y = volume % of soil water = 25

z = volume % of soil solids = 50
d = density of soil solids = 2.5
w = weight % of soil organic
           carbon in oven dry soil
K_w = water–air ratio
K_{oc} = soil adsorption constant

This is a hypothetical case with an idealized soil consisting of 25% air, 25% water, and 50% solids. It assumes that equilibrium is established, thus permitting calculation of the fractional portion of chemical residing in the air of this hypothetical soil. If one assumes that volatility is proportional to the percentage of chemical in the soil air, then it is logical to evaluate the potential for volatility by comparing the percentages for unknown compounds with those of compounds of known volatilization rates. By this process one can estimate soil volatilization rates of unknown compounds if there is information available on vapor pressures, water solubilities, sorption constants, and volatilization rates of standards.

Equilibrium expressions combining K_w with K_{oc} as in Eq. 6-9 may not be the only important indicator of volatility from soil. There is evidence from recent work by McCall and Swann (Ref. 6-58) that under certain situations vapor pressure may be the best initial indicator. This possibility exists for initial losses of surface applied chemicals before equilibration with the water and the soil solids has been completed. Data from McCall and Swann for nitrapyrin and trifluralin are reproduced in Table 6-4. If one compares the ratios of the volatility rate constants of nitrapyrin and trifluralin with the ratios of their vapor pressures and percentage in soil air values, it appears that, at first, the best indicator for rate of loss is vapor pressure, but that later the best indicator becomes the percentage in the soil air. The data also suggest that when nitrapyrin and trifluralin are incorporated into soil the best initial and subsequent indicator of volatility is the calculated percentages of the materials in the soil air.

More recently Swann *et al* (Ref. 6-74) described a simpler expression for volatility by assuming that volatilization rates from wet soil surfaces are directly proportional to vapor pressure but inversely proportional to water solubility and the soil adsorption constant K_{oc}. Their relationship is expressed as

$$k \propto V_p/[(S)(K_{oc})] \qquad\qquad \text{(Eq. 6-10)}$$

where

k = volatility rate constant
V_p = vapor pressure in mm Hg @ 25°C
S = water solubility in ppm @ 25°C
K_{oc} = soil adsorption constant

TABLE 6-4. Comparison of Nitrapyrin and Trifluralin Volatility
Rate Constants.

Condition	Rate Constant k		Ratio of N/T
	Trifluralin (T)	Nitrapyrin (N)	
Initial rates ($k \times 10^3$ min^{-1}):			
Glass surface	2.16	51.5	23.8
Commerce soil surface	1.49	14.61	9.8
Barnes soil surface	0.65	8.28	12.7
Commerce soil surface (air dry)	0.20	1.44	7.2
Vapor pressure ratio	—	—	35.2
Rate at 4 to 8 hr ($k \times 10^4$ min^{-1}):			
Commerce soil surface	8.14	12.28	1.5
Barnes soil surface	4.95	10.93	2.2
Commerce soil incorporated	2.60	5.34	2.0
Barnes soil incorporated	0.62	0.80	1.3
Percentage in soil air $\times 10^3$	0.82	1.55	1.9

Source: McCall, P. J., and Swann, R. L., "Nitrapyrin Volatility from Soil," *Down to Earth*, **34**, pp. 21–27, 1978.

In a way this combination of vapor pressure, water solubility, and adsorption constant can also be viewed as representing the amount of chemical in the air phase of an air–water–soil system. When the quantity $V_p/[(S)(K_{oc})]$ for several compounds is plotted against their volatility rates on a log–log basis, a good correlation is obtained. Nine chemicals—nitrapyrin, lindane, trifluralin, chlorpyrifos, diuron, carbofuran, dinoseb, DDT, and atrazine—varying greatly in rate of volatility were used to develop the correlation. The correlation coefficient obtained was 0.96, and the regression equation modified to express rate directly in days required for 50% loss was:

$$t/2 = [1.58 \times 10^{-8} (K_{oc})(S)]/V_p \qquad \text{(Eq. 6-11)}$$

So far this study has only been carried out in one soil. It remains to be seen how applicable Eq. 6-11 is in other soils after the proper corrections are made for differences in organic carbon contents.

Comparisons such as these have rarely been used in soil volatility studies and, therefore, precise expressions that best represent volatilization from soil have yet to be determined. It may be that no single expression will hold for all environmental situations, but that there will be a discrete set of expressions for certain types of environments. One obvious basis for environmental types is the distinction between surface-applied chemicals and soil-incorporated ones.

The most comprehensive studies of volatilization of chemicals from soils were conducted by Spencer and coworkers at the University of California at Riverside. They prepared a very useful review on this subject (Ref. 6-72), covering actual measurements of volatility under field and laboratory situations. In it they describe the myriad of parameters controlling rates of volatility from soil; these are summarized in Table 6-5.

Table 6-5 is designed to impress upon the reader that volatilization from soil is a very complicated process controlled by a large number of variables. Values for many of these variables are constantly changing, and so it is difficult to make quantitative predictions of volatilization under field situations. Too many parameters of the specific environment must be continuously measured. In our own work on volatility, this dilemma was resolved by concentrating on more general objectives aimed at predictions that would describe trends. This topic is returned to in Section 6-2.3, where interpretation of data is discussed.

TABLE 6-5. Factors Influencing Rates of Volatility of Chemicals from Soil.

Category	Factor
I. Influence on vapor density:	
	1. Saturation vapor pressure
	2. Solubility in water
	3. Temperature
	4. Sorption to surfaces
	5. Concentration of chemical
	6. Soil moisture
	7. Soil properties related to sorption
II. Influence on movement away from evaporating surface:	
	8. Air flow
	9. Diffusion rate
	10. Molecular weight
	11. Vapor density
	12. Temperature
III. Influence on movement to evaporating surface:	
	13. Diffusion rate
	14. Vapor density
	15. Mass water flow
	16. Soil bulk density
	17. Soil moisture

Source: Based (in part) on Spencer, W. F., Farmer, W. J., and Cliath, M. M., "Pesticide Volatilization," *Residue Reviews*, 49, pp. 1–47, 1973.

The work of Spencer and coworkers merits additional attention with regard to the methodology for measurement of soil volatility in the laboratory. Their basic apparatus is reproduced in Fig. 6-2. With it they have been able to provide excellent information on vapor pressures of pure chemicals alone (Ref. 6-69) and in combination with soil (Refs. 6-69, 6-70). Using the chlorinated hydrocarbon lindane, they have shown that vapor densities in soil are directly correlated with the concentration of chemical in soil solution (Ref. 6-70). At least for lindane Henry's Law is obeyed in a soil-water system, lending credence to the use of water–air ratios and soil adsorption constants for estimation of soil volatilization potential (Eqs. 6-9, 6-10, 6-11).

This concludes the discussion of volatility and physical properties and their relation to the migration of chemicals in general. We have neither attempted to cover factors specifically influencing the migration of chemicals, nor all of the information that has been generated on soil volatility. Reviews by Hamaker and Thompson (Ref. 6-34), Weber (Ref. 6-83), Osgerby (Ref. 6-61), Hance (Ref. 6-36), Spencer et al. (Refs. 6-71, 6-72), Plimmer (Ref. 6-63), and Hartley (Ref. 6-38) summarize the literature very well, and the reader is referred to them for more information.

Soil Degradation. This piece of information is extremely important because it makes it possible to estimate the exposure of living things not only to parent

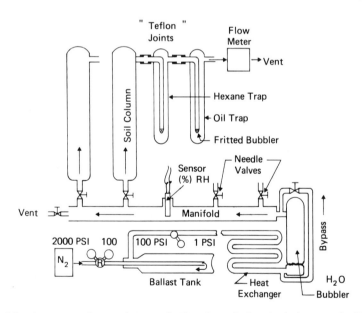

Fig. 6-2. Apparatus for studying volatilization of chemicals from soil. *Source*: Spencer, W. F., and Cliath, M. M., "Vapor Density of Dieldrin," *Environ. Sci. Technol.*, 3, pp. 670–674, 1969.

chemicals but also to their breakdown products. In general, stable chemicals tend to migrate from the placement site regardless of their physical properties. Thus stability tends to increase the potential for widescale exposure. In contrast, unstable materials are less apt to migrate and generally have a low exposure potential.

To be meaningful and suitable for risk analysis, soil degradation experiments should have the following objectives:

1. To determine the expected average rate of breakdown of parent in soil under reproducible and well-defined moisture/temperature conditions.
2. To characterize the variation in breakdown rates from one soil to another.
3. To characterize the interaction of decomposition rate with soil moisture, soil temperature, and initial soil concentration of chemical.
4. To identify significant decomposition products and characterize their inherent stability in soil.

This kind of information can help provide an idea of what to expect from a chemical with respect to its destruction under field conditions. The objectives are designed in such a way as to provide data on rates of soil degradation that can be used in a variety of ways, ranging from categorization of a chemical's soil stability in a general sense to estimation of environmental concentrations in specific situations by means of computer modeling. Because stringent control of incubation conditions is required to accomplish these objectives, field studies are all but ruled out, leaving laboratory experiments under controlled environments as the primary tool for investigation. Because of the way the data will ultimately be used, these studies must characterize only the degradation of a chemical and not be confounded with losses arising from volatility and other forms of movement that cause disappearance but not degradation. Soil degradation kinetics is complex enough by itself without adding the burden of several additional dissipation mechanisms operating at the same time.

In order to restrict studies to degradation alone, they must be performed in closed systems under controlled conditions. These systems generally are of the diffusion flask type described by Stotzky (Ref. 6-73) and by Bartha and Pramer (Fig. 6-3). Typically, they consist of a main soil compartment and a secondary compartment for absorption of carbon dioxide. Since the flasks are closed to the atmosphere, some means of aeration must be provided. Methods have ranged from periodic manual introduction of fresh air (Ref. 6-3) to automatic replenishment by generation from a carbon dioxide trapping solution of oxygen-producing barium peroxide (Ref. 6-15).

For our own work, Bartha and Pramer's flask was redesigned to allow oxygen to be replaced automatically by connection to a positive pressure oxygen manifold. The subsequent design is shown in Fig. 6-4. We retained the two-compartment concept, one for soil and one for carbon dioxide but replaced the rubber

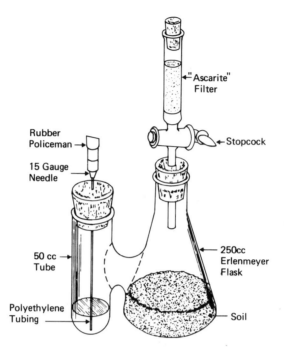

Fig. 6-3. Biometer flask. *Source:* Bartha, R. and Pramer, D. "Features of a Flask and Method for Measuring the Persistence and Biological Effects of Pesticides in Soil," *Soil Sci.,* 100, pp. 68-70, 1965.

stoppers with glass so that the chemical under study contacts only glass surfaces. This is because rubber is an excellent sink for many nonpolar organic chemicals, and its presence will not permit quantitative recovery. The second major change was to enlarge the carbon dioxide trap to accommodate the use of dilute sodium hydroxide instead of the concentrated solutions used normally. This circumvents problems of soil drying because of water transfer to concentrated caustic. It also allows easy measurement of radioactive carbon dioxide by counting an aliquot of the dilute caustic directly in one of the several commercial liquid scintillation cocktails now available and designed for handling water solutions. With this design we have been able to maintain good accountability of chemical, normal oxygen levels, and constant soil moisture levels over incubation periods lasting for a year or more.

Use of what are called open systems is very common in soil degradation studies. They can provide useful information on degradation rates if volatilization of the chemical from the soil is negligible. Interpretation of the data becomes much more complicated when volatilization occurs.

Many kinds of open systems have been used. They range from simple open containers filled with treated soil and incubated on lab benches to capped con-

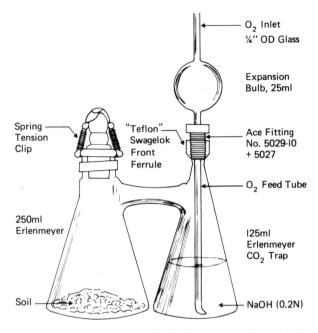

Fig. 6-4. Diagram of soil incubation flask. *Source*: Laskowski, D. A., Swann, R. L., and Regoli, A. J., "Incubation System for Studying Soil Degradation of Chemicals," Presentation Amer. Chem. Soc. National Meeting, Anaheim, California, 1978.

tainers aerated intermittently or with a steady stream of air and incubated under controlled environmental conditions. The containers have consisted of such things as greenhouse pots, bottles, jars, and other suitably sized holders for soil. One apparatus meriting special attention is the continuous air flow system described by Bartholomew and Broadbent (Ref. 6-4) and Goring *et al.* (Ref. 6-28). It is shown in Fig. 6-5. As pictured, the system is not capable of recovering volatilized chemical, but Kearney and Kontson (Ref. 6-44) solved this problem by adding a polyurethane foam trap to the air stream as it leaves the soil incubation flask.

Regardless of the system used, it must be recognized that the data generated do not simulate actual field environments. The system is designed to minimize migration losses and to be carried out under constant conditions of temperature and moisture. Actual field environments are not like this. Moisture and temperature are in a constant state of change and migration losses contribute to the loss patterns observed in the field. Extrapolating the results of laboratory degradation studies to field situations can be very misleading unless these differences are taken into account. Generally laboratory data suggest greater field residuality than what actually occurs.

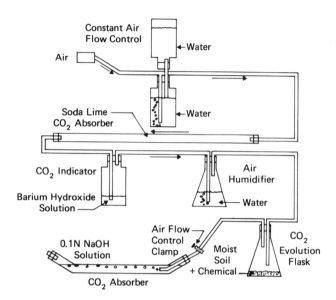

Fig. 6-5. Continuous air flow system for soil incubation of chemicals. *Source*: Goring, C.A.I., Griffith, J. D., O'Melia, F. C., Scott, H. H., and Youngson, C. R., "The Effect of TORDON on Microorganisms and Soil Biological Processes," *Down to Earth*, 22, pp. 14-17, 1967.

When differences from the field situation are taken into consideration laboratory studies do a creditable job of suggesting actual field results. This fact has been demonstrated very nicely by Walker (Ref. 6-68, 6-78-6-82) for the pesticides simazine, atrazine, propyzamide, linuron, metamitron, trifluralin, metribuzin, chlorthal-dimethyl, prometryne, asulam, and napropamide. Loss of these materials from field plots was predicted quite well from laboratory data when field moisture and temperature patterns were measured and taken into consideration. Agreement between calculated and observed values was not as close when field conditions conducive to migration were present.

Another basic consideration in conducting laboratory degradation studies is the variability among soils in their ability to degrade chemicals (see Table 6-6). This variation occurs not only in degradation rate but also in the distribution of breakdown products. Rate differences may be more than tenfold and are totally unpredictable with today's level of knowledge. In the work by Walker on computer modeling this problem was overcome by using soils for the laboratory studies that came from the same field plots used to establish the field soil dissipation curves. As can be seen by the soil variations shown in Table 6-6, if this were not done it would have been impossible to model the field dissipation patterns. Soil variation is one of the most frustrating problems a modeler of

TABLE 6-6. Variation in Rate of Decomposition Among Surface Soils for Several Chemicals.

Chemical	# Soils	Range of Observed Differences Among Soils	Reference
Crotoxyphos	3	36×	6-46
Linuron	4	2×	6-53
Methomyl	2	2×	6-39
Glyphosate	4	19×	6-66
Aldicarb	2	2×	6-65
Carbofuran	4	25×	6-24
Diazinon	4	2×	6-25
Thionazin	4	7×	6-25
Methidathion	4	3×	6-23
Nitrilotriacetate	11	80×	6-75
Nitrapyrin	10	6×	6-48
Picloram	13	19×	6-35
Propyzamide	5	2×	6-81

Note: Only experiments that expose soils to chemicals under identical conditions are cited.

terrestrial environments will encounter because it is so unpredictable and at present so poorly defined.

Those engaged in soil degradation studies have not yet paid much attention to differences between soils. Far too many times conclusions have come from studies based upon a single "fresh garden soil" (whatever that means) or at best from one or two additional soils. One soil taken from the population of roughly 12,000 soils found in the U. S. is unlikely to provide a statistically valid estimate of soil degradation. If there is to be meaningful extrapolation to field conditions from laboratory data, there must be an assessment of how many soils are needed for evaluation of rates of degradation. This is not possible with the currently available data, like that shown in Table 6-6, since it is much too scattered to provide a reliable measure of variability among soils. It is possible that future work aimed at specific evaluation of soil variation will show differences less than the two- to 80-fold magnitudes shown in Table 6-6. Characterization of soil variability and evaluation of the number of soils required in degradation studies are research problems that need solution before we can rely on general extrapolation of laboratory soil degradation data to actual field situations.

A third point to remember when conducting laboratory degradation studies is the problem of collection and storage of soils (see Ref. 6-64). Soil is a dynamic system of fragile biochemical activity, and any disturbance to the soil is apt to change its activity. Therefore, chemicals degraded biochemically could be affected by soil handling methods, and breakdown rates or transformation pathways could be altered. Although the quantitative and qualitative significance of these possibilities are not known, there are enough incidents reported in the literature to establish the importance of the problem.

It is well known that drying of soils destroys microorganisms (Ref. 6-1). Bacteria seem to be more susceptible than other species. If the decomposition of a chemical is mediated mainly by bacteria, then drying could influence the subsequent kinetics of a chemical's destruction. Freezing also kills microbes, and gram-negative bacteria seem more sensitive than other microbial types (Ref. 6-9). In general, microbial activity declines during hot, dry summer months and increases in spring and fall (Ref. 6-1). Consequently, soils might possibly show seasonal differences in degradability as well as effects from drying and storage under frozen conditions. The mere act of storage under any condition apparently influences the biochemistry of soils; this influence increases with the length of time the soil is stored (Ref. 6-64).

Unfortunately, too little is known about these effects on degradation of chemicals to comprehend their signficance. They have been noted to influence the degradation of the herbicide propanil in soil (Refs. 6-2, 6-50, 6-51), but this is only for one chemical and not a measure of the overall signficance of the problem. Ideally, fresh soil collected in late spring or early fall and used immediately would be the most suitable for soil degradation studies. This is not always a practical procedure, however. An alternative procedure adopted by many experimenters is collection of the top three to four inches of soil in polyethylene bags fitted into cardboard fiberpak containers. The soil is then stored in the fiberpaks at collection moisture and 4 to 5°C for periods of up to a year. No one knows if this is the best method or even if it is free of significant artifacts, but until information to the contrary is developed it can be regarded as an adequate standard practice.

Changes in soil degradation characteristics during storage do need to be studied so they may be placed in proper perspective. They probably do not represent an unmanageable problem since Walker in his series of modeling studies (Refs. 6-68, 6-78–6-82) successfully described field dissipation for 11 different chemicals from laboratory degradation data, even though some of the soils were air dried before being used in the laboratory studies.

Climatic factors such as temperature and moisture, and the concentration of chemical applied to the soil can modify the rates of degradation of chemicals. Their effects must be determined if laboratory data are to be successfully extrapolated to field environments.

The Arhenius equation provides a fairly satisfactory expression for the relationship between temperature and rate of degradation, as shown below:

$$\log k = (-\Delta E / 2.303\, RT) + A \qquad\qquad \text{(Eq. 6-12)}$$

where

k = rate constant at temperature T
ΔE = activation energy, in calories/mole
R = molar gas constant, 1.987 calories/deg-mole
T = absolute temperature, in degrees K
A = constant

Most experimenters assume that first-order kinetics apply to soil decomposition and therefore use first-order rate constants. This assumption was reasonably successful in Walker's modeling exercises with his 11 compounds.

The effects of moisture on degradation have been mathematically described on a more empirical basis than temperature. Walker (Ref. 6-77) used the expression:

$$H = aMc^{-b} \qquad \text{(Eq. 6-13)}$$

where

H = half-life
Mc = soil moisture content
a, b = constants

A more universal version of Eq. 6-13 suitable for soils might involve expressing the moisture as a percentage of field capacity rather than actual moisture content. Field capacity is defined as the water retained by soil after equilibration with gravity and relates to the availability of water in soils. Since the availability of water varies from soil to soil, due to differences in texture, organic matter content, and soil structure, the use of moisture contents in Eq. 6-13 does not reflect the availability of water. It is necessary to compare the two methods of expressing moisture, selecting the best method for incorporation in an equation that suitably expresses the effect of moisture on rates of degradation.

The effects of concentration on rates of degradation in soil can be very complicated. Clearly some chemicals are good potential energy sources for some of the soil microorganisms. When high concentrations of these chemicals are applied to soil, the microorganisms adapt to their presence and develop the capacity to degrade them rapidly (Ref. 6-30). The kinetics of decomposition resemble what might be expected for the disappearance of a nutrient from a broth culture inoculated with a microorganism that readily adapts itself to use the nutrient as an energy source.

However, the great majority of the chemicals reaching soil are at such low concentrations or such poor food sources that decomposition occurs primarily through cometabolic processes (Ref. 6-30). The kinetics of these processes are quite different from a situation where the chemical is being used as a readily available energy source.

The kinetics of degradation of chemicals in soil is generally assumed to be first order as a first approximation. However, the actual situation is more complicated. Two kinds of processes influence the observed kinetics. The first process is the actual decomposition of that portion of the chemical available for degradation, and the second process the migration of the chemical in and out of the solid matrix of the soil. The chemical in this matrix is generally unavailable for degradation.

The degradation process for water-soluble, relatively poorly sorbed chemicals appears to follow Michaelis–Menten kinetics (Ref. 6-30). In this process, rates of degradation tend to be independent of concentration at high concentrations (zero-order kinetics) and proportional to concentration at low concentration (first-order kinetics). However, the migration of the chemical into the solid matrix of the soil continuously decreases the amount of chemical available for degradation (Ref. 6-33).

These two forces act on the degradation rate to produce certain patterns that are well illustrated in recent work carried out in our laboratory by Swann with picloram (see Fig. 6-6). At relatively high concentrations of picloram the slope of the initial decay curve is much shallower than at low concentrations, as would be expected if degradation is taking place according to Michaelis–Menten kinetics and results of such studies are plotted on semi-log paper.

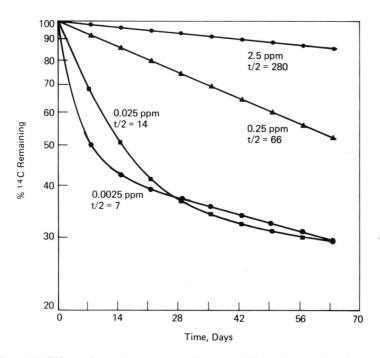

Fig. 6-6. Effect of starting concentration on soil degradation of picloram.

However, the shapes of the curves that would be expected if *only* Michaelis–Menten kinetics acted on degradation are different from the actual curves observed (see Fig. 6-7). At the higher concentrations, a curve that bends downwards would be expected (zero-order kinetics) but straight lines were apparently obtained (first-order kinetics). At the lower concentrations, straight lines were expected (first-order kinetics) but in fact curves that gradually decreased in slope were actually obtained.

The deviation from ideal can be explained at both low and high concentration by the impact of migration into the soil matrix. In both instances this migration continuously removed material from solution and reduced rates of degradation *below* what might be expected from Michaelis–Menten kinetics. The impact expected and observed is a bending of the degradation curve upwards so that at high concentrations there appears to be first-order kinetics, but at low concentrations the rate of degradation slows down more than might be expected from first-order kinetics.

The migration phenomenon has been discussed by Hamaker (Ref. 6-32) and modeled by Hamaker and Goring (Ref. 6-33) for the low-concentration case (first-order kinetics expected) (Fig. 6-8). No attempt has as yet been made to develop a model that applies over the entire range of Michaelis–Menten kinetics.

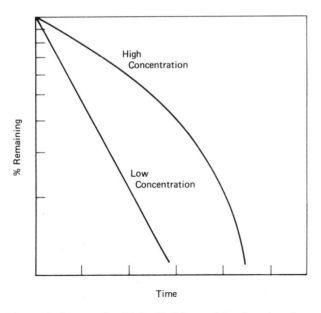

Fig. 6-7. Theoretical curves for Michaelis–Menten kinetics when data are plotted as semilog plots.

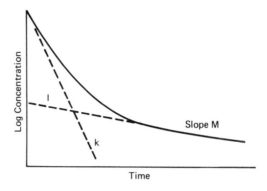

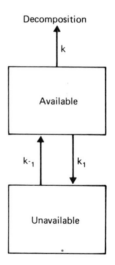

Fig. 6-8. Model for soil decomposition. *Source*: Hamaker, J. W., and Goring, C.A.I., "Turnover of Pesticide Residues in Soil," in D. D. Kaufman, G. G. Still, G. D. Paulson, and S. K. Vandal, (Eds.), *Bound and Conjugated Residues*, ACS Symp. Ser. 29, Amer. Chem. Soc., Washington, D. C., pp. 219-243, 1976.

Indeed, for practical purposes this would not ordinarily be necessary. For most chemicals the actual concentrations that occur in the soil water (taking into account initial sorption) are so low that degradation for the most part seems to follow first-order kinetics.

In Fig. 6-8 the two-compartment model developed by Hamaker and Goring (Ref. 6-33) is described by a series of first-order reaction rates. With the aid of a computer, values for each of the constants can be estimated by least-squares curve-fitting techniques. For cases where a computer is not available, Hamaker and Goring (Ref. 6-33) also describe an estimation method employing the constants shown in Fig. 6-8. These are the rate constant k from the initial rapid portion of the degradation curve, the slope m of the linear portion of the slow region of the curve, and the y intercept value of I, from the slow portion projected to time zero.

This brings to a close the discussion on degradation of chemicals in soil. We would like to leave the reader with the thought that these studies should be designed to provide average decomposition rates of parent, to identify the products formed during the decomposition process, and to characterize the manner in which soil, climatic, and concentration factors influence rates and breakdown products. If done properly, the information will allow the analysis of environmental exposure.

6-2.3 Data Interpretation

Field Monitoring. The most difficult part of environmental chemistry is the actual interpretation of data. It is easy to carry out elaborate experiments and generate numbers, but it is quite another matter to evaluate the results in terms of risk analysis. In the past, risk was assessed by conducting monitoring programs designed to demonstrate the fate of chemicals under field conditions. For example, it was learned through monitoring that DDT, dieldrin, PCB, and other chlorinated hydrocarbons are widely distributed in the environment and are capable of accumulating in the biotic compartment to levels that may be destructive to certain kinds of biota. The properties of these materials now serve as benchmarks for judging the potential behavior of other compounds. In contrast, there are other chemicals like 2, 4-D that have also been used extensively but have not concentrated in any of the environmental compartments. The properties of 2, 4-D have proved to be much more environmentally acceptable than those of the chlorinated hydrocarbons.

Comprehensive monitoring programs were initiated by the Federal government in 1963. These programs include the monitoring of residues in humans, air, water, soil, fish, estuarine organisms, and food from the market place (Refs. 6-13, 6-18-6-20, 6-22, 6-41, 6-67, 6-85, 6-87, 6-88). The information gathered is slowly becoming available. Data on soil residues has been published for the years 1969 (Ref. 6-86), 1970 (Ref. 6-16), 1971 (Ref. 6-11), and 1972 (Ref. 6-12). Water data has been published for the years 1964–1968 (Ref. 6-52), 1966–1968 (Ref. 6-54), and 1969–1971 (Ref. 6-76). Fish residue levels were reported in 1971 (Ref. 6-40) and again in 1978 (Ref. 6-8); human residue levels during fiscal years 1970-1974 were reported in 1977 (Ref. 6-47), total diet data for 1972–1973 in 1976 (Ref. 6-42), and residue levels in wild ducks during 1972–73 in 1976 (Ref. 6-84). In addition, Edwards (Ref. 6-21) has provided an excellent review summarizing the environmental distribution of such persistent chlorinated hydrocarbons as DDT.

The most striking observation from all of these field studies is that nature has a tremendous capacity for transporting and dissipating chemicals, especially if they happen to be stable. When given enough time and chemical stability, nature does distribute chemicals between each of her compartments, so that sooner or later the entire system becomes permeated with the chemical at some level of concentration. This level rapidly decreases in most compartments, approaching the lower limits of detectability with the latest of analytical techniques. The actual levels are dictated by input of chemical, rate of migration, partitioning between compartments, and stability within each environmental compartment.

Unfortunately the precipitousness of decline in field residues, often observed for even the DDT family of compounds, has served to allay the concerns of more than one researcher. In the case of DDT, sufficient passage of time and continued use of the chemical showed the falseness of this security, and made researchers realize that observed declines simply involved transport to other places. Such experiences with DDT and, more recently, PCBs provide compelling incentive for recognizing that all of the environmental compartments shown earlier in Fig. 6-1 are quite closely interconnected. They may not yet be in equilibrium because transport processes on a large scale require time for equilibration to take place. Eventually, however, the equilibrium will be approached.

Nevertheless, it is now quite clear that because equilibrium is not quickly achieved in the environmental system, fate analysis cannot be accomplished solely by monitoring. The monitoring task becomes very tedious and time consuming, requiring many experiments in time and location before recognizable patterns are obtained. The costliness and the frustrations arising from the monitoring process has in recent years stimulated the development of a more fundamental approach based on the measurement of the physical and chemical properties of chemicals. Information of this kind can be used in two ways. One is by comparison of the properties of older benchmark chemicals with those of new unknown compounds having no previous history of usage. The second method is by direct modeling of chemical behavior using computers. Monitoring can and will continue to play a role in the evaluation of chemicals, but now that there are the additional tools of benchmark comparison and computer modeling, the monitoring should primarily serve to verify the predictions of benchmark comparisons and computer modeling.

Benchmark Comparison. Data interpretation by benchmark comparison is accomplished by comparing a series of fundamental properties of chemicals whose environmental behavior is unknown with the standards whose environmental behavior is known. No assumptions are made with regard to the actual environment into which the chemical is placed. Nor is there any attempt to generate absolute values for exposure levels, soil or water contamination levels, vaporization losses, etc., since these numbers are constantly changing in the field situation, depending on usage pattern and environmental conditions. Instead, the method assumes that chemicals respond environmentally in direct proportion to the numerical values for their physical/chemical properties and that a relative ranking within a list does not change from one environment to another. In other words, those chemicals having the greatest tendency to act in a particular fashion in one environment will also do so in a totally different environment, even though the extent of the action may be completely different for the two environments.

This approach works in the following manner. The discussion centers around exposure that occurs on-site, and exposure resulting from movement away from point of introduction to off-site locations. The term "on-site" refers to those portions of the environment serving as targets for initial introduction of the chemical, and exposure in this part of the environmental system deals with rates of introduction and dissipation. Off-site exposure involves total production volume, migration away from the target site, and the direction of migration between the environmental compartments. Because this chapter is on terrestrial environments, we will assume that soil is the initial target. It will play a central role in the evaluation process. Soil does not have to occupy this position; one of the other environmental compartments could serve just as well, depending upon the needs and views of the investigator.

The first step in the analysis is to assemble information on key environmental properties for a series of chemicals (see Table 6-7). The values used in Table 6-7 were the best available but undoubtedly will be improved with time.

The second step is to combine these properties in ways that will quantify key environmental characteristics. To do this we call the reader's attention to the original compartmental concept of the environment shown earlier in Fig. 6-1. This model shows the four compartments—air, water, soil, and biota—to be in equilibrium with each other. These equilibria are represented by the constants K_{ow}, K_{oc}, and K_w. Since volatility has already been shown to be proportional to the amount of chemical in the soil air, as described by the expression $V_p/[(S)(K_{oc})]$ (Ref. 6-74), extension of this line of reasoning to include expressions for the chemical content of the water and soil phases lead to the following proportionalities:

$$\text{Amount in Water} \propto S/[(V_p)(K_{oc})] \qquad \text{(Eq. 6-14)}$$

$$\text{Amount in Soil} \propto [(K_{oc})(S)]/V_p \qquad \text{(Eq. 6-15)}$$

Using these combinations of properties and the following assumptions we then develop a benchmark system for ranking compounds:

1. Total movement within or from soil is inversely proportional to rate of decomposition in soil, i.e., given equal physical properties, stable chemicals migrate more than unstable ones.

TABLE 6-7. Environmental Properties for Selected Chemicals.

Chemical	Soil $t/2$, Days	V_P [a]	S [b]	K_{oc}	K_{ow}
Alachlor	7	2.2×10^{-5}	240	190	830
Atrazine	130	3.0×10^{-7}	33	150	480
Carbofuran	80	2.0×10^{-5}	420	10	40
Chlordane	3500	1.0×10^{-5}	0.056	38000	380000
Chlorpyrifos	60	1.9×10^{-5}	0.3	6100	98000
Diazinon	30	1.4×10^{-4}	40	580	3800
Dicamba	20	2.0×10^{-5}	4500	0.4	130
1, 3-D	10	2.5×10^{1}	2700	26	120
2, 4-D	4	6.0×10^{-7}	900	60	37
DDT	3800	1.9×10^{-7}	0.0017	240000	960000
Dieldrin	1000	4.9×10^{-6}	0.022	8400	680000
Dinoseb	30	5.0×10^{-5}	50	120	4900
Diuron	200	1.9×10^{-6}	42	150	94
Endrin	4300	2.0×10^{-7}	0.024	8100	220000
EPTC	7	2.0×10^{-2}	380	280	610
Heptachlor	2000	3.0×10^{-4}	0.030	24000	540000
Lindane	600	3.2×10^{-5}	0.15	1300	160000
Malathion	1	4.0×10^{-5}	150	930	780
Methyl Parathion	7	1.0×10^{-5}	57	9800	2700
Monuron	170	5.0×10^{-7}	230	80	29
Nitrapyrin	10	2.8×10^{-3}	40	560	2600
Parathion	15	3.8×10^{-5}	24	4800	6400
Phorate	14	8.4×10^{-4}	50	3200	3000
Picloram	180	6.8×10^{-7}	430	13	2
Propachlor	7	2.3×10^{-4}	610	420	560
Triclopyr	40	1.3×10^{-6}	430	27	3
Trifluralin	70	1.3×10^{-4}	0.6	14000	220000

[a] Vapor pressure in mm Hg @ 25°C whenever possible.
[b] Water solubility in ppm @ 25°C whenever possible.

2. Rate of movement by leaching through soil is directly proportional to the quantity of chemical in the water of an air/water/soil system (Eq. 6-14).
3. Rate of movement by volatilization from soil is directly proportional to the quantity of chemical in the air of an air/water/soil system (Eq. 6-10).
4. On-site exposure is directly proportional to bioconcentration, frequency of application, and level of application, but inversely proportional to rate of decomposition and amounts of chemical in the air and water of an air/water/soil system. The latter becomes directly proportional to amounts sorbed to soil solids, and assumes that migration away from the target area is proportional to all chemical not associated with soil.
5. Off-site exposure is inversely proportional to decomposition rates in soil, water, and air, and directly proportional to total production volume, bio-

concentration, and amounts of chemical in the air and water of an air/water/soil system. Amounts in the air and water are inversely proportional to amounts adsorbed to soil solids.

From these assumptions we can derive a series of mathematical equations that express the environmental behavior of chemicals on a relative basis. These equations are shown below.

$$\text{Leaching Potential} = S/[(V_p)(K_{oc})] \qquad \text{(Eq. 6-16)}$$

$$\text{Leaching Index} = [(S)(t/2(]/[(V_p)(K_{oc})] \qquad \text{(Eq. 6-17)}$$

$$\text{Volatility Potential} = V_p/[(S)(K_{oc})] \qquad \text{(Eq. 6-18)}$$

$$\text{Volatility Index} = [(V_p)(t/2)]/[(S)(K_{oc})] \qquad \text{(Eq. 6-19)}$$

$$\text{On-Site Exposure} = [(t/2)(K_{oc})(S)(K_{ow})(F)(R)]/V_p \qquad \text{(Eq. 6-20)}$$

$$\text{Off-Site Exposure} = [(t/2)(V_p)(K_{ow})(V)]/[(K_{oc})(S)] \qquad \text{(Eq. 6-21)}$$

In addition to the terms defined earlier for Eq. 6-10, $t/2$ represents soil degradation half-life at $25°C$ whenever possible, K_{ow} the octanol–water coefficient represents bioconcentration, F is the frequency of application, R is the concentration of chemical introduced, and V stands for volume of material manufactured. The distinction made between migration potential and migration index operates so that the latter always includes the effects of the degradation rate in addition to the rate of migration. Thus, "index" describes the overall extent of migration.

With these expressions it is now possible to calculate numerical values that can be used comparatively to express relative trends in volatilization, leaching, and exposure to the biota. The results of such calculations for the list of chemicals in Table 6-7 are shown ranked in descending order in Tables 6-8————6-10. Frequency of application, application rate, and total volume manufactured were assumed to be equal and therefore did not enter into the comparisons.

Chemicals with the greatest propensity for migration or exposure are found at the top of the lists in Tables 6-8 through 6-10, while those demonstrating the least occupy the bottom positions. Very quickly one can develop some impression of a chemical's potential environmental behavior from such lists and relate personal experiences with known chemicals to unknown materials of similar rankings.

This method of benchmark ranking might seem crude. One might even wonder whether it is a futile exercise in number manipulation, or whether there is real substance to such rankings. To provide at least a partial answer, actual field experience is considered, along with environmental monitoring information collected for the last 30 years.

The monitoring studies cited at the beginning of Section 6-2.3 show clearly that chlorinated hydrocarbons like DDT, dieldrin, chlordane, and endrin have permeated the biotic compartment of our environment to higher concentrations

TABLE 6-8. Relative Ranking of Leaching Through Soil.

Leaching Potential[a]		Leaching Index[b]	
Chemical	Value	Chemical	Value
Dicamba	5.6×10^8	Dicamba	1.1×10^{10}
Picloram	4.9×10^7	Picloram	8.8×10^9
2, 4-D	2.5×10^7	Monuron	9.8×10^8
Triclopyr	1.2×10^7	Triclopyr	4.9×10^8
Monuron	5.8×10^6	Carbofuran	1.7×10^8
Carbofuran	2.1×10^6	2, 4-D	1.0×10^8
Atrazine	7.3×10^5	Atrazine	9.5×10^7
Diuron	1.5×10^5	Diuron	2.9×10^7
Alachlor	5.7×10^4	Alachlor	4.0×10^5
Dinoseb	8.3×10^3	Dinoseb	2.5×10^5
Propachlor	6.3×10^3	Endrin	6.4×10^4
Malathion	4.0×10^3	Propachlor	4.4×10^4
Methyl Parathion	5.8×10^2	Diazinon	1.5×10^4
Diazinon	4.9×10^2	Methyl Parathion	4.1×10^3
Parathion	1.3×10^2	Malathion	4.0×10^3
EPTC	6.8×10^1	Lindane	2.2×10^3
Nitrapyrin	2.6×10^1	Parathion	2.0×10^3
Phorate	1.9×10^1	Chlordane	5.3×10^2
Endrin	1.5×10^1	Dieldrin	5.3×10^2
1, 3-D	4.2	EPTC	4.8×10^2
Lindane	3.6	Nitrapyrin	2.6×10^2
Chlorpyrifos	2.6	Phorate	2.6×10^2
Dieldrin	5.3×10^{-1}	Chlorpyrifos	1.6×10^2
Trifluralin	3.3×10^{-1}	DDT	1.4×10^2
Chlordane	1.5×10^{-1}	1, 3-D	4.2×10^1
DDT	3.7×10^{-2}	Trifluralin	2.3×10^1
Heptachlor	4.2×10^{-3}	Heptachlor	8.3

[a] From Eq. 6-16.

[b] From Eq. 6-17; includes effect of stability on migration.

than any of the other chemicals in the list. As these chemicals reside at the top of the rankings for both on-site and off-site exposure, this observation agrees well with the exposure assessments in Table 6-10. Similarly, the fumigant 1, 3-D at the bottom of the on-site exposure list is seldom if ever found in samples of biota taken from actual treated areas. Field experience indicates this material preferentially migrates into the air from soil; this is in keeping with the leaching and volatility rankings of Tables 6-8 and 6-9. Off-site exposure to 1, 3-D is estimated to be greater than on-site mainly because of the ready migration to the air. This might suggest a need for vapor phase decomposition studies with 1, 3-D so that the exposure rating can be readjusted for degradation in this compartment. In setting up the model for benchmark comparisons, air was treated

TABLE 6-9. Relative Ranking of Volatility from Soil.

Volatility Potential[a]		Volatility Index[b]	
Chemical	Value	Chemical	Value
1, 3-D	3.6×10^{-4}	1, 3-D	3.6×10^{-3}
Heptachlor	4.2×10^{-7}	Heptachlor	8.3×10^{-4}
EPTC	1.9×10^{-7}	Lindane	9.8×10^{-5}
Lindane	1.6×10^{-7}	Dieldrin	2.7×10^{-5}
Nitrapyrin	1.3×10^{-7}	Chlordane	1.6×10^{-5}
Dieldrin	2.7×10^{-8}	Endrin	4.4×10^{-6}
Trifluralin	1.5×10^{-8}	DDT	1.8×10^{-6}
Dicamba	1.1×10^{-8}	EPTC	1.3×10^{-6}
Chlorpyrifos	1.0×10^{-8}	Nitrapyrin	1.3×10^{-6}
Dinoseb	8.3×10^{-9}	Trifluralin	1.1×10^{-6}
Diazinon	6.0×10^{-9}	Chlorpyrifos	6.2×10^{-7}
Phorate	5.3×10^{-9}	Carbofuran	3.8×10^{-7}
Carbofuran	4.8×10^{-9}	Dinoseb	2.5×10^{-7}
Chlordane	4.7×10^{-9}	Dicamba	2.2×10^{-7}
Endrin	1.0×10^{-9}	Diazinon	1.8×10^{-7}
Propachlor	9.0×10^{-10}	Phorate	7.4×10^{-8}
Alachlor	4.8×10^{-10}	Diuron	6.0×10^{-8}
DDT	4.7×10^{-10}	Picloram	2.2×10^{-8}
Parathion	3.3×10^{-10}	Atrazine	7.9×10^{-9}
Diuron	3.0×10^{-10}	Propachlor	6.3×10^{-9}
Malathion	2.9×10^{-10}	Parathion	4.9×10^{-9}
Picloram	1.2×10^{-10}	Monuron	4.6×10^{-9}
Triclopyr	1.1×10^{-10}	Triclopyr	4.5×10^{-9}
Atrazine	6.1×10^{-11}	Alachlor	3.4×10^{-9}
Monuron	2.7×10^{-11}	Malathion	2.9×10^{-10}
Methyl Parathion	1.8×10^{-11}	Methyl Parathion	1.3×10^{-10}
2, 4-D	1.1×10^{-11}	2, 4-D	4.4×10^{-11}

[a] From Eq. 6-18.

[b] From Eq. 6-19; includes effect of stability on migration.

only as a vehicle for transport and not as media in which degradation took place. Any degradation that does occur in this compartment would serve to reduce exposure proportionately. The same is true for water when it serves as the transport vehicle.

Although the extremes in behavior are readily discerned from existing field data, differences between compounds in the middle area of the rankings are not so readily apparent from field experience. This is because these compounds tend to be newer materials and have thus not been studied as widely in the field as have materials such as the chlorinated hydrocarbons.

There are other potential combinations of properties that might provide a more suitable ranking method than those described by Eqs. 6-16 through 6-21.

For instance, it is possible that off-site exposure correlates as well or better with the combination of production volume, soil stability, and K_{ow} given in Eq. 6-22.

$$\text{Off-site Exposure} = (t/2)(K_{ow})(V) \qquad \text{(Eq. 6-22)}$$

This expression takes into account the migration of materials caused by erosion (This was disregarded totally in Eq. 6-21.) but still maintains that ex-

TABLE 6-10. Relative Ranking of Exposure to a Series of Soil-Applied Chemicals[a].

On-Site[b]		Off-Site[c]	
Chemical	Value	Chemical	Value
DDT	7.8×10^{18}	Heptachlor	4.5×10^2
Endrin	9.2×10^{17}	Dieldrin	1.8×10^1
Chlordane	2.8×10^{17}	Lindane	1.6×10^1
Dieldrin	2.6×10^{16}	Chlordane	6.3
Heptachlor	2.6×10^{15}	DDT	1.7
Methyl Parathion	1.1×10^{15}	Endrin	9.7×10^{-1}
Atrazine	1.0×10^{15}	1, 3-D	4.3×10^{-1}
Trifluralin	1.0×10^{15}	Trifluralin	2.4×10^{-1}
Lindane	5.9×10^{14}	Chlorpyrifos	6.1×10^{-2}
Chlorpyrifos	5.7×10^{14}	Nitrapyrin	3.3×10^{-3}
Parathion	2.9×10^{14}	Dinoseb	1.2×10^{-3}
Monuron	1.8×10^{14}	EPTC	8.0×10^{-4}
Diuron	1.2×10^{13}	Diazinon	6.9×10^{-4}
Diazinon	1.9×10^{13}	Phorate	2.2×10^{-4}
Dinoseb	1.8×10^{13}	Parathion	3.2×10^{-5}
2, 4-D	1.6×10^{13}	Dicamba	2.9×10^{-5}
Alachlor	1.2×10^{13}	Carbofuran	1.5×10^{-5}
Phorate	8.0×10^{12}	Diuron	5.7×10^{-6}
Propachlor	4.4×10^{12}	Atrazine	3.8×10^{-6}
Picloram	3.0×10^{12}	Propachlor	3.5×10^{-6}
Malathion	2.7×10^{12}	Alachlor	2.8×10^{-6}
Triclopyr	1.1×10^{12}	Methyl Parathion	3.4×10^{-7}
Carbofuran	6.7×10^{11}	Malathion	2.2×10^{-7}
Dicamba	2.3×10^{11}	Monuron	1.3×10^{-7}
Nitrapyrin	2.1×10^{11}	Picloram	1.0×10^{-7}
EPTC	2.3×10^{10}	Triclopyr	1.3×10^{-8}
1, 3-D	3.4×10^6	2, 4-D	1.6×10^{-9}

[a] Exposure encompasses effects of compartmental partitioning, migration, stability and bioconcentration.

[b] From Eq. 6-20.

[c] From Eq. 6-21.

posure is a combination of migration, bioconcentration, and production volume. This is accomplished by allowing the soil half-life in Eq. 6-22 to stand for total migration potential through recognition that stable chemicals migrate more extensively than do unstable ones.

Even if Eqs. 6-16 through 6-21 are not flawless, they are clearly very useful for a benchmark approach to environmental impact assessment and do focus attention on the key environmental properties of chemicals. The biggest advantage of the benchmark method is the fact that only a few environmental parameters need be considered in the evaluation process because relative rather than absolute results are obtained and compared. Due to the heterogeneity of terrestrial environments and their totally unattenuated exposure to the whims of nature, the absolute fate and impact of a chemical is extremely difficult to predict. This hurdle can be sidestepped by using the benchmark method to estimate a chemical's probable behavior and effect on the environment.

Computer Modeling. The disadvantage of the benchmark approach is that it does not provide information on actual compartmental concentrations. Although it can provide estimates of a chemical's potential to become distributed in various compartments of the environment, much more precise descriptions of a chemical's concentration at any place in the environment at any time can be obtained with sophisticated computers. Using mathematical relationships developed from laboratory studies, computer modeling techniques can predict dissipation under field conditions.

Modeling the dissipation of a chemical in a terrestrial system requires specific information on rate processes and mechanisms. The following fundamental processes must be considered: the rates of decomposition and diffusion of the chemical in soil, and the rate of mass transport of the chemical in soil or water. The decomposition rate depends on various degradative mechanisms (microbial, hydrolytic, catalytic) that can be integrated together to give a single rate constant. The diffusion rate is a function of the soil properties (moisture, texture, organic carbon) and the physical properties of the chemical (molecular weight, vapor pressure, soil adsorption coefficient). The rate of mass transport depends on a combination of rate processes associated with the hydrology of the environment, i.e., percolation rate of water through the soil profile, rate of runoff, and rate of evapotranspiration.

These processes are all involved in dissipation from soil. In addition, the way in which the chemical is applied may suggest the need for more information. For surface-applied chemicals, the rate of vaporization from the surface, the rate of photodegradation on soil or plant surfaces, and wind erosion must also be incorporated into the model. It has also been demonstrated that with foliar applied chemicals, approximately 50–70% of the material can remain on the foliage. Therefore, a transfer rate from foliage to soil and the surrounding atmosphere must be estimated.

Finally, climatic conditions (temperature, rainfall, snowfall, wind, solar radiation) must be measured or estimated in order to take these factors into account in the rate expressions of the model. For example, degradation of the chemical in soil is influenced by temperature and moisture. Walker's successful simulations of persistence in the field, as described earlier, comes to mind in this instance.

There have been several models developed to describe the fate of chemicals in the environment. One in particular, the Agricultural Runoff Management Model (Ref. 6-17) has taken a comprehensive approach to modeling agricultural watersheds. At present, it represents the best effort of modeling a terrestrial system.

The ARM Model simulates the hydrologic, sediment production, pesticide, and nutrient processes on the land surface and in the soil profile that determine the quantity and quality of agricultural runoff. Preliminary indications are that the model can adequately represent the major processes affecting agricultural runoff and thereby provide a useful tool for evaluation of the fate of chemicals applied to terrestrial systems. The model incorporates the rate processes previously discussed and has been successful in simulation of runoff and sediment. More work is needed in the simulation of pesticide dissipation; predictive pesticide degradation and attenuation models are required. Although the ARM model focuses on runoff, the parameter inputs generally describe all the important processes which pertain to non-point source-applied chemical in the terrestrial environment. Thus this type of model has the potential to be altered to emphasize chemical concentration in any part of the terrestrial environment.

One serious problem with this type of model is that it requires extensive input of meteorological data in order to compare simulated results with actual field data. Supplying actual meteorological data can be a time-consuming and expensive proposition. For certain chemicals of high interest the effort might be worthwhile; however, for the evaluation of new chemicals or widely used chemicals the model is too costly and too inefficient.

Another problem with models of this type is that they rest on the assumption that simple kinetic processes apply to various degradative mechanisms. This is not always the case, especially with breakdown in soil. First-order kinetics are generally able to describe fairly well the breakdown of chemical for the first half-life but often the last 10–20% degrades at a substantially slower rate. This phenomenon has been observed in many laboratory studies and is generally thought to be caused by migration of the chemical into the solid soil matrix. Regardless of the mechanisms, it will be difficult to model until the process is better understood.

Moreover, we do know that there can be a substantial loss of many chemicals through volatilization. Further clarification of the mechanisms regulating loss by volatilization is needed. Currently, soil moisture, percent organic carbon, wind speed, temperature, the physical properties of the chemical, and other parameters listed in Table 6-5 are known to affect the rate of loss. However, until more background research is completed, volatilization will be a difficult process to model.

Therefore, although the type of modeling exemplified by the ARM model has great potential, much work is needed before chemical dissipation can be predicted with an accuracy that would be considered acceptable.

Perhaps a better approach to evaluating exposure potential to chemicals in terrestrial systems would be to combine benchmark and modeling. If the basic philosophy of the benchmark concept, ranking and comparing chemicals based on their physical and chemical properties, is put into a modeling framework, then a useful tool for predicting exposure could evolve. This would require the determination of key laboratory parameters on a sufficient number of chemicals, combination of these parameters in order to relate various environmental processes to each other, and then prediction of exposure in different compartments of the environment for specific environmental situations. We envision that a series of specific environmental situations could be constructed and used for general environmental modeling of all chemicals. Such an approach would elevate modeling from its current role as an expression of monitoring results to its future role as predictor of environmental behavior, impact, and ultimate fate.

REFERENCES

6-1. Alexander, M., *Introduction to Soil Microbiology*, John Wiley & Sons Inc., New York, 1961.

6-2. Bartha, R., "Altered Propanil Biodegradation in Temporarily Air-Dried Soil," *J. Ag. Food Chem.*, **19**, pp. 394–395, 1971.

6-3. Bartha, R. and Pramer, D., "Features of a Flask and Method for Measuring the Persistence and Biological Effects of Pesticides in Soil," *Soil Sci.*, **100**, pp. 68–70, 1965.

6-4. Bartholomew, W. V., and Broadbent, F. E., "Apparatus for Control of Moisture, Temperature, and Air Composition in Microbiological Respiration Experiments," *Soil Sci. Soc. Amer. Proc.*, **14**, pp. 156–160, 1949.

6-5. Black, C. A., Evans, D. D., White, J. L., Ensminger, L. E., and Clark, F. E. (Eds.), *Methods of Soil Analysis*, 2 Vols., Amer. Soc. Agron., Madison, WI, 1965.

6-6. Briggs, G. G., "A Simple Relationship Between Soil Adsorption of Organic Chemicals and Their Octanol/Water Partition Coefficients," Proc. 7th British Insect. Fung. Conf., pp. 83–85, 1973.

6-7. Buckman, H. O., and Brady, N. C., *The Nature and Properties of Soils*, 6th ed., MacMillan Co., New York, 1960.

6-8. Butler, P. A., and Schutzmann, R. L., "Residues of Pesticides and PCBs in Estuarine Fish, 1972-76–National Pesticide Monitoring Program," *Pest. Monit. J.,* **12**, pp. 51–59, 1978.

6-9. Campbell, C. A., Biederbeck, V. O., and Warder, F. G., "Influence of Simulated Fall and Spring Conditions on the Soil System: III. Effect of Method of Simulating Spring Temperatures on Ammonification, Nitrification, and Microbial Populations," *Soil Sci. Soc. Amer. Proc.,* 37, pp. 382–386, 1973.

6-10. Call, F., "The Mechanism of Sorption of Ethylene Dibromide on Moist Soils," *Jour. Sci. Food Agric.,* 8, pp. 630–639, 1957.

6-11. Carey, A. E., Gowen, J. A., Tai, H., Mitchell, W. G., and Wiersma, G. B., "Pesticide Residue Levels in Soils and Crops, 1971–National Soils Monitoring Program (III)," *Pest. Monit. J.,* 12, pp. 117–136, 1978.

6-12. Carey, A. E., Gowen, J. A., Tai, H., Mitchell, W. G., and Wiersma, G. B., "Pesticide Residue Levels in Soils and Crops from 37 States, 1972–National Soils Monitoring Program (IV)," *Pest. Monit. J.,* 12, pp. 209–229, 1979.

6-13. Carver, T. C., "Estuarine Monitoring Program," *Pest. Monit. J.,* 5, p. 53, 1971.

6-14. Chiou, C. T., Freed, V. H., Schmedding, D. W., and Kohnert, R. L. "Partition Coefficient and Bioaccumulation of Selected Organic Chemicals," *Environ. Sci. Technol.,* 11, pp. 475–478, 1977.

6-15. Cornfield, A. H. "A Simple Technique for Determining Mineralization of Carbon During Incubation of Soils Treated with Organic Materials," *Plant Soil,* 14, pp. 90–93, 1961.

6-16. Crockett, A. B., Wiersma, G. B., Tai, H., Mitchell, W. G., Sand, P. F., and Carey, A. E., "Pesticide Residue Levels in Soils and Crops, FY-70–National Soils Monitoring Program (II)," *Pest. Monit. J.,* 8, pp. 69–97, 1974.

6-17. Donigian, A. S., Jr., Beyerlein, D. C., Davis, H. H., Jr., and Crawford, N. H., "Agricultural Runoff Management (ARM) Model Version II: Refinement and Testing," *EPA Ecological Research Series,* No. EPA-600/3-77-098, 1977.

6-18. Duggan, R. E. and Cook, H. R., "National Food and Feed Monitoring Program," *Pest. Monit. J.,* 5, pp. 37–43, 1971.

6-19. Duggan, R. E., Lipscomb, G. Q., Cox, E. L., Heatwole, R. E., and Kling, R. C., "Pesticide Residue Levels in Foods in the United States from July 1 1963 to June 30, 1969," *Pest. Monit. J.,* 5, pp. 73–212, 1971.

6-20. Dustman, E. H., Martin, W. E., Heath, R. G., and Reichel, W. L., "Monitoring Pesticides in Wildlife," *Pest. Monit. J.,* 5, pp. 50–52, 1971.

6-21. Edwards, C. A., *"Persistent Pesticides in the Environment,"* 2nd ed., CRC Press, Cleveland, OH, 1973.

6-22. Feltz, H. R., Sayers, W. T., and Nicholson, H. P., "National Monitoring Program for the Assessment of Pesticide Residues in Water," *Pest. Monit. J.,* 5, pp. 54–62, 1971.

6-23. Getzin, L. W., "Persistence of Methidathion in Soils," *Bull. Environ. Cont. Tox.,* 5, pp. 104–110, 1970.

6-24. Getzin, L. W., "Persistence and Degradation of Carbofuran in Soil," *Environ. Entomol.,* 2, pp. 461–467, 1973.

6-25. Getzin, L. W., and Rosefield, I., "Persistence of Diazinon and Zinophos in Soils," *J. Econ. Ent.,* 59, pp. 512–516, 1966.

6-26. Goring, C. A. I., "Physical Aspects of Soil in Relation to the Action of Soil Fungicides," *Ann. Rev. Phytopathol.,* 5, pp. 285–318, 1967.

6-27. Goring, C. A. I., "Agricultural Chemicals in the Environment: A Quantitative Viewpoint," In Goring, C. A. I., and Hamaker, J. W., (Eds.), *Organic Chemicals in the Soil Environment,* Vol. 2 Marcel Dekker, Inc., New York, pp. 793–863, 1972.

6-28. Goring, C. A. I., Griffith, J. D., O'Melia, F. C., Scott, H. H., and Youngson, C. R., "The Effect of TORDON on Microorganisms and Soil Biological Processes," *Down To Earth,* 22, pp. 14–17, 1967.

6-29. Goring, C. A. I., and Hamaker, J. W. (Eds.), *Organic Chemicals in the Soil Environment,* 2 Vols., Marcel Dekker, Inc., New York, 1972.

6-30. Goring, C. A. I., Laskowski, D. A., Hamaker, J. W., and Meikle, R. W., "Principles of Pesticide Degradation in Soil," in *Environmental Dynamics of Pesticides,* R. Haque and V. H. Freed (Eds.), Plenum Press, New York, pp. 135–172, 1975.

6-31. Guenzi, W. D. (Ed.), *Pesticides in Soil and Water,* Soil Sci. Soc. Am. Inc., Madison, WI, 1974.

6-32. Hamaker, J. W., "Decomposition: Quantitative Aspects," in *Organic Chemicals in the Soil Environment,* Vol. 1 C. A. I. Goring and J. W. Hamaker (Eds.), Marcel Dekker, Inc., New York, pp. 253–340, 1972.

6-33. Hamaker, J. W., and Goring, C. A. I., "Turnover of Pesticide Residues in Soil," in *Bound and Conjugated Residues,* ACS Symp. Ser. 29, D. D. Kaufman, G. G. Still, G. D. Paulson, and S. K. Bandal (Eds.), Amer. Chem. Soc., Washington, D. C., pp. 219–243, 1976.

6-34. Hamaker, J. W., and Thompson, J. M., "Adsorption," in Goring, Ref. 6-29, pp. 49–143.

6-35. Hamaker, J. W., Youngson, C. R., and Goring, C. A. I., "Rate of Detoxification of 4-Amino-3, 5, 6-Trichloropicolinic Acid in Soil," *Weed Research,* 8, pp. 46–57, 1968.

6-36. Hance, R. J., "Influence of Sorption on the Decomposition of Pesticides," in *Sorption and Transport Processes in Soils,* SCI Monographs 37, Soc. Chem. Ind., London, pp. 92–104, 1970.

6-37. Hartley, G. S., "Physiochemical Aspects of the Availability of Herbicides in Soil," in *Herbicides and the Soil,* E. K. Woodford and G. R. Sagar (Eds.), Blackwell, Oxford, pp. 63–78, 1960.

6-38. Hartley, G. S., "Evaporation of Pesticides," in *Pesticidal Formulations Research, Physical and Colloidal Chemical Aspects,* Adv. Chem. Ser. 86, pp. 115–134, 1969.

6-39. Harvey, J., Jr. and Pease, H. L., "Decomposition of Methomyl in Soil," *J. Agric. Food Chem.,* 21, pp. 784–786, 1973.

6-40. Henderson, C., Inglis, A., and Johnson, W. L., "Organochlorine Insecticide Residues in Fish–Fall 1969 National Pesticide Monitoring Program," *Pest. Monit. J.,* 5, pp. 1–11, 1971.

6-41. Inglis, A., Henderson, C., and Johnson, W. L., "Expanded Program for Pesticide Monitoring of Fish," *Pest. Monit. J.,* 5, pp. 47–49, 1971.

6-42. Johnson, R. D., and Manske, D. D., "Pesticide Residues in Total Diet Samples (IX)," *Pest. Monit. J.,* 9, pp. 157–169, 1976.

6-43. Kearney, P. C., and Kaufman, D. D. (Eds.), *Herbicides, Second Edition,* 2 Vols., Marcel Dekker, Inc., New York, 1975.

6-44. Kearney, P. C., and Kontson, A., "A Simple System to Simultaneously Measure Volatilization and Metabolism of Pesticides from Soils," *J. Agric. Food Chem.,* **24,** pp. 424–426, 1976.

6-45. Kenaga, E. E., and Goring, C. A. I., "Relationship Between Water Solubility, Soil Sorption, Octanol-Water Partitioning, and Bioconcentration of Chemicals in Biota," ASTM Third Aquatic Toxicology Symposium, New Orleans, Oct. 15–20, 1978.

6-46. Konrad, J. G., and Chesters, G., "Degradation in Soils of Ciodrin, an Organophosphate Insecticide," *J. Agric. Food Chem.,* 17, pp. 226–230, 1969.

6-47. Kutz, F. W., Yobs, A. R., Strassman, S. C., and Viar, J. F., Jr., "Effects of Reducing DDT Usage on Total DDT Storage in Humans," *Pest. Monit. J.,* 11, pp. 61–63, 1977.

6-48. Laskowski, D. A., and Regoli, A. J., "Influence of the Environment on N-SERVE Stability," *Agron, Abstracts,* p. 97, 1972.

6-49. Laskowski, D. A., Swann, R. L., and Regoli, A. J., *"Incubation System for Studying Soil Degradation of Chemicals,"* Presentation Am. Chem. Soc. National Meeting, Anaheim, California, 1978, unpublished.

6-50. Lay, M. M., and Ilnicki, R. D., "Peroxidase Activity and Propanil Degradation in Soil, *Weed Research,* 14, pp. 111–113, 1974.

6-51. Lay, M. M. and Ilnicki, R. D., "Effect of Soil Storage on Propanil Degradation," *Weed Research,* 15, pp. 63–66, 1975.

6-52. Lichtenberg, J. J., Eichelberger, J. W., Dressman, R. C., and Longbottom, J. E., "Pesticides in Surface Waters of the United States–a 5-Year Summary, 1964-1968," *Pest. Monit. J.,* 4, pp. 71–86, 1970.

6-53. Lode, O., "Decomposition of Linuron in Different Soils," *Weed Research,* 7, pp. 185–190, 1967.

6-54. Manigold, D. B., and Schulze, J. A., "Pesticides in Selected Western Streams–A Progress Report," *Pest. Monit. J.,* 3, pp. 124–135, 1969.

6-55. Mackay, D., and Cohen, Y., "Prediction of Volatilization Rate of Pollutants in Aqueous Systems," in *Symposium on Non-Biological Transport and Transformation of Pollutants on Land and Water: Processes and Critical Data Required for Predictive Description*, National Bureau of Standards, Gaithersburg, Maryland, 1976.

6-56. Mackay, D., and Leinonen, P. J., "Rate of Evaporation of Low-Solubility Contaminants from Water Bodies to Atmosphere," *Env. Sci. Technol.,* 9, pp. 1178–1180, 1975.

6-57. Mackay, D., and Wolkoff, A. W., "Rate of Evaporation of Low-Solubility Contaminants from Water Bodies to Atmosphere," *Environ. Sci. Technol.,* 7, pp. 611–614, 1973.

6-58. McCall, P. J., and Swann, R. L., "Nitrapyrin Volatility from Soil," *Down To Earth,* 34, pp. 21–27, 1978.

6-59. McLaren, A. D., and Peterson, G. H. (Eds.), *Soil Biochemistry,* Marcel Dekker, Inc., New York, Vol. 1, 1967.

6-60. McLaren, A. D., and Skujinš, J. (Eds.), *Soil Biochemistry,* Marcel Dekker, Inc., New York, Vol. 2, 1971.

6-61. Osgerby, J. M., "Sorption of Unionized Pesticides by Soils," in *Sorption and Transport Processes in Soils,* SCI Monographs 37, Soc. Chem. Ind., London, pp. 92–104, 1970.

6-62. Paul, E. A., and McLaren, A. D. (Eds.), *Soil Biochemistry*, Marcel Dekker, Inc., New York, Vol. 3, 1975.

6-63. Plimmer, J. R., "Volatility," in *Herbicides*, 2nd Ed., P. C. Kearney and D. D. Kaufman (Eds.), New York, Marcel Dekker, Inc., pp. 891–934, Vol. 2, 1976.

6-64. Pramer, D., and Bartha, R., "Preparation and Processing of Soil Samples for Biodegradation Studies," *Environ. Letters*, 2, pp. 217–224, 1972.

6-65. Richey, F. A., Jr., Bartley, W. J., and Sheets, K. P., "Laboratory Studies on the Degradation of (the Pesticide) Aldicarb in Soils," *J. Agric. Food Chem.*, 25, pp. 47–51, 1977.

6-66. Rueppel, M. L., Brightwell, B. B., Schaefer, J., and Marvel, J. T., "Metabolism and Degradation of Glyphosate in Soil and Water," *J. Agric. Food Chem.*, 25, pp. 517–528, 1977.

6-67. Schechter, M. S., "Revised Chemicals Monitoring Guide for the National Pesticide Monitoring Program," *Pest. Monit. J.*, 5, pp. 68–71, 1971.

6-68. Smith, A. E., and Walker, A., "A Quantitative Study of Asulam Persistence in Soil," *Pestic. Sci.*, 8, pp. 449–456, 1977.

6-69. Spencer, W. F., and Cliath, M. M., "Vapor Density of Dieldrin," *Environ. Sci. Technol.*, 3, pp. 670–674, 1969.

6-70. Spencer, W. F., and Cliath, M. M., "Desorption of Lindane from Soil as Related to Vapor Density," *Soil Sci. Soc. Am. Proc.*, 34, pp. 574–578, 1970.

6-71. Spencer, W. F., and Cliath, M. M., "Vaporization of Chemicals," in *Environmental Dyanmics of Pesticides*, R. Haque and V. H. Freed (Eds.), Plenum Press, New York, pp. 61–78, 1975.

6-72. Spencer, W. F., Farmer, W. J., and Cliath, M. M., "Pesticide Volatilization," *Residue Reviews*, 49, pp. 1–47, 1973.

6-73. Stotzky, G., "Microbial Respiration," in *Methods of Soil Analysis*, C. A. Black, D. D. Evans, J. L. White, L. E. Ensminger, and F. E. Clark (Eds.), Am. Soc. Agron., Madison, WI, Vol. 2, pp. 1550–1572, 1965.

6-74. Swann, R. L., McCall, P. J., and Unger, S. M., *"Volatility of Pesticides From Soil Surfaces,"* 177th National Meeting Am. Chem. Soc., 1979.

6-75. Tiedje, J. M., and Mason, B. B., "Biodegradation of Nitrilotriacetate (NTA) in Soils," *Soil Sci. Soc. Am. Proc.*, 38, pp. 278–283, 1974.

6-76. Truhlar, J. F., and Reed, L. A., "Occurrence of Pesticide Residues in Four Streams Draining Different Land-Use Areas in Pennsylvania, 1969–71," *Pest. Monit. J.*, 10, pp. 101–110, 1976.

6-77. Walker, A., "Use of a Simulation Model to Predict Herbicide Persistence in the Field," *Proc. Eur. Weed Res. Coun. Symp. Herbicides-Soil*, pp. 240–250, 1973.

6-78. Walker, A., "A Simulation Model for Prediction of Herbicide Persistence," *J. Environ. Quality*, 3, pp. 396–401, 1974.

6-79. Walker, A., "Simulation of Herbicide Persistence in Soil I. Simazine and Prometryne," *Pestic. Sci.*, 7, pp. 41–49, 1976.

6-80. Walker, A., "Simulation of Herbicide Persistence in Soil II. Simazine and Linuron in Long-Term Experiments," *Pestic. Sci.*, 7, pp. 50–58, 1976.

6-81. Walker, A., "Simulation of Herbicide Persistence in Soil III. Propyzamide in Different Soil Types," *Pestic. Sci.,* 7, pp. 59–64, 1976.

6-82. Walker, A., "Simulation of the Persistence of Eight Soil-Applied Herbicides," *Weed Research,* 18, pp. 305–313, 1978.

6-83. Weber, J. B., "Interaction of Organic Pesticides with Particulate Matter in Aquatic and Soil Systems," in *Fate of Organic Pesticides in the Aquatic Environment,* Adv. in Chem. Series III, Washington, D. C., Am. Chem. Soc., pp. 55–120, 1972.

6-84. White, D. H., and Heath, R. G., "Nationwide Residues of Organochlorines in Wings of Adult Mallards and Black Ducks, 1972–73," *Pest. Monit. J.,* 9, pp. 176–185, 1976.

6-85. Wiersma, G. B., Sand, P. F., and Cox, E. L., "A Sampling Design to Determine Pesticide Residue Levels in Soils of the Conterminous United States," *Pest. Monit. J.,* 5, pp. 63–66, 1971.

6-86. Wiersma, G. B., Tai, H., and Sand, P. F., "Pesticide Residue Levels in Soils, FY 1969– National Soils Monitoring Program," *Pest. Monit. J.,* 6, pp. 194–228, 1972.

6-87. Yobs, A. R., "The National Human Monitoring Program for Pesticides," *Pest. Monit. J.,* 5, pp. 44–46, 1971.

6-88. Yobs, A. R., "National Monitoring Program for Air," *Pest. Monit. J.,* 5, p. 67, 1971.

7

Mathematical modeling for prediction of chemical fate

Shonh S. Lee*

SRI International**
Menlo Park, California

7-1 INTRODUCTION

There is a continuing need for accurate predictions of the environmental impacts of materials before decisions can be made regarding their production, use, and disposal. Generally, the earlier an assessment is obtained, the less the expense. Mathematical models are extremely attractive in this regard, for they offer relatively rapid and cheap assessments of potential hazards in the early stages of product or process development. Although no one model is likely to meet the needs of regulators, producers, and academicians, reliable models at least offer explicit sets of assumptions and predictions that are useful as bases for negotiation.

This chapter presents an overview of concepts pertinent to modeling the transport and transformation of specific chemicals

---

*With editorial review by Dr. Buford R. Holt of the U. S. Geological Survey, Palo Alto, California.

**Current Address: 267 Pulido Road, Danville, CA 94526.

in aquatic environments. Similar models can be prepared for the terrestrial and atmospheric environments (see Chapters 5 and 6). Concepts are introduced and discussed in the context of a deterministic model that includes descriptions of individual transformation rates and transport mechanisms. There is reference made to sections in eleven other chapters of this book that pertain to modeling.

7-2 EMISSION AND ENVIRONMENTAL FACTORS

Emission Sources and Rates. The source of the chemical first needs to be addressed. Two types of emission sources are generally considered: point and nonpoint sources. Definitions are ambiguous because a nonpoint source may become a point source as the area to be modeled becomes very large. Often, discharges from industrial plants or municipal wastewater handling facilities are treated as point sources, while surface runoffs from agricultural land and urban areas are nonpoint or diffuse sources.

The emission rate can be determined from the concentration in the discharge water and the water discharge rate using the formula:

$$E_i = C_i \times Q \qquad \text{(Eq. 7-1)}$$

where E_i and C_i are the emission rate and its concentration of chemical i, respectively, and Q is the carriage–water discharge rate. However, concentrations of specific chemicals are often unavailable, making it necessary to use indirect estimates of emission rate. Equation 7-1 is then replaced by

$$E_i = e_i P_i \qquad \text{(Eq. 7-2)}$$

where e_i is the emission factor, expressed in kg per kg of product, and P_i is the production rate of the chemical i, respectively. Reported emission factors for point sources can be found in Ref. 3-1, but should be used cautiously because emission factors usually vary with production processes and are often difficult to estimate.

Emissions from nonpoint sources are estimated differently, usually from models that allow for variations in soil characteristics, potency factors, and the manner in which the chemicals are applied (Refs. 3-11, 3-12). Section 3-5 presents a sophisticated model for this purpose.

Hydrodynamics and Hydrology. The flow pattern in the studied environment is important for projection of distribution of the chemical. This flow pattern can be described by the equation of motion, which in verbal form is

$$\left\{ \begin{array}{c} rate\ of \\ momentum \\ accumulation \end{array} \right\} = \left\{ \begin{array}{c} rate\ of \\ momentum \\ in \end{array} \right\} - \left\{ \begin{array}{c} rate\ of \\ momentum \\ out \end{array} \right\} + \left\{ \begin{array}{c} sum\ of \\ forces\ acting \\ on\ system \end{array} \right\} \qquad \text{(Eq. 7-3)}$$

Expressed in vector form, it is possible to solve this equation either analytically or numerically for simple cases in which tributaries can be considered as point sources. There, the velocity will be estimated to be the flowrate divided by the cross-sectional area. However, if a large tributary is studied, detailed velocity profiles are needed for hydrological data and topographic information of the tributary. Thus, solution of the equation then becomes difficult even though, in principle, with sufficient initial and boundary conditions, the vector form of the equation can be solved by either the finite difference or finite element method.

An alternative is to focus on the conservation of mass, using the equation of continuity, which can be used to describe transport in the studied environment when it is written as

$$\frac{\partial L}{\partial t} + \bar{\mu}\,\frac{\partial L}{\partial X} = (D_m + e)\,\frac{\partial^2 L}{\partial X^2} \qquad \text{(Eq. 7-4)}$$

where $\bar{\mu}$ is the velocity vector, D_m is molecular diffusivity, and e is eddy diffusivity. When advection becomes important, as in the case of a fast flowing environment, the right-hand part of Eq. 7-4 can be deleted. In contrast, in a stagnant impoundment, the second term in the left-hand side of equation can be omitted. To solve the above equation, $\bar{\mu}$ information is calculated from Eq. 7-3.

Environmental Factors. The receiving environments to be modeled often are referenced to a watershed, river, or lake, each of which has geographic and hydrologic characteristics profoundly affecting the persistence of chemicals in the environment. Patterns of insolation and seasonability of mixing, for example, markedly vary in the environment. Additionally, the quality of the aquatic environment is important to the fate of chemicals. For example, pH greatly affects the hydrolysis rate of chemicals and the sediment content strongly influences chemical deposition. Generally, water quality factors of greatest interest in an evaluation of the persistence of chemicals are temperature, pH, transparency, alkalinity, salinity, microbial population and physiology, and sediment load.

Water temperature varies the chemical reaction rates according to the Van Hoff Law, where the reaction rate will be double as temperatures rise every $10^{\circ}C$. Geographical and hydrological properties of the environment generally govern the physical transport of the less volatile chemicals. A shallow and wide impoundment, for example, favors evaporative loss and uniform vertical distribution of chemicals, because of high ratio of vertical mixing.

7-3 TRANSPORT AND TRANSFORMATIONS

Most chemicals continuously undergo chemical, physical, and biological transformations after having entered the environment. Through these transformations, most chemicals eventually degrade to innocuous materials (e.g., carbon dioxide, water, methane), although occasionally the derivatives may have a higher toxicity (e.g., nitrosoamines). These transformations are governed by their reaction kinetics and their thermodynamic properties. The principal processes of concern are hydrolysis, oxidation, and photolysis.

As described by Lassiter, Baughman, and Burns (Ref. 7-18), the dynamics of a trace organic pollutant in the environment can be represented as a sum of contributing processes:

$$-\frac{dS}{dt} = V + I_D + I_S + R_h + A + M \pm Se + D - L \qquad \text{(Eq. 7-5)}$$

where S is the quantity or concentration of the pollutant in the system; the processes are volatilization (V), direct photolysis (I_D), sensitized photolysis (I_S), hydrolysis (R_h), breakdown by photoautotrophs (A), microbial degradation (M), exchanges with sediment reservoirs (Se), dilution (D), and loadings of the pollutant into the system (L).

Hydrolysis. Hydrolysis of chemical compounds usually results in introduction of a hydroxyl function (-OH) into the chemicals. The kinetics of hydrolysis can be expressed as

$$R_h = k_h[S] = k_B[OH^-][S] + k_A[H^+][S] + k_N[H_2O][S] \qquad \text{(Eq. 7-6a)}$$

where k_h is the measured first-order rate constant at a given pH. Equation 7-6a can be rewritten as

$$R_h = k_h[S] = (k_B[OH^-] + k_A[H^+] + k_N'[H_2O])[S] \qquad \text{(Eq. 7-6b)}$$

It is clear that when k_B and/or $k_A \neq 0$, k_h will depend on pH. From the auto-protolysis equilibrium,

$$[H^+][OH^-] = K_w \simeq 10^{-14} \qquad \text{(Eq. 7-7)}$$

Equation 7-6b may be rewritten

$$k_h = \frac{k_B K_w}{[H^+]} + k_A[H^+] + k_N'[H_2O] \qquad \text{(Eq. 7-8)}$$

The contribution of each term to k_h will depend on the acidity (or pH) of the solution. Three regions may be defined.

$$\text{Acid}\left(\text{i.e., } k_A[H^+] > k_N + \frac{k_B K_w}{[H^+]}\right): \quad \log k_h = \log k_A + \log[H^+] \qquad \text{(Eq. 7-9)}$$
$$= \log k_A - \text{pH}$$

$$\text{Base}\left(\text{i.e., } \frac{k_B K_w}{[H^+]} > k_N + k_A[H^+]\right): \quad \log k_h = \log k_B K_w - \log[H^+] \qquad \text{(Eq. 7-10)}$$
$$= \log k_B K_w + \text{pH}$$

$$\text{Neutral}\left(\text{i.e., } k_N > k_A \, [H^+] + \frac{k_B K_W}{[H^+]}\right): \quad \log k_h = \log k_N \qquad \text{(Eq. 7-11)}$$

The dependence of k_h on the pH of the solution is conveniently shown by a plot of $\log k_h$ as a function of pH (Fig. 7-1). Equations 7-9 to 7-11 show that in the pH range where the base-catalyzed process is dominant, a slope of +1 is found; a slope of –1 is found in the acid-catalyzed region. The neutral hydrolysis is pH independent and shows a slope of zero.

Values of k_h obtained in the laboratory must be adjusted to environmentally relevant temperatures before they can be used in the model. The Arrhenius equation

$$k_h = A \exp(-E/RT) \qquad \text{(Eq. 7-12)}$$

is used, where E is the activation energy and A is a constant.

Photolysis. Often in natural waters the transformation of chemical compounds occurs through direct photolysis by absorption of light at wavelengths above 290 nm or through photosensitized reactions. Although the kinetics and mechanisms of direct photolysis of chemical compounds can usually be evaluated using present theory and experience, the details of sensitized photolysis in which other organics in natural waters act as sensitizers are largely undefined.

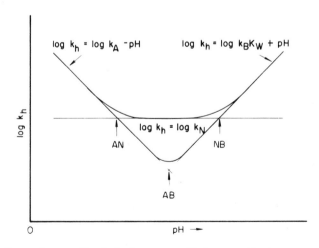

Fig. 7-1. Dependence of hydrolysis rate on pH. *Source: Environmental Pathways of Selected Chemicals in Freshwater Systems,* SRI International, U.S. Environmental Protection Agency Report EPA-600/7-78-074.

The rate of adsorption of light is dependent on the molar absorbability ϵ (also called the molar extinction coefficient), and light intensity, I_λ. This can be expressed as

$$I_A = \epsilon I_\lambda \, [S] = k_a \, [S] \qquad \text{(Eq. 7-13)}$$

where $k_a = \epsilon I_\lambda$. The rate of direct photolysis of a chemical (rate constant k_p) is then obtained by multiplying I_A by the quantum yield ϕ, which is the efficiency for converting the absorbed light into chemical reaction, measured as the ratio of moles of substrate transformed to einsteins of photons absorbed.

$$-\frac{dS}{dt} = k_a \phi \, [S] = K_p \, [S] \qquad \text{(Eq. 7-14)}$$

where K_p, equal to $k_a \phi$, is the photolysis rate constant. It can be estimated in two ways. Obviously, it depends upon light intensity, penetration, and scattering. Equation 7-14 is a first-order kinetic equation; the half-life for photolysis is given by

$$\left(t_{\frac{1}{2}} \right)_\text{p} = \frac{ln2}{k_p} \qquad \text{(Eq. 7-15)}$$

For model simulation, the average value of k_p for a full days' photolysis is usually used. When short half-times are being investigated, better projections in photolysis can be obtained by fluctuating k_p with the time of day.

However, in some aquatic environments, rates of photolysis may differ significantly from those measured in pure water, owing to the presence of naturally occurring light absorbers, quenchers, or sensitizers. In water, naturally occurring materials such as humic or fulvic acids, which have high optical densities, may absorb sunlight and effectively screen the chemical from being photolyzed. The presence of particulate materials in water may result in light scattering. In both cases, the photolysis rate of the chemical would be slower than in pure water because the physical processes reduce the light available for reaction.

These processes can occur through light absorption by a natural substance, which then interacts with the chemical. In the photosensitized reaction, the excited-state energy from the sensitizer is transferred to the chemical, which then undergoes reaction, the sensitizer maintaining its own identity. In the photoinitiated reaction, the natural substance that absorbed the light reacts with the chemical and both materials are transformed. If either process is more rapid than direct photolysis of the chemical, the rate of photolysis will be accelerated. However, in natural waters with significant optical densities, an acceleration of photolysis due to either mechanism may be somewhat offset by the screening capability of the water.

Because the presence of natural water can either accelerate or retard the photolysis of a chemical, half-lives based on pure water photolyses must be interpreted with some caution.   If experiments in natural waters give faster photolysis rates than in pure water, the photolysis rate in pure water can serve as a conservative value (i.e., maximum half-life).   When the photolyses in natural waters are slower than in pure water, the half-life estimate obtained for the pure water photolysis should be used with the qualification that in some cases longer half-lives may occur and that more experiments may be needed to determine how much slower the photolysis is likely to be.

Free Radical Oxidation. Oxidation of chemical compounds by free radical processes may be important under some environmental conditions.   The most general reaction scheme involves initiation, oxidation, and termination reactions. Oxidation process can be expressed as a pseudo-first-order kinetics

$$\gamma_{ox} = k'_{ox} \, [RO_2 \cdot] \, [S] = k_{ox} [S] \qquad \text{(Eq. 7-16)}$$

where k'_{ox} is a pseudo-first order rate constant and $[RO_2 \cdot]$ is peroxyl radical concentration in the environment.

For purposes of environmental assessment, $[RO_2 \cdot]$ in the aquatic environment is assumed to be approximately 10^{-9} M. This concentration combined with an experimental value of k_{ox} places a probable upper limit on the oxidation.

Volatilization. Volatilization, or movement of pollutants from water into the atmosphere, is an important consideration in determining the fate of many chemicals.   Volatilization is strongly influenced by air–water contact conditions and the properties and concentration of the chemical.   Compounds of low molecular weight, high vapor pressure, and low water solubility have been shown to volatilize rapidly.   However, some high-molecular-weight, low-solubility compounds such as DDT also volatilize at an environmentally significant rate because competing transport/transformation mechanisms are slow.

Many equations have been developed to estimate the volatilization rate.   The principal difficulties with these equations are either related to measurement of the required parameters or the fact that they bear little relation to real environments. Tsivoglou has made the important observation that the ratio of the volatilization rates of several low-molecular-weight gases from water is constant over a wide range of turbulence conditions (Refs. 7-15, 7-16).   Thus, for compounds A and B, the ratio

$$\frac{k_v^A}{k_v^B} = \text{constant} \qquad \text{(Eq. 7-17)}$$

It is convenient to choose the compound under test for A and oxygen for B. The law of microscopic reversibility requires that the rate of volatilization equal the rate of dissolution into the liquid for identical conditions. The term, oxygen reaeration rate, is commonly used to express the rate at which oxygen from the atmosphere dissolves in oxygen-deficient water. The oxygen reaeration rate is defined by

$$\frac{d[O_2]}{dt} = k_v^O ([O_2]_{sat} - [O_2])$$ (Eq. 7-18)

where $[O_2]$ is the oxygen concentration, $[O_2]_{sat}$ is the oxygen concentration when the water is saturated, and k_v^O is the oxygen reaeration rate constant. The oxygen reaeration rate has the additional advantage of having been measured for many different water bodies. Values for ponds, rivers, and lakes were taken in one study as 0.19, 0.96, and 0.24 day^{-1}, respectively (Ref. 7-13).

If the ratio of the substrate volatilization rate to the oxygen reaeration rate constant can be measured in the laboratory, it is possible to estimate the volatilization rate of the substrate in a real water body for which the oxygen reaeration rate is known:

$$(k_v^S)_{\text{water body}} = (k_v^O)_{\text{water body}} (k_v^S/k_v^O)_{\text{laboratory}}$$ (Eq. 7-19)

Hill et al. (Ref. 7-6) used this procedure successfully to estimate the volatilization rate constants for vinyl chloride.

Sorption. Sorption of organic substrates on or in sediments and biota can be a very important phenomenon in the aquatic environment. The sediment can act as a sink for sorbed materials, removing them from the water column. However, the sorbed chemicals are often released (desorbed) from the sediments at a later time. In this way, sorbed material can become a source of pollution.

Two types of equations are often used to describe sorption/desorption mechanisms—the Langmuir and Freundlich isotherms. First, S_w and S_s are defined as

$$S_w = \frac{\text{weight substrate in solution}}{\text{milliliters solution}}$$

$$S_s = \frac{\text{weight substrate sorbed}}{\text{grams sorbent}}$$

at equilibrium. The substrate weights must be in the same units (e.g., ng, μg). For a dilute aqueous solution, 1 ml of solution equals 1 g of solution, and

$$S_w = \frac{\text{weight substrate in solution}}{\text{grams solution}}$$

The Langmuir isotherm equation is defined as

$$S_s = \frac{abS_w}{1 + bS_w}$$ (Eq. 7-20)

where a and b are constants and

$$a = X_c$$

$$b = \frac{K_p}{X_c}$$

where X_c is the sorption capacity of the sorbent and K_p is a partition coefficient. Data for gas–solid sorption generally and data for organic substrates sorbed on clay minerals usually fit the Langmuir isotherm. However, natural sediments are not homogeneous—sorbed complex organic material such as humic substances is already present on the clay particles—and sorption by natural sediments usually fails to fit the Langmuir isotherm.

Data for sorption of multiple substrates from solution on nonuniform surfaces generally fit the Freundlich isotherm, which is defined as the following empirical equation:

$$S_s = KS_w^{1/n} \qquad \text{(Eq. 7-21)}$$

At low substrate concentrations, n is often very nearly equal to 1. If $n = 1$ and S_s and S_w are in the same units, K becomes a partition coefficient, as defined by

$$S_s = K_p S_w \qquad \text{(Eq. 7-22)}$$

In some cases, hystersis is observed in process of desorption. This phenomenon indicates that pollutants adsorbed onto sediments are not completely released back to the water column during the desorption process.

Biodegradation. Most organic chemicals can be decomposed by microbial metabolic mechanisms. Microorganisms are highly susceptible to frequent enzymatic reorientation in response to environmental change or alteration in substrate availability. Natural waters normally contain many types of microorganisms, depending on the water body, season, and organic substrates being introduced.

Generally, biodegrading systems are described by classical kinetic expressions in which the organic chemicals are the growth-rate-limiting carbon source. The Monod kinetic equations (Refs. 7-9, 7-14) describe the microbial growth rate as

$$\frac{dX}{dt} = \mu X \qquad \text{(Eq. 7-23)}$$

$$\mu = \frac{\mu_{\max} S}{K_s + S} \qquad \text{(Eq. 7-24)}$$

where S and X are the concentrations of substrates and biomass per unit volume, respectively, μ is the specific growth rate, $\mu_{\max}$ is the maximum growth rate, and

K_s is the concentration of chemical supporting a half-maximum growth rate. The chemical decomposition rate is expressed as the following equation:

$$-\frac{dS}{dt} = \frac{\mu}{Y} X = \frac{\mu_{max}}{Y} \frac{SX}{(K_s + S)} = k_b \frac{SX}{(K_s + S)} \qquad \text{(Eq. 7-25)}$$

where Y is the cell yield. For many of the more common substrates, K_s is on the order of 10^{-1} $\mu g/ml$ (Ref. 7-12) and this, in general, is considerably higher than the concentration that would be expected for chemicals in natural waters. If $S \ll K_s$, Eq. 7-25 reduces to

$$-\frac{dS}{dt} = \frac{\mu_{max}}{YK_s} XS = K_{b2} \ XS \qquad \text{(Eq. 7-26)}$$

where K_{b2} is a second-order rate constant equal to

$$K_{b2} = \frac{\mu_{max}}{YK_s} = \frac{k_b}{K_s} \qquad \text{(Eq. 7-27)}$$

If the microbial mass is assumed to be constant, which is likely in a given ecological stabilized environment, then Eq. 7-26 can be rewritten as

$$-\frac{dS}{dt} = K_{b2}XS = K_{b1}S$$

The above equation becomes a first-order degradation kinetic.

7-4 CONTINUOUS-FLOW MODELS

Many environmental systems can be conceptualized as a chemical reactor. Flows are continuously coming in and going out. Plug-flow and completely mixed flow systems are considered to be the two extremes. In the latter system, the incoming material is mixed uniformly upon entering the system. In the plug-flow system, on the other hand, incoming flow (e.g., carriage water) retains its identity after entering the system; no mixing takes place. However, plug-flow and completely mixed flow systems are the limiting cases; actual flow patterns lie between these idealized cases.   In plug-flow situations, many of the first-order equations previously cited directly apply.

If the environment is considered as a single chemical reactor, the mass balance equation can be written as follows

$$V \frac{dC}{dt} = M_0 - QC - VkC \qquad \text{(Eq. 7-28)}$$

where V is volume. If a pulse input is assumed, $C_{(t=0)}$ can be approximated to be $M/\bar{V}$, where $\bar{V}$ is equal to mean volume. Equation 7-28 can be rewritten for $t > 0$

$$V\frac{dC}{dt} = -(Q + Vk)C$$

or

$$\frac{dC}{dt} = (Q/\bar{V} + k)C \qquad \text{(Eq. 7-29)}$$

Integrating Equation 7-29, we obtain

$$\ln C = -(Q/\bar{V} + k)t + \text{constant}$$

or

$$\ln C/C_0 = -(Q/\bar{V} + k)t \qquad \text{(Eq. 7-30)}$$

The half-life can be estimated by the following equation:

$$t_{\frac{1}{2}} = \frac{\bar{V} \ln 2}{(Q + \bar{V}k)} \qquad \text{(Eq. 7-31)}$$

where a first-order kinetic reaction rate is assumed. In the case of a zero or low flowrate, Eq. 7-31 becomes

$$t_{\frac{1}{2}} = \frac{\bar{V} \ln 2}{\bar{V}k} = \frac{\ln 2}{k} \qquad \text{(Eq. 7-32)}$$

A mass-balance equation to describe changes of the chemicals in a completely mixed reactor is

$$\bar{V}\frac{dC}{dt} = QC_0 - QC - \bar{V}r \qquad \text{(Eq. 7-33)}$$

where Q is the influent (or effluent) flow, C is the chemical concentration, $\bar{V}$ is the reactor volume, and r is reaction rate. To determine the half-lifes of the chemicals in the reactor (assuming inlet concentration becomes zero), Eq. 7-33 can be rearranged:

$$\bar{V}\frac{dC}{dt} = -QC - \bar{V}r \qquad (t \geq 0) \qquad \text{(Eq. 7-34)}$$

$$t_{\frac{1}{2}} = \frac{\bar{V} \ln 2}{Q} \qquad \text{(Eq. 7-35)}$$

where $C \longrightarrow 0$ as $t \longrightarrow \infty$; hence $C_0 = r\bar{V}/Q$. $\qquad$ (Eq. 7-36)
From the above equations, one will note that the fate of chemicals in natural waters depends not only upon the transformation route but also upon the transport mechanism.

A single reactor system is not sufficient to describe the detailed changes of chemicals in the environment in a spatial sense. In a large system, more sophisticated models are required to achieve a better understanding of the fate of chemicals. The environment can be visualized as a system composed of many single reactors or compartments in series. Each compartment represents a reach in the river or lake, in which complete mixing is assumed. The mass-balance equation for chemicals in each compartment can then be written as follows:

$$\frac{dV_iC_i}{dt} = Q_{i-1}C_{i-1} - Q_iC_i - \sum_j V_ir_j \qquad \text{(Eq. 7-37)}$$

or, to be more general,

$$\frac{dV_iC_i}{dt} = \sum_k Q_kC_k - \sum_l Q_lC_i - \sum_j V_ir_j \qquad \text{(Eq. 7-38)}$$

where the first term on the right-hand side is the total of incoming streams to compartment $i;$ note that one of the incoming streams is from the upstream adjacent compartment. Similarly, the second term on the right-hand side of Eq. 7-38 is the total stream withdrawn from the compartment $i;$ note that one of the outgoing streams is Q_i, which flows to the adjacent $(i + 1)$th compartment. The degradation term includes all possible degradation and transfer processes as discussed in the previous section. Here, a pseudo-first-order kinetics is assumed for all processes.

As discussed before, characteristics of sediments are important in determination of the fate of chemicals—size distribution and organic content are two key factors. However, these two factors can be ignored for the purpose of a first-cut assessment. Scouring and settling are two important transport mechanisms. The former is defined as the bottom sediments being returned to the water column. This is determined by the hydrodynamics of the water body at the particular time under consideration, and will occur when the shear force of water movement is high enough. The settling rate of sediments depends upon the bouyant forces and velocity. In turbulent rivers, sediments remain in the water column instead of being deposited on the bottom. For modeling purposes, scouring and settling are considered, respectively, as percent of sediments returned to and settled from the water body. Therefore, the mass balance equation for the sediment in the compartment i becomes

$$\frac{dV_iS_i}{dt} = \sum_k Q_kS_k - \sum_i Q_iS_i \qquad \text{(Eq. 7-39)}$$

Scouring rate and sedimentation rate, respectively, are included in the first and second terms on the right-hand side of Eq. 7-39.

Until this point, excepting adsorption and desorption, most degradation and transfer processes have been formulated in Eq. 7-38 and 7-39. A partition coefficient is often used to describe adsorption and desorption at a steady state condition; it indicates the chemical distribution between the water and the sediment. If a dynamic rate equation for adsorption and desorption can be derived, it can be simply attached to the right-hand side of both Eq. 7-38 and 7-39. The new equation will become a coupling equation. For simulation results, these two equations have to be solved simultaneously.

Adsorption and desorption are considered to reach an equilibrium state within a very short period of time in comparison to the time interval to be used for simulation of the environment system. If partition coefficient, P_s is defined in the following equation:

$$P_s = \frac{m_{wi}M_{si}}{m_{si}M_{wi}} \qquad \text{(Eq. 7-40)}$$

where m_{wi} is the mass of water and m_{si} is the mass of suspended sediments in compartment i. M_{si} and M_{wi} are, respectively, the masses of chemical constituent in the suspended sediment phase and the water phase.

If no biodegradation or chemical transformation processes are taking place on the surface or inside the adsorbant, the total mass of the chemical constituent in the compartment will not be changed by adsorption/desorption. Therefore, the relationship of the total mass of the chemical constituent in the aqueous phase (M_{wi}) and the suspended sediment (M_{si}) before (prime) and after (not prime) the adsorption/desorption process can be written as:

$$M_{wi} + M_{si} = M_{wi} + M_{si} \qquad \text{(Eq. 7-41)}$$

The mass distribution after adsorption, therefore, can be calculated by solving Eq. 7-40 and 7-41. If bioadsorption is also considered, the partition coefficient, P_b, similar to Eq. 7-40 can be written as follows:

$$P_b = \frac{m_{bi}M_{wi}}{m_{wi}M_{bi}} \qquad \text{(Eq. 7-42)}$$

where m_{bi} is the mass of microorganisms and M_{bi} is the mass of the chemical constituents adsorbed on the surface of microorganisms, respectively.

Equation 7-42, therefore, can be rewritten to accommodate the bioadsorption term:

$$M_{wi} + M_{si} + M_{bi} = M_{wi} + M_{si} + M_{bi} \qquad \text{(Eq. 7-43)}$$

The mass distribution of chemical among aqueous phase, solid phase, and microbial population can be calculated by solving Eqs. 7-39, 7-42, and 7-43.

7-5 DATA REQUIREMENTS

The fate of chemicals in the aquatic environment can be theoretically simulated by the equations in the foregoing section(s), provided all input data are available. In this section, input data will be considered in the following groups: system data, physical parameters, and degradation data. The model requires a set of system data appropriate for computer execution, e.g., simulation period, simulation interval, and control index for specific programming purposes. System data are determined by specific objectives of the model from either a programming or a simulation point of view.

Physical parameters define the studied environment, e.g., such as number of compartments, size of each compartment, location of incoming tributaries and outflows, inflow and outflow rates, and interflows. Most of these data can be condensed from historical hydrographic information. First, the studied environment should be compartmentalized according to study objectives. A rule of thumb is to set compartment boundaries in regard to chemical input points. In some cases, shoreline compartments are desired to track off-shore transport in a large water body. At the central core of the lake, large compartments are designed. It is believed, in general, that the relative rates of water movements in the central core of the lake are slower than at the shoreline. This would mean that chemical constituents have much faster rates of change along the shoreline than in the central part of the lake. In estuaries, small compartments are designed where pollution discharges are expected. Once compartments are defined, the size of each compartment can be estimated. Inflows and outflows can be estimated from hydrologic data. Interflows connecting the compartments are much more difficult to estimate. Rigorously, a hydrodynamic model can be used to estimate the interflows. However, this approach usually requires as much an effort to execute as a fate assessment model. Circulation information can be used to estimate the interflows between compartments. Note that the aforementioned data are time dependent, unless for a short-term assessment, where it is sufficient to take the information as an average. Sometimes, seasonal averages are used to compensate for drastic changes of hydrodynamic properties.

Degradation information include those kinetic constants for various degradation processes. The kinetic data can be obtained from either published papers or laboratory experiments. Data obtained from the literature often refer to a certain environmental or experimental condition. Further interpretation is necessary for model simulation; the data should be adjusted in accordance to environmental conditions. Rate constants are temperature and pH dependent, partition coefficients depend on organic contents of sediments, and photolysis rates are determined by turbidity of the water. Again, rate constants are time dependent, especially on a seasonal basis.

7-6 SPECIFIC APPLICATIONS

Applications of models to environmental risk assessment are described in articles cited in the References to this chapter, especially Refs. 7-1 to 7-5, 7-8, 7-10, 7-11, 7-13, 7-17, and 7-18. In addition, modeling is discussed to some extent in almost every chapter of this book, especially:

Sections 1-3 and 1-5 (general considerations)
Sections 2-2 to 2-4 (physical properties for models)
Section 3-5 (model for estimating product entry rate and cumulative quantity)
Section 4-2 (aquatic fate studies)
Sections 5-1 to 5-6 (atmospheric fate studies)
Section 6-2 (assessment of terrestrial exposure)
Section 8-5 (parametric analysis of microcosms)
Sections 11-2 and 11-3 (risk assessment of synthetic-fuel wastewater)
Sections 14-1 to 14-5 (nonlinear regression model for predicting bioaccumulation and ecosystem effects of a pesticide)
Section 15-2 (fate model for setting pollutant limit values for soils)
Sections 17-2 and 17-3 (hazard evaluation of solid wastes)

REFERENCES

7-1. Branson, D. R., "Predicting the Fate of Chemicals in the Aquatic Environment from Laboratory Data," *Estimating the Hazard of Chemical Substances to Aquatic Life,* J. Cairns, K. Dickson, and A. Maki (Eds.), STP 657, Am. Soc. for Testing and Materials, Philadelphia, PA, 1978.

7-2. Brown, R. A., and Weiss, F. T., *Fate and Effects of Polynuclear Aromatic Hydrocarbons in the Aquatic Environment,* American Petroleum Institute Publication #4297, Washington, D.C., 1978.

7-3. Eschenroeder, A., "Multimedia Modeling of the Fate of Environmental Chemicals," presented at Am. Soc. for Testing and Materials' 4th Symposium on Aquatic Toxicity, Chicago, IL, October 16–17, 1978.

7-4. Gillett, J., Hill, J., Jarvinen, A., and Schoor, W., *A Conceptual Model for the Movement of Pesticides through the Environment,* U. S. Environmental Protection Agency Ecological Research Series, EPA-660/2-74-024, 1974.

7-5. Haque, R. and Freed, V. H., *Behavior of Pesticides in the Environment: Environmental Chemodynamics, Residue Reviews,* F. A. Gunther (Ed.), Springer-Verlag, New York, 1974.

7-6. Hill, J., *et al., Dynamic Behavior of Vinyl Chloride in Aquatic Ecosystems,* EPA-600-13-76001, U. S. Environmental Protection Agency, Washington, D. C., 1967.

7-7. Lyman, W. J., Harris, J. C., and Nelken, L. H., *Research and Development of Methods for Estimating Physicochemical Properties of Organic Chemicals of Environmental Concern,* Final Report Phase 1, February, 1979, Contract No. DAMD17-78-C-8073, prepared for U. S. Army Medical Research and Development Command by A. D. Little, Inc., Cambridge, MA.

7-8. Mill, T., "Data Needed to Predict the Environmental Fate of Organic Chemicals," presented at Symposium on Environmental Fate and Effects, American Chemical Society, Miami, Florida, September, 1978. Proceedings to be published by Ann Arbor Science.

7-9. Monod, J., "The Growth of Bacterial Cultures," *Am. Rev. Microbial.* **3**, pp. 371–394, 1949.

7-10. Neely, W. B., "A Preliminary Assessment of the Environmental Exposure to be Expected from the Addition of a Chemical to a Simulated Aquatic Ecosystem," *International Journal of Environmental Studies,* **13**, pp. 101–108, 1979.

7-11. Neely, B., "An Integrated Approach to Assessing the Potential Impact of Organic Chemicals in the Environment," *Analyzing the Hazard Evaluation Process,* K. Dickson, A. Maki, and J. Cairns (Eds.), published by Water Quality Section of the American Fisheries Society, Washington, D. C., 1979.

7-12. Pirt, S. J., *Principles of Microbe and Cell Cultivation,* J. Wiley & Sons, Inc., New York, 1975.

7-13. Smith, J. H., Mabey, W. R., Bohonos, N., Holt, B., Lee, S., Chou, T., Bomberger, D., and Mill, T., *Environmental Pathways of Selected Chemicals in Freshwater Systems,* SRI International, U. S. Environmental Protection Agency Report EPA-600/7-78-074; available as PB-274548 from National Technical Information Service, Springfield, Virginia 22161, 1978.

7-14. Stumm-Zollinger, E., and Harris, R. H., in *Kinetics of Biologically Mediated Aerobic Oxidation of Organic Compounds in Aquatic Environments,* S. L. Faust and J. V. Hunter (Eds.), Marcel Dekker, Inc., New York, pp. 555–598, 1971.

7-15. Tsivoglou, E. C., "Tracer Measurement of Atmospheric Reaeration," *J. Water Pollution Control Fed.,* **37**, pp. 1343–1362, 1965.

7-16. Tsivoglou, E. C., *Measurement of Stream Reaeration,* U. S. Department of the Interior, Washington, D. C., 1967.

7-17. *Prescreening for Environmental Hazards – A System for Selection and Prioritizing Chemicals,* EPA/560/1-77-002 (NTIS PB 267093), A. D. Little, Inc., 20 Acorn Park, Cambridge, MA, April, 1977.

7-18. Lassiter, R. R., Baughman, G. L., and Burns, L. A., "Fate of Toxic Organic Substances in the Aquatic Environment." *State-of-the-Art in Ecological Modeling,* S. E. Jorgensen (Ed.), Int. Soc. for Ecological Modeling, Copenhagen, Denmark, Vol. 7, pp. 219–246, 1978.

8

Model ecosystems

P.H. Pritchard, Ph.D.

Environmental Scientist
U.S. Environmental Protection Agency
Environmental Research Laboratory
Sabine Island
Gulf Breeze, Florida

8-1 INTRODUCTION

The inventiveness and success of chemical and energy industries in developing new pesticides and industrial chemicals have resulted in an exponential increase in their production. The mere use of such chemicals in their designated manner leads to unavoidable global contamination—the burden of which is only beginning to be perceived. Our poor perception of this burden is a consequence, in part, of a failure in risk analysis programs to comprehend that predictions of environmental catastrophe depend on the ability to integrate the results of a variety of single parameter laboratory tests into an environmentally meaningful hazard assessment. There is, accordingly, increased social and scientific pressure to make assessments on the waste assimilatory capacity of the world's ecosystems (i.e., their ability to accept pollution loading without adverse environmental effects). As Cairns states (Ref. 8-32); "Our present population size and distribution, as well as the current life style in industrialized countries, will require that we make some use of environmental assimilative capacity. It does not require that we degrade the environment in doing so."

It is now obvious that existing integration capabilities for describing the limits of this capacity, are inadequate and consequently hinder attempts to abate pollution. For example, the highly acclaimed restriction on the use of several organochlorine insecticides focused attention on our lack of knowledge about assimilatory capacity and our inability to accurately assess the fate and effects of synthetic organic compounds (xenobiotics) in the environment. This, in turn, resulted in the so-called zero discharge philosophy that, in many cases, was erroneously interpreted to mean that the elimination of pollution in one environmental area would result in a net environmental impact of zero. In retrospect, zero discharge pollution control often resulted in pollution being displaced to another location or form, where the consumption of power, technology, and chemicals to alleviate the problem were potentially worse than the orginal effect.

Extensive research is being undertaken to remold (through chemical synthesis) persistent organochlorine compounds. Newly developed pesticides maintain their highly effective control of target organisms but become more degradable in the environment (Ref. 8-29). Similar efforts have been applied to development of novel pesticides, such as those that inhibit chitin synthesis in insects (Ref. 8-186). Underlying such efforts is the assumption that adequate techniques and methods are available for accurately predicting the fate and effects of these new pesticides in the real world.

The evolution of chemicals that have a finite lifetime in the environment magnifies the problems of accurately predicting actual fate and effects from laboratory studies. For example, if a pesticide has an adverse effect on a non-target species, the time required to eliminate the effect will be critical in predicting the assimilatory capacity of the environment. Thus, current problems facing industry and regulatory agencies become, in general terms, a matter of generating environmentally meaningful information in the laboratory in short periods of time and then integrating this information into a realistic environmental assessment.

The Toxic Substances Control Act has focused attention on the manufacturer or producer of toxic chemicals; clearance from the U. S. Environmental Protection Agency (EPA) must now be obtained before the chemical can be manufactured and distributed. This mandate has forced both the EPA and the manufacturer to develop methods and protocols to assess the waste assimilatory capacity for any specified environment. Unfortunately, results generated by existing test methods have only recently been examined for their extrapolability to "real world" situations.

Expectedly, a great deal of interest has been directed toward the use and application of laboratory test systems that encompass complex aspects of natural environments. Much interest has focused around the use of model ecosystems, as evidenced by recent workshops, symposia, and reviews (Refs. 8-53, 8-54, 8-124, 8-200). Unfortunately, the term "microcosm" or "model ecosystem" is vague and beset by a myriad of interpretations. Experimental studies using microcosms are highly varied and often quite meaningless because of poor experimental design and confusion as to how, when, and where results can be applied to natural environments.

However, the microcosm approach has an excellent possibility of addressing and even answering many questions related to risk assessment and waste assimilatory capacity. This chapter selectively reviews some of the research investigations that employ the microcosm or model ecosystem approach. It attempts to define important parameters of a microcosm study and to characterize precautions, limitations, and developments in microcosm technology as they apply to risk assessment. The views expressed are the author's and they do not necessarily represent the policy of the U. S. Environmental Protection Agency. As a result of my training and convictions as an environmental scientist, the discussion will be slanted toward aquatic systems. Papers reviewed for this chapter specifically mentioned microcosms or model ecosystems. Numerous other laboratory investigations, however, can be considered as microcosm studies, although they are not so labeled. My coverage is by no means comprehensive, but includes general sources of information on microcosms (Refs. 8-14, 8-22, 8-26, 8-43, 8-53, 8-54, 8-124, 8-196, 8-200).

8-2 NATURAL ECOSYSTEMS AND THE MICROCOSM APPROACH

8-2.1 Properties of Natural Ecosystems

Ecosystems operate around both biotic and abiotic components that regulate and drive processes such as nutrient cycling, organism successions, homeostatic control, and energy flow. Integration of these components constitutes the ecosystem. Natural ecosystems, regardless of their size, are highly complex entities that exhibit structural and metabolic orderliness. Emmel (Ref. 8-57) points out that "the interrelationship between the components of an ecosystem is not haphazard; it has a definite history of development, a particular spatial orientation, an involvement of time in operation of the systems, and an involvement of specific energetic sequences."

Ecosystems generally evolve toward a mature, stable condition in which the accumulation of biomass remains constant. This condition signifies that the resulting communities can process a variety of inputs without greatly affecting the system (Refs. 8-131, 8-151). If autotrophic succession in aquatic environments is used as an example (Ref. 8-44), the ratio of the rate of photosynthesis to the rate of respiration stabilizes around unity in a mature ecosystem. Photosynthetically, young ecosystems characteristically have low structural diversity but high production, which makes their systems inefficient. Older or mature ecosystems, on the other hand, demonstrate higher efficiency as a result of higher diversity in autotrophic organization. The mature system, therefore, maintains a certain temporal homeostasis both structurally and metabolically. In natural settings, evolution toward a homeostatic state is slow and significantly affected by seasonal and environmental changes.

Ecosystems also typically demonstrate an ability to maintain or self-sustain their homeostatic stage, which has been referred to as the stability of an ecosystem (Ref. 8-194). Stability may be defined as the capability of a system to maintain within certain bounds, its structural and functional continuity in the face of pressure exerted by external or internal perturbations (Refs. 8-151, 8-152). The ability of a natural ecosystem to restore its structure following acute or chronic disturbance (natural or human-induced) is termed resilience (Refs. 8-33, 8-92). Similarly, the property of an ecosystem to resist displacement (structurally and functionally) from some equilibrium is referred to as inertia (Refs. 8-31, 8-194).

Despite their stability, natural ecosystems are not characterized by invariant processes. A constant state of flux exists; the rates of environmental processes vary considerably in time and over boundaries. Again, climatic seasonal and physical conditions of ecosystems are the major factors that regulate these fluxes; they are not controllable, per se, by man.

From a risk assessment standpoint, information derived about the environmental impact of pollutants often require, an in-depth understanding of the workings of natural ecosystems, including the aspects of homeostasis, stability, and system flux. The inherent variability or flux within these natural systems makes risk assessment extremely difficult and time consuming. Solutions to pollution problems necessitate short operational time scales. Consequently, two basic experimental approaches have arisen to fill the need for shortcuts: First, the use of a laboratory system to study ecosystem processes makes it possible to examine structural and functional aspects under defined and standardized climatic and physical conditions, thereby greatly diluting the effects of certain uncontrollable parameters typically associated with field studies. Second, the most influential environmental factors can be assessed by ultimately testing different sets of climatic and environmental conditions in the laboratory systems. It is from these approaches to environmental studies that the "microcosm approach" has evolved. The concept is not new in itself but its contemporary meaning reflects a new and timely emphasis on laboratory studies of ecosystem processes.

8-2.2 The Microcosm Approach

The microcosm approach potentially covers a large variety and scale of laboratory studies. It is used for investigations of some portion of an ecosystem, and it assumes that appropriate analogies or simulations of that ecosystem can be obtained in laboratory systems. Presumably, by controlling certain ecosystem variables, such as seasonal changes and variations in the physical parameters of an environment, and by carrying out process analysis using ancillary experiments of simpler design, a more rapid and complete interpretation of the processes occurring in nature can be obtained.

Further, it also indirectly stipulates that, as an analogy to some part of an ecosystem, the same homeostatic and stability measures typical of natural ecosystems can be demonstrated in laboratory systems. This means that if microcosm studies are to be environmentally meaningful, considerations of stability, resilience, and inertia should be incorporated into the design features, the systems analysis, and the data interpretation of these laboratory systems. The creation of microcosms in the laboratory undoubtedly causes a disruption of perturbation of the environmental sample. This could mean a breakdown in the ecosystem stability and/or inertia; recovery of the system to some equilibrium then must be checked. If the wrong ecosystem parameter is chosen to monitor this recovery, a false indication of inertia or stability will be obtained. This could greatly affect the utility of the microcosm approach and it demonstrates the necessity of studying and analyzing laboratory systems with the same conceptual foundations employed in the study of natural ecosystems.

Finally, from a regulatory and a risk assessment standpoint, if the microcosm approach is to be valid, it is necessary that a degree of observational confirmation with the real world be provided. Because of the simulation potential, it should be possible, within certain constraints, to extrapolate laboratory results to actual environmental situations. Microcosm studies are far from simple experimental approaches; they are as much a part of basic ecological research as research in support of regulation.

8-2.3 Advantages and Disadvantages

From an experimental standpoint, the microcosm approach offers several other major advantages not readily obtainable with field studies:

1. Their small size permits replication and the use of appropriate controls.
2. The chemical composition of the media within the systems and physical parameters (pH, salinity, temperature, etc.) can all be varied within an experimental regime and the subsequent effects monitored.
3. The systems can be perturbed in a variety of ways.
4. The trophic structure of the systems can be manipulated, within limits, to provide greater similarity to natural situations.
5. The imports and exports to the systems can be controlled.
6. The systems circumvent large-scale environmental contamination for the examination of pollutant effects and are cost- and time-effective approaches

to assessments of the environmental risks of chemical and biological pollution.

7. The difficulty in accessibility and containment of a field study can be supplanted in certain situations by a microcosm study.

Microcosms also have a number of limitations that require consideration in both system design and interpretation of results:

1. Containerization of environmental components will, on a pragmatic laboratory scale, lead to the exclusion of certain higher trophic levels (i.e., their existence in a laboratory system cannot naturally be sustained).
2. Containerization of environmental components can lead to structural and functional changes that are more related to system design rather than to the environment from which the samples were taken.
3. The cost and time expenditure for a microcosm study are often initially high but tends to even itself out with continued study and application.
4. Biological recruitment into microcosms must be carefully modeled and controlled because natural systems depend heavily on recruitment for maintenance of organic diversity under stressed conditions.
5. Laboratory microcosms invariably have unnaturally high surface-to-volume ratios, due primarily to large vessel wall surfaces.

Many advantages and disadvantages will be discussed as part of other sections in this chapter.

8-3 DEFINITIONS AND CRITERIA FOR MICROCOSM STUDIES

8-3.1 Microcosm Definitions

The microcosm approach as described in Section 8-2.2 is a relatively simple concept. There are numerous justifications, based on problems in studying natural ecosystems, for this approach. Consequently, the nature of a microcosm study will be based on the questions asked by the investigator who must consider the following: Are there any criteria that would limit and define the design of a microcosm? Does it have definable features? Are there any consistent characteristics that would have general application to a whole range of hypothesis testing exercises.

   In this section, I will discuss a few general performance criteria, that can be applied to a variety of microcosm studies and hopefully afford a degree of uniformity and consistency to this experimental approach. This does not necessarily mean that I want to standardize microcosm design or its operation.

It does mean, however, that I would like to see the *conceptual ideas* behind a microcosm study articulated into performance criteria which would dictate when an experimental undertaking is classified a microcosm study.

The actual definition of a microcosm is quite complex and confusing, due to the very general nature of the term and its connotations. For example, consider several of the available definitions. The one recently adopted at the Terrestrial Microcosm Workshop (Ref. 8-200) states that a microcosm "is a controlled, reproducible laboratory system that attempts to simulate the situation (i.e., processes and interactions of components) in a portion of a (terrestrial) ecosystem." Abbott (Ref. 8-1) simply described them as "miniaturized ecosystem(s)." Beyers (Ref. 8-14), in discussing the problems of ecosystem boundaries, stated, "one solution to the boundary problem is to enclose an ecosystem in some kind of container to isolate a functional ecological unit from the rest of the biosphere. It is this deliberately isolated ecosystem that has come to be called a microcosm or microecosystem." Leffler (Ref. 8-123), at a presentation of a paper at the symposium, "Microcosms in Ecological Research," defined microcosms as "small living models of ecosystem processes." "Models" referred to any general representation of an ecosystem and "small" referred to dimensions and simplicity. To Jassby *et al.* (Ref. 8-108), "the word microcosm, when used in an ecological context, refers to a collection of chemicals and organisms within well-defined spatial boundaries, generally under controlled physical conditions, and in a volume convenient for laboratory study, i.e., much smaller than ecosystems of interest in nature." Moreover, they are "appropriate experimental objects for the investigation of systemic properties of naturally occurring ecosystems and the delineation of various details concerning trophic interactions, nutrient cycles, and certain other topics." Cooke (Ref. 8-43) indirectly defines microcosms as closed experimental laboratory systems, such as flasks or beakers, in which the "assemblage of organisms is natural and is usually obtained from nature as a unit" such that the organisms are allowed to. . .reorganize and reassemble under (controlled) laboratory conditions. . . ." Finally, a literal translation of the word microcosm, i.e., "anything regarded as a world in miniature," implies that all structures and functions of a tropical rain forest can be miniaturized to fit into a terranium, which, of course, is ludicrous.

Despite its somewhat metaphoric qualities and its degree of indefiniteness, the term is now commonly applied to a variety of experimental problems and laboratory systems. Unfortunately, from the cited definitions, it would appear a microcosm study could be defined as any study containing a component whose complexity cannot be immediately described.

The indefiniteness of the microcosm definition, as well as a lack of experimental criteria for establishing or designating a laboratory system as a microcosm, has permitted many types and levels of experimentation to be included. For example, in a study by Sugiura *et al.* (Ref. 8-177) on the toxic effects of copper ions to a synthetic community of bacteria, protozoa, and algae, the test system was arbitrarily labeled a microcosm. If their investigation is considered a microcosm study, then the same must be done for a vast number of similar studies (Ref. 8-145), even though the use of microcosms is not mentioned.

In a similar vein, Allan and Brock (Ref. 8-3) used simple test tubes to cultivate mixtures of heterotrophic microorganisms from sewage at particular temperatures. Their emphasis was on the assemblages of microorganisms as the microcosm and not on the laboratory system used in the cultivation. Other investigators have also emphasized this community approach, such as fungal microcosms (Ref. 8-65) and coral reef microcosms (Ref. 8-50). Another example of the confusion resulting from the microcosm approach is exemplified in the fate study reported by Juengst and Alexander (Ref. 8-110), who studied the effects of environmental conditions on the degradation of DDT in a "model marine ecosystem;" as no criteria or reasons are set forth in the development of the model system, it is difficult to know whether this type of study is any better than similar studies using marine sediment and water (Ref. 8-2) that are not labeled as model ecosystems, or microcosms.

8-3.2 Operational Criteria

What constitutes a reasonable definition of a microcosm and what operational limits can be established to give it some specificity? There seems to be sufficient literature to make it possible to apply certain limits. In an attempt to impose some of these limits and to encourage discussion in this area, I advance the following operational criteria which I feel should be addressed, at least in part, in any microcosm study.

Criterion #1. A microcosm study must involve a laboratory test system in which a component, in as natural a state as possible, of a real world system is brought into the laboratory and kept under standard conditions of light, temperature, humidity, aeration, pH, Eh, and other similar factors.

This criterion is developed for two reasons. First, it is designed to discourage nonlaboratory from being labeled microcosms. This is necessary because nonlaboratory studies, which would include a great variety of field studies, provide such an endless array of experimental systems that it would completely dilute the use of the term microcosm. Further, nonlaboratory studies cannot be climatically controlled or placed under standard operating conditions, which, of course, is one of the main conceptual parameters of the microcosm approach (see Section 8-2.2). Thus, experimental investigations centered around manmade streams, enclosures, impoundments, ponds, etc., should not be termed microcosm studies. Perhaps a more appropriate label would be incarcerated ecosystems, as they often represent attempts to house or encompass as much of the function and structure of a natural system as possible. This is not necessarily the intent of a microcosm study from my viewpoint. Instead, these nonlaboratory studies represent a source of ecological information that can function as verification data bases for many microcosm systems (see Section 8-6).

Second, the underlying inference of this criterion is that a microcosm harbors a level of complexity that cannot be readily duplicated, at least not initially in the laboratory, by assembling component parts. It may ultimately be that the synthesis of a quasicomplex system, using individual components or monocultures, will provide a sufficient (depending on those questions being asked) analogy to a real world situation; however, this has not been experimentally demonstrated in most instances. Until reliable quasicomplex systems can be developed, a true microcosm study will be the next best laboratory procedure for modeling the complexities of nature. For example, from a fate standpoint, it may often be very difficult (due to the slow evolution of new knowledge and new techniques) to isolate from an environmental sample a microorganism that will metabolize a particular xenobiotic (a manmade chemical). However, since the introduction of the xenobiotic to a natural environmental sample leads to its metabolism, there are obviously some important interactions residing in the natural sample that cannot be duplicated in the synthetic system. In order for the effect of this interaction to be studied in the laboratory, the sample must be transported, in as intact a form as possible, to the laboratory and studied in an experimental system.

In other words, many processes occurring in natural ecosystems presumably cannot be modeled in the laboratory by making composites of individuals. One of the virtues of a microcosm study, therefore, is the ability to bring natural assemblages into the laboratory for study even though the interactions and composition of the assemblages may not be known. For many microcosms, the use of natural soil or sediment (as opposed to clean sand or sterile silt, etc.) may readily satisfy this level of complexity; these materials would contain a large variety of types and numbers of microorganisms and invertebrates that could never be assembled on an individual basis. Thus, the criterion as stated precludes, to a certain extent, the use of synthetic community studies (see Section 8-7) as microcosms.

Criterion #2. Microcosms are ecosystems unto themselves that attempt to simulate a natural environmental situation. They must accordingly be calibrated with the environmental situation to be simulated.

Three important points in regard to this criterion need to be emphasized. First, it is basically shortsighted to believe that the transport of an environmental sample to the laboratory and its containerization in a microcosm will result in continued duplication of its original condition. A variety of studies have examined and characterized these changes (Refs. 8-15, 8-44, 8-164, 8-195). In general, these changes rapidly approach and maintain, for a finite period of time, a homeostatic state in which much of the structure and function of the system become balanced. Although this balanced state does not reflect the condition of the original sample, it is both unique to the microcosm and probably representative of a state that could exist in the environment under consideration. In discussing this aspect relative to stream microcosms, Warren and Davis (Ref. 8-191) stated, "Closed or partially closed laboratory stream ecosystems are ecosystems of a sort, probably not stream ecosystems, certainly not pond ecosystems, but ecosystems having some of the characteristics of each. Though these systems appear to have no natural counterpart, populations of many different species do live in them for relatively long periods of time; and these populations, in finding places in which to live, competing for energy and materials, consuming one another, elaborating protoplasm, and perhaps even reproducing, must do so in interactive ways that in some respects resemble life in nature. Despite this representativeness rather than duplication, it is very likely the ecological state observed in the microcosm contains functions and structures similar to the natural system." The degree of similarity and precision required in modeling natural environmental conditions will be a research question.

Second, it is important to recognize that a microcosm represents only a small portion or situation of the natural system. It should not be considered as a general representation of an entire ecosystem, that is, it is conceptually trivial to label a microcosm as, for example, a "model estuary" or "model forest." Certain trophic levels and trophic interactions are not amenable to study in a microcosm. In fact, it is reasonable to speculate that microcosm studies encompassing trophic levels above small invertebrate animals or small rooted plants are not feasible because containerization forces a restrictive condition upon the higher trophic levels that has no correlation in the real world.

Thirdly, it is crucial in a microcosm study to calibrate the laboratory system and the results it produces against the environmental situation under consideration. Probably the most severe shortcoming of past and present microcosm studies is the almost total absence of an attempt to determine the differences and similarities in structure and function between laboratory systems and field sites. It cannot be assumed, *a priori*, that natural components, when brought into the laboratory for study, will behave in the same fashion as they did in their natural setting. Thus, if this calibration step is not performed, any differences between laboratory results and field results will basically be uninterpretable, particularly from the viewpoint of extrapolation. Unfortunately, calibration is probably one of the most complicated and poorly researched aspects of microcosm studies (see Section 8-6).

It is also important to emphasize that a microcosm and the results it generates must be calibrated against system design as well. For example, a particular microcosm result may make no sense in relation to an actual environmental situation simply because the wrong design of a laboratory system was chosen. Extensive efforts to calibrate or parameterize the effects of *scaling* and *containerization* should be considered in initiating any microcosm program. The extent of calibration, be it for simulation potential or system design optimization, will depend on the question being asked of the microcosm study.

Criterion #3. The integrity of a microcosm study cannot stand solely on the merits of the microcosm system. The microcosm system (that is, the actual container and accessories used to study the environmental sample) should be considered only as a tool in a research undertaking, not as the research undertaking itself.

The most immediate utility of a microcosm study is to reveal processes and effects that cannot be observed in simpler experiments or in some types of field studies. Since the microcosm is a laboratory representation of a complex milieu of an environment, it will provide information based on the complex aggregation of various actions and interactions occurring in the environmental sample under study. The information obtained is basically an end result from which little knowledge can be gained about the various processes and interactions involved. As a consequence, the microcosm results cannot be extrapolated to the environment without kinetic information about the relevant processes and their response to a variety of environmental conditions. Therefore, specific processes and their kinetics should be studied throughout the microcosm experiments—an exercise I call process analysis (see Section 8-5). These parameterization-type experiments will, of course, be considerably simpler than the microcosm; they will provide information on pathways, mechanisms, and limiting conditions that are critical to the interpretation of the microcosm results. This process information will potentially lead to a better picture of what goes on in the natural environment and will provide a much better basis for extrapolations to real world situations.

Criterion #4. A microcosm study should be considered initially as a verification tool, not as a screening tool. In my view, any study that uses a test system in which an arbitrary but consistent mixture of environmental material (e.g., water and sediment) is incubated in a container of arbitrary size and shape without any further attempt to compensate for containerization effects or deviations from natural situations is not a microcosm study. Rather, it is a study that simply uses *actual media* as part of a standardized test method. Such a study is useful on a relative basis to estimate factors such as biodegradability, toxicity, or bioconcentration potential of a pollutant as long as the test system is run in the same way each time.

I reserve the microcosm system as the verification tool (see Figure 8-15). Its results are the *integration* of a variety of processes associated with the natural complexities of the environment. It will be the experimental focus for other investigations to describe and elucidate the individual processes and interactions occurring in a real world setting. From such information, either new screening tests will be developed or current screening tests will be validated. An actual media test system may, for some experimental questions, turn out to be a perfectly good microcosm, but this remains to be shown. A further discussion of actual media screening test systems can be found in Section 8-8.

8-3.3 Summary Definition

In summary, I have proposed four operational criteria as a guide for a microcosm study. Hopefully, these criteria will prevent a variety of laboratory experimentations from incorrectly being considered microcosm studies, as they have been in the past, due to the broad generalizations inherent in earlier definitions.

Accordingly, a general definition of a microcosm study encompassing these criteria is stated as follows:

A microcosm is defined as a calibrated laboratory simulation of a portion of a natural environment in which environmental components, in as undisturbed a condition as possible, are enclosed within definable physical and chemical boundaries and studied under a standard set of laboratory conditions. By calibrated, two things are meant: first, the simulation should be checked by comparing common properties of the microcosm (functional processes, for example) with similar properties in the environment; second, the conditions in the microcosms or the results generated must be checked for their relation to the system design.

If a microcosm study is part of a method for predicting occurrences in a natural environment, four test phases should be implemented:

1. The microcosm study itself, checked against the design of the experimental container and observed under several environmental conditions.
2. Supplemental experimentation on individual processes thought to occur in the complex system; process analysis.
3. Calibration of the microcosm with the natural conditions of the environment being modeled; field calibration.
4. Mathematical modeling of the microcosm study and the applicable environment to enhance interpretation and extrapolation.

Adherence to such criteria makes it obvious that few existing experimental investigations would qualify as microcosm research. Very few microcosm studies cited in the literature are extensive enough to provide information on the test phases listed above; in many cases, information is given only about the micro-cosm system itself while interpretation tools (modeling) and parameterization type experiments are ignored. Therefore, this chapter focuses on a series of topics (following sections) that, when taken as a whole, exemplify the attributes of a comprehensive microcosm study. Presumably as these attributes become experimentally tested and verified, a significant degree of commonality in microcosm studies will surface, allowing the application of microcosm technology to become simpler and standardized.

8-4 System Design Features

In writing this chapter, it was anticipated that a careful review of microcosm and related literature would reveal a series of general criteria in microcosm design that could be recommended to investigators interested in this approach, irre-spective of the basic experimental questions being asked. It was assumed that enough reports of microcosm studies would make it possible to generalize regarding when and how certain experimental design features should be applied. However, there is such a large diversity in systems labeled microcosms that the generaliza-tions are difficult. For example, the 13,000-liter marine ecosystem research laboratory microcosm tanks of Pilson (Ref. 8-160) and the 300-ml bottles of Cooke (Ref. 8-44) are both used as systems to study the dynamics of phyto-plankton populations. It is difficult to form generalizations in system design from these studies because there are no base-line studies for comparisons. It also is evident that the system design in most microcosm studies basically depends on the arbitrary decision of the investigator. Rarely is much effort de-voted to determining if the design could be improved or whether its design had an effect on the results being obtained.

In the section below I suggest some experimental constraints for microcosm design that may be useful in a microcosm program. In some areas, suggestions are put forth as a possible incentive for future research in that area.

8-4.1 Size and Scale Reduction

The size or scaling of a microcosm depends heavily on the tophic level of the ecosystem to be housed in the system. It is obvious that certain higher trophic levels can never be studied in a laboratory, regardless of the microcosm size. Three general criteria can be used in deciding what trophic levels can be ac-commodated.

First, it is desirable that the microcosm be large enough to sustain a plant or animal community using the renewable food sources inherent in the microcosm system. That is, if the culturing of plants and/or animals in a laboratory system of a specific size (or shape) cannot be maintained as part of a balanced eco-system, consisting of producers, consumers, and decomposers, then it is possible that the microcosm size may be limiting.

Second, if a particular trophic level irreversibly upsets the biological balance (say, through the total elimination of a prey species) or the physical continuity (disruption of substratum structure such as through burrowing) of the microcosm, then the system is probably too small. Generally, this situation results from cases where the feeding habits of an animal are so intense, relative to the avail-able food sources, that they virtually obliterate some biological or physical component of the microcosm, often leading to extreme situations of diversity and ecosystem stability. For example, in developing synthetic communities, Gillette and Gile (Ref. 8-73) and Cole *et al.* (Ref. 8-39) (see Section 8-7) added animals to the system, knowing that their existence was finite. In the former case, a meadow vole introduced into a rather lush crop of rye grass and alfalfa totally denuded the system of its vegetation. In the latter case, saltmarsh caterpillars and voles did much the same thing to corn and other plants. Al-though the experimental questions being asked of these synthetic communities are not affected by this overgrazing phenomena, it is nonetheless illustrative of the complicated problems that can be associated with attempts to incorporate higher trophic levels into microcosms. Brockway *et al.* (Ref. 8-28) reported other problems associated with snails in their aquarium-type microcosms—the consumption of periphytic algal populations by the snails introduced a functional instability into their systems and were consequently removed. Restrictive size of the experimental containers may have been one of the factors contributing to the exaggerated effects of the snail populations.

Third, the size chosen for a microcosm must be large enough to unambiguously demonstrate that the results generated are a consequence of some biological event and not a consequence of the normal variations in measured parameters. A reduc-tion in scale from field to laboratory is likely to cause an increased variability in the data generated and a greater sensitivity of certain ecosystem parameters to perturbation. This will be due, in part, to the inherent decreases in diversity that accompany scale reduction and that have been shown to affect stability (Refs. 8-47, 8-132).

It would appear, based on the small amount of published information, that lower trophic levels in aquatic environments are minimally affected by scaling particularly from a functional standpoint (Refs. 8-69, 8-159). Laboratory sys-tems from 5 to 500 liters (with or without sediment) should provide adequate experimental conditions for modeling the ecology of trophic levels in aquatic environments up to and including zooplankton and benthic invertebrate com-munities. This microcosm size range provides for good species diversity and appropriate successional responses (see for example, Refs. 8-15, 8-28, 8-43,

8-108, 8-164). It will also accommodate a large variety of fate and effects studies, eventually producing results that can be extrapolated to natural situations. In some cases, it is possible that even smaller size systems can be employed; some fate studies, for example, can be successfully studied in sediment-water systems containing only a few liters of water (Pritchard and Bourquin, unpublished results).

Giddings and Eddlemon (Ref. 8-69) showed that sediment-water microcosms from 7 and 70 liters generally behave in the same manner relative to each other for parameters like pH, dissolved oxygen, inorganic nutrients, dissolved organic carbon (DOC), algal biomass, and arsenic distribution. Perez et al. (Ref. 8-159) reported that experimental laboratory systems of 150 liters basically mimic conditions seen in an equivalent field situation. Zooplankton densities were statistically the same in both systems. However, phytoplankton densities in the laboratory system were consistently higher than those observed in the field. No explanation was given for this discrepancy other than to say that it was not attributable to differences in water turbulence or light intensities. Kemmerer (Ref. 8-114) and McLaren (Ref. 8-144) demonstrated that natural phytoplankton populations develop in small laboratory tanks in a manner that agrees with populations seen in surrounding lake water.

Another critical scaling parameter in aquatic systems is the question of whether the ratio of sediment surface area to overlying water volume needs to be rigidly simulated in the microcosm. The question is difficult to answer because few studies directly address this topic. For many environmental settings, it is logistically impossible to mimic the proper ratio in the laboratory. Perez and his colleagues (Ref. 8-159), in attempting to model a portion of Narragansett Bay, used a 150-liter tank and suspended a box of sediment (of appropriate surface area) just below the water surface to assure proper light penetration to the sediment. However, the effects of a larger or smaller sediment surface area were not tested.

It is very complicated to circumvent the environmentally unnatural effects of wall surface area of the microcosm tank or container. The vessel walls, in fact, provide such a large surface area that the amount of sediment surface area could become insignificant. It is virtually certain that the container walls will be covered rapidly with layers of bacteria and algae, and their presence could definitely affect the dynamics and structure of the ecosystem contained therein. A potential remedy to this problem may be the continual scraping of the walls; however, this is a technical headache. It must be remembered that the biomass residing on the walls may simply be redistributed within the system, thus potentially producing other unpredictable consequences. At the very least, the wall growth can be tolerated if one attempts to determine its' role in or its' effect on the results generated from the microcosm.

In terrestrial studies, many of the same constraints of aquatic systems apply. Terrestrial microcosms of a size amenable to laboratory study apparently will provide data, such as leaching from soil columns (Ref. 8-52) or CO_2 evolution (Ref. 8-6) reflective of natural situations. Generally, soil masses from 0.1 kg to several hundred kilograms are suggested. Despite a state-of-the-art review in terrestrial microcosms (Ref. 8-200), very little information is available relative to the effects of scale; in fact, this problem was minimally addressed in this review. Jackson *et al.* (Ref. 8-103) described 45 X 45 X 25 cm box cores that appear to be ideal systems for most studies: they are large enough to handle leaf litter, a variety of rooted vegetation, leaching and CO_2 evolution measurements, and subsampling (using mini cores) and small enough to allow replicates and tests with a variety of environmental parameters. Their studies (Refs. 8-101, 8-104) demonstrated these systems do model real world situations.

8-4.2 Higher Trophic Level Herbivores and Carnivores

A number of microcosm studies have shown that laboratory systems inoculated with natural components rapidly evolved into a state of both successional and functional equilibrium (see Refs. 8-15, 8-28, 8-43, 8-44, 8-69, 8-108) in the absence of zooplankton and phytoplankton grazers, such as small fish and crustaceans. If these studies are to be considered representative of natural successional events, then it is important to eventually include the effects of higher trophic level herbivores and carnivores on the equilibrium state observed in the microcosm study. Several studies have shown that the effects of grazing on primary productivity and zooplankton communities may be substantial. Cooper (Ref. 8-45) found that reduction in producer standing crops by a herbivore caused an increase in primary productivity. Hargrave (Ref. 8-85) found that a deposit-feeding amphipod significantly affected the productivity of sediment microflora, while Paine and Vadas (Ref. 8-154) have shown similar grazer effects on primary productivity of attached benthic communities. Thus, it would seem important to assure that some grazer component be incorporated into a microcosm study and that scaling factors be adjusted to accommodate the grazing effect.

However, other studies suggest that small zooplankton grazers, even though they can potentially be incorporated into laboratory systems, may still have pronounced and unnatural effects on processes occurring therein. For example, McConnell (Ref. 8-142) and Goodyear *et al.* (Ref. 8-75), using small pools, reported that small herbivore fish, *Tilapia mossambica* and *Gambusia affinis* respectively, can reproduce in laboratory tanks at a rate related to the natural gross primary productivity of these systems. Jassby *et al.* (Ref. 8-107), using 700-liter tanks, demonstrated a very significant effect of two mosquito fish (*Gambusia affinis*) on the organic structure of their microcosms. The fish initially consumed the dominant large cladoceran zooplankter, *Simocephalus vetulus*, and the large cyclopoid copepod, *Cyclops vernalis*, causing approximately a tenfold decrease in their concentration. A small cladocern, *Alona gutlata*, which was not a major source of food for the fish, became the dominant member of the zooplankton community. The change in dominance, coupled with no increase in other crustaceans (which represents an inherent limitation of this microcosm system), provoked a release of grazing pressure on the phytoplankton and bacterial communities; subsequently, rotifer populations expanded since their food source was increased and the grazing by cladocerans reduced. These results reflect conditions typical of heavily stocked fish ponds and not natural systems (Ref. 8-107). Indeed, even with a 700-liter tank, the two fish represented a concentration of 5 mm^3 liter^{-1} which would only rarely be encountered in a natural mesotrophic system.

Flint and Goldman (Ref. 8-61) also demonstrated the importance of grazing pressure on benthic periphyton communities by crayfish, again questioning the validity of microcosms that permit zooplankton and plankton communities to operate in the absence of grazing pressure. In their studies, primary productivity of diatom communities (developed on styrofoam substrates) was increased by crayfish grazing in both in situ experiments and in 15-liter laboratory aquaria. However, the effects were dependent on the crayfish biomass; crayfish biomass greater than 203 g/m^2 decreased productivity while a biomass greater than 69 g/m^2 was required to keep standing crops of the macrophyte, *Myriophyllum*, at natural levels. In addition, Flint and Goldman indicated that the fecal material from the crayfish may be a signficant contributor of both organic and inorganic nutrients to the ecosystem. Thus, grazing effects may be an important factor in considering microcosm design. If the effects of grazers are shown to be important, but scaling prohibits incorporation of a biological entity to meet this end, then it may be possible to physically design or manipulate a microcosm system in which grazing effects are mimicked artificially.

8-4.3 Inoculum

There is accumulating evidence (Refs. 8-24, 8-52, 8-105, 8-162) that the microcosm study should commence with a sample taken directly from the environment, thereby incorporating as many natural components as possible. Selective removal of some component (i.e., by coarse filtering) or mechanical agitation of the sample, appears to promote results that differ from an undisturbed sample. In most cases, if the inoculum can be extracted from the environment as an intact core, the results generated will be the most representative of field conditions. This has been demonstrated for soil samples (Refs. 8-52, 8-105) and for sediment–water systems (Ref. 8-162). Jackson et al. (Ref. 8-102) have shown that homogenized soil cores have significantly greater leaching rates of Ca than either intact cores or field plots. Giddings and Eddleman (Ref. 8-69) have shown that clean sand rather than actual sediments in a microcosm produced lower concentrations of inorganic nutrients in the water column, far less algal growth, low pH and dissolved oxygen (DO) values, and much less binding of arsenic to particulates.

8-4.4 System Operation

Many operational parameters of a microcosm system are chosen at the discretion of the investigator. However, several helpful pointers are in order. First, I recommend the use of a continuous-flow mode for aquatic-type microcosms to prevent a buildup of toxic end products. Under continuous flow conditions a number of biological and environmental parameters are time independent, thus aiding in data interpretation. A continuous-flow mode, will also provide a realistic dilution of sediment-originated inorganic nutrients. Regarding this last point, since a microcosm system will generally underestimate the normal depth of water over the sediment (logistically too complicated) an unnatural buildup of inorganic nutrients may occur in the water column if a static (no flow) system system is employed. This can be prevented by a continuous-flow operation and, in fact, flow can be adjusted until nutrient concentrations in the water column equal field levels. Flowrates giving retention times of 5 to 20 hrs are probably sufficient for proper dilution of these nutrients; however, nutrient loading rates from sediment into the water column relative to natural environmental situations could serve as a criterion for setting flowrates.

A variety of stirring and mixing devices have been employed in microcosm systems but the most common is simple aeration using a bubbler. This approach seems to properly maintain DO levels, allows aeration without great agitation of the sediment–water interface, and maintains an environmentally realistic oxidized zone at the sediment bed surface. However, it is questionable whether aeration provides a continuously mixed reactor that is desirable from the standpoint of data analysis and mathematical modeling. Two alternatives to aeration by bubbling, with similar attributes, are stirrers (Figure 8-1) that were developed at the U. S. EPA Environmental Research Laboratory (ERL) in Gulf Breeze, FL. In each case, the stirrers maintain rapid mixing within the water column and stabilize reaeration rates at the water surface, permitting detection of diurnal dissolved oxygen concentrations. The devices also allow mixing in the presence of rooted macrophytes and can be used on relatively large volume systems (up to 20 liters). For larger tanks, rotating paddles below the water surface should be employed.

Lighting systems for most microcosms consist of 40-W cool-white florescent lights that generate between 500 to 15,000 lux of light in a 12-hr light–dark cycle. They do not, however, completely simulate the sunlight energy spectrum.

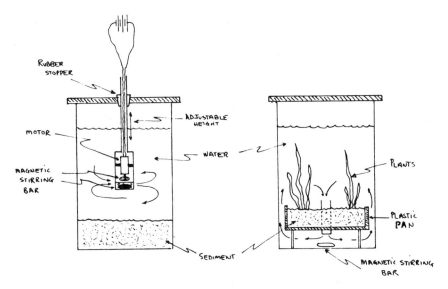

Fig. 8-1. Schematic diagram of stirring apparatus developed for use in sediment/ water microcosms. *Source:* Pritchard, P.H., unpublished data, U.S. EPA ERL, Gulf Breeze, Florida (1980).

8-4.5 Pretest Incubation

A general trend in microcosm studies is the incubation of laboratory systems for weeks to months before any testing is commenced. It is assumed that freshly constituted systems need time to settle and adjust to both the perturbation involved in inoculating the microcosm and to the standardized conditions of the laboratory. Undoubtedly, in some cases, pretest incubation is necessary (Refs. 8-43, 8-69). Very few papers, however, document changes occurring during the pretest period. As evidenced in several aquatic microcosms (see Section 8-6.8), successional responses occur during this pretest incubation, eventually leading to a stable or equilibrium condition considerably different from what had been typical of the original sample. There is no guarantee that any systems receiving the sample inoculum will evolve through the same successional pattern and attain the same ultimate diversity. Pond studies of Hall *et al.* (Ref. 8-81) showed that, even though the ponds were originally structured identically, it was very difficult to predict the manner or the degree in which they diverged in their development and maturation. Further, it may be difficult to extrapolate microcosm results, after extended incubation periods, to the field in certain situations. Microcosm results may show greater correlation or perhaps less variability if the microcosm is established and tested after only minimal incubation periods (2-3 weeks). This assumes that adjustments to disturbances caused by inoculation occur rapidly; it is not unreasonable to expect intact cores to behave accordingly. Fate studies of pesticides in aquatic microcosms (Ref. 8-162) indicate that this assumption may be correct. At the very least, microcosm researchers should consider short-term incubations as potentially the best way to obtain direct simulation of any environmental situation.

8-5 PROCESS ANALYSIS

Although a microcosm system may be shown to be an excellent analogy to a real world situation, the results or data produced from a microcosm study are only as good as the background information required to interpret the microcosm results. This background information has to be derived from simpler more defined experiments, what I call process analysis. As stated in Section 8-3.2, process analysis will provide the necessary information on the processes, rates, mechanisms, interactions, and environmental influences that affect results observed in the microcosms. The critical nature of this process analysis is evident in risk assessment programs where ecosystem mathematical models are used to predict environmental impact. The mathematical model will require both a listing of the rate constants of major processes involved in the fate of the particular compound and some verification or guide as to the integration of these processes and their respective rate constants. The microcosm, in principle, will supply information on the final result of this integration aspect, but will not necessarily produce information to help elucidate and quantitate the major processes involved and the effects of environmental conditions (such as pH, DO, salinity, and temperature) on the processes.

Relatively few microcosm studies have been accompanied by process analysis; the standard procedure has been to set up a microcosm, operate it, and then simply report the results, remaining vague as to how results can or will be extrapolated to a natural situation. Extrapolation, in fact, can only be attempted after a process analysis study. As examples of the importance of process analysis, I will describe briefly three case studies that employ process analysis to some extent. Each study, however, is incomplete relative to the operational criteria of a microcosm study I set forth in Section 8-3.2, but is illustrative of the potential experimental design that can be applied to microcosm studies and risk assessment.

The first example comes from the work of Witkamp and his colleagues (Refs. 8-158, 8-197, 8-198, 8-199) who examined the decomposition of leaf litter and mineral cycling in forest ecosystems. Their basic experimental concept was to first assemble simple replicate soil–litter systems: clear plastic boxes with drain spouts containing approximately 200 g of sterile washed sand and 3 to 5 g dry wt. of one-month-old sterilized leaf litter previously tagged with ^{137}Cesium (Refs. 8-158, 8-199). Boxes were incubated in a greenhouse (30°C) under ambient light conditions. Each soil–litter system was made progressively more complex by adding various combinations of leaf litter, microflora, millipedes, snails, plants, and soil. The distribution of radioactivity in each compartment after a specific incubation period was then determined as a function of the combinations. A wetting cycle provided a system leachate.

The interesting aspect of these experiments is the measured effects of the various compartmental combinations on the dynamics and decay of leaf litter (weight loss). Figure 8-2 gives examples of the relative distribution of radio-cesium within these compartments. Millipedes accounted for the greatest direct affect on turnover of ^{137}Cs, possibly because of their role in detachment of leaf fragments. Interestingly, systems containing both millipedes and snails did not have as great an effect on cesium transfer, leaf weight loss, and CO_2 evolution as did the sum of the effects observed in systems with millipedes and snails alone (Ref. 8-199).

A kinetic analysis of the cesium movement was performed by analog computer simulation modeling (Ref. 8-158) and relatively good fits were obtained with first-order approximations. In this way, analysis of the transfer pathways, the rate constants for these pathways and the transfer mechanisms were more quantitative. Information can, therefore, be generated about time to equilibrium, steady state concentrations, concentration factors, input and output fluxes, turnover rates, and stability. For example, their studies showed that material turnover in the soil–litter systems, as a whole, increased as the number of compartments increased. By examining rate constants for various pathways linking each component (Refs. 8-158, 8-197), it was possible to draw additional inferences about the relative sensitivity and stability of various links within the system. A conceptual diagram of these results is shown in Fig. 8-2 (insets).

Witkamp refers to all of his experiments as microcosm studies, including the simple experiments with only one or two components. According to my operational definition, Witkamp's experiments did not employ a microcosm system. Instead, they are examples of a type of process analysis that could be applied to the results generated from a microcosm. In fact, under my definition, Witkamp has never studied a microcosm system, i.e., excising an intact portion of soil and leaf litter from a natural setting and bringing it back to the laboratory for study under standardized conditions. It would be interesting to know whether an integration of all the component experiments (i.e., using his computer model) would provide the same picture of mineral cycling as that observed in a microcosm system. Witkamp, in fact, acknowledges (Ref. 8-197) that rates obtained from his component studies are faster than the rates seen in field studies, although the transfer mechanisms and pathways are common to both systems.

Another example of an experimental study representative of process analysis is the work of Coleman and his colleagues (Refs. 8-4, 8-38, 8-41, 8-42, 8-89). A series of factorially designed experiments were employed to investigate the

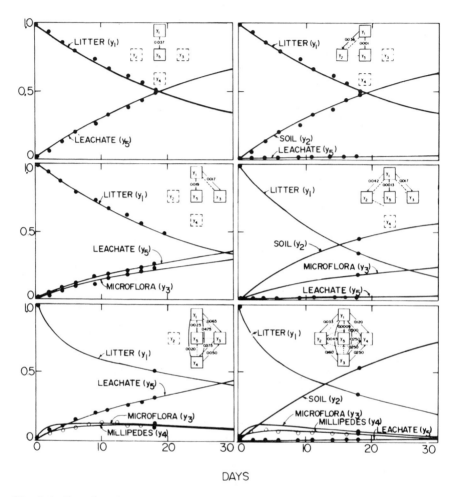

DAYS

Fig. 8-2. Fit of analog computer curves to data showing changes in relative distribution of (y) radiocesium in microcosms with various biotic components. (Y1-5) (a-f, respectively). Means of three replicates are shown as solid circles. Open circles represent estimates of relative cesium concentration in millipedes based on radioassays of animals that died. Insets show compartment combinations and rate constants (day $^{-1}$) for each indicated transfer route. Pathways over which no cesium was transferred are illustrated by broken arrows. *Source:* Patten, B.C., and Witkamp, M., Systems analysis of ^{134} cesium kinetics in terrestrial microcosms. *Ecology,* 48, pp. 813-824, 1967. (Reprinted with permission of Duke University Press, Durham, NC.)

effects of trophic interactions (modeled as a simple food web consisting of plant exudate, bacteria, amoebae, and nematodes) on the mineralization of organically immobilized carbon, nitrogen, and phosphorous in soil. A conceptual model of the environmental situation under study is shown in Fig. 8-3. It was initially

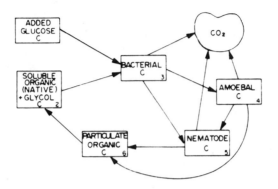

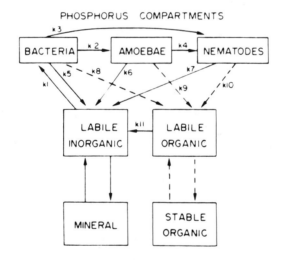

Fig. 8-3. Conceptual model of major carbon and phosphorous compartments and the corresponding nutrient flows in soil microcosm containing bacteria, amoebae and nematodes. *Source:* Coleman, D.C., Cole, C.V., Hunt, H.W., Klein, D.A., Trophic interactions in soil as they affect energy and nutrient dynamics. I. Introduction, *Microb. Ecol.,* 4, pp. 345-349, 1978. (Reprinted with permission of Springer Verlag, New York.)

hypothesized in their study that the integration of all the component studies would provide a complete picture of the whole (Ref. 8-41).

The experimental systems used in these studies consisted of flasks with 20 g of propylene-oxide sterilized soil inoculated with pure cultures of bacteria (*Pseudomonas cepacia* was chosen to represent a microorganism that responded to releases of high nutrient root exudate, Ref. 8-89), amoebae, and nematodes in various combinations. Population dynamics as a function of glucose additions (to simulate root exudate) were followed over a 24-day period (Ref. 8-4). Because of the simple nature of the experimental design, large numbers of flasks were set up at one time to observe factors such as biological treatment, carbon levels, incubation time, and replication. Extended studies also examined flows of biomass and changes in metabolic activity as monitored by respired CO_2 release (Ref. 8-42). It was established, by examining phosphorous movement in the system (Ref. 8-38), that once phosphorous was immobilized by bacteria it was rapidly remineralized by the saprophagic grazing of the amoeba and subsequently released as inorganic phosphorous. These results are illustrated in Fig. 8-4.

Overall, this approach is excellent for gaining component information about a complex process in soil. However, there is no information to suggest that the relationship established between various compartmental combinations in these mechanistic studies actually exists in a natural situation. This information could be obtained in either field studies and/or microcosm experiments, using intact samples of natural soil (that is, coinciding with my operational definition). Field studies, although complicated and time consuming, can be performed on soil plots; phosphorous dynamics and population responses could be followed as a function of either actual root exudate or simulated root exudate. On the other hand, a microcosm experiment would provide a fine intermediate testing system between the simple experiments and the field studies. For example, it could indicate whether the rate and extent of phosphorous mineralization was the integrated result of a large number of interactions taking place in the soil sample or simply the result of the bacterium–amoeba–nematode interaction. This approach demonstrates how the component experiments provide quantitative baseline information on the supposed transfer pathways or mechanisms involved, and how the microcosm provides a verification tool for the integration of these component experiments.

A third microcosm study that also illustrates the need for process analysis was conducted by Procella and his colleagues (Refs. 8-161, 8-175) on the Great Salt Lake in Utah. Due to the lake's extreme conditions, the biology is relatively simple and, therefore, easier to model in microcosms. The northern end of the Great Salt Lake contains a very low diversity of biological organisms—a red alga, *Dunaliella salina*, a green alga, *D. viridis*, the brinefly *Ephydra gracilis*, the brineshrimp *Artemia salina*, and several types of halophilic bacteria. The red alga is the primary planktonic form serving as the main food source for the brine shrimp that, in turn, are grazed upon by the larval stages of the brinefly. The halophilic bacteria presumably feed on the exudates of live and dead algae and the arthropods (Ref. 8-175).

The relationship between these organisms was studied with microcosms using water and sediment from the lake. The microcosms (Fig. 8-5) were constructed of lucite plastic cylinders (75 cm high by 15.5 cm in diameter); a 15-cm layer of sediment was contained in the bottom, covered by approximately 9 liters of

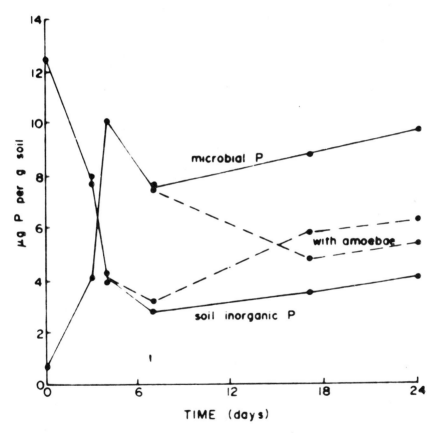

Fig. 8-4. Effects of amoebae on levels of bicarbonate-extractable inorganic phosphorous and microbial phosphorous during a 24-day incubation in soil microcosms. Data plotted are means of treatments with and without amoebae ($N =$ 12). *Source:* Cole, C.V., Elliott, E.T., Hunt, H.W., and Coleman, D.C., Trophic Interactions in Soils as they Affect Energy and Nutrient-Dynamics. V. Phosphorous transformations, *Microb. Ecol.,* 4, pp. 381-387, 1978 (Reprinted with permission of Springer Verlag, New York.)

lake water. The water in the systems was mixed only during biweekly medium exchanges (although a mixing apparatus was designed into the system for other experiments), in which approximately 900 ml of fresh media (mineral salts, constituted according to composition found in lake) were added through an inlet port just above the sediment. This displaced an equal volume of spent medium at an outlet port near the top of the water column. Chemical and biological

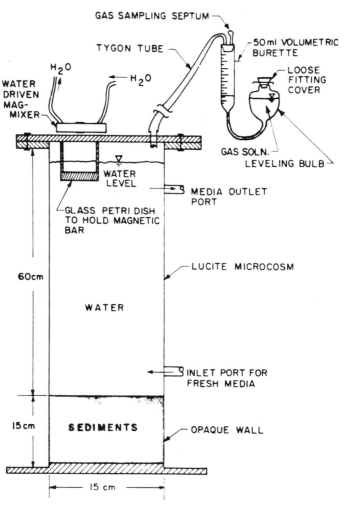

Fig. 8-5. Design of experimental microcosm used to study nitrogen cycling in the Great Salt Lake. *Source:* Stube, J.C., Post, F.J., Porcella, D.B., Nitrogen cycling in microcosms and application to the biology of the northern arm of the Great Salt Lake. U.S. Dept. of Interior, Office of Water Res. and Tech., PRJ SBA-016-1, 1976. (Reprinted with permission of the author.)

analyses were then performed on the effluent water. The microcosm system was also designed so that quantities of O_2, N_2, CO_2, CH_4, and $CH_2=CH_2$ could be determined in the exiting gases. Several sets of microcosms were operated simultaneously under light–dark (16-hr light, 8-hr dark) conditions and exposed to regimes of NO_3^{-1}, NH_4^{+1}, urea, and glutamic acid enrichment. Concomitantly, field samples were taken periodically analyzed chemically and biologically in the same way as microcosm samples.

These microcosm studies revealed a variety of responses. *D. viridis*, the green alga, was specifically stimulated by nitrate whereas the red alga, *D. salina*, was simulated by ammonia (also verified in axenic culture studies). This corresponded with actual conditions in the lake, where *D. salina* is the predominant algal species and ammonia is the predominant form of nitrogen. The microcosm studies revealed that supplementation of the systems with glutamic acid stimulated the pink halophilic bacteria to the point of producing enough biomass to turn the microcosm pink. This colored condition is also typical of observations seen in the lake itself; the high levels of amino acids found in the lake (6 to 7 mg/liter) may account for the stimulation of halophilic bacteria. A simplified scheme of the interrelationships observed in the microcosm studies is shown in Fig. 8-6. This scheme is based on the major assumption that bacterial growth and metabolism are stimulated by algal excretion products and then the bacteria stimulate the predominant algal species through the production of ammonia.

This study, although it is representative of what I would consider a good microcosm study, has not yet provided a complete picture of the interactions occurring in the Northern basin of the Great Salt Lake because process analysis was not complete enough. It would be interesting to determine the conditions in the lake that lead to the natural production of amino acids; several variations and simulations could be attempted in the microcosm. Likewise, process analysis on the individual factors that stimulate both algae and bacteria would be useful in verifying proposed interactions. For example, results from the microcosm suggest that algal excretion products were the primary stimulation for the bacterial populations. Pure culture studies of *D. salina* could be used to determine (both qualitatively and quantitatively) if algal extracellular products would support the growth of halophilic bacteria. The environmental factors that control this production and stimulation could also be studied through process analysis.

This microcosm study again emphasizes the need for processes analysis. It also illustrates the type of study that could be set up for a risk assessment program in any aquatic environment. The knowledge gained by performing both the microcosm experiment and process analysis will provide an excellent, although partial (a mathematical model and other calibration experiments are needed), base-line study that could be used to assess the fate and effects of a variety of pollutants in aquatic systems.

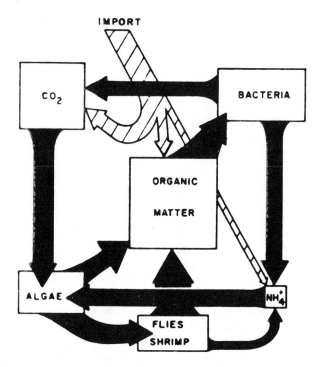

Fig. 8-6. Conceptual model of the microbial community interactions in the Great Salt Lake. Organic matter and ammonia are the key nutrients; nitrogen fixation, nitrification and denitrification are absent in the lake ecosystem. Invertebrates play a role through grazing and contribute to the organic and ammonia pools. The "IMPORT" includes organic and inorganic material from the south arm of the lake. *Source:* Stube, J.C., Post, F.J., Porcella, D.B., Nitrogen cycling in microcosms and application to the biology of the nothern arm of the Great Salt Lake. Dept. of Interior, Office of Water Res. and Tech., PRJ SBA-016-1, 1976. (Reprinted with permission of the author.)

8-6 ECOSYSTEM MEASUREMENTS FOR FIELD CALIBRATING MICROCOSMS

Earlier in this chapter (Section 8-3.2), I stated that if a microcosm was to simulate some environmental situation and be used for valid extrapolations from laboratory to field, it would be necessary to make an effort to demonstrate how similar or dissimilar the microcosm is from the field situation, i.e., simulatory capability. The simulatory capability of the microcosm is the key step in making extrapolations for risk assessment programs. The degree of simulation determines the confidence in the extrapolation exercise and the accuracy of the predictions in risk assessment.

It is a difficult experimental problem to determine whether a microcosm behaves like the portion of the environment it is attempting to simulate. In my view, the term "verification" (Ref. 8-200) implies more absoluteness than can be experimentally obliged, that is, the processes and organic structures in a microcosm cannot be expected to be identical with their real world equivalent. However, it is reasonable to expect that the biology and ecology of a microcosm system will *approach* some of the conditions of a natural system and the same perturbations to each system will result in similar outcomes from each. Therefore, I contend that it is better to *calibrate* a microcosm with the natural system; that is, they should behave in a similar manner and contain many of the same processes and structures.

Presumably, certain ecosystem-level measurements can be employed as practical and realistic calibration checks for simulation capability, but information is limited regarding which measurements to use and under what conditions. Unquestionably, the success of microcosm studies in supporting and supplementing risk assessment programs will depend in large part in the ability to select and test the proper measurements for this calibration process. These measurements also provide experimental probes for assessing other aspects of microcosm behavior:

1. They can be used to monitor microcosm equilibrium periods. For example, after an environmental sample is brought into the laboratory and containerized in a microcosm, a variety of adjustments by the indigenous communities will occur, eventually approaching some stable equilibrium. Knowledge about the time course and extent of this equilibration process can be provided by ecosystem-level measurements.

2. They can be employed to determine whether the biology of the system tends to deteriorate or evolve to some end point that is environmentally unrealistic. This would also supply information of divergence in community structure, changes in diversity and adjustments to sensitivity, resilence, resistance to specific stresses.

3. They can provide a good base line upon which replicability and reproducibility of microcosms can be assessed.

Several common ecosystem-level measurements will be discussed in this section. Most measurements are discussed in the context of aquatic systems, due to the author's bias. However, the general ecological concepts behind these measurements readily span all types of ecosystems and the microcosms designed to study them. It is also important to point out that the measurements chosen for field calibration may vary, depending on the questions being asked by the investigator and the type of risk assessment being performed.

Most of the examples involve functional rather than structural measures mainly because the former are simpler and can be performed by untrained personnel over long incubation periods both in the lab and the field. However, it should be remembered that functional continuity between any two systems can be due to drastically different ecosystem structures, much of which could stem from large artifacts caused by system design. Therefore, although functional aspects of microcosm behavior seem to be the first choice for comparisons with natural situations, considerably more research, particularly in numerous aspects of basic ecology and the effects of laboratory containerization and scaling, are required before our confidence in the simulatory nature of microcosms is beyond reproach.

8-6.1 *P/R* Ratios

A common functional ecosystem measurement often used to judge the successional maturity of a microcosm is the ratio between autotrophic growth as measured by daytime oxygen production (P) and the heterotrophic community respiration as measured by nighttime oxygen uptake (R) (Refs. 8-28, 8-44, 8-68). Generally, any containerized environmental sample, maintained under constant laboratory conditions (including light-dark circles), initially shows a P/R ratio of > 1, indicating a net increase in algal biomass. With continued incubation, however, the autotrophic and heterophic metabolism become balanced and the P/R ratio approaches one. A typical example is shown in Fig. 8-7. In some cases a large organic structure is supported under this balanced condition (Ref. 8-44). The P/R ratio remains quite stable over long incubation periods indicating minimal changes in net community production (Refs. 8-44, 8-69). The ratio can also be shown to be sensitive to a number of environmental variables (Refs. 8-16, 8-43, 8-46) and to ecosystem stress (Ref. 8-68). Many types of natural ecosystems show a similar bioenergetic pattern (Ref. 8-15, 8-151): early stages of ecological succession are characterized by P/R ratios of less than or more than one, while in mature stages P/R ratios tend to reach unity. Thus, if a microcosm after a particular equilibration period has a $P/R = 1$, it can be assumed that the energy being fixed by the autotrophic communities is counterbalanced by energy being consumed for maintenance of the heterotrophic communities; thus, the containerized ecosystem behaves energetically like real world situations.

The production-to-respiration ratio is also relatively easy to measure, both in the laboratory and the field, by either the pH method (based on changes in CO_2 concentration) (Ref. 8-17) or the dissolved oxygen method (Ref. 8-1). A number of studies have shown that replicate microcosms have good coefficients of variation (15%) for P/R ratios (Refs. 8-1, 8-28, 8-76); despite the fact that production and respiration values themselves may differ substantially from one microcosm to another, their ratios remain fairly close.

Caution, however, should be exercised because the P/R ratio concept only reflects systems that are functionally similar and, to some extent, self-sustaining.

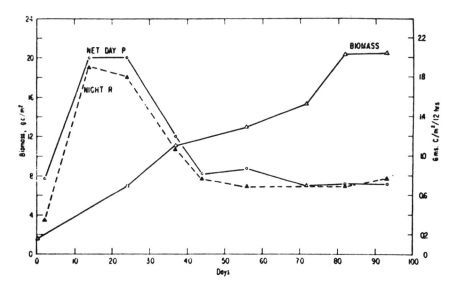

Fig. 8-7. Pattern of net community photosynthesis, night respiration, and biomass changes during autotrophic succession in laboratory microecosystems. Micro-ecosystems consisted of 400-ml beakers with 300 ml of nitrate- and phosphate-supplemented fish tank water inoculated with a farm pond water sample. During initial incubation period, algal blooms produced a high ratio of gross community photosynthesis to biomass, i.e. a small amount of biomass is very productive. As incubation continues (no additional nutrient), production efficiency decreased providing a large unproductive biomass. *Source:* Cooke, G.D., The pattern of autotrophic succession in laboratory microcosms, *Bioscience,* 17, pp. 717-721, 1967. (Reprinted with permission of American Institute of Biological Sciences, Arlington, VA.)

It says nothing about community structure. It is possible that two microcosms have good correlation in *P/R* ratios but drastically different population com-positions, even to the point that producer and the consumer communities are heavily dominated by one or two species. The question being addressed by the microcosm study will undoubtedly determine how critical the community diversity and structure will be; from the standpoint of functionality, however, the *P/R* ratio provides a good indication that energy flow in the microcosms is occurring in an environmentally realistic fashion.

8-6.2 Dissolved Organic Carbon Concentration

Closely coupled to the *P/R* ratio concept is the measurement of dissolved organic carbon (DOC) concentrations resulting from secretions by living organisms and partial decomposition of dead organisms. Transformation of organic carbon, of course, is part of the organic nutrient cycling process in any ecosystem, nutrient release being the major growth-controlling factor for the autotrophic community. As such, DOC would be expected to vary as a function of both autotrophic and heterotrophic activity. In fact, in a mature microcosm under standard laboratory conditions (including light–dark cycle), DOC concentrations should also approach an equilibrium state. Brockway *et al.* (Ref. 8-28) have observed this condition.

Actual DOC concentrations also reflect the eutrophic condition of a natural system (Ref. 8-190). Due to the fact that a microcosm may contain excessive wall surface area for algal growth, DOC may be considerably elevated over natural conditions. This could have a significant effect on metabolic turnover and, consequently, could provide a misleading evaluation of risk assessment, particularly in fate studies of xenobiotics. Thus, knowledge of DOC can provide a potentially good measure of comparability between field and laboratory systems.

8-6.3 Recycling of Inorganic Nutrients

Primary productivity in many aquatic systems is controlled by nutrient re-generative fluxes (Ref. 8-195). The recycling of nitrogen, for example, in nearshore (Ref. 8-79) and estuarine (Ref. 8-91) waters has shown that a large portion of the nitrogen required to maintain standing crops and productivity rates of phytoplankton comes from *in situ* (in the absence of external nutrients inputs) mineralization of organic matter. Aquatic sediments provide an important site for nitrogen remineralization. Microcosms that involve sediment-water systems are likely to have nitrogen remineralization processes and rates that could serve as a tractable ecosystem parameter to compare laboratory with field situations. It is quite possible for microcosm design (aeration systems, sediment perturbation processes, sediment surface to water volume ratios, flowrates) to affect sediment structure and function, ultimately leading to an effect on nitrogen recycling and thereby producing an ecosystem situation entirely unlike the field situation being modeled. For example, it is known that in stagnant basins (which could be equated to the conditions of a closed aquarium-type microcosm) restricted circulation lowers overall nutrient turnover rates and causes a buildup of nutrients in the water column (Ref. 8-176). A similar situation could arise in a microcosm study. Thus, an appreciation of the nitro-gen dynamics in an aquatic microcosm could provide a basic understanding of a microcosm's overall behavior.

It would appear from several exemplary studies (Refs. 8-7, 8-78, 8-79, 8-184) that simply measuring nitrate, nitrite, and ammonia nitrogen in the water column will provide data on nitrogen cycling comparable with similar data obtained in the field. Several *in situ* methods for measuring nitrogen flux in the field are available (Refs. 8-62, 8-168).

Rates and extents of ammonification and nitrification can be estimated in several ways: the use of a specific inhibitor of nitrification, such as with "N-serve" (Ref. 8-202), is a good example.

The fate of nitrate nitrogen, in part, depends on how much NO_3^{-1} diffuses back into anaerobic sediments where conversion to N_2 through denitrification can occur (Ref. 8-36). Several methods exist for directly measuring denitrification in marine sediments (Refs. 8-172, 8-173), soils (Ref. 8-202), and lake water sediments (Refs. 8-35, 8-180).

Various studies have shown that these particular reactions of the nitrogen cycle are sensitive to a variety of environmental parameters and provide excellent ecosystem parameters for field calibration of a microcosm. Graetz et al. (Ref. 8-78), Terry & Nelson (Ref. 8-180), and Vankessel (Ref. 8-184), have shown, for example, that factors such as aeration in the water column and redox potential in the sediments critically control these transformations. These factors then must be controlled and monitored in a microcosm to assure that the nitrogen cycle is being properly simulated.

Thus, it would appear that a knowledge of the nitrogen dynamics or the dynamics of other nutrients such as phosphorous, silicone, manganese, etc. in a microcosm will not only provide a basis for comparison with real world situations (rates of remineralization, rates of denitrification), but could also project the metabolic status of sediment/water systems.

8-6.4 Carbon Dioxide Evolution

The measurement of CO_2 evolution rates from soil and sediment is a tested and reliable technique for assessing general biological activity. It is also a very sensitive measurement and is often affected by the manner in which the sample is taken and the subsequent incubation conditions. As with the other techniques described above, CO_2 evolution is a universal measure that does not distinguish the relative contribution of the different living components in the sample, i.e., microorganisms, plant roots, and invertebrate animals. The technique is applicable to field measurements, as shown by Witkamp (Ref. 8-199), Domsch et al. (Ref. 8-51), and Reiners (Ref. 8-165).

An example of the application of CO_2 evolution to soil systems is the soil-litter microcosm developed by researchers at Battelle Laboratories (Refs. 8-6, 8-103). Intact soil box cores from a hardwood forest were placed in fiberglass boxes and kept under standard laboratory conditions. Small blackened boxes are inverted over alkaline traps sitting on the soil surface. To measure CO_2 production at periodic intervals, the amount of CO_2 trapped in the alkali was determined by titration; 2.60 mg of CO_2/day was released from control microcosms, and this rate remained constant for approximately 20 months of incubation. CO_2 efflux was independent of soil moisture and weakly dependent on temperature. The measurement, therefore, appears to be a useful technique to characterize the behavior of a microcosm and to compare it with those observed CO_2 effluxes in the field. Similar types of measurements should also be possible for marine tidal flat areas and sediment–water systems (Ref. 8-49).

Harrison *et al.* (Ref. 8-87) have suggested that the mineralization (CO_2 production) of ^{14}C labeled glucose in sediment will suffice as a representative measure of the *in situ* metabolic activity of microorganism and other biota in sediments (it can be applied to soil as well). It is important in performing this test to employ several concentrations of substrate so that kinetic analysis can be performed properly. The technique provides an interesting tool for field calibrating microcosms.

8-6.5 Oxygen Uptake

Rates of oxygen consumption by soils and sediments are indicative of the metabolic activity of the microorganism contained therein and as such it is directly related to the oxidation of organic matter and nutrient regeneration occurring in this medium. Included in these measurements of oxygen uptake is the oxygen consumed during the chemical oxidation of inorganic materials. Available methods for measuring oxygen consumption provide the possibility of using this functional parameter as a means of calibrating microcosms with field studies.

Most oxygen uptake measurements in the laboratory use intact cores of water and sediment (Refs. 8-48, 8-55, 8-86, 8-106, 8-156). Pamatmat (Ref. 8-156), for example, devised a multiple coring system, operated from shipboard, that will extricate undisturbed intact sediment/water cores. For inshore areas, cores can be taken by hand, using glass tubes and rubber stoppers (Ref. 8-162). Once the cores are brought into the laboratory, they can be fitted with DO probes (coupled with miniature stirring devices), and the rate of oxygen depletion from the water column followed. The laboratory system used by Pamatmat (Ref. 8-156) is shown in Fig. 8-8. Controls for chemical oxidation, which often constitute the major source of O_2 consumption, can be obtained by sterilizing the cores.

Part of the advantage of using oxygen uptake measurements for field calibration of microcosms is the availability of excellent methods for obtaining *in situ* measurements of oxygen uptake by aquatic sediments and soils. Techniques usually involve bell-jar systems (Refs. 8-156, 8-171) fitted with dissolved oxygen recording devices. Pamatmat (Ref. 8-156) has provided good evidence to show that intact shipboard cores give oxygen uptake rates very similar to *in situ* measurements taken at the sampling site. He also demonstrated that the cores exhibit constant oxygen uptake rates over incubation periods of several hours, but with longer incubation periods, uptake rates become dependent on O_2 concentrations in the overlying water. Continuous-flow cores could prob-

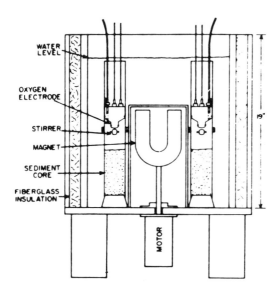

Fig. 8-8. Cross-sectional diagram of waterbath and sealed sediment cores used to measure oxygen uptake rates. *Source:* Pamatmat, M.M., Oxygen consumption by the seabed. IV. Shipboard and laboratory experiments, *Limnol. Oceanogr.* 16, pp. 536-550, 1971. (Reprinted with the permission of American Society of Limnology and Oceanography, Ann Arbor, MI.)

ably be used to monitor O_2 uptake over long periods. Disturbance of the top layers of sediment has been shown to'stimulate O_2 uptake due to exposure of the anaerobic sediment zones, but original uptake values can be regained after the systems are allowed to settle for brief periods (Refs. 8-84, 8-156).

8-6.6 Reduction Processes in Sediment

A functional process that is closely related to oxygen uptake measurements is the measurement of reductive metabolism in anaerobic environments, using oxidation–reduction indicators. The technique involves the biological reduction of tetrazolium salts to formazan by the flavoproteins, quinones, or cytochromes of the electron transport system (ETS) in biological systems. Theoretically, the reduction of tetrazolium can be used as an indicator of overall biological oxidation potential because electron transfer in communities is measured regardless of the respiratory and fermentative status of its members. Methods for examining ETS activity have been described by Olanczuk-Neyman and Vosjan (Ref. 8-149), Zimmerman (Ref. 8-204), and Pamatmat and Bhagwat (Ref. 8-155). An example of the use of this technique is shown in Fig. 8-9 (Ref. 8-187). ETS activity was

Natural Tidal Flat Sediment

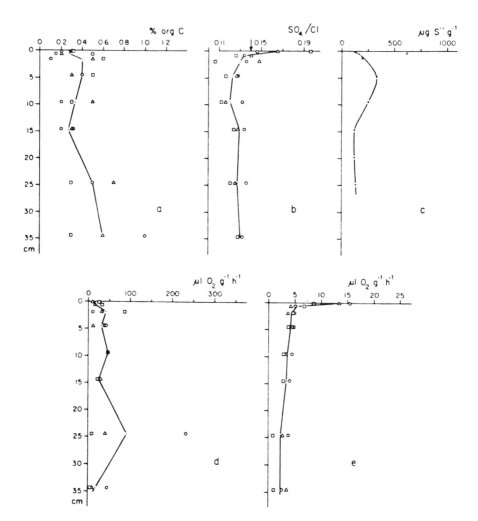

Figure 8-9. Vertical distributions in an artificial tidal flat system and in natural tidal flat sediment from the Dutch Wadden sea: a. organic carbon content (% of dry sediment); b. sulphate-chlorinity ratio in the interstitial water; c. sulfide content (μg S$^-$g^{-1}); d. oxygen utilization (μl O$_2$ g^{-1} h^{-1}) and e. ETS activity (μl O$_2$ g^{-1} h^{-1}). Determinations in 3 separate cores; lines connect mean values. *Source:* Vosjan, J.H., Olanczuk – Neyman, K.M., Vertical distribution of mineralization processes in a tidal sediment. *Netherland J. Sea Research,* 11, pp. 14-23, 1977. (Reprinted with permission, from *Netherlands J. Sea Research.*)

Artificial Tidal Flat Sediment

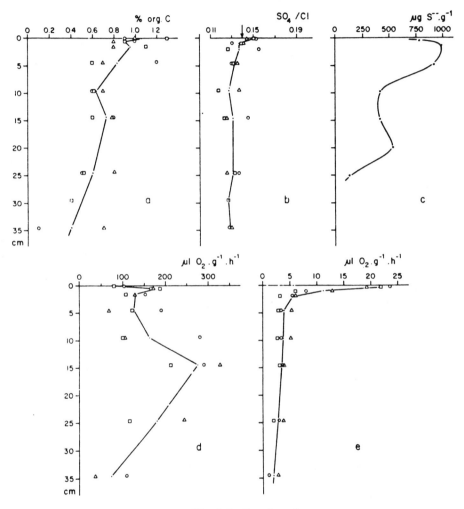

Fig. 8-9. Continued.

compared with other measurements (O_2 uptake, % organic matter, sulfide, and SO_4) and with natural and laboratory derived tidal flat sediment systems originating from different depths above the sediment surface. No major differences are apparent between the natural and artificial systems. With the ETS activity measurement, it is possible to demonstrate that most of the mineralization of organic matter occurred in the deeper anaerobic layers of the sediment. This fact has also been verified by Jorgensen and Fenchel (Ref. 8-109), who employed sulfate reduction as a mineralization monitor. In all cases, ETS activity revealed considerably more mineralization potential in sediments than in the overlying water. These results have an impact on the design of a microcosm, because if sediment depth is insufficient (i.e., allowing for only a surface oxidized zone) then mineralization will be reduced and nutrient cycling into the water column will become container dependent, rather than environment dependent, and will affect the biology in the overlying water. Therefore, it is important that a microcosm system be developed whose ETS activity level approaches that seen in natural systems. Pamatmat and Bhagwat (Ref. 8-155) have demonstrated, using direct microcalorimetry, that measures of ETS activity in undisturbed sediment–water cores agreed with *in situ* measurements. Their results also suggest that ETS activity may be the most accurate assessment of metabolic activity in sediments.

8-6.7 Nitrogen Fixation

With the tremendous wealth of information that now exists on nitrogen fixation in aquatic and terrestrial environments, it seems likely that this functional parameter could furnish another potential field calibration tool for microcosms. Using aquatic systems as an example, studies have demonstrated in several situations that it is possible to get *in situ* measurements with bell jar-like systems (Ref. 8-20) or sealed cores (Ref. 8-88). A variety of methods are also available for studies in laboratory systems (Refs. 8-83, 8-88, 8-121, 8-129, 8-202). However, the acetylene assay for N_2 fixation is complicated in sediment containing systems because the measurement of N_2 fixation is dependent on the diffusion of acetylene and its reduction product, ethylene, to and from the active sediment sites (Ref. 8-121). Herbert (Ref. 8-88), for example, has revealed large differences between *in situ* and laboratory measurements, presumably due to gas diffusions. Attempts to rectify this problem invariably result in disturbance or disruption of sediment–water interfaces, giving stimulated uptake rates that cannot be unequivocally attributed to the natural biology of the system (Ref. 8-60). However, refinements of the test may enhance its application in field calibrating microcosms.

8-6.8 Community Succession

One of the more common ecosystem structural measurements applied to microcosm studies for the purposes of field calibration is the study of autotrophic and heterotrophic organism succession. Even though successional events are community originated, the successional patterns, the rates of change, and eventual limits in successional development are a function of the physical system. Consequently, in a microcosm study, the size, shape, and inputs of a microcosm determine the physical environment and ultimately the structure of succession. Thus, a microcosm is an ecosystem unto itself, which, from an organic structural standpoint, may or may not have a corollary in the real world. However, the successional trends in a microcosm probably do have excellent similarities to naturally occurring ecological succession.

One of the most comprehensive studies of succession in a microcosm type system has been carried out by Jassby et al. (Ref. 8-108). Their study is a potential model for examining function and structure in laboratory systems, particularly for possible risk assessment. In their experiment, large tanks containing a 4-cm layer of acid-washed, river sand covered with 700 liters of water were supplemented with inorganic nutrients that would support algal growth and inoculated with a 3.5-liter sample of water from a eutrophic lake. The systems were then incubated for 150 days under constant aeration and a 12-hr light–dark cycle. Fluctuations in population densities of phytoplankton, protozoa, rotifers, and crustaceans were periodically monitored, as were changes in levels of pH, inorganic carbon, organic carbon, NH_4^{+1}, $NO_3^{-1} + NO_2^{-1}$, inorganic phosphorous, and total phosphorous. An example of the successional pattern observed in the study is shown in Fig. 8-10. The results can be divided into two major phases: a primary phase (0–80 days) in which a major bloom of phyto- and zooplankton occured and a secondary phase (80–120 days) in which population densities remained relatively constant. As would be expected in a closed system, the initial presence of high nutrient levels and low phytoplankton density produced a bloom of phytoplankton productivity, as has been observed in numerous other studies (e.g., Refs. 8-43, 8-107). Calculations and control experiments based on expected productivity from the available inorganic nutrients indicated that the standing crop was being regulated by the grazing activities of rotifers and crustaceans.

Studies of food sources for zooplankton are greatly complicated by the presence of periphyton on the walls, representing an unnaturally high source of productivity in the microcosm (approximately 9–10 times higher than water column productivity). This effect might be reduced by increasing the ratio of water volume (liters) to effective wall height (centimeters) (Ref. 8-108).

This type of successional study can provide good base-line data to compare with field situations. However, it may prove difficult to confirm similarities between laboratory and field studies because there is no information to demonstrate its reproducibility. It is also conceptually unclear at what point in the succession the comparisons should be made. Although the successions may not be totally representative of natural conditions the successional trends within the microcosms are probably very similar (under conditions of minimal recruitment), and as a result they can form the basis of effects testing, which could lead to valuable risk assessment predictions.

An excellent example of why caution should be exercised in investigating successional changes in laboratory systems is the work of Reed (Ref. 8-164), which illustrates the effects of system design and it's physical complexity on species diversity. A series of wide-mouth jars containing 3 liters of pond water were used to study effects of three variables (in eight combinations) in system design on species diversity. The variables involved the complexity or simplicity of the physical configuration (layer of washed river sand with a clay flower pot, compared with a beaker of rice or a beaker of glass tubing pieces), the stability or unstability of the temperature, and whether fertilizer was added (a commercial fish emulsion that supplied NO_3^{-1} and PO_4^{-3} along with organic matter). Each system was maintained under a 12-hr light–dark regime and incubated for 20 weeks. Periodic samplings revealed that the experimental variables produced

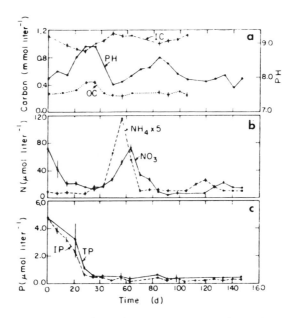

Fig. 8-10. Chemical and biological measurements performed on water samples taken from a microcosm system containing 400 liters of nutrient-supplemented lake water inoculated with an eutrophic lake water sample. Vertical lines indicate the standard error for duplicate measurements. (a) pH, inorganic carbon (IC), and organic carbon (OC); (b) NH_4 and $NO_3 + NO_2$; (c) inorganic phosphorus (IP) and total phosphorus (TP); (d) total volume of phytoplankton; (e) total volume of protozoa, rotifera, and crustacea; into cladocera, copepoda, and ostracoda; resolution of the cladocera into component species. *Source:* Jassby, A., Dudzik, M., Rees, J., Lapan, E., Levy, D. and Harte, J., "Production Cycles in Aquatic Microcosms." U.S. EPA Ecological Research Series #600/ 7-77-097, 1977. (Reprinted with permission of the author.)

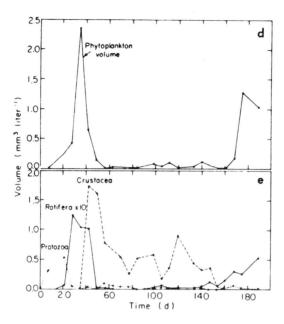

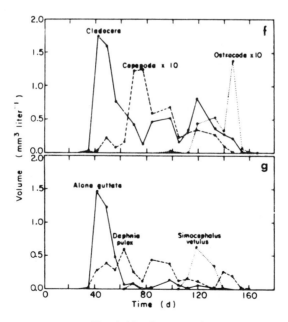

Fig. 8-10. Continued.

significantly different types of species diversity among phytoplankton, microbenthos, periphyton, macrobenthos, and microcrustacea. The fertilized systems, for example, had less diversity than unfertilized systems, a situation similar to eutrophic lake conditions. On the other hand, physical complexity increased species diversity in the unfertilized system but further decreased it in the fertilized system. Stable temperature conditions also led to greater diversity in some cases. These studies demonstrate that the system design readily effects the organic structure and this approach could provide another means of field calibrating a microcosm.

8-7 SYNTHETIC COMMUNITY STUDIES

With the increasing production and use of pesticides and other toxic chemicals, and with major gaps in our knowledge about the relationship between chemical structure and biodegradability or ecological magnification, several research laboratories have recognized the need to study laboratory ecosystems for the purposes of generating information to fill this knowledge gap. Prominent in this endeavor was the work of Metcalf and his colleagues (Ref. 8-134), who developed a unique laboratory system capable of screening chemicals for their potential biodegradability and ecological magnification. This system was called a "model ecosystem" by Metcalf but has been termed a laboratory microcosm on numerous occasions. In fact, his systems are synthetic communities in which both the kind and number of species added to the system depended on the decision of the investigator. There was little attempt to actually simulate a natural situation, except in the most general sense. Instead, the systems were designed for a high degree of homology and replicability so that each new compound tested in the model ecosystem would experience the same set of conditions, thus permitting valid relative correlations and comparisons.

For several years now, the Metcalf approach has been the model for laboratory studies designed to supplement risk assessment programs. As a result of the large number of compounds screened by Metcalf, the model ecosystem concept has dominated microcosm technology and new conceptual approaches to laboratory microcosm experimentation, especially those concerned with extrapolating laboratory results to the field, were slow to evolve. A careful discussion of the application of the Metcalf approach and other related studies to risk assessment therefore is merited.

8-7.1 The Metcalf Model Ecosystem

The basic "aquatic/terrestrial" Metcalf system consists of 10×10×20 in. glass aquaria containing a sloping shelf of washed white sand (Fig. 8-11). Approxi-

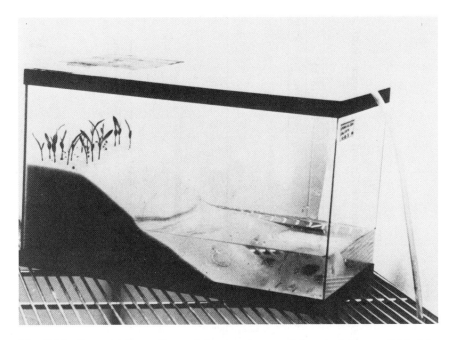

Fig. 8-11. Terrestrial-aquatic model ecosystem used for studying pesticide bio-degradability and ecological magnification. *Source:* Metcalf, R.L., "A Model Ecosystem for the Evaluation of Pesticide Biodegradability and Ecological Magnification," *Outlook on Agr.,* 7, pp. 55-59, 1972. (Reprinted with permission of ICI, England.)

mately 7 liters of aerated reference water (from another laboratory aquarium) containing *Oedogonium carciacum, Daphnia magna, Culex pipiens*, and *Physa sp.* covers the bottom portion of the slope. On the upper part of the sand (the terrestrial part) *Sorghum vulgare* seeds are planted. After 20 days of incubation, when the *Sorghum* leaves are approximately 10 cm high, a radiolabeled pesticide is applied to the *Sorghum* leaves at a rate equivalent to a field application. In order to disperse the labeled pesticide, the saltmarsh caterpillar, *Estigmene aire*, is allowed to eat the sorghum; its defecation products and the decay of dead caterpillars transports the pesticides to the soil. At the end of 30 days, three *Gambusia affinis* are added to the aquatic part and allowed to feed on the *Culex* for three more days. The selection of each of the biological components has been discussed by Metcalf (Ref. 8-135). At the end of this incubation period, the aquarium is completely dismantled and the quantity of pesticide and its degradation products in all of the biological components and in the water and sand are determined.

A picture of the system is shown in Figure 8-11. The potential food chain pathways are shown in Table 8-1.

TABLE 8-1. Ecological magnification (E. M.) and biodegradability index (B. I.) of PCBs and DDE compared with water, their solubility and partition coefficients

Chemical	Ecological Magnification (E. M.)				Biodegradability Index (B. I.)			
	Alga	Snail	Mosquito	Fish	Alga	Snail	Mosquito	Fish
Tri-Cl-PCB	7,315	5,795	815	6,400	0.30	0.17	0.35	0.60
Tetra-Cl-PCB	17,997	39,439	10,562	11,863	0.015	0.082	0.076	0.060
Penta-Cl-PCB	5,464	59,629	17,345	12,152	0.029	0.027	0.0134	0.019
DDE	11,251	36,342	59,390	12,037	0.069	0.049	0.033	0.050

Chemical	H_2O Solubility (ppb)	Partition Coefficient
Tri-Cl-PCB	16	7,803
Tetra-Cl-PCB	16	8,126
Penta-Cl-PCB	19	16,037
DDE	1.3	18,893

Source: Metcalf, R. L., Sanborn, J. R., Lu, Po-Yung, and Nye, D., "Laboratory Model Ecosystem Studies of the Degradation and Fate of Radiolabeled Tri-, Tetra-, and Pentachloro-biphenyl Compared with DDE," Arch. of Environ. Contam. and Toxicol., 3, pp. 151–165, 1975 (Reprinted with permission of Springer-Verlag, New York.)

Metcalf has emphasized that his system can be adjusted to fit certain environmental situations. For example, the pesticide can be injected directly into soil as a similation for a preemergent herbicide (Refs. 8-120, 8-134) or it can be added to the water component to simulate toxic effluent. In a similar manner, the physical and biological components of the system can be varied (Ref. 8-135); the elimination of the terrestrial component, including the rooted plants, provides a model aquatic system (Refs. 8-128, 8-138). The *Sorghum* can be replaced with other agriculturally important crops, such as rice (Ref. 8-120), or the aquatic portion can be removed completely to provide a totally terrestrial system (Ref. 8-39). In the latter case, vermiculite contained in the bottom of a 19-liter wide-mouth jar served as the soil and caterpillars, an earthworm, a slug, an isopod, and an omnivoric prairie vole, *Microtus ochrogaster,* served as the principal food-chain linkage with the plants (corn). This system was later modified to include a flooded soil component as well (Ref. 8-39).

Metcalf and his colleagues have provided an impressive data base (Refs. 8-39, 8-134, 8-141). Their model ecosystem studies offered early warning systems to screen pesticides, plasticizers (Refs. 8-136, 8-137, 8-140, 8-167), chlorinated organics (Refs. 8-128, 8-137), and other types of chemicals (Ref. 8-139), and thereby provided *relative* estimates of environmental impact. Yu and his colleagues (Ref. 8-203) also generated similar data sets on herbicides. As Metcalf states, "the model ecosystem approach is essentially comparative and pesticides (for example) are evaluated vis-a-vis certain well-established compounds, such as DDT, dieldrin, and 2,4-D, whose real world environmental toxicology has been studied in great detail. A variety of structural modifications of a potentially useful pesticide nucleus, e.g., triazine, diphenyl ethane, pyrethroid, or growth regulators may be compared in detail to permit selection of derivatives which optimize target-site effectiveness with appropriate environmental safety" (Ref. 8-141).

Several specific measurements and observations can be obtained from the Metcalf systems; the most important of these are the types and distribution of chemical and biological degradation products, the determination of a biodegradability index (Ref. 8-137), (B.I., the ratio of the amount of solvent-unextractable material, or polar products, found in water or animal tissue to the amount of solvent-extractable material, with B.I. decreasing with decreasing ratio), and the determination of the ecological magnification (E.M., ratio of parent compound and degradation products in organisms to the amount in water; not to be confused with bioconcentration factors). Very useful information, therefore, can be gained about comparative metabolism both in terms of compound structure *potential* environmental degradation pathways, persistence of degradation products, and the relationship of intrinsic molecular properties (benchmark characteristics, Refs. 8-27, 8-147) of a xenobiotic to fate and ecological magnification.   This experimental versatility can be illustrated by two exemplary studies.   First, model ecosystem studies revealed chemical substitutions and deletions in the DDT (Ref. 8-141) and the PCB molecule, (Ref. 8-139) that were most susceptible to biodegradation and ecological magnification. A summary of the results with DDT is shown in Table 8-2.   The results agree well with microbiological investigations of others (Ref. 8-2) and of Focht and Alexander (Ref. 8-63) who performed similar studies, using pure cultures of bacteria. In a second example, the correlation between ecological magnification and water solubility (and/or octanol–water partition coefficients) was firmly established (Refs. 8-128, 8-134, 8-138).   The linearity of this correlation is shown in Fig. 8-12.   This agrees with similar correlations using standard single species bio-concentration tests (Ref. 8-80).

8-7.2 Drawbacks of the Synthetic Community Approach

The Metcalf approach, as a potential research tool in risk assessment programs, is not without its drawbacks. Several points need to be iterated.

(1) The analytical chemistry involved in the use of a model ecosystem study is enormous and requires several well-trained chemists and a large outlay of time and money.   The examination of the fate of ten insecticides and fungicides in the terrestrial simulation model ecosystem required 14,000 quantitative and qualitative analyses (Ref. 8-40). This heavy analytical burden greatly restricts

TABLE 8-2. Model ecosystem characterization of biodegradability (B. I.) and ecological magnification (E. M.) of DDT analogues in *Gambusia*

$$R^1 \!\!-\!\!\bigcirc\!\!-\!\!\overset{\overset{\text{H}}{|}}{\underset{\overset{|}{R_3}}{C}}\!\!-\!\!\bigcirc\!\!-\!\!R^2$$

R^1	R^2	R^3	E. M.	Fish B. I.
Cl	Cl	CCl_3	84,500	0.015
Cl	Cl	$HCCl_2$	83,500	0.054
CH_3O	CH_3O	CCl_3	1,545	0.94
CH_3	CH_3	CCl_3	140	7.14
CH_3S	CH_3S	CCl_3	5.5	47
Cl	CH_3	CCl_3	1,400	3.43
CH_3	C_2H_5O	CCl_3	400	1.20
CH_3O	CH_3S	CCl_3	310	2.75
CH_3O	CH_3O	$C(CH_3)_3$	1,636	1.04
Cl	Cl	$HC(CH_3)NO_2$	112	3.27

Source: Metcalf, R. L., "Model Ecosystem Studies of Bioconcentration and Biodegradation of Pesticides," in *Pesticides in Aquatic Environments*, M. A. Q. Khan (Ed.), Plenum Press, New York, 1977. (Reprinted with permission of Plenum Press, New York.)

on a relative basis within the *same* experimental design. In this way, the arbitrary nature of the system is potentially circumvented.

However, this approach hinges on several major assumptions. First, in many attempted simulations, one organism is primarily responsible for the initial step in transport and distribution of the pollutant within the system, i.e., the caterpillar in the terrestrial–aquatic system and the vole in the terrestrial system. As a consequence, in order to insure replicability and reproducibility with each test system run, great care must be exercised in handling and introducing the animals into the systems. A particular set of caterpillars failing to feed efficiently on the *Sorghum* leaves could curtail the distribution of a pesticide or alter its time course. In some tests (Ref. 8-40), the initial inoculum of caterpillars died and new ones were added until all of the *Sorghum* leaves were consumed, altering the distribution pattern of the test compound in the various biological components. Second, if any members of the synthetic community die during a test, it is perplexing to decide on a course of action. Regardless of whether the deaths are ignored or new animals are added to return the population to its original complement, the availability of the pesticide to various compartments in the system is likely to change. In either case, the results could greatly affect the ecological magnification (since the exposure concentration is variable and unpredictable) or the extent of degradation, particularly if degradation was tightly linked to the aquatic portion. This was reflected in studies on diflubenzuron by Metcalf *et al.* (Ref. 8-140), who reported that contamination of the water was variable between three separate tests. It is quite possible that experimental error in the values obtained from a model ecosystem study are large enough to negate certain conclusions on fate and transport. Unfortunately, statistical analysis is not a routine part of the Metcalf experiments; this may be due, in part, to the large time factor involved in each study. Isensee (Ref. 8-98), using an altogether different synthetic community, has shown significant statistical variations.

(3) A decision to embark on a risk assessment program involving the Metcalf approach should be considered carefully because the same results might be obtained by much simpler and less costly experiments. For example, in Metcalf's ecological magnification experiments, the source and availability of a xenobiotic, or its degradation product (i.e., directly from soil or from food), to a particular organism is not known in the model ecosystem study; therefore, the interrelationships between the species and the surrounding milieu lead to biomagnification information that is of no consequence in making an environmental prediction. Branson (Ref. 8-27) pointed out that a standard bioconcentration factor test (a more straight forward test in which uptake by a single organism is optimized for a pesticide) is far more valuable in environmental decision-making than Metcalf's ecological magnification value simply because Metcalf cannot account for the route and extent of exposure. This is illustrated in Table 8-3. Note that ecological magnification values did predict the potential for biological accumulation to the same extent as single species bioconcentration

TABLE 8-3. Comparisons of biomagnification values predicted from model ecosystem studies by Metcalf and from octanol–water partition coefficients

Compound	Predicted Bioconcentration Factors[a]		Aquatic Ecosystems[b]	
	BCF	Rank in Series	Ecological Magnification	Rank in Series
Pentachlorophenol	750	1	296	2
Chlorobenzene	46	2	650	1
Benzene	19	NR[c]	NR	NR
Benzoic acid	14	3	21	4
Nirtobenzene	13	3'	29	3
3, 5, 6-trichloro-2-pyrindinol	11	3''	16	4
Phenol	8	NR	NR	NR
Aniline	4	4	6	6

[a] BCF were calculated from Neely *et al.* (Ref. 8-147) using $\log_{10} = 0.56 \log P + 0.124$. Source reference contains information of $\log P$ values used.

[b] Values are those reported by Lu and Metcalf (Ref. 8-128) from model ecosystem studies.

[c] NR, not ranked because data on ecological magnification not available.

Source: Branson, D. R., "Predicting the Fate of Chemicals in the Aquatic Environment from Laboratory Data," in *Estimating the Hazard of Chemical Substances to Aquatic Life*, J. Cairns, K. L. Dickson, and A. W. Maki (Eds.), ASTM STP 657, Philadelphia, pp. 55-70, 1978. (Reprinted with the permission of ASTM, Philadelphia, PA.)

tests. It should also be pointed out that if the intent is to screen toxic chemicals to determine which structural derivatives are environmentally safe on a relative basis, the *same* information can be obtained by performing single-species bioconcentration tests, which are simpler and less expensive.

(4) Use of the term "aquatic" is quite misleading. Since tap water and terrestrial soils are the major components, the contribution of microorganisms and benthic invertebrates from actual aquatic ecosystems is totally absent. Their presence (which would be simple to incorporate into the system) could have a significant effect on the bioaccumulation values either by modifying biodegradation pathways and kinetics or by altering toxicant releases from bed solids through bioturbation processes. It is misleading to report slow biodegradation of a carbamate insecticide in an aquatic community model ecosystem study similar to Metcalf's system (Ref. 8-112) when indigenous micro- and macroflora of actual aquatic environments were not employed. The same argument also applies to the use of clean sand as a biological model of soil.

(5) Metcalf's biodegradability index is based on the ratio of the nonpolar material (parent compound and immediate degradation products) in water to the polar degradation products remaining in the water. Although the existence of unextractable products probably represents an extensively degraded parent compound, it is an ambiguous measure of biodegradation. It is not clear what factors control or permit accumulation of these unextractable products or whether it is valid to assume that polar product formation is independent of chemical type. Some chemicals that are extensively degraded to CO_2 may receive misleadingly low biodegradation index values; CO_2 production from parent compound in Metcalf's systems is not determined. A similar situation can also occur through irreversible (solvent unextractable) bound residues or conjugate formation in soil or sediment. Branson (Ref. 8-27) has compared biodegradation rankings of several chemicals, using biodegradation index information from Lu and Metcalf and 5-day BOD values from Veith and Comstock (Ref. 8-185) (Table 8-4). As can be seen, rankings, in some cases, are not consistent; this may reflect the problems associated with the definition of biodegradability index.

TABLE 8-4. Comparisons of biodegradability determined from BODs (5 day) and from biodegradation index (B. I.) values obtained through model ecosystem studies

Compounds	BOD		Aquatic Ecosystems[a]	
	% of Theory	Rank in Series	B. I.	Rank in Series
Phenol	76	NR[b]	NR	NR
Aniline	68	1	1.78	2
Benzoic acid	66	1'	2.97	1
Benzene	45	NR	NR	NR
Chlorobenzene	24	2	0.014	6
Nitrobenzene	0	3	0.023	5
Pentachlorophenol	0	3	0.338	3
3, 5, 6-Trichloro-2-pyridinol	0	3	0.311	4

[a] From Lu and Metcalf (Ref. 8-128): B. I. $= \dfrac{\text{polar (thin layer chromatography origin)}}{\text{nonpolar (total }^{14}\text{C minus polar)}}$

[b] Not ranked because aquatic ecosystem data were unavailable for comparison.

Source: Branson, D. R., "Predicting the Fate of Chemicals in the Aquatic Environment from Laboratory Data," in *Estimating the Hazard of Chemical Substances to Aquatic Life*, J. Cairns, K. L. Dickson, and A. W. Maki, (Eds.), ASTM STP 657, Philadelphia, pp. 55-70, 1978 (Reprinted with the permission of ASTM, Philadelphia, PA.)

Similar concerns with the Metcalf systems have been noted by Bondietti *et al.* (Ref. 8-22). However, their criticisms are, in part, based on "what if" questions—what if the species of animals are changed, what if it rains on the soil, what if something else eats the *Sorghum* and then flies away, what if the compound is converted to CO_2? These questions are only germane if one views the Metcalf system as a simulation of an actual environmental situation. It is not—it is only a conceptual simulation and, therefore, incapable of producing conditionally absolute information about environmental processes and effects. As such, it is meaningless to force the question of how certain variables in system design and operation affect a model ecosystem data set. The system represents a unified screening approach based on relative measures and thus is independent of "what if" inquiries.

8-7.3 Other Types of Synthetic Community Studies

Several other types of synthetic community test systems have been developed and examined for particular applications. Tsuge *et al.* (Ref. 8-183) have modified the Metcalf system to monitor $^{14}CO_2$ released by the degradation of radio-labeled parent compound. Isensee and his colleagues (Refs. 8-99, 8-100) have used an aquatic community system in which the principal feature is the monitoring of the fate of a toxic material by introducing it to the test systems adsorbed on soil particles. This permits a more predictable method of distributing the toxic material throughout the system. It also models toxicant input into the system as it might occur during erosion of pesticide contaminated soil.

The Isensee test system consists of an aquarium in which the adsorbed ^{14}C radiolabeled material is spread over the bottom and covered with 80 liters of continuously aerated tap water. The soil also is covered with pea gravel and a screen in order to maintain an unsuspendable bed. If desired, tanks can be vertically bisected with screens to permit spatial separation of predacious components of the animal community. Aquarium walls are light-shielded to reduce proliferation of algae. The synthetic community consists of fingerling catfish (*Ictalurus*) or *Gambusia*, crayfish (*Procambaru*) snails (*Physa*), a crustacean (*Daphnia*), algae (*Oedogonium*), and duckweed (*Lemna*). Water in the tanks is chemically analyzed at 2-day intervals, and on the 20th day all organisms are harvested and analyzed for residues.

Another example of an elaborate synthetic community test system is the terrestrial microcosm chamber developed by Gillette and Gile (Ref. 8-73). These are large chambers (1 $\times$ 0.75 $\times$ 0.61 m) operated under standardized conditions of light (16-hr days), temperature, air movement, and rainfall. The soils consist of a mixture of an organic matter source, 20-grit sea sand, and illite clay all layered on a bed of river gravel and course sand. The systems are designed to monitor the movement of a pesticide through the atmosphere, soil, biota, and ground water (obtained by collection of simulated rainfall through the soil and out the gravel bottom). The biotic components are numerous in type and include alfalfa, ryegrass, nematodes, earthworms, pillbugs, meal worms, crickets, snails, and, finally, a gray-tailed vole.

Many of the problems and advantages associated with the Metcalf system also apply to other synthetic community studies; analytical efforts are massive and, therefore, slow. The movement of materials among components of the food chain is largely unknown because component studies on the food chain dynamics have not been performed and large numbers of runs with reference toxicants must be completed before good comparative data can be obtained from each system. At their present stage of the development, these systems represent potential screening tools. It would appear, however, that many system variables, such as air flow control, water balance, population dynamics, pesticide disposition, replication, and so on, need to be standardized before the systems are useful as screens and are able to supplant simpler screening methods. This is discussed in more detail in the proceedings of a terrestrial microcosm workshop held in Corvallis, Oregon (Ref. 8-200). Favorable points raised in Section 8-7.2 are counterbalanced by the inappropriateness of such synthetic community studies for producing information that can be extrapolated to actual environmental settings and their inability to fulfill the operational criteria defined for microcosms in Section 8-3.

8-8 FATE AND TRANSPORT STUDIES

Two approaches often are employed to assess the fate of a xenobiotic in the environment, and it is important to carefully distinguish them.

The first approach seeks to determine what I loosely call the biodegradability of a compound. The second approach attempts to determine (or predict) what is referred to as the estimated exposure concentration (EEC), or simply the concentration of a chemical to which some target organism may be exposed at some environmental point in time and space. These approaches are by no means mutually exclusive and, as our technology and conceptual basis in risk assessment improves, they may eventually be unified into a single approach, probably that of an EEC determination. However, it is important to outline the differences between the two approaches and indicate where microcosm studies play a major role.

8-8.1 Biodegradability Testing—Applications and Limitations

The determination of biodegradability includes the use of a test method or system in which the rate of disappearance of the parent compound (usually half-life) is determined. There are, however, several self-imposed limitations and assumptions to this biodegradability testing.

(1) Although the test system or method is initially designed to be "environmentally realistic," the question of how the test results vary from an *actual* environmental situation is never pursued; they are simply accepted for their qualitative value. The results have, *a priori*, no relationship to actual rates of degradation in the environment. In fact, pursuit of such a relationship, as by calibrating the test system with natural environmental conditions, will, in most cases, jeopardize the practicality of the test. These biodegradability tests, therefore, do not represent microcosm studies, but rather are what I call an actual media tests, meaning that biodegradation is assessed by using actual samples of environmental material.

(2) Test results on a new chemical are interpreted *relative* to previous results on other chemicals. It is only appropriate to assess the biodegradability of a chemical on the basis of its ranking with other chemicals, i.e., whether one chemical is more or less degradable than another chemical. It is not appropriate to make quantitative judgments about how *fast* the chemical will degrade in the environment as there is no information provided (nor is it necessarily intended to do so) on how representative the biodegradability test is of the environment of concern.

(3) The utilitarian aspects of a relative biodegradability test are tremendous, and it is probably correct to say that the EPA and other regulatory groups base most of their decisions regarding various chemicals on empirical biodegradability tests much in the same way as standard toxicity tests are presently used for water quality criteria evaluation. The ability to determine if a compound takes a few hours or days to disappear, as opposed to several weeks or months, is important in predicting persistence and is of immense value to a regulatory agency and to the chemical industry in their development of new chemicals. However, there are "gray area" compounds for which a simple yes/no answer is not possible; procedures other than a simple biodegradability test are required.

(4) It is entirely possible that, after a certain degree of research, a biodegradability test, in fact, may be shown to be quite reliable for determining a biodegradation rate directly applicable to a specific environmental situation. An example of such an approach is the work of Baughman and his colleagues (Ref. 8-157), which is further discussed below. However, a biodegradability test should not be considered initially for its extrapolability to real life situations, but for its pragmatic potential in forecasting the persistence of a chemical and providing qualitative information on aspects such as degradation products, microbial interactions, effects of anoxia, and effects of different environmental samples.

8-8.2 Examples of Biodegradability Testing Methods

Several examples of biodegradability testing methods are in order. The methods commonly used in soil testing for pesticide degradation have been reviewed by Kaufman (Ref. 8-111) and Howard *et al.* (Ref. 8-94). Two basic types have been developed, opened and closed. The closed system is exemplified by the soil biometer flask test developed by Bartha and Pramer (Ref. 8-10). This test method for tracking biodegradation uses radiolabeled compounds and follows mineralization to carbon dioxide. It has been standardized in the EPA's "Pesticide Guidelines." Moreover, this test method has excellent versatility—degradation products and pathways can be determined, bound residues can be observed, multiflask experiments can be run for replication, and the test can be run under anaerobic conditions. However, the oxygen content of these systems is difficult to control, and reductive conditions may develop that would affect biodegradation (Ref. 8-77).

Problems of unambiguously distinguishing between $^{14}CO_2$ and other volatile ^{14}C were addressed and experimentally compensated for by Marvel et al. (Ref. 8-133) and Kearny and Konstson (Ref. 8-113). A logical extension of the soil biometer flask was the incorporation of a gas flowthrough device to provide a constant movement of air through the system. Alkaline traps were employed to monitor exiting $^{14}CO_2$ produced from the degradation of a radiolabeled parent compound. These systems are typified by those developed by Kearney and Konstson (Ref. 8-113) and Murthy and Kaufman (Ref. 8-146). Kaufman (Ref. 8-111) noted that degradation rates in flow-through shake flasks are faster than in static biometer flask.

Generally, the results generated from these test methods agree at least qualitatively with results obtained with field studies using soil plots amended with test chemical. However, not enough work has been done to demonstrate that rates of degradation in the laboratory biodegradability tests are equivalent to those observed in the field. Until this equivalency can be demonstrated, the need for continued terrestrial microcosm development is important.

Methods for aquatic environments typically involve river die-away tests—the disappearance of parent compound (as determined by a variety of analytical procedures) in a shake flask containing natural water samples with or without sediment is employed to measure biodegradability. This test was developed to study the biodegradation of surfactants and has been employed extensively to study the biodegradability of a number of other detergents (Ref. 8-178), pesticides (Ref. 56), and polyglycols (Ref. 72). Several other related tests, involving among other things shake flasks and standardized sewage or activated sludge inoculums, have also been developed and extensively utilized (Refs. 8-30, 8-59, 8-72, 8-94).

Another offshoot of the river die-away tests is a biodegradability test that examines fate of xenobiotics in estuarine systems under conditions of minimal disturbance to the sediment water interface. A sediment-core fate testing system, for example, has been developed at the U. S. EPA Environmental Research Laboratory, Gulf Breeze (Refs. 8-24, 8-25, 8-162). It is referred to as "eco-core" and is schematically depicted in Fig. 8-13. It generates fate infor-

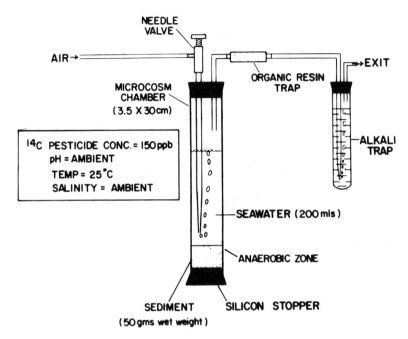

Fig. 8-13. Schematic diagram of eco-core biodegradation test system used for biodegradation studies in estuarine salt marsh environments. *Source:* Pritchard, P.H., Bourquin, A.W., Frederickson, H.L., and Maziarz, T., "System Design Factors Affecting Environmental Fate Studies in Microcosms," in *Proceedings of the Workshop: Microbial Degradation of Pollutants in Marine Environments,* A.W. Bourquin and P.H. Pritchard (Eds.), U.S. EPA Ecological Research Series, 600/9-79-012, pp. 251-272, 1979.

mation based on an integration of all the fate processes occurring in the sediment–water sample. Thus, not only are data supplied on hydrolysis, sorption, biodegradation, and volatilization but they are supplied on other known and unknown fate processes that significantly contribute to the overall fate and detoxification of the pesticide. The system is easily set up and analyzed; therefore, a large number of cores can be employed, permitting a variety of environmental parameters to be tested (e.g., sediment type, sediment source, enrichment) (Ref. 8-162).

The eco-cores show very good replicability. Cores taken simultaneously from the same general sampling area behave similarly. At low pesticide concentrations, the eco-core does not deteriorate with time. A variety of experiments (sediment–water ratios, intact versus mixed sediment, aeration mixing) have shown that system design features are optimum (Ref. 8-162). The effects of variations in the physical and biological conditions of sediment–water samples, taken at different times in the year, can be normalized by making relative comparisons with a standard reference compound.

Figure 8-14 illustrates the typical results obtained from eco-core experiments employing intact cores from a salt marsh. It shows that methyl parathion disappears from the warer column (half-life = 29 days) due to chemical and biological hydrolysis to paranitrophenol and due to biological (nonspecific) reduction to amino methyl parathion, a relatively nontoxic intermediate. This latter product accumulates and then slowly disappears. Table 8-5 gives an example of the total budget from one core in this experiment. Note the level of polar products (solvent unextractable from water), indicating possible biological oxidation to water soluble products, and the level of sediment bound residue (radioactivity on sediment remaining after extraction).

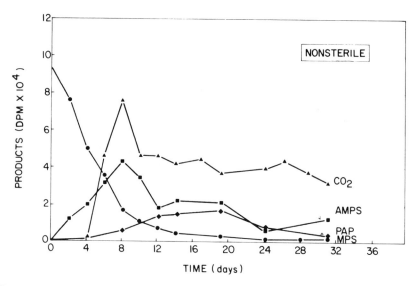

Fig. 8-14. Degradation products detected in methylene chloride extractable material from water overlying sediment in sterile and nonsterile ecocore test systems spiked with methyl parathion (0.5 mg/liter). MPS = methyl parathion, AMPS = aminomethyl parathion, PAP = p-aminophenol. Products are expressed as dmp/ml of water column. $^{14}CO_2$ is total produced averaged per day in collection period. *Source:* Pritchard, P.H., Bourquin, A.W., Frederickson, H.L., and Maziarz, T., "System Design Factors Affecting Environmental Fate Studies in Microcosms," in *Proceedings of Workshop: Microbial Degradation of Pollutants in Marine Environments,* A.W. Bourquin and P.H. Pritchard (Eds.), U.S. EPA Ecological Research Series 600/9-79-012, pp. 251-272, 1978.

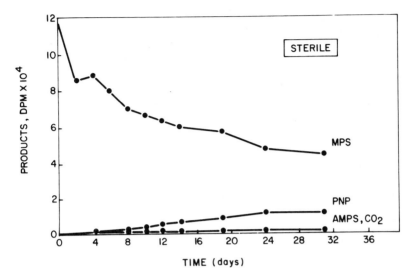

Fig. 8-14. Continued.

Thus, this sediment–coring system provides a biodegradability test system that that can be applied to a variety of aquatic situations. Although this system has been previously labeled a microcosm (Ref. 8-25), very little attempt has been made in our laboratory to determine how it actually relates to a natural estuarine setting. The decision at the present time is to use it as a biodegradability test system and to utilize the data only on an intracompartative basis. Research with other microcosm systems in our laboratory (see Section 8-8.5) may eventually validate the similarity or differences between eco-core and natural environmental situations.

8-8.3 Research Development in Biodegradability Testing

In tests both for soil and aquatic environments, the usefulness of the results depends on how they are incorporated into risk assessments. In some cases, such as with detergents, the BOD loading to an environment is the major concern, and thus total mineralization of the chemical will have to be considered. With pesticides, on the other hand, toxicology is of far greater concern than the total carbon input to the environment. Therefore, the mechanism of detoxification is important and may involve a very simple analytical assessment, such as the extent of hydrolysis.

One of the more serious shortcomings of these actual media tests is the variation of bacterial activity and biomass from one environmental sample to another, even when samples are taken from the same field site. It is possible to

TABLE 8-5. Material balance of radioactivity in water and sediment taken from an eco-core initially spiked with ^{14}C-MPS (0.5 mg/liter) and analyzed after 32 days of incubation

Type of Sample	Percentage of total ^{14}C	
	Sterile	Nonsterile
Carbon dioxide	0.1	23.7
Overlying water	62.2	41.3
Extracted water[a]	20.1	28.4
Extractable products[a]		
MPS	62.7	3.9
PNP	16.3	2.2
AMPS	0.4	48.0
PAP	0.6	12.0
BASE	0.0	4.4
Detritus		
Extractable	6.0	1.6
Bound	1.2	14.6
Sand		
Extractable	1.0	0.4
Bound	0.8	2.2
Interstital water	23.8	11.6
Core wall washes	3.5	4.3
Resin trap	0.0	0.0

[a] Percentage figures are percent of total ^{14}C in overlying water. MPS = methyl parathion, PNP = para nitrophenol, AMPS = amino methyl parathion, PAP = para aminophenol, base = radioactive material remaining at base of TLC plate after development.

Source: Pritchard, P. H., Bourquin, A. W., Frederickson, H. L., and Maziarz, T., "System Design Factors Affecting Environmental Fate Studies in Microcosms," in Proceedings of Workshop; Microbial Degradation of Pollutants in Marine Environments, A. W. Bourquin, and P. H. Pritchard, (Eds.), U. S. EPA Ecological Research Series 600/9-79-012, pp. 251-272, 1978.

somewhat compensate for these variations, somewhat, by running a standard reference chemical with each new chemical tested. The microbial biomass problem, however, can be tackled in another way, as proposed by Baughman and his colleagues (Ref. 8-157). They conducted river die-away experiments to determine the kinetics for the disappearance of parent compound. The rate of disappearance of three different pesticides was found to be pseudo-first order with respect to substrate (Ref. 8-157). These rates, of course, were a function of compound structure but the overall *rate constant* for any particular compound in any environmental sample was the same as long as the samples were normalized for microbial biomass. Table 8-6 presents a summary of their data for chloropropham, malathion, and 2, 4-D. Note that the pseudo-first-order rate constants do not vary by more than a factor of 4. This work implies that it is

TABLE 8-6. Second-order degradation rate constants for three
compounds in water

Compound	k 1 org^{-1} hr^{-1} [a]	M [b]	$T(^\circ C)$ [c]
2, 4-D butoxyethyl ester	$(5.4 \pm .50) \times 10^{10}$	31	18.8 ± 6.6 (1-29)
Malathion	$(4.4 \pm .66) \times 10^{11}$	14	21.1 ± 7.8 (2-27)
Chlorpropham	$(2.4 \pm .42) \times 10^{14}$	11	19.1 ± 5.6 (7-28)

[a] Mean across all sites and all concentrations ± relative standard deviation.

[b] Number of sites for which k was determined.

[c] Average temperature in degrees Centigrade ± standard deviation. Values in parentheses are the range of temperature.

Source: Baughman, G. L., Paris, D. F. and Steen, W. C., "Quantitative Expression of Biotransformation Rates," in *Biotransformation and Fate of Chemicals in the Environment,* American Society of Microbiology, in press. (Reprinted with permission of American Society of Microbiology, Washington, D. C.)

potentially possible to reduce the experimental description of the biodegradation of a particular compound to simple rate constants that may be applicable over a large range of environmental variables. As long as substrate concentration is kept low, a biodegradation rate constant from a river die away experiment is as good as one from a microcosm study or any other study if the test systems can be normalized for microbial biomass. Their work also shows that the higher overall biodegradation rates frequently seen in sediments still represents the same rate constant but higher biomass. The rate constant approach is certainly a provocative concept capable of potentially expediting risk assessment and the extrapolation of laboratory data to the field. It also will certainly stimulate and orient a large data gathering effort.

However, some caution is in order. The rate constant approach not only depends heavily on our technology for accurately measuring microbial biomass (which is anything but ideal), but also on the assumption that all members of a microbial population are active on the test chemical and each is active to the same extent. Other investigations by Wright (Ref. 8-201) have shown that specific activity measurements of heterotrophic activity (^{14}C glucose uptake per bacterial cell) vary considerably from one environmental sample to another. Similarly, work with hydrocarbon degradation in aquatic environments has shown that oil-polluted areas are composed of microbial populations with a higher percentage of hydrocarbon oxidizing bacteria than populations from nonpolluted areas (Refs. 8-166, 8-188). This could mean, therefore, that differences in degradation rates between water samples may not be due to differences in total biomass but to differences in the percent degraders in that microbial population. Determining the percent degraders in a population may eventually prove to be a very elusive and complicated limitation of this approach.

It is also possible that the degradation rate of certain chemicals may be influenced by acclimation of the microbial populations during the experiment, which, of course, means a change in the percent degraders. Work in our laboratory (Spain *et al.*, unpublished results) shows that acclimation does, in fact, occur at low substrate concentrations (100 ppb) and that is occurrence is a function of the environmental conditions, not the associated microbial biomass.

8-8.4 Exposure Concentration Estimations—Role of Microcosms

In the preceding discussion, I emphasized that biodegradability tests are of tremendous practical consequence in determining relative measures of persistence. However, there arises a question as to how laboratory data can be extrapolated to field situations for risk assessment, that is, how definitive exposure concentration estimation (ECE) can be determined. As stated earlier, the biodegradability testing methods are not necessarily designed with this purpose in mind, although future research may reveal that they are quite applicable to ECE determinations, particularly if a kinetic approach is utilized. Our alternatives for experimental methodology to carrying out extrapolation-type work covers several areas: mechanistic studies, field studies, monitoring, and microcosms. It is not the purpose of this chapter to review the first three categories other than to point out their close interrelation and that validation sources for extrapolation procedures will undoubtedly come from these areas. The role of microcosm research in performing extrapolation of laboratory results to the field again must be discussed. It is clear that very few in-depth microcosm studies in fate assessment have been performed and that their potential for bridging the extrapolation gap has been only sparingly addressed.

First, the role of microcosms in fate work and exposure concentration estimations must be defined before published studies are reviewed. Ideally for an exposure concentration estimation, some type of fate and transport mathematical model of the environment or ecosystem in question must be available; models such as those developed by Lassiter *et al.* (Ref. 8-118), Neely, *et al.* (Ref. 8-148) and Cleseri *et al.* (Ref. 8-34) serve as examples. These models demand rate information about each of the major fate processes (biodegradation, hydrolysis, photolysis, bioaccumulation, etc.) affecting a particular chemical. The major question is how to integrate the rates of these fate processes in order to predict overall fate. Secondarily, an ECE will require knowledge of how environmental characteristics of a particular ecosystem compartment affect each fate process and their integrated outcome.

The simplest and most straightforward integration approach is to simply sum the rate constants of these processes, much in the manner described by Baughman and Lassiter (Ref. 8-12) and later more fully developed as the model EXAMS (Ref. 8-118) by the U. S. EPA Environmental Research Laboratory in Athens, Georgia. From the standpoint of risk assessment, it is important to know if all the important fate processes were considered and, if so, whether they were integrated in a way that mimics the overall fate observed in a natural situation. The microcosm, if it can adequately simulate a real-world condition, can supply this information; it generates results that are indicative of natural integration of these fate processes. If, for example, the rate of disappearance of parent compound in the microcosm agrees with that predicted by the mathematical model, a validation step has been made. For ECE determinations then, it is simply a matter of determining the rate constants for the individual fate processes, using simple tests (such as biodegradation rate constants determined in river die-away tests), and then plugging these values into the model. The microcosm has then technically served its immediate purpose.

However, if there is disagreement between the model's prediction and the microcosm results then presumably (if the microcosm is assumed to be a realistic analogy to the environment) either an important fate processes was overlooked or the integration of the rate constants for the fate processes was simply not an additive function.

An example of this latter case is indicated in some preliminary work from my laboratory in which the fate of methyl parathion in estuarine systems was examined (Ref. 8-162). Results indicate that, at certain times, substantial amounts of methyl parathion may be reduced to the nontoxic amino methyl parathion form, as shown in Fig. 8-15 using the eco-core test system. If reduction is a significant fate process, then an evaluative model such as described by

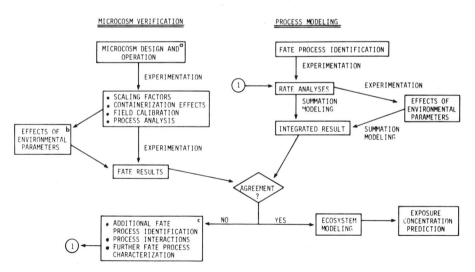

Fig. 8-15. Flow diagram depicting the use of microcosms as verification tools. The ability to make exposure concentration predictions depends on confidence in the information generated in boxes a, b, and c.

Baughman and Lassiter (Ref. 8-12) would be incomplete, not having considered a reducing environment. The major research problem then is to determine if reduction is a significant process in the environment (which can be answered with a microcosm study) and to derive a rate constant for reduction processes as a function of specific environmental parameters.

Thus, the role of microcosms in ECE determination is summarized in Fig. 8-18. It is also important to mention that the utility of exposure concentration estimations for subsequent risk assessment will depend heavily on the availability of rate constants for each of the important fate processes. This is because the rate constant approach is the simplest and most direct way of ultimately providing ECE determinations, particularly as a function of various environmental conditions. For this reason, microcosms are at best considered as supplemental verification tools for integrating these rate constants into the environmental models.

8-8.5 Types of Microcosm Studies Used for Fate Analysis—Terrestrial

The limited availability of good microcosms for studying chemical fate in terrestrial environment is largely due to an arbitrary establishment of system design, without checking its effect on the ensuing results and its effect on simulation capabilities. The situation is also complicated by the fact that there is no common denominator employed in these studies that would allow comparison between laboratory data and field studies; certainly the determination and use of rate constants could be a very appropriate normalizing factor.

Surprisingly, the availability of field-calibrated microcosms to study fate in terrestrial environments is quite limited despite the great many laboratory soil biodegradation tests that have been performed and their considerably greater accessibility to field studies than that possessed by aquatic environments. Two major aspects need to be considered in the design of terrestrial microcosms. First, soils are dominated by rooted vegetation, creating a unique environment, the plant rhizosphere, which can critically affect the fate of a chemical introduced near the plant roots. The rhizosphere, for example, can have significant effects on pH, nutrient composition, oxygen and CO_2 concentrations and on microbial activity in the soil both in time and space (Ref. 8-11). Second, one of the major fate processes affecting chemicals is leaching through soil. Leaching, unfortunately, is a difficult process to analyze and quantitate in the field and thus makes the calibration of microcosms considerably more tenuous. These aspects are generally ignored in standard soil biodegradation tests, thus necessitating microcosm-like studies to supplement these simpler tests.

Several types of microcosm-like systems are available to study leaching of organics and their transformation products. Lichtenstein et al. (Refs. 8-125, 8-126) have developed synthetized soil/plant systems that model typical field plot situations (Fig. 8-16). In the most recent design (Ref. 8-125), stainless-steel troughs (10 cm high, 22 cm deep, and 10 cm wide) were filled with soil and spiked by directly mixing a ^{14}C radiolabeled pesticide in the soil. Corn was then planted and allowed to grow for 22 days. During this incubation period, the plants and soil were periodically moistened with an artificial rain. Runoff water was collected in aquaria containing samples of lake mud, a floating water

Fig. 8-16. Two types of "plant-soil" systems used by Lichtenstein and his colleagues to study the fate of pesticides in terrestrial environments. These systems permit studies on potential pesticide transport through soil percolation and soil run-off. *Source:* in *Terrestrial Microcosms,* workshop report, J.W. Gillett and J.M. Witt (Eds.), National Science Foundation, Washington, D.C., pp. 00-00, 1900. (Reprinted with permission of the author.)

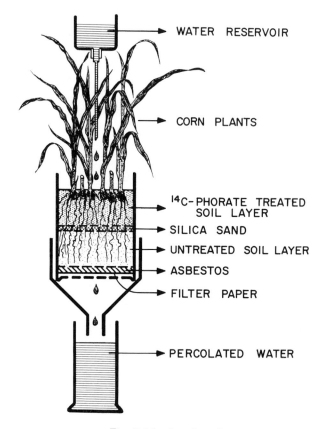

WATER   RESERVOIR

CORN   PLANTS

^{14}C-PHORATE   TREATED
   SOIL   LAYER

SILICA   SAND

UNTREATED   SOIL   LAYER

ASBESTOS

FILTER   PAPER

PERCOLATED   WATER

Fig. 8-16. Continued.

fern, and guppy fish. These aquaria were designed to model soil runoff into lakes. At the end of an arbitrary time period the soil and aquatic components, and biological species contained therein, were chemically analyzed for concentrations of parent compound and transformation products.

As with the systems developed by Metcalf and Isensee (see Section 8-7), the Lichtenstein system is synthetized and it is not shown how the system actually models the terrestrial environment under consideration. The system thus has many of the problems associated with synthetic community systems (see Section 8-7.2). However, the terrestrial component of the Lichtenstein system is potentially useful in studying the factors that affect leaching and transformation of pesticides in soil. Without further calibration with actual field situations, this microcosm approach cannot be used to predict, despite Lichtenstein's inference to the contrary (Ref. 8-125), how much of the parent compound or its degradation products will be released into *aquatic environments* by soil runoff.

Other less intricate types of leaching soil systems have been described by Best and Weber (Ref. 8-13), Hill and Arnold (Ref. 8-90), and Draggon (Ref. 8-52). Their systems provide for recovery of radiolabeled volatile products. The system developed by Draggon (Fig. 8-17) also uses intact soil cores that are considerably better representations of actual terrestrial situations than the artificially structured or synthesized systems used in the other studies. Draggon (Ref. 8-52) has shown that there are considerable differences in arsenic transport (leaching) between homogenized and intact soil cores. The mere fact that Draggan has attempted to calibrate his system against its own design makes this leaching test the state-of-the-art choice for providing good qualitative fate information.

An extended application of the soil-coring concept was described by Ausmus and Jackson (Refs. 8-102, 8-103). Their laboratory system was designed to

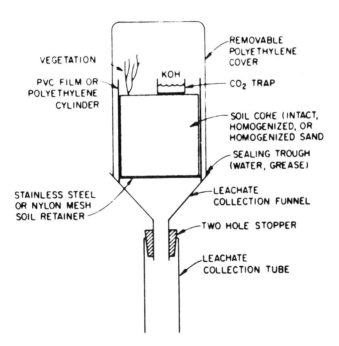

Fig. 8-17. Diagram of soil core microcosm. Intact cores are 5 cm in diameter and 5 cm in depth. *Source:* Draggan, S., "Effects of Substrate Type and Arsenic Dosage Level on Arsenic Behavior in Grassland Microcosms." Part I: Preliminary results on ^{74}As transport, in *Terrestrial Microcosms and Environmental Chemistry,* proceedings of symposium, Corvallis, Oregon, J.M. Witt and J.W. Gillett (Eds.), National Science Foundation, Washington, D.C., pp. 102-110, 1977. (Reprinted with permission of author.)

model a portion of a forest ecosystem and consisted of intact forest floor block sections (45 × 45 × 25 cm) in fiberglassed boxes. Each section contained an *Acer rubrum* sapling (2 m in height) and associated ground flora. A typical setup is shown in Fig. 8-18. Each box was incubated under greenhouse conditions and dosed with water to provide both rain simulation and a soil leachate taken from the bottom of the boxes. These systems were designed for ecosystem stress studies, but would be excellent for fate studies as well. They represent systems that could be used for scaling studies, and the existence of base-line studies in nutrient cycling (Ref. 8-104) and soil carbon metabolism (total gaseous carbon lost) (Ref. 8-6) provides good information for field calibration studies.

8-8.6 Types of Microcosm Studies Used for Fate Analysis—Aquatic

Artificial streams have been used by several investigators to help understand organic matter transformations of natural communities and trophic interactions. Most of these studies have been designed to gain information on the ecology of streams rather than the fate and effects of selected toxicants. However, stream microcosms are one of the few laboratory test systems that have been calibrated for their relationship to natural stream conditions. As such, they represent experimental systems with good baseline data sets and thus have good potential for fate and effects studies. Typically, stream microcosms are designed to model a large portion of the stream, such as a pool and riffle sections, in miniature, rather than a small section of just a pool or a riffle area (Refs. 8-115, 8-130, 8-143). As a consequence, they are constructed *de novo* in the laboratory by transporting material from the actual stream into the lab and then allowing the containerized system to colonize and stabilize over a period of months to years. The model streams are complicated and expensive to design and construct; they are usually large troughs or channels with mixing and pumping devices, monitoring instruments and elaborate lighting systems. McIntyre *et al.* (Ref. 8-143), Lauff and Cummings (Ref. 8-119), and Bott *et al.* (Ref. 8-23) have developed methods for removing and sampling portions of the model streams, using trays embedded in the bottom sediment. Figure 8-19 shows the stream channel microcosm developed by McIntire (Ref. 8-193). It is typical of many other studies: note the trays in the benthic zone. His studies, as well as others, have basically shown it to be a very suitable system for studying the structural characteristics of benthic algal communities, the effects of current velocity, leaf-litter decomposition.

Ponds and littoral zones in lakes have also been commonly modeled in microcosm systems. These studies use tanks or aquaria in various structured

Fig. 8-18. Forest microcosm unit used to study the behavior of heavy metals in forest ecosystems. *Source:* Jackson, D.R., Selvidge, W.J., and Ausmus, B.S., "Behavior of Heavy Metals in Forest Microcosms: Transport and Distribution Among Components," *Water, Air,* and *Soil Pollut.,* 10, pp. 3-11, 1978. (Reprinted with permission of Reidel Publishing, Dordrecht, Holland.)

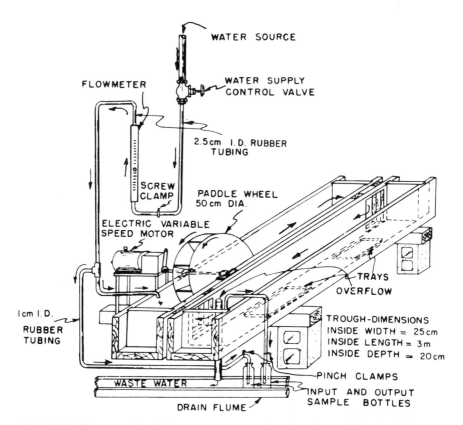

Fig. 8-19. Diagram of a laboratory streams, showing the paddle wheel for cir-culating the water between the two interconnected troughs and the water ex-change system. *Source:* McIntire, C.D., "Structural Characteristics of Benthic Algal Communities in Laboratory streams," *Limnol. and Oceanogr.*, 9, pp. 92-102, 1965. (Reprinted with permission of Duke University Press, Durham, NC.)

forms. Simsiman and Chesters (Ref. 8-170), for example, studied the persistence of the herbicide diquat in a system they called a "stimulated lake impound-ment." It consisted of large Plexiglas tanks (90 cm high, 30 cm in diameter) containing approximately 8 liters of sediment and 60 liters of water. Prior to the herbicide application, two plants—elodea (*Elodea canadensis*) and watermil-foil (*Myriophyllum spicatum*)—were allowed to grow in high densities. Although the study qualitatively revealed that the fate of diquat was dependent on the ex-tent of its adsorption to plant material (where microbial degradation of the herbicide can occur) rather than its adsorption to clay minerals (where the herbi-cide becomes unavailable for microbial attack), it is difficult to extrapolate to actual lake conditions because there was no attempt in the laboratory study to show whether the plant biomass was realistic relative to the amount of water and sediment. As often happens in these impoundment experiments, enrichment occurs, producing artificially high biomass. This could lead to the erroneous conclusion that diquat was not persistent due to its binding to plant material when, in fact, in the natural setting, the role of plant biomass in adsorption may be considerably less, thereby forcing more of the diquat to the sediments and increasing persistence.

Huang (Ref. 8-95) provides an example of a 38-liter aquarium tank study designed to study the sorption of pesticides to sediment. Although the system (pure clay as sediment, no mixing, constant temperature) might appear to be an analogy to existing conditions in a lake, it is, in fact, far too artificial for that purpose. Work in the Environmental Research Laboratory, Gulf Breeze (J. Connolly, unpublished data) has shown that very accurate and applicable sorption–desorption kinetics can be obtained in simple stir-flask systems using water volumes of 500 ml. This points to the importance of carefully defining the experimental question and then employing the simplest study system; it avoids the tendency of saying a system is analogous to a real world situation when, in fact, the appropriate comparisons have not been performed.

A study by Sikka and Rice (Ref. 8-169) on the fate of the herbicide endothall in farm ponds is one of the few attempts to correlate laboratory results with results from the field. A small, spring fed, farm pond (0.1 acre surface area and average depth of 4 ft) was sprayed with a water solution of endothall at a rate of 2 mg/liter. Periodic samples of water and sediment were analyzed for parent compound and the results compared with similar samples taken from a 38-liter aquarium containing 27 liters of water and a 4-cm layer of sediment from the same pond. The aquarium system was allowed to equilibrate for 7 days and then dosed with 2 ppm endothall. Figure 8-20 compares the rate of disappearance of parent compound from water and its rate of accumulation in sediment for each type of system. It is difficult to compare the two systems because biodegradation rate constants were not derived (no biomass available). It appears, however, that rates of disappearance in the aquarium were considerably faster than those in the pond. The temperature differences (1 to 4°C) may have been responsible for this, but they are unlikely to have caused the 20-day difference in time required to lower the concentration of endothall below detectable limits. Supporting evidence indicated that the ponds may have been previously acclimated to endothall degradation. The more rapid response of the aquarium system could have been due to this previous acclimation; either a large increase in biomass (and endothall degraders) could have occurred because of the containerization effect, thereby affecting biodegradation rates. It may have also been the result of a sediment surface area to volume ratio which may have grossly misrepresented natural pond conditions. This study emphasizes the

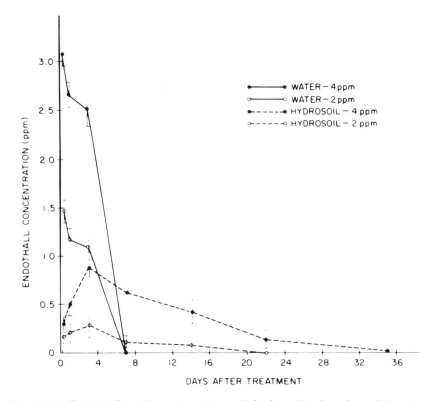

Fig. 8-20. Endothall residues in water and hydrosoil taken from laboratory aquaria and from a farm pond. Aquaria were treated with 2 and 4 ppm of the herbicide and farm pond was treated with 2 ppm. The bars represent the range of duplicate values. *Source:* Sikka, H.C. and Rice, C.P., "Persistence of Endothall in Aquatic Environments as Determined by Gas-Liquid Chromatography," *J. Agr. Food Chem.,* 21, pp. 842-846, 1973. (Reprinted with permission of American Chemical Society, Washington, D.C.)

critical need for system design calibration and the need to make laboratory-field correlations using a common experimental parameter such as biodegradation rate constants.

The problems of studying oil degradation in marine environments has prompted a number of microcosm-like experiments. For studies which model pelagic environments, it appears that small vessels containing natural water samples are sufficient test systems. For example, in reviewing studies using laboratory shake flasks, Bartha and Atlas (Ref. 8-9) concluded that weathering effects on crude oil were basically similar to observations made in the field (Refs. 8-19, 8-71, 8-93, 8-96, 8-150). They have also attempted to predict biodegradation rates for crude oil

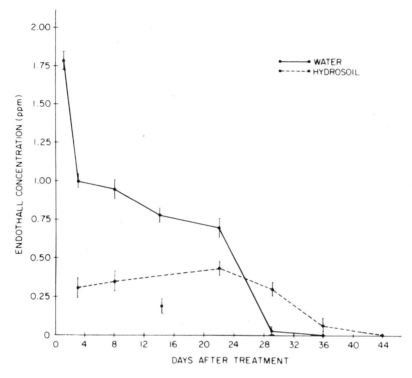

Fig. 8-20. Continued.

by using flow-through seawater tanks with miniature slicks of Sweden crude oil contained within vertical columns (Ref. 8-5). After a two-week lag, presumably due to the oil's toxicity, degradation (as measured by weight loss) commenced at a linear rate of approximately 10 g m^{-3} day^{-1}. Other investigators have used the mineralization of ^{14}C hexadecane as an indication of biodegradation potential in seawaters; rates from 0.001 to 0.050 g m^{-3} day^{-1} of hydrocarbon degraded (assuming 50% of the original hydrocarbon was converted to CO_2) have been observed (Ref. 8-9).

All of these results reflect degradation of the n-alkane fractions, rather than the branched or aromatic hydrocarbon components that degrade at considerably slower rates (Ref. 8-122). These studies also reflect the numerous complications involved in studying oil degradation in laboratory systems—the principal substrate, of course, is immisible with water, N and P often become rapidly growth-limiting in static systems, enrichment of specific alkane-utilizing microbial populations often occurs, and it is very difficult to model, in the laboratory, the physical–chemical processes (dissolution, sinking volatilization, emulsification, adsorption to particulate, photolysis, etc.) that occur in natural systems. The latter processes probably affect the fate of oil at sea more than biodegradation (Ref. 8-9). Even if good biodegradation rates were obtained in a laboratory model system, it would be difficult to apply this information to an environmental situation because the availability of good monitoring data on the actual fate of oil would be limited.

Based on these and other field and laboratory studies, Gibbs (Ref. 8-67) designed a special semiclosed continuous-flow apparatus that provides a method for measuring the oxygen uptake caused by the biodegradation of crude oil under nutrient limitations typical of the marine environment being considered. Although this apparatus has not been directly calibrated with the environment, it represents a microcosm approach capable of producing significant estimations of oil degradation rates in nature. A schematic diagram of the apparatus is shown in Fig. 8-21. His studies showed that the availability of nitrogen was the major controlling factor in biodegradation, with approximately 4 moles of nitrogen being required for the oxidation (theoretically derived from O_2 uptake) of 1 mg of crude oil. For English channel waters, this meant that the amount of oil oxidized in the system was dependent on the turnover rate for nitrogen; by extrapolation, the seawater was able to oxidize approximately 30 g of the saturate fraction of the oil per cubic meter per day. This is only one of a few studies in which a laboratory rate has been extrapolated to a natural setting. It was made possible by the fact that a specific normalizing factor, nitrogen limitation, could be used to make the comparison. The results, of course, are dependent on the validity of the nitrogen-limited situation seen in the test system.

Estuarine marsh and coastal area are characterized as nutrient-rich metabolically active transition zones that incorporate biological and chemical contributions from terrestrial and river runoff with similar indigenous contributions. Many estuarine areas are also the breeding grounds for large numbers of aquatic animals. As a result, interest in the toxic effects of certain organic chemicals has stimulated new questions about their fate, particularly in relation to the possibility that the rates and extents of biodegradation in estuarine environment cannot be deduced from studies in any generalized aquatic environment. Relatively few studies, however, have attempted to bridge the gap between laboratory and field studies in this particular setting and, consequently, microcosm-like studies are also uncommon.

In our laboratory, we are developing a continuous-flow sediment–water microcosm to learn more about the fate toxic organics in estuarine marsh areas (Refs. 8-24, 8-162). This system (Fig. 8-22) consists of a 1000-ml glass vessel containing 550 ml water over a 3-cm layer of detritus that, in turn, lies above a 3.5-cm layer of sand. The system is designed for continuous-flow operation; seawater enters from a head box containing natural seawater and from a carboy containing sterilized artificial seawater and the compound of interest (usually ^{14}C labeled). Both sources are pumped at equal rates, and the rate is slow enough to permit a change in concentration of parent compound between inflow and outflow (in this case a 48-hr retention time). The glass vessel is fitted with a Plexiglas lid that supports monitoring probes (pH, Eh, dissolved oxygen, temp) and a special stirring apparatus that keeps the water column completely mixed but does not resuspend the sediments (Fig. 8-1). The system is gastight in order to trap any radiolabeled CO_2 released from the metabolism of the parent compound. It is also designed so that water and sediment can be readily sampled

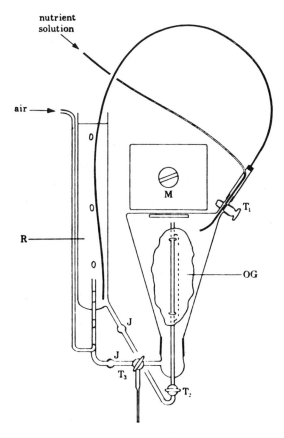

Fig. 8-21. Respirometer vessel used to determine oil degradation rates in sea-water as a function of nutrient concentration. M, magnetic stirrer, OG, oil on glass-fiber filter paper; T_1, T_2, stopcocks; T_3, "T bore" stopcock; J, ball in socket joints; R, reservoir. *Source:* Gibbs, C.F., "Quantitative Studies on Marine Biodegradation of Oil. I. Nutrient limitation at 14°C," *Proc. R. Soc. Lond. B. Biol. Sci.,* 188, pp. 61-82, 1975. (Reprinted with permission of the Royal Society, London, England.)

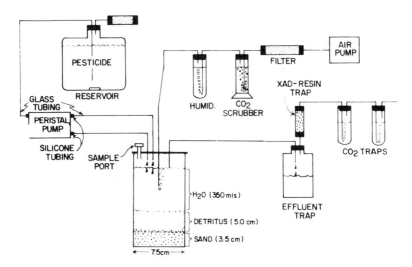

Fig. 8-22. Schematic diagram of continuous-flow microcosm used to study the fate of toxic organics in estuarine saltmarsh environments. *Source:* Pritchard, P.H., Bourquin, A.W., Grederickson, H.L., and Maziarz, T., "System Design Factors Affecting Environmental Fate Studies in Microcosms," in *Proceedings of the Workshop: Microbial Degradation of Pollutants in Marine Environments,* A.W. Bourquin and P.H. Pritchard (Eds.), U.S. EPA Ecological Research Series, Washington, D.C., pp. 251-272, 1978.

from the vessel by a minicoring device. It has the basic advantages of a continuous-flow system (Section 8-4.4) and alleviates, to some extent, selection pressure artifacts by continually resupplying microorganisms to the reactor vessel. The small size of systems allow replicates and sterile controls (formalin treated) to be run.

After these microcosms have been set up, approximately 6 to 7 days are allowed for the systems to settle and stabilize. By monitoring dissolved oxygen concentrations in the water column, it can be shown that sustaining and reproducible diurnal changes (12-hr light–dark cycle) are operating by this time. Further work with DO levels, invertebrate activity in the sediment, and nutrient cycling will permit calibration with field studies in the marsh.

In order to follow the fate of a chemical in this system, a mathematical model has been developed to help interpret the results. A schematic of the conceptual basis behind the model is shown in Fig. 8-23. The results from an experiment with methyl parathion are shown in Fig. 8-24; a reasonably good fit of the model to the points was obtained by measuring the amount of radioactivity in the water column (parent compound was ^{14}C ring labeled) in both sterile and active systems. This was made possible by performing process analysis experiments to determine hydrolysis rate and partitioning to sediments. In the active systems, low levels of radioactivity in the water could only be accounted for in the model by assuming an increase of binding to sediment above and beyond simple equilibrium partition. Process analysis experiments have shown that binding of radioactivity to sediment occurs through methyl parathion transformation.

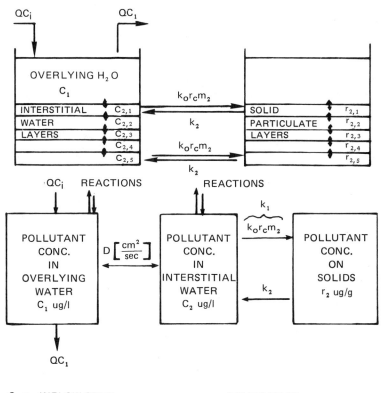

C_i = INFLOW CONC.
Q = FLOW RATE
C = CONC. IN WATER
r = CONC. ON SOLIDS
k_1 = ADSORPTION RATE

m = SOLIDS MASS
k_o = PARTITION COEFFICIENT
k_2 = DESORPTION RATE
D = DIFFUSION COEFFICIENT

Fig. 8-23. Conceptual framework for a diffusion model of a small continuous-flow sediment-water microcosm.

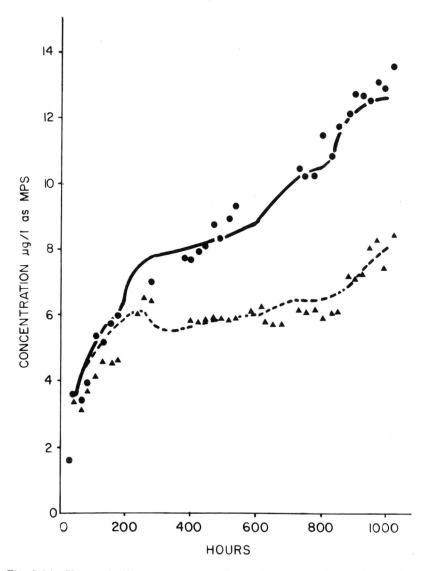

Fig. 8-24. Changes in the concentration of parent compound (methyl parathion— MPS) with time in sterile (circles) and nonsterile (triangles) sediment-water continuous flow microcosms. Points represent actual data and lines are derived from the diffusion model in Fig. 8-22. Waviness in lines is due to variations in inflow concentration. The close fit of the line with data points from the nonsterile systems was obtained by assuming extensive biologically mediated binding to sediments.

Continuing tests to determine relationship of this laboratory system to a field situation and to examine the effects of system design (Ref. 8-162) indicate that valid fate information derived in the laboratory can potentially be extrapolated to estuarine environments.   At the very least, the tests will provide critical information on the fate processes that affect toxic organics in an estuarine marsh situation and on the mechanism by which these processes are integrated together.

8-9 TOXICOLOGICAL EFFECTS STUDIES IN MICROCOSMS

The application of microcosm studies to toxicological hazard evaluation of chemicals entering aquatic and terrestrial environments is still in its infancy. This slow growth can be attributed primarily to the considerable success of standard toxicological testing procedures involving acute lethality, embryo–larval testing (both mortality and growth), and full life-cycle chronic tests. These tests have proven to be cost- and time-effective, amenable to standardization and reproducibility, and eminently defensible from a litigation standpoint. Moreover, for pragmatic reasons, these tests have concentrated on the use of monocultures of plants and animals that can be raised in a laboratory and on test protocols that are optimized for data analysis rather than for ecological simulation. Consequently, the development and application of more complex testing systems (like microcosms), encompassing natural multispecies communities, has been slow to materialize and, in some cases, been looked upon with skepticism (Ref. 8-27). However, at the same time, it has been recognized that the present bank of toxicity testing protocols has a number of limitations that should be scientifically and experimentally addressed in the near future in order to build the integrity of environmental risk assessment programs. These limitations, as they apply to microcosm studies, are discussed in the following paragraphs.

(1) The common application of the "limiting factor concept" (Ref. 8-18) fosters the assumption that if a toxicological effect can be demonstrated on a species in isolation, particularly from the standpoint of the worst possible case (that is, the species providing a toxic response at the lowest toxicant concentration), then the effect will also be reproduced on that species (or an equivalent) in its natural situation. Lane and Levins (Ref. 8-117), for example, have discussed the shortcomings of this concept and have shown that it is an invalid assumption under some conditions.   Toxicological testing schemes clearly require more studies in which toxicological responses of individual species are measured as a function of their intertwinement with natural communities under environmentally realistic parameters. Ostensibly, the microcosm approach (see Section 8-2) and perhaps a whole gradation of approaches in between microcosms and bioassays, could provide for this need.

(2) In most toxicological tests, a constant toxicant concentration and standardized conditions of water quality are employed to expose the test species to the toxicant. In actual environmental situations, however, water quality will vary, and exposure concentration will be intermittent and graded, depending on certain fate and transport processes. It is possible to model, in microcosm systems, some or many of these "natural" exposure conditions, thus, potentially improving toxicological testing.

(3) In some cases, emphasis has been placed on the importance of sublethal toxicological effects, even to the point of supplanting routine mortality tests. At best, this concept suffers from a potpourri of uncoordinated approaches and is difficult to experimentally formulate into a test. This is because quantitation of the adverse effect is complicated and natural systems possess large compensatory powers, particularly through recruitment and recovery functions (environmental adaptability), which tend to readily dilute the effect. Thus, in order to validate the predictability of a sublethal test (and, in fact, even a more direct toxicological test), at some point in time it becomes mandatory to undertake extensive field studies involving in-depth ecological parameterizations or to develop microcosm systems or special types of synthetic communities that simulate natural ecosystems to whatever degree possible and permit laboratory examination of toxicological effects under "natural" conditions.

Two complications seem to be responsible for the slow application of microcosm studies to effects testing. First, because of containerization effects, laboratory microcosms cannot satisfactorily harbor certain higher trophic levels, particularly those with an existing toxicological data base. Even in cases where a microcosm is able to accommodate a higher trophic level organism under environmental conditions similar to its own natural habitat, there are likely to be severe sampling limitations; the sampling of organisms from the microcosm may have more effect on the laboratory system than the toxicant itself. This will provide a poor data base for statistical analyses of mortality and growth. From a population dynamics standpoint, any experienced toxicologist will probably have a good appreciation of which organism or ecosystem complexity can and cannot be adequately modeled in laboratory microcosms. Many of the studies cited in this chapter point to the fact that nutrient cycling, productivity, aspects of diversity, and structure and truncated energy flow can be potentially modeled in microcosms. With continued validation studies, these ecological aspects could form the basis for a variety of new toxicity tests that involve more complex and more natural situations than present bioassay protocols. Consequently, the application of microcosms to effects testing can have two very useful outcomes: first, new or sensitive multispecies test systems and protocols may come forth and, second, our confidence in making accurate (and quantitative) hazard assessment predictions may be greatly enhanced.

The second reason microcosm technology has not been more vigorously applied to toxic effects studies is the need to undertake extensive efforts in standardization and statistical analysis. Consequently, the microcosm approach is relegated to a slow questioning acceptance by the present scientific community engaged in methods development in toxicity testing, because of the inherent complex nature of this approach. Rigid control of parameters such as reproducibility, replicability, organism selection, technical expertise, and time/cost effectiveness will be difficult to accommodate in the conceptual basis around which microcosms are designed and applied. Attempts at providing standardization have resulted in the synthetic community approach as typified by the work of Metcalf, Isensee, Gillette, and others (see Section 8-7). The limitations of this approach have been discussed in Section 8-7.2 and it is with some reluctance that I even label these systems microcosms. If a synthetic community approach is required to accomplish standardization, considerable caution is warranted because a microcosm system that is simply the additive effect of simple bioassays is inadequate; the individual bioassays are, of course, simpler and less expensive to execute. The microcosm approach should either verify the environmental significance of individual bioassay studies (meaning that the microcosm would no longer be needed once verification was established) or provide a laboratory representation of the interrelationships between the species that are affected by the toxicant and indicate how, in fact, to design and implement further tests on the role of recovery and effects of environmental parameters on these interrelationships.

The real virtue of the microcosm approach, as I see it, is to verify the environmental significance of existing studies and methods. That is, can we environmentally verify at the appropriate trophic level a toxic effect in time and space against a variety of environmental parameters using a laboratory model of an ecosystem situation? Or, in another light, can the dynamics and structure of laboratory microcosm, assuming it has good simulation capabilities, be the focal point in which to detect and develop *new* toxicity test measures? These two aspects, in my view, describe the most useful role of microcosms. Microcosm studies, in conjunction with field studies and monitoring information, should provide new simple, replicable, and cost-effective bioassays, or should verify extant toxicity tests for their extrapolation capacity. Accordingly, the routine (standardized) toxicity test will always be the screening tool, not the microcosm. The microcosm will represent one of several research tools for verification and enhancement of extrapolatibility of the screening tests.

The types of microcosm-like systems available for toxicity testing are highly varied and numerous. Many potential examples have already been cited in this chapter (i.e., many systems designed for fate studies are equally applicable to effects studies). The major challenge, as indicated earlier, is to demonstrate that a microcosm is environmentally realistic and that its results can be extrapolated back to the environment. Other examples of microcosm studies specifically applied to toxicology studies are discussed below. This is a cursory examination of the available studies and it is biased toward aquatic systems, but the review is designed to be illustrative of the approaches employed and their potential application.

A number of microcosm-like studies have developed from a need to characterize the toxic effects of herbicides and pesticides to various estuarine communities. Davis *et al.* (Ref. 8-49) developed tidal flat microcosms constructed of plastic tanks (51 cm long, 51 cm wide, 97 cm deep). Several small intact box cores from a salt marsh area were inserted to form a contiguous salt marsh sediment surface in the bottom of the tank. Macrophytes (*Spartina alterniflora*), diatoms, and benthic invertebrates associated with the box cores were consequently included in the microcosms. Several larger animals (snails, mussels, fiddler crabs, blood worms) were added to the tanks and the systems allowed to stabilize for 17 days while being subjected to a simulated tidal flux (using variable differential head pressure between the microcosm and a reservoir tank of water). Various ecosystem parameters measured in these microcosms eventually settled down and became quite stable. Plant diseases of the *Spartina* were reduced to acceptable levels and the crabs, after their population density was adjusted, also appeared to function normally. Good similarities were obtained in comparisons with field studies. Atrazine had similar low level toxic and bioconcentration effects in both the laboratory and field systems (Ref. 8-49).

Hansen (Ref. 8-82) and Tagatz *et al* (Ref. 8-179) developed a very useful microcosm-like system to study the effects of pesticides on developing estuarine animal communities. Rather than extract a preexisting community structure from the environment and bring it into the laboratory for testing, they initiated the study with clean sand and then allowed larval forms in the raw inflowing seawater to colonize the sand both in the presence and absence of the pesticide. A diagram of the apparatus is shown in Fig. 8-25. At present, this test is being developed as a standard bioassay (using relative toxicity as predictive tool), but it is possible to use this system for extrapolation purposes and risk assessment because a variety of secondary effects (nutrient cycling, food sources, sediment turnover) could be related to the reduction in colonization caused by a toxicant. The system can potentially be calibrated with the environment, tested for the effects of various environmental parameters, and used to predict population dynamics of the invertebrates contained therein, particularly in their ability to recover from the initial toxic effects. Finally, it is one of the few systems with a built-in recruiting system (continual inoculation), which is typical of many environmental situations and is a process which often rapidly dilutes a particular toxic response.

A very critical aspect of any aquatic environment is the nutrient cycling processes that are mediated by bacterial communities both in the sediment and on plant surfaces. For many areas, where macrophytes are common, the death of the plants is rapidly followed by microbial decomposition, a process that can release significant amounts of nitrogen and phosphorous into the water column.

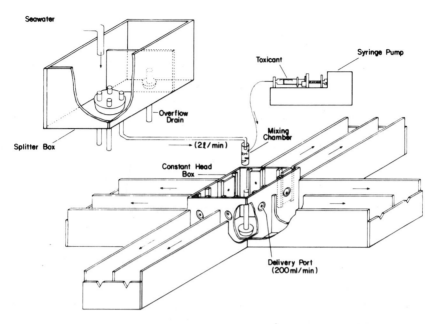

Fig. 8-25. Apparatus used to test the effect of toxic chemicals on composition of estuarine animal settling communities. *Source:* Tagatz, M.E., "Effects of Sevin on Development of Experimental Estuarine Communities," *J. Toxicol. Environ. Htl.* 5, pp. 643-651, 1979.

There appears to be a very close relationship between plant senescence and microbial activity (Refs. 8-64, 8-74, 8-116, 8-163) that consequently provides an excellent system to study effects of certain toxicants. Ramsay (Ref. 8-163), for example, studied the heterotrophic activity (using ^{14}C glucose mineralization) of bacterial epiphytes on *Elodea* and *Chara* leaves in the presence and absence of the herbicide, paraquat. He showed that unnatural nutrient releases from herbicide-treated plants significantly stimulated bacterial activity as measured by maximum uptake rates and substrate affinities. Although his system did not employ a microcosm approach per se, it provided an experimental tool that could be applied to toxicity studies of macrophytes grown in the laboratory under simulated natural conditions. Strange (Ref. 8-174) studied the microbially mediated nutrient cycling phenomena associated with dead plant material in large outdoor pools designed to allow monitoring of CO_2 production and O_2 consumption, as well as N and P levels. His results, using the herbicide combination, endothall/diaquat, could be easily accommodated in a laboratory microcosm. Christian *et al.* (Ref. 8-37) examined the microbial community associated with *Spartina* roots and its sensitivity to a variety of chemical and physical perturbations. Noteworthy is the use of community adenylate energy charge as a monitor of system state. This approach also appears to have an application to microcosm studies.

In a somewhat related study, Giesy (Ref. 8-70) has studied the toxic effects of cadmium on leaf litter decomposition in a model stream channel at the EPA-ERDA stream microcosm facility at the Savannah River project in Aiken, South Carolina. Although, by my definition, this is not a microcosm study (not a laboratory system) but an outdoor artificial stream study, the experimental details seem highly applicable to the development of a smaller laboratory system that can be compared to the results of the artificial stream channel (that is, to field calibration). Giesy's toxicology study shows that 5 and 10 μg/liter Cd in the water, flowing through the stream channel over a 28-day incubation period, is enough to significantly decrease microbial colonization of the leaf-litter material and thereby decrease weight loss. Overall, it would appear that the association of natural microbial populations with leaf and plant surfaces is a natural interrelationship that is not only sensitive to toxicants but represents a complex ecosystem process that can be modeled and studied in a laboratory microcosm.

A large number of effect studies have been carried out on freshwater ponds. As an example, Hurlbert et al (Ref. 8-97) have made an extensive population study of predacious and herbivorous insects, tadpole shrimp, planktonic crustaceans, rotifers, and phytoplankton contained in small ponds exposed to the organophosphate pesticide, Dursban. This investigation would be an excellent base-line study around which to develop microcosms in an attempt to reproduce much of the same toxicological results in the laboratory. If good correlation could be obtained between the microcosm and pond studies, then the experimental details for the design and construction of a field-calibrated microcosm of a pond would be available for other toxicological tests and experiments. Large outdoor tank facilities such as those at the Naval Undersea Center in Hawaii (Ref. 8-58) and at the University of Rhode Island (MERL) (Ref. 8-160) provide potentially similar base-line studies.

With terrestrial systems, effects studies are encompassed in the excellent work of Jackson and Ausmus at Battelle's Columbus Laboratory (Ref. 8-105) and Draggon (Ref. 8-52) at Oak Ridge National Laboratory. Their systems were described in Section 8-6. The concept behind their microcosm studies is that nutrient cycling in the soil involves a close functional couple between the autotrophs and the heterotrophs responsible for transformation of dissolved organic matter—a couple that cannot be studied with isolated compartments. The transfer of materials between these compartments is apparently mediated by pore water movement. Thus, for every soil sample, certain nutrients and certain nutrient concentrations can be normally detected in the soil water. Stress to this nutrient recycling regime, therefore, should appear as changes in type and concentration of these nutrients. Soil cores were brought into the laboratory to measure this effect and moistened to produce a leachate to be analyzed for various nutrients, particularly Ca^{+2} and NO_3^{-1}. Originally, small cores (5 cm diameter, 10 cm depth) were utilized by encasement in shrinkable PVC plastic sheeting (Refs. 8-53, 8-105) (see Fig. 8-21). For forest floor cores, after 4 weeks of preequilibration (Ca efflux constant), arsenic at 100 μg cm^{-2} caused significant increases in Ca and NO_2 efflux whereas similar grassland cores were considerably less sensitive. Other indicators, such as ATP concentration and various enzymatic activities, failed to show an effect at these arsenic levels (Ref. 8-105). Draggon (Ref. 8-53) has shown that arsenic leaching from the cores is affected by homogenization of the soil and by soil type. He concluded that intact cores were most representative of the environment of concern.

Studies of Jackson and Ausmus have now been expanded to accommodate large box cores that permit inclusion of a rooted tree sapling in the system (Ref. 8-102). Transport and distribution of heavy metals in forest ecosystems can be modeled (Ref. 8-102)—nutrient cycling (NO_3^{-1} and Ca^{+2}) (Ref. 8-103) and CO_2 evolution (Ref. 8-6) were demonstrated to be good indicators of heavy metal stress to these systems. Results from these microcosm systems have also been compared to exposed field plots (Refs. 8-101, 8-104). In some cases, the effects of arsenic dosage had on Ca^{+2}, DOC, NH_4^{+1}, PO_4^{-3} leachate concentrations did not coincide between field plots and microcosm. This is illustrated in Fig. 8-26. Although these differences between field and laboratory were partly attributed to a dilution of arsenic stress in field plots (greater arsenic dispersal), it is still likely that the containerization effect increased the susceptibility of the soil cores to arsenic stress. Regardless, this does not detract from the usefulness of the microcosm approach and, in fact, signifies an experimental problem that should be characterized and elucidated before good simulation can be obtained.

Several other types of soil microcosm have been developed for effects tests. Weber developed a "Dual-Pot" method, using styrofoam cups (Ref. 8-193). This system allowed the exposure of only the plant roots to a herbicide. Lighthart and Bond (Refs. 8-21, 8-127) developed soil–litter microcosms that are unique in the respect that community carbon dioxide and oxygen dynamics in a temperature and moisture-controlled chamber can be monitored. It is interesting that their soil systems took approximately 7–10 days to stabilize in regard to O_2 uptake and CO_2 production. These systems also showed significant effects on gas dynamics due to cadmium stress.

As pointed out in regard to aquatic systems, a large amount of literature documents effects of certain toxicants on various types of terrestrial ecosystems, the details of which could provide an excellent data base for designing and calibrating microcosm studies. A typical example is the work of Barrett (Ref. 8-8), in which the plants, arthropods, and mammalian populations and the litter de-

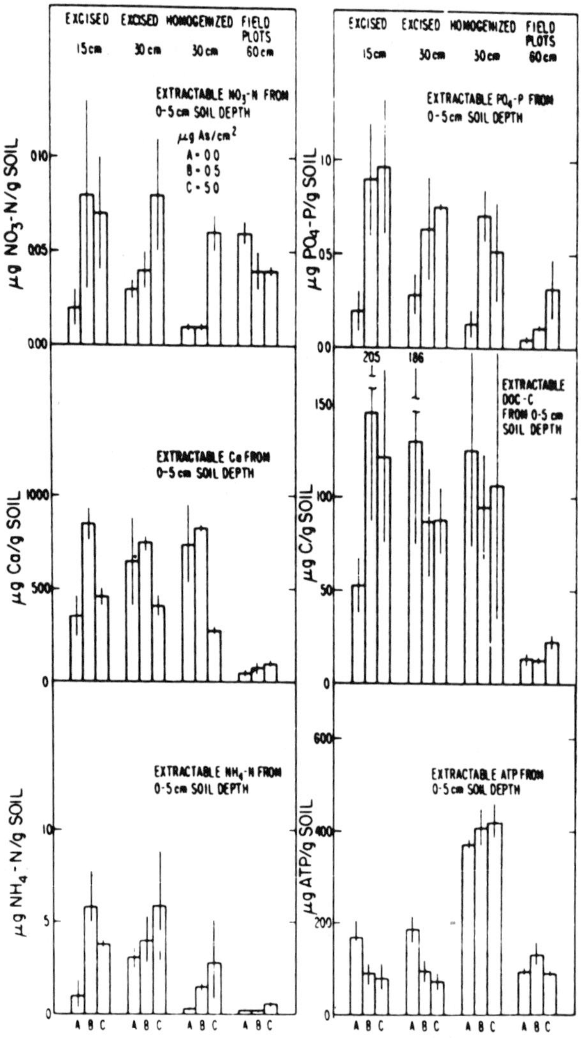

Fig. 8-26. Extractable nutrients from microcosms (excised, homogenized) and field plots treated with arsenic (± 1 SE, n = 3). *Source:* Jackson, D.R. and Ausmus, B.S., "Effects of Arsenic on Nutrient Dynamics of Grassland Microcosms and Field Plots," *Water, Air, Soil Pollut.,* 11, pp. 13-21, 1978. (Reprinted with permission of Reidel Publishing Co., Dordeecht, Holland.)

composition of semi-enclosed field plot were monitored in detail to determine the effects *and* the recovery after exposure to the carbamate pesticide, Sevin. Other examples involving plant communities (Refs. 8-182, 8-189) and leaf litter decomposition in a red maple wood lot (Ref. 8-192) are also appropriate for this approach.

8-10 SUMMARY

The application of microcosm technology to risk assessment programs is basically an exercise in determining if the sum of the parts can be made to equal the whole. Our ability to test a wide range of environmental situations and system variables and our proficiency at quantitatively translating the information obtained into new test protocols and risk assessment predictions will more or less determine the success of the microcosm approach. Although field verification of microcosm results is the ultimate challenge before us, it is important to realize that microcosm development will have to progress hand in hand with our scientific knowledge of the individual processes and the interactions within the ecosystem under study. The current recognition of this fact in the conceptual advancement of risk assessment programs is, in itself, cause for excitement and encouragement. As Orians (Ref. 8-153) has stated, "Ecology shares with all sciences the difficult problems associated with moving from microanalysis to macroanalysis of complex systems. A key part of the problem involves the means by which we select aggregate variables and macrodescriptors that capture the essence of the system we wish to describe. Meaningful aggregate variables must be based upon and be reasonable extensions of known or suspected processes at lower levels. One of the great challenges in ecology over the next decade will be to build conceptual bridges between micro- and macroecology. A predictive science of community ecology will have to be based on underlying processes, but without dealing with the details of those processes. We have been slow to develop ways of integrating the two areas because, until very recently, we had little microecology worth using as building blocks for imaginative macroecological models. This is no longer true, and there is much to be done."

It is also important to reemphasize that microcosm studies, if implemented properly, are interesting pivotal points within risk assessment programs. On the one hand, they provide a continuous source of ecological data that will ultimately result in greater and greater comprehension of ecosystem structure and function and a closer application to risk assessment questions. On the other hand, they continually verify the validity and environmental significance of the current bank of simple fate and toxicity testing protocols, such as biodegradability tests and single species bioassays. In the face of the massive effort required to ascertain potential environmental risk of literally thousands of new chemicals, these simple test protocols will be critical to our efforts. Our confidence in these protocols will be enhanced by microcosm studies. Thus, our adeptness at implementing microcosm studies is a goal of great consequence. I hope this chapter will provide an initial framework upon which to meet this goal.

Finally, the use of microcosms in risk assessment can often unwittingly delude an investigator into making wrong deductions or misleading assumptions. As Warren and Davis (Ref. 8-191) asked: "Do our conceptual constructs draw our thinking nearer to the natural relationships we seek to understand, or do they take our thinking in another direction? This is an important question that only experience with natural systems can ever answer and then, if at all, only very partially. Because of their elements of realness, laboratory models can perhaps be misleading. A healthy skepticism, if not a sense of humor, is required with regard to our conceptual constructs."

REFERENCES

8-1. Abbott, W., "Microcosm Studies on Estuarine Waters I. The Replicability of Microcosms," *J. Water Pollut. Control Fed.*, 38:258-270, 1966.

8-2. Albone, E.S., Eglinton, G., Evans, N.C., Hunter, J.M., and Rhead, M.M., "Fate of DDT in Seven Estuary Sediments," *Environ. Sci. Technol.*, 6:914-919, 1972.

8-3. Allan, S. and Brock, T., "The Adaptation of Heterotrophic Microcosms to Different Temperatures," *Ecology*, 49:343-346, 1968.

8-4. Anderson, R.V., Elliot, E.T., McClellan, J.F., Coleman, D.C., Cole, C.V., and Hunt, H.W., "Trophic Interactions in Soils as They Affect Energy and Nutrient Dynamics. III. Biotic Interactions of Bacteria, Amoebae, and Nematodes," *Microb. Ecol.*, 4:361-371, 1978.

8-5. Atlas, R.M. and Bartha, R., "Simulated Biodegradation of Oil Slicks Using Oleophilic Fertilizers," *Environ. Sci. Technol.*, 7:538-541, 1973.

8-6. Ausmus, B.S., Dodson, G.J., and Jackson, D.R., "Behavior of Heavy Metals in Forest Microcosms. III. Effects on Litter-soil Carbon Metabolism," *Water, Air, Soil Pollut.*, 10:19-26, 1978.

8-7. Austin, E.R., and Lee, G.R., "Nitrogen Release from Lake Sediments," *WPCF*, 45:870-879, 1973.

8-8. Barrett, G.W., "The Effects of an Acute Insecticide Stress on a Semienclosed Grassland Ecosystem," *Ecology*, 49:1019-1035, 1968.

8-9. Bartha, R., and Atlas, R.M., "The Microbiology of Aquatic Oil Spills, *Adv. Appl. Microbiol.*," 22:225-266, 1977.

8-10. Bartha, R., and Pramer, D., "Feature of a Flask and Method for Measuring the Persistence and Biological Effects of Pesticides in Soil," *Soil Sci.*, 100:68-70, 1965.

8-11. Bartha, R. and Hsu, T.S., "Accelerated Mineralization of Two Organophosphate Insecticides in the Rhizosphere," *Appl. Environ. Microbiol.*, 37:36-41, 1979.

8-12. Baughman, G.L. and Lassiter, R.R., "Prediction of Environmental Pollutant Concentration," in *Estimating the Hazard of Chemical Substances to Aquatic Life*, J. Cairns, K.L. Dickson, and A.W. Maki, (Eds), Am. Soc. for Testing and Materials. Philadelphia, PA, STP 657, pp. 35-54, 1978.

8-13. Best, J.A. and Weber, J.B., "Disappearance of s-Triazines as Affected by Soil pH Using a Balance-Sheet Approach," *Weed Sci.*, 22:364-373, 1974.

8-14. Beyers, Robert J., "The Microcosm Approach to Ecosystem Biology," *Am. Biol. Teach.*, 26:491-497, 1964.

8-15. Beyers, R.J., "The Metabolism of Twelve Aquatic Laboratory Microecosystems," *Ecol. Monogr.*, 33:281-306, 1963.

8-16. Beyers, R.J., "Relationship Between Temperature and the Metabolism of Experimental Ecosystems," *Science*, 136:980-982, 1962.

8-17. Beyers, R.J., and Odum, H.T., "The Use of Carbon Dioxide to Construct pH Curves for the Measurement of Productivity," *Limnol. Oceanogr.*, 4:499-502, 1959.

8-18. Blackman, F.F., "Optimal and Limiting Factors," *Ann. Bot.*, 19:287-295, 1905.

8-19. Blumer, M., and Sass, J., "Oil Pollution Persistence and Degradation of Spilled Fuel Oil," *Science*, 176:1120-1122, 1972.

8-20. Bohool, B.B. and Wiebe, W.J., "Nitrogen-Fixing Communities in an Intertidal Ecosystem," *Can. J. Microbiol.*, 24:932-938, 1978.

8-21. Bond, H., Lighthart, B., Shimabuku, R. and Russel, L., "Some Effects of Cadmium on Coniferous Forest Soil and Litter Microcosms," *Soil Sci.*, 21:278-287, 1976.

8-22. Bondiett, E.A., Draggan, S., Traballa, J.R. and Witherspoon, J.P., "State-of-the-art and Recommended Testing for Environmental Transport of Toxic Substances," U.S. Environmental Protection Agency, Environ. Sciences Div. Publication No. 893, Oak Ridge National Laboratory/EPA-1, 1974.

8-23. Bott, T.L., Preslan, J., Finlay, J. and Brunker, R., "The Use of Flowing-Water Microcosms and Ecosystem Streams to Study Microbial Degradation of Leaf Litter and Nitrilotriacetic Acid," *Dev. Ind. Microbiol.*, 18:171-184, 1976.

8-24. Bourquin, A.W., Hood, M.A., and Garnas, R.L., "An Artificial Microbial Ecosystem for Determining Effects and Fate of Toxicants in a Salt-Marsh Environment," *Dev. Ind. Microbiol.*, 18:185-191, 1977.

8-25. Bourquin, A.W., Garnas, R.L., Pritchard, P.H., Wilkes, F.G., Cripe, C.R. and Rubinstein, N.I., "Interdependent Microcosms for the Assessment of Pollutants in the Marine Environment," *Int. J. Environ. Stud.*, 13:131-140, 1979.

8-26. Bourquin, A.W. and Pritchard, P.H., "Microbial degradation of pollutants in marine environments workshop," U.S. EPA, Ecological Research Series, EPA 600/9-79-012, Cincinnati, OH, 351 pp., 1979.

8-27. Branson, D.R., "Predicting the Fate of Chemicals in the Aquatic Environment from Laboratory Data," in *Estimating the Hazard of Chemical Substances to Aquatic Life*, J. Cairns, K.L. Dickson, A.W. Maki (Eds.), Am. Soc. Testing and Materials, Philadelphia, PA, pp. 55-70, 1978.

8-28. Brockway, D.L., Maudsley, J. and Lassiter, R.R., "Development, Replicability and Modeling of Naturally Deriving Microcosms," *Int. J. Environ. Stud.*, 13:149-158, 1979.

8-29. Brooks, G.T., "Chlorinated Insecticides: Retrospect and Prospect," in *Pesticide Chemistry in the 20th Century,* J.R. Plimmer (Ed.), American Chemical Society, Washington, D.C., 1977.

8-30. Bunch, R.L. and Chambers, C.W., "A biodegradability test for organic compounds," *J. Water Pollut. Control Fed.,* 39:181–184, 1967.

8-31. Cairns, J., Dickson, K.L. and Herricks, E.E., "A Challenge for Action: Symposium Analysis," in *Recovery and Restoration of Damaged Ecosystems,* J. Cairns, K.L. Dickson, and E.E. Herricks, (Eds.), Univ. of Virginia Press, Charlottesville, VA, pp. 522–525, 1977.

8-32. Cairns, J., Dickson, K.L., Slocomb, J.P., Almeida, S.P. and Ev, J.K.T. "Automated Pollution Monitoring with Microcosms," *Int. J. Environ. Stud.,* 10:43–49, 1976.

8-33. Clapham, W.B., *Natural Ecosystems,* Macmillan, New York, 1971.

8-34. Clesceri, L.S., Park, P.A. and Bloomfield, J.A., "General Model of Microbial Growth and Decomposition in Aquatic Ecosystem," *Appl. Environ. Microbiol.,* 33:1047–1058, 1977.

8-35. Chan, Y.K., and Knowles, R., "Measurement of Denitrification in Two Freshwater Sediments by an In Situ Acetylene Inhibition Method," *Appl. Environ. Microbiol.,* 37:1067–1072, 1979.

8-36. Chen, R.L., Keeney, D.R. and Konrod, J.F., "Nitrification in sediments of selected Wisconsin lakes," *J. Environ. Qual.,* 1:151–154, 1972.

8-37. Christian, R.R., Bancroft, K. and Wiebe, W.J., "Resistance of the Microbial Community Within Salt Marsh Soils to Selected Perturbations," *Ecology,* 59:1200–1210, 1978.

8-38. Cole, C.V., Elliott, E.T., Hunt, H.W., and Coleman, D.C., "Trophic Interactions in Soils as They Affect Energy and Nutrient Dynamics," V. Phosphorus Transformations, *Microb. Ecol.,* 4:381–387, 1978.

8-39. Cole, L.K., Metcalf, R.L., and Sanborn, J.R., "Environmental Fate of Insecticides in Terrestrial Model Ecosystems," *Intern. J. Environ. Stud.* 10:7–14, 1976.

8-40. Cole, L.K., and Metcalf, R.L., "Predictive Environmental Toxicology of Pesticides in the Air, Soil, Water, and Biota of Terrestrial Model Ecosystems," in *Terrestrial Microcosms and Environmental Chemistry,* J.M. Witt and J.W. Gillett (Eds.), National Science Foundation, Washington, D.C., pp. 57–73, 1978.

8-41. Coleman, D.C., Cole, C.V., Hunt, H.W., and Klein, D.A., "Trophic Interactions in Soils as They Affect Energy and Nutrient Dynamics. I. Introduction" *Microb. Ecol.* 4:345–349, 1978.

8-42. Coleman, D.C., Anderson, R.V., Cole, C.V., Elliot, E.T., Woods, L., and Campion, N.R., "Trophic Interactions in Soils as They Affect Energy and Nutrient Dynamics. IV. Flows of Metabolic and Biomass Carbon," *Microb. Ecol.,* 4:373–380, 1978.

8-43. Cooke, G.D., "Aquatic Laboratory Microsystems and Communities," in *Structure and Function of Fresh Water Microbial Communities,* J. Cairns (Ed.), Virginia Polytechnic Institute, Blacksburg, VA, pp. 47–86, 1974.

8-44. Cooke, G.D., "The Pattern of Autotrophic Succession in Laboratory Microcosms," *Bioscience,* 17:717–721, 1967.

8-45. Cooper, D.C., "Enhancement of Net Primary Productivity by Herbivore Grazing in Aquatic Laboratory Microcosms," *Limnol. Oceanogr.,* 18:31–37, 1973.

8-46. Copeland, B.S., "Evidence for Regulation of Community Metabolism in a Marine Ecosystem," *Ecology*, 46:563–564, 1969.

8-47. Crowley, P.H., "Effective Size and the Persistence of Ecosystems," *Oecologia*, 35:185–195, 1978.

8-48. Dale, T., "Total Chemical and Biological Oxygen Consumption of the Sediments in Lindaspollene, Western Norway," *Marine Biol.*, 49:333–341, 1978.

8-49. Davis, D.E., Weete, J.D., Pillai, G.G., Plumley, F.G., Everest, J.W., Truelove, B., and Diner, A.M., "Atrazine Fate and Effects in a Salt Marsh," U.S., EPA Ecological Research Series, EPA-600/3-79-111, Cincinnati, OH, 84 pp., 1979.

8-50. DiSalvo, L.H., and Ross, H.C. "Oxygen Metabolism of Some Reef Substrate Microcosms," *Hydro-lab J*, 3:25–30, 1974.

8-51. Domsch, K.H., Anderson, J.P.E., and Ahlers, R., "Method for Simultaneous Measurement of Radioactive and Inactive Carbon Dioxide Evolved from Soil Samples During Incubation with Labeled Substrates," *Appl. Microbiol.*, 25:819–824, 1973.

8-52. Draggan, S., "Effects of Substrate Type and Arsenic Dosage Level on Arsenic Behavior in Grassland Microcosms. Part I: Preliminary Results on [74] As Transport," in *Terrestrial Microcosms and Environmental Chemistry*. J.M. Witt and J.W. Gillett, (Eds.), National Science Foundation, Washington, D.C., pp. 102–110, 1977.

8-53. Draggan, S., "The Role of Microcosms in Ecological Research," *Int. J. Environ. Stud.*, Vol. 13, Special Issue, pp. 83–182, 1979.

8-54. Draggan, S., "The Role of Microcosms in Ecological Research," *Int. J. Environ. Stud.*, Vol. 10, Special Issue, pp. 1–70, 1976.

8-55. Eberly, W.B., "The Use of Oxygen Deficit Measurements as an Index of Eutrophication in Temperate Dimictic Lakes," *Vern. Int. Verein. Limnol.*, 19:439–441, 1975.

8-56. Eichelberger, J.W. and Lichtenberg, J.J., "Persistence of Pesticides in River Water," *Environ. Sci. Technol.*, 5:541–544, 1971.

8-57. Emmel, T.C., *"Ecology and Population Biology,"* W.W. Norton and Company, New York, p. 5, 1973.

8-58. Evans, E.C., III, "Microcosm Responses to Environmental Perturbants. An Extension of Baseline Field Survey," *Helgol. Wiss. Meeresunters.* 30:178–191, 1977.

8-59. Fischer, W.K., Gerike, P., and Smid, R.D., "Combination of Methods for Successive Testing and Evaluation of the Biodegradability of Synthetic Substances as Organic Complexing Agents by Means of Generally Valid Nonspecific Parameters," *Wasser und Alowasser Forschung*, 15:99–118, 1974.

8-60. Flett, R.J., Hamilton, R.D., and Campbell, N.E.R., "Aquatic Acetylene Reduction Techniques Solutions to Several Problems," *Can. J. Microbiol.*, 22:43–51, 1976.

8-61. Flint, R.W., and Goldman, C.R., "The Effects of a Benthic Grazer on the Primary Productivity of the Littoral Zone of Lake Tahoe," *Limnol. Oceanogr.*, 20:935–944, 1975.

8-62. Focht, D.D., and Verstraete, W., "Biochemical Ecology of Nitrification and Denitrification," in *Advances in Microbial Ecology*, M. Alexander (Ed.), Plenum Press, New York, 1:135–214, 1977.

8-63. Focht, D.D. and Alexander, M., "DDT Metabolites and Analogs: Ring Fission by *Hydrogenomonas*," *Science*, 170:91–92, 1970.

8-64. Gallagher, J.L. and Pfeiffer, W.J., "Aquatic Metabolism of the Communities Associated with Attached Dead Shoots of Salt Marsh Plants," *Limnol. Oceanogr.,* 22:562–565, 1977.

8-65. Gessner, R.O., Boos, R.D., and Sieburth, J. McN., "The Fungal Microcosm of the Internodes of *Spartina alterniflora*," *Marine Biol.,* 16:269–273, 1972.

8-66. Gibbs, C.F., "Quantitative Studies on Marine Biodegradation of Oil. I. Nutrient Limitation at 14°C," *Proc. R. Soc. Lond. B., Biol. Sci.* 188:61–82, 1975.

8-67. Gibbs, C.F., and Davies, S.J., "The Effects of Weathering on Crude Oil Residues Exposed at Sea," *Water Res.,* 9:275–285, 1975.

8-68. Giddings, J., and Eddlemon, G.K., "Photosynthesis/Respiration Ratios in Aquatic Microcosms Under Arsenic Stress," *Water, Air, Soil Pollut.,* 9:207–212, 1978.

8-69. Giddings, J., and Eddlemon, G.K., "The Effects of Microcosm Size and Substrate Type on Aquatic Microcosm Behavior and Arsenic Transport," *Arch. Environ., Contam. Toxicol.,* 6:491–505, 1977.

8-70. Giesy, J.P., "Cadmium Inhibition of Leaf Decomposition in an Aquatic Microcosm," *Chemosphere,* 6:467–475, 1978.

8-71. Gilbert, P.D., and Higgins, I.J., "The Microbial Degradation of Crude Mineral Oils at Sea," *Journal Gen. Microbiol.,* 108:63–70, 1978.

8-72. Gilbert, P.A., and Watson, G.K., "Biodegradability and Its Relevance to Environmental Acceptability," *Tenside,* 14:171–177, 1977.

8-73. Gillett, J.W., and Gile, J.D., "Pesticide Fate in Terrestrial Laboratory Ecosystems," *Int. J. Environ. Stud.,* 10:15–22, 1976.

8-74. Golterman, H.L., "Role of Phytoplankton in Detritus Formation," in *Detritus and Its Role in Aquatic Ecosystems*, F.M. Anderson and A. Macfayden (Eds.), Blackwell Scientific Publ., Oxford, England, pp. 89–103, 1972.

8-75. Goodyear, C.P., Boyd, C.E., and Beyers, R.J., "Relationships Between Primary Productivity and Mosquitofish (*Gambusia affinis*) Production in Large Microcosms," *Limnol. Oceanogr.,* 17:445–450, 1972.

8-76. Gorden, R.W., Beyers, R.J., Odum, E.P., and Eagon, R.G., "Studies on a Simple Laboratory Microecosystem: Bacterial Activities in a Heterotrophic Succession," *Ecology,* 50:86–100, 1968.

8-77. Goswami, K.P., and Koch, B.L., "A Simple Apparatus for Measuring Degradation of ^{14}C-Labeled Pesticides in Soil," *Soil Biol. Biochem.,* 8, pp. 527–528, 1976.

8-78. Graetz, D.A., Keeney, D.R., and Aspiras, R.B., "The Status of Lake Sediment–Water Systems in Relation to Nitrogen Transformations," *Limnol. Oceanogr.,* 18, 908–917, 1973.

8-79. Haines, E.B., "Nitrogen Pool in Georgia Coastal Waters," *Estuaries,* 2, 34–39, 1979.

8-80. Haque, R., Kearney, P.C., and Freed, V.H., "Dynamics of Pesticides in Aquatic Environments," in *Pesticides in Aquatic Environments*, M.A.Q. Khan (Ed.), Plenum Press, New York, pp. 39–52, 1977.

8-81. Hall, D.J., Cooper, W.E., and Werner, E.E., "An Experimental Approach to the Production Dynamics and Structure of Freshwater Animal Communities," *Limnol. Oceanogr.,* 15:839–928, 1970.

8-82. Hansen, D.J., "Aroclor^R 1254: "Effect on Composition of Developing Estuarine Animal Communities in the Laboratory," *Marine Sci.,* 18:19–33, 1974.

8-83. Hanson, R.B., "Comparison of Nitrogen Fixation Activity in Tall and Short *Spartina alterniflora* Salt Marsh Soils," *Appl. Environ. Microbiol.,* 33:596–602, 1977.

8-84. Hargrave, B.T., "Stability in Structure and Function of the Mud Water Interface," *Vern. Int. Verein. Limol.,* 19:1073–1079, 1975.

8-85. Hargrave. B.T., "The Effect of a Deposit-Feeding Amphipod on the Metabolism of Benthic Microflora," *Limnol. Oceanogr.,* 15:21–30, 1970.

8-86. Hargrave, B.T., "Similarity of Oxygen Uptake by Benthic Communities," *Limnol. Oceanogr.,* 14:801–805, 1969.

8-87. Harrison. M.J., Wright, R.T., and Morita, R.Y., "Method for Measuring Mineralization in Lake Sediments," *Appl. Microbiol.,* 21:698–702, 1971.

8-88. Herbert, R.A., "Heterotrophic Nitrogen Fixation in Shallow Estuarine Sediments," *J. Exp. Mar. Biol. Ecol.,* 18:215–225, 1975.

8-89. Herzberg, M.A., Klien, A., and Coleman, D.C., "Trophic Interactions in Soils as They Affect Energy and Nutrient Dynamics. II. Physiological Responses of Selected Rhizosphere Bacteria," *Microb. Ecol.,* 4:351–359, 1978.

8-90. Hill, I.R., and Arnold, D.J., "Transformation of Pesticides in the Environment – The Experimental Approach," in *Pesticide Microbiology,* I.R. Hill, and S.J.L. Wright, (Eds.), Academic Press, London, pp. 222–238, 1978.

8-91. Ho, C.L. and Barrett, B.B., "Distribution of Nutrients in Louisiana's Coastal Waters Influenced by the Mississippi River," *Estuarine Coastal Mar. Sci.,* 5:173–196, 1977.

8-92. Holling, C.S., "Resilience and stability of ecological systems," *Annual Rev. Ecol. Syst.,* 4:1–24, 1973.

8-93. Horowitz, A. and Atlas, R.M., "Continuous Open Flow-Through Systems as a Model for Oil Degradation in the Arctic Ocean," *Appl. Environ. Microbiol.,* 33:647–653, 1977.

8-94. Howard, P.H., Saxena, J. and Sikka, H., "Determining the Fate of Chemicals," *Environ. Sci. Technol.,* 12:398–407, 1978.

8-95. Huang, J., "Effect of Selected Factors on Pesticide Sorption and Desorption in the Aquatic System," *J. WPCF,* 1739–1748, 1971.

8-96. Hughes, D.E., and McKenzie, P., "The Microbial Degradation of Oils in the Sea," *Proc. R. Soc. London,* B189:375–390, 1975.

8-97. Hurlbert, S.H., Mulla, M.S., and Willson, H.R., "Effects of an Organophosphorus Insecticide on the Phytoplankton, Zooplankton, and Insect Populations of Fresh-Water Ponds," *Ecol. Monogr.,* 42:269–299, 1972.

8-98. Isensee, A.R., "Variability of Aquatic Model Ecosystem-Derived Data," *Int. J. Environ. Stud.,* 10:35–41, 1976.

8-99. Isensee, R., and Jones, G.E., "Distribution of 2, 3, 7, 8-Tetrachlorodibenzo-p-dioxin (TCDD) in Aquatic Model Ecosystem," *Environ. Sci. Technol.,* 9:668–672, 1975.

8-100. Isensee, A.R., Kearney, P.C., Woolson, E.A., Jones, G.E., and Williams, V.P., "Distribution of Alkyl Arsenicals in Model Ecosystem," *Environ. Sci. Technol.,* 7:841–845, 1973.

8-101. Jackson. D.R., and Levin, M., "Transport of Arsenic in Grassland Microcosms and Field Plots," *Water, Air, Soil Pollut.*, 11:3-12, 1979.

8-102. Jackson, D.R., Ausmus, B.S., and Leving, M., "Effects of Arsenic on Nutrient Dynamics of Grassland Microcosms and Field Plots," *Water, Air, Soil Pollut.*, 11:13-21, 1979.

8-103. Jackson, D.R., Selvidge, W.J., and Ausmus, B.S., "Behavior of Heavy Metals in Forest Microcosms: I, Transport and Distribution Among Components," *Water, Air, Soil Pollut.*, 10:3-11, 1978.

8-104. Jackson, D.R., Selvidge, W.J., and Ausmus, B.S., "Behavior of Heavy Metals in Forest Microcosms: I. Effects on Nutrient Cycling Processes," *Water, Air, Soil Pollut.*, 10: 13-18, 1978.

8-105. Jackson, D.R., Washburne, C.D., and Ausmus, B.S., "Loss of Ca and NO -N from Terrestrial Microcosms as an Indicator of Soil Pollution," *Water, Air, Soil Pollut.*, 8:279-284, 1977.

8-106. James, A., "The measurement of benthal respiration," *Water Res.*, 8:955-959, 1974.

8-107. Jassby, A., Rees, J., Dudzik, M., Levy, D., and Lapan, E., "Trophic Structure and Modifications by Planktivorous Fish in Aquatic Microcosms," U.S. EPA Environmental Research Series, EPA-600-7-77-096, Cincinnati, OH, PB-274, 1977.

8-108. Jassby, A., Dudzik, M., Rees, J., Lapan, E., Levy, D., and Harte, J., "Production Cycles in Aquatic Microcosms," U.S. EPA Ecological Research Series, EPA-600/7-77-077, Cincinnati, OH, 1977.

8-109. Jorgensen, B.B., and Fenchel, T., "The Sulfur Cycle of a Marine Sediment Model System," *Marine Biol.*, 24:189-201, 1974.

8-110. Juengst, F.W., and Alexander, M., "Effect of Environmental Conditions on the Degradation of DDT in Model Marine Ecosystems," *Marine Biol.*, 33:1-6, 1975.

8-111. Kaufman, D.D., "Approaches to Investigating Soil Degradation and Dissipation of Pesticides," in *Terrestrial Microcosms and Environmental Chemistry*, J. Witt and J. Gillette, (Eds.), National Science Foundation, Washington, DC, pp. 41-54, 1977.

8-112. Kazano, Hsakawa, H., and Tomizawa, C., "Fate of 3, 5-Xylyl Methyl Carbamate Insecticide (XMC) in a Model Ecosystem," *Appl. Entomol. Zool.*, 10:108-115, 1975.

8-113. Kearney, P.C., and Kontson, A., "A Simple System to Simultaneously Measure Volatilization and Metabolism of Pesticides from Soils," *J. Agric. Food Chem.*, 24:424-426, 1976.

8-114. Kemmerer, A.J., "A Method to Determine Fertilization Requirements of a Small Sport Fishing Lake," *Trans. Am. Fish. Res. Bd. Can.*, 97:425-428, 1968.

8-115. Kevern, N.R., and Ball, R., "Primary Productivity and Energy Relationships in Artificial Streams," *Limnol. Oceanogr.*, 10:74-87, 1965.

8-116. Knaver, G.A., and Ayers, A.V., "Changes in Carbon, Nitrogen, Adenosine, Triphosphate, and Chlorophylla in Decomposing *Thalassia testudinum* leaves," *Limnol. Oceanogr.*, 22:408-414, 1977.

8-117. Lane, P., and Levins, R., "The Dynamics of Aquatic Systems; the Effects of Nutrient Enrichment on Model Plankton Communities," *Limnol. Oceanogr.*, 22:454-471, 1977.

8-118. Lassiter, R.R., Baughman, G.L., and Burns, L.A., "Fate of Toxic Organic Substances in the Aquatic Environment," in *State-of-the-Art in Ecological Modeling*, S.E. Jorgensen (Ed.), *Int. Soc. Ecol. Modeling*, 7:219-245, 1978.

8-119. Lauff, G.A., and Cummins, K.W., "A Model Stream for Studies in Lotic Ecology," *Ecology*, 45:188-191, 1964.

8-120. Lee, A., Lu, P., Metcalf, R.L., and Hsu, E., "The Environmental Fate of Three Dichlorophenyl Nitrophenyl Ether Herbicides in a Rice Paddy Model Ecosystem," *J. Environ. Qual.*, 5:482-486, 1976.

8-121. Lee, K.K. and Watanabe, I., "Problems of the Acetylene Reduction Technique Applied to Water-Saturated Paddy Soils," *Appl. Environ. Microbiol.*, 34:654-660, 1977.

8-122. Lee, R., and Ryan, C., Microbial Degradation or Organochlorine Compounds in Estuarine Waters and Sediments, in *Microbial Degradation of Pollutants in Marine Environments*, A.W. Bourquin and P.H. Pritchard (Eds.), U.S. EPA Ecological Research Series, EPA-600/9-79-012, Cincinnati, OH, pp. 443-450, 1978.

8-123. Leffler, J., personal communication, 1979.

8-124. Levandowsky, M., "Multispecies Cultures and Microcosms," *Marine Ecology*, O. Kinne (Ed.), John Wiley and Sons, New York, Vol. III, pp:1399-1452, 1977.

8-125. Litchtenstein, E.P., Liang, T.T., and Fuhremann, T.W., "A Compartmentalized Microcosm for Studying the Fate of Chemicals in the Environment," *J. Agric. Food Chem.*, 26:948-953, 1978.

8-126. Litchtenstein, E.P., Fuhremann, T.W., and Schulz, K.R., "Translocation and Metabolism of (^{14}C) Phorate as Affected by Percolating Water in a Model Soil-Plant Ecosystem," *J. Agric. Food Chem.*, 22:991-996, 1974.

8-127. Lighthart, B. and Bond, H., "Design and Preliminary Results from Soil/Litter Microcosms," *Int. J. Environ. Stud.*, 10:51-58, 1976.

8-128. Lu, P. and Metcalf, R.L., "Environmental Fate and Biodegradability of Benzene Derivatives as Studies in a Model Aquatic Ecosystem," *Environ. Health Perspect.*, 10: 269-284, 1975.

8-129. Macgregor, A.N., and Keeney, D.R., "Acetylene-Reduction Assay of Anaerobic Nitrogen Fixation by Sediments of Selected Wisconsin Lakes," *J. Environ. Qual.*, 2: 438-440, 1973.

8-130. Maki, A.W., and Johnson, H.E., "Evaluation of a Toxicant on the Metabolism of Model Stream Communities," *J. Fish. Res. Board Can.*, 33:2740-2746, 1976.

8-131. Margalef, R., "Diversity, Stability, and Maturity in Natural Ecosystems," in *Unifying Concepts in Ecology*. W.H. VanDobben and R.H. Lowe – McConnell, Dr. W. Junk, B.V. Publishers, The Hague, pp. 150-160, 1975.

8-132. Margalef, R., "On Certain Unifying Principles in Ecology," *Am. Nat.*, 97:357-374, 1963.

8-133. Marvel, J.T., Brightwall, B.B., Malik, J.M., Sutherland, M.L., and Rueppel, M.L., "A Simple Apparatus and Quantitative Method for Determining the Persistence of Pesticides in Soil," *J. Agric. Food Chem.*, 26:1116-1120, 1978.

8-134. Metcalf, R.L., "Model Ecosystem Studies of Bioconcentration and Biodegradation of Pesticides," in *Pesticides in Aquatic Environments*, M.A.Q. Khan (Ed.), Plenum Press, New York, pp. 127-144, 1977.

8-135. Metcalf, R.L., "Development of Laboratory Model Ecosystems as Early Warning Elements of Environment Pollution," in *Proceedings of the Annual Conference on Environmental Toxicology*, Aerospace Medical Research Laboratory, TR-74-125, 17-16, 1974.

8-136.   Metcalf, R.L., Booth, G.M., Schuth, C.K., Hansen, D.J., and Lu, P.Y., Uptake and Fate of di-2-Ethylhexyl phthalate in Aquatic Organisms and in a Model Ecosystem, *Environ. Health Perspect.*, 4:27-34, 1973.

8-137.   Metcalf, R.L., Kapoor, I.P., Lu, P., Schuth, K., and Sherman, P., "Model Ecosystem Studies of the Environmental Fate of Six Organochlorine Pesticides," *Environ. Health Perspect.*, 4:35-44, 1973.

8-138.   Metcalf, R.L., and Lu, D., "Partition Coefficient and Environmental Fate of Xenobiotic Compounds," in *Terrestrial Microcosms and Environmental Chemistry,* J. Witt, and J. Gillette, (Eds.), National Science Foundation, Washington, DC, pp. 23-34, 1977.

8-139.   Metcalf, R.L., Sanborn, J.R., Lu, Po-Yung, and Nye, D., "Laboratory Model Ecosystem Studies of the Degradation and Fate or Radiolabeled Tri-, Tetra-, and Pentachloro-Biphenyl Compared with DDE," *Arch. Environ. Contam. Toxicol.*, 3:151-165, 1975.

8-140.   Metcalf, R.L., Lu, P., and Bowlus, S., "Degradation and Environmental Fate of 1-(2, 6-Difluorobenzoyl)-3-(4-chlorophenyl) Urea," *J. Agric. Food Chem.*, 23:359-364, 1975.

8-141.   Metcalf, R.L., Sangha, G.K., and Kapoor, I.P., "Model Ecosystem for the Evaluation of Pesticide Biodegradability and Ecological Magnification," *Environ. Sci. Technol.*, 5:709-713, 1971.

8-142.   McConnell, W.J., "Relationship of Herbivore Growth to Rate of Gross Photosynthesis in Microcosms," *Limnol. Oceanogr.*, 10:539-543, 1965.

8-143.   McIntire, C.D., "Structural Characteristics of Benthic Algal Communities in Laboratory Streams," *Limnol. Oceanogr.*, 9:92-102, 1965.

8-144.   McLaren, I.A., "Primary Production and Nutrients on Ogac Lake, a Land Locked Fjord on Baffin Island," *J. Fish. Res. Bd. Can.*, 26:1561-1567, 1969.

8-145.   Mitchell, R., and Chet, I., "Indirect Ecological Effects of Pollution," in *Water Pollut. Microbiol.*, R. Mitchell (Ed.), John Wiley and Sons, New York, Vol. 2, 1978.

8-146.   Murthy, N.B.K. and Kaufman, D.D., "Degradation of Pentachloronitrobenzene (PCND) in Anaerobic Soils," *J. Agric. Food Chem.*, 26:1151-1156, 1978.

8-147.   Neely, W.B., and Blau, G.E., "The Use of Laboratory Data to Predict the Distribution of Chlorpyrifos in a Fish Pond," in *Pesticides in Aquatic Environments,* M.A.Q. Khan (Ed.), Plenum Press, New York, pp. 145-166, 1977.

8-148.   Neely, W.B., Blav, G.E. and Alfrey, T., "Mathematical Models Predict Concentration Time Profiles Resulting From Chemical Spill in a River," *Environ. Sci. Technol.*, 10: 72-76, 1976.

8-149.   Olanczuk-Neyman, K.M., and Vosjan, J.V., "Measuring Respiratory Electron-Transport System Activity in Marine Sediment," *Neth. J. Sea. Res.*, 11:1-13, 1977.

8-150.   Olivieri, R., Baccnin, P., Robertiello, A., Oddo, N., Degen, L., and Tonolo, A., "Microbial Degradation of Oil Spills Enhanced by a Slow Release Fertilizer," *Appl. Environ. Microbiol.*, 31:629-637, 1976.

8-151.   Odum, E.P., *Fundamentals of Ecology,* N.B. Saunders Co., Philadelphia, PA, pp. 251-274, 1971.

8-152.   Orians, G.H., "Diversity, Stability, and Maturity in Ecosystems," in *Unifying Concepts in Ecology,* W.H. VanDobbens, and R.H. Lowe – McConnell, (Eds.), Dr. W. Junk, B.V. Publishers, The Hague, pp. 139-150, 1974.

8-153.   Orians, G.H., "Micro and Macro in Ecological Theory," *Biol. Sci.*, 30:79, 1980.

8-154. Paine, R.T., and Vadas, R.L., "The Effects of Grazing by Sea Urchins *Strongylocentrotus* spp. on Benthic Algal Populations," *Limnol. Oceanogr.,* 14:710–719, 1969.

8-155. Pamatmat, M.M., and Bhagwat, A.M., "Anaerobic Metabolism in Lake Washington Sediments," *Limnol. Oceanogr.,* 18:611–627, 1973.

8-156. Pamatmat, M.M., and Fenton, O., "An Instrument for Measuring Subtidal Benthic Metabolism *In Situ,*" *Limnol. Oceanogr.,* 13:537–540, 1968.

8-157. Paris, D.F., Steen, W.C., and Baughman, G.L., "Prediction of Microbial Transformation of Pesticides in Natural Waters." Abstract 175th National Meeting of Am. Chem. Soc., Anaheim, CA, 1978.

8-158. Patten, B.C., and Witkamp, M., "Systems Analysis of ^{137}Cesium Kinetics in Terrestrial Microcosms," *Ecology,* 48:813–824, 1967.

8-159. Perez, K.T., Morrison, G.M., Lackie, N.F., Oviatt, C.A., Nixon, S.W., Buckley, B.A., and Heltshe, J.F., "The Importance of Physical and Biotic Scaling to the Experimental Simulation of a Coastal Marine Ecosystem," *Helgol. Wiss. Meeresunters,* 30:144–162, 1977.

8-160. Pilson, M.E.Q., "An Adventure in Studying Marine Ecosystems," *Maritimes,* 22:12–15, 1978.

8-161. Porcella, D.B., Adams, V.D., Cowan, P.A., Austrheim, S., Holmes, W.F., Hill IV., J., Grenner, W.J., and Middlebrooks, E.J., "Nutrient Dynamics and Gas Production in Aquatic Ecosystems: The Effects and Utilization of Mercury and Nitrogen in Sediment-Water Microcosms," U.S. Dept. Inter. Office of Water Res. and Technol., PRW-Q, 121-1, 1975.

8-162. Pritchard, P.H., Bourquin, A.W., Frederickson, H.L., and Maziarz, T., "System Design Factors Affecting Environmental Fate Studies in Microcosms," in *Proceedings of the Workshop: Microbial Degradation of Pollutants in Marine Environments,* A.W. Bourquin and P.H. Pritchard (Eds.), U.S. EPA Environmental Research Series, EPA-600/9-79-012, Cincinnati, OH, pp. 251–272, 1978.

8-163. Ramsay, A.J., "The Use of Autoradiography to Determine the Proportion of Bacteria Metabolizing in an Aquatic Habitat," *J. Gen. Microbiol.,* 80:363–373, 1974.

8-164. Reed, C., "Species Diversity in Aquatic Microecosystem," *Ecology,* 59:481–488, 1978.

8-165. Reiners, W.A., "Carbon Dioxide Evolution from the Floor of Three Minnesota Forests," *Ecology,* 49:471–483, 1968.

8-166. Robertson, B., Arhelger, S., Kinney, P.J., and Button, D.K., "Hydrocarbon Biodegradation in Alaskan Waters," in *Microbial Degradation of Oil Pollutants.* D.G. Ahearn and S.P. Meyers (Eds.), LA State Univ. Publ. No. LSU-SG-73-01, Baton Rouge, LA, pp. 171–184, 1973.

8-167. Sanborn, J.R., Metcalf, L., Bruce, N., and Lu, Po-Yung, "The Fate of Chlordane and Toxaphene in a Terrestrial-Aquatic Model Ecosystem," *Environ. Entomol.,* 5:533–538, 1975.

8-168. Schippel, F.A., and Bagander, L.E., "Chemical Dynamics of Baltic Sediments, an *in situ* Investigation Method," *Oikos,* 15:63–67, 1973.

8-169. Sikka, H.C., and Rice, C.P., "Persistence of Endothall in Aquatic Environment as Determined by Gas-Liquid Chromatography," *J. Agric, Food. Chem.,* 21:842–846, 1973.

8-170. Simsiman, G.V., and Chesters, G., "Persistence of Diquat in the Aquatic Environment," *Water Res.,* 10:105–112, 1976.

8-171. Smith, K.L., Burns, K.A., and Teal, J.M., "*In Situ* Respiration of Benthic Communities in Castle Harbor, Bermuda," *Marine Biol.,* 12:196–199, 1972.

8-172. Sorensen, J., "Capacity for Denitrification and Reduction of Nitrate to Ammonia in a Coastal Marine Sediment," *Appl. Environ. Microbiol.,* 35:301–305, 1978.

8-173. Sorensen, J., "Capacity for Denitrification and Reduction of Nitrates to Ammonia in a Coastal Marine Sediment," *Appl. Environ. Microbiol.,* 35:301–305, 1978.

8-174. Strange, R.J., "Nutrient Release and Community Metabolism Following Application of Herbicide to Macrophytes in Microcosms," *J. Appl. Ecology,* 13:889–897, 1976.

8-175. Stube, J.C., Post, F.J., and Procella, D.B., "Nitrogen Cycling in Microcosms and Application to the Biology of the Northern Arm of the Great Salt Lake," U.S. Dept. Int., Office of Water Res. and Technol., PRJSBA-016-1, 1976.

8-176. Suess, E., "Nutrients Near the Depositioned Interface," in *The Benthic Boundary Layer,* I.N. McCave, (Ed.), Plenum Press, New York, pp. 57–79, 1976.

8-177. Sugiura, K., Sato, S., and Goto, M., "Toxicity Assessment Using an Aquatic Microcosm," *Chemosphere,* 2:113–118, 1976.

8-178. Swisher, R.D., *Surfactant Biodegradation,* Marcel Dekker, Inc., New York, pp. 1–275, 1970.

8-179. Tagatz, M.E., Ivey, J.M., and Lehman, H.K., "Effects of Sevin on Development of Experimental Estuarine Communities," *J. Toxicol. Environ. Health,* 5:643–651, 1979.

8-180. Terry, R.E., and Nelson, D.W., "Factors Influencing Nitrate Transformation in Sediments," *J. Environ. Qual.,* 4:549–554, 1975.

8-181. Tomizawa, C., and Kazano, H., "Some Aspects of a Model Ecosystem Fate of Carbon, Nitrogen, and Phosphorus in a Model Ecosystem," *Appl. Entomol. Zool.,* 12:27–34, 1977.

8-182. Tomkins, D.J., and Grant, W.F., "Effects of Herbicides on Species Diversity of the Plant Communities," *Ecology,* 58:398–406, 1977.

8-183. Tsuge, S., Kazano, H., and Tomizawa, C., "Some Devices in a Model Ecosystem for a Volatile Compound and Its Application to Carbaryl and *p, p'*-DDT," *J. Pestic. Sci.,* 1:307–311, 1976.

8-184. VanKessel, J.F., "Factors Affecting the Denitrification Rate in Two Water–Sediment Systems," *Water Res.,* 11:259–267, 1977.

8-185. Veith, G.D., and Comstock, V.M., "Apparatus for Continuously Saturating Water with Hydrophobic Organic Chemicals," *J. Fish Res. Board Can.,* 32:1849–1851, 1979.

8-186. Verloop, A., and Ferrell, C.D., "Benzoylphenyl Ureas – A New Group of Larvicides Interfering with Chitin Deposition," in *Pesticide Chemistry in the 20th Century,* J.R. Plimmer (Ed.), American Chemical Society, Washington, D.C., pp. 237–270, 1977.

8-187. Vosjan, J.H. and Olanczuk-Neyman, K.M., "Vertical Distribution of Mineralization Processes in a Tidal Sediment," *Neth. J. Sea Res.,* 11:14–23, 1977.

8-188. Walker, J.D., and Colwell, R.R., "Role of Autochthonous Bacteria in the Removal of Spilled Oil from Sediment" *Environ. Pollut.,* 12:51–56, 1977.

8-189. Walsh, G.E., Miller, C.W., and Heitmuller, P.T., "Uptake and Effect of Dichlobenil in a Small Pond," *Bull. Environ. Contamin. Toxicol.,* 6:279-288, 1971.

8-190. Wangersky, P.J., "Production of Dissolved Organic Matter," in *Marine Ecology,* Vol. IV: Dynamics, O. Kinne (Eds.), John Wiley and Sons, New York, pp. 115-220, 1978.

8-191. Warren, C.E. and Davis, G.E., "Laboratory Stream Research: Objectives, Possibilities, and Constraints," *Ann. Rev. Ecol. Syst.,* 2:111-144, 1971.

8-192. Weary, G.C., and Merriam, H.G., "Litter Decomposition in a Red Maple Wood Lot Under Natural Conditions and Under Insecticide Treatment," *Ecology,* 59:180-184, 1978.

8-193. Weber, J.B., "Soil Properties, Herbicide Sorption, and Model Soil Systems," in *Research Methods in Weed Science,* B. Truelove (Ed.), Southern Weed Science Society, Auburn Printing Co., Auburn, AL, pp. 59-72, 1977.

8-194. Westman, W.E., "Measuring the Inertia and Resilience of Ecosystems," *Bioscience,* 28:705-710, 1978.

8-195. Whittaker, R.H., "Experiments with Radiophosphorus Tracer in Aquarium Microcosms," *Ecol. Monogr.,* 81:158-188, 1961.

8-196. Witherspoon, J.P., Bondietti, E.A., Draggon, S., Taub, F., Pearson, P., and Trabokla, J.R., "State-of-the-Art and Proposed Testing for Environmental Transport of Toxic Substances," U.S. EPA Ecological Research Series, EPA-560/5-76-001, Cincinnati, OH, 1976.

8-197. Witkamp, M., "Microcosm Experiments on Element Transfer," *Intern. J. Environ. Studies,* 10:59-63, 1976.

8-198. Witkamp, M., and Frank, M.L., "Effects of Temperature, Rainfall, and Fauna on Transfer of ^{137}Cs, K, MG, and Mass in Consumer–Decomposer Microcosms," *Ecology,* 51:465-474, 1970.

8-199. Witkamp, M., and Frank, M.L., "Cesium-137 Kinetics in Terrestrial Microcosms," in Proceedings of Second National Symposium, Conf. 670503, Oakridge National Lab., Oak Ridge, TN, pp. 635-643, 1967.

8-200. Witt, J.M., and Gillett, J.W., "*Terrestrial Microcosms and Environmental Chemistry*," in Proceedings of Symposium, Corvallis, OR, National Science Foundation, Washington, D.C., 147 pp. 1977.

8-201. Wright, R.T., "Measurement and Significance of Specific Activity in the Heterotrophic Bacteria of Natural Waters," *Appl. Environ. Microbiol.,* 36:297-305, 1978.

8-202. Yoshinari, T., Hynes, R., and Knowles, R., "Acetylene Inhibition of Nitrous Oxide Reduction and Measurement of Denitrification and Nitrogen Fixation in Soil," *Soil Biol. Biochem.,* 9:177-183, 1977.

8-203. Yu, C., Booth, G.M., and Larsen, J.R., "Fate of Triazine Herbicide Cyanazine in a Model Ecosystem," *J. Agric. Food Chem.,* 23:1014-1015, 1975.

8-204. Zimmerman, A.P., "Electron Transport Analysis as an Indicator of Biological Oxidations in Freshwater Sediments," *Verh. Int. Verein. Limnol.,* 19:1518-1523, 1975.

9

Diseases caused
by chemicals

Ahmed Nasr, M.D., Ph.D.

Assistant Director
Health, Safety, and Human Factors Laboratory
Eastman Kodak Company
Rochester, New York

It has been said that the essence of wisdom is the ability to make the right decision from insufficient evidence. Determining whether or not a particular chemical causes disease in man is frequently a complex issue. Unless dealing with chemicals that have been in use for many years, such as lead and mercury and their compounds, the evidence is incomplete. Even for such chemicals, the effects of long-term low-level exposure are still debatable. In recent years, scientific issues in this area have been pervaded by public sentiment and concern (Ref. 9-1), and by several government regulations.

Regarding government regulations, several laws were enacted in the past few years. In 1970, the Occupational Safety and Health Act, the Poison Packaging Prevention Act, and the Clean Air Act were passed. These were followed in 1972 by the Consumer Product Safety Act, the Federal Water Pollution Control Act, and an amendment of the Federal Insecticide, Fungicide, and Rodenticide Act (1947). The Toxic Substances Control Act and the Resource Conservation and Recovery Act appeared in 1976. In order to comprehend and comply with these regulations, health professionals and managers in industry have had to consult frequently with lawyers within and outside their organizations. This is fraught with difficulties (Ref. 9-2) and is but one of the dilemmas that are being faced in dealing with issues concerning the effects of chemicals on human health.

Obviously, this chapter is not intended to be an exhaustive treatise on diseases caused by chemicals. It will deal only with basic concepts, some of which are complex and controversial. In trying to simplify complicated issues, one may unwittingly become dogmatic. If indeed some of the statements that will follow seem too categorical, those qualified to question dogma are at no risk to be misled by it.

The philosophy and teachings of occupational health constitute useful bases when considering the matter of exposure of humans to chemicals via water, air, and soil. Exposures of wokers are at a higher level and responses are more easily measured; our experience in this area is much greater. Related topics of particular interest are exposure modes, disease types, carcinogenesis, toxicity testing, and epidemiology.

9-1 GENERAL CONSIDERATIONS

9-1.1 Chemicals: How Many Are There?

As of November 1977, Chemical Abstracts Service computer registry of chemicals contained slightly more than four million distinct entities. The Environmental Protection Agency (EPA) estimates that there may be as many as 50,000 chemicals in everyday use, not including pesticides, pharmaceuticals, and food additives. The EPA also estimates that there may be as many as 1500 different ingredients in pesticides. The Food and Drug Administration (FDA) estimates that there are about 4000 active ingredients in drugs and about 2000 other compounds used as excipients to promote stability and inhibit the growth of bacteria. The FDA also estimates that there are about 2500 additives used for nutritional value and flavoring, and 3000 chemicals used to prolong product life. The best estimate thus is that there are about 63,000 chemicals in common use (Ref. 9-3). It is estimated that about 20,000 chemicals are produced in the United States, 2000 of which are starting materials of major industrial significance. It is also estimated that 600 to 800 chemicals are entered annually into commerce (Ref. 9-4).

9-1.2 Harmful Effects Of Man-Made Versus Natural Chemicals

It is a misconception that man-made chemicals are more toxic than those that occur naturally. Man competes only feebly with nature in making highly toxic substances. It has been estimated that seven ounces of botulinum toxin are sufficient to destroy the entire population of the earth. Crampton and Charlesworth (Ref. 9-5) described the occurrence of natural toxins in food. Aflatoxins, nitrosamines, and safrole are some of the "natural" food carcinogens they listed. It may be surprising to some that all parts of the potato plant, with the exception of the tuber, are toxic. Castor beans, lily-of-the-valley, foxglove, rhubarb, and narcissus are but a few examples of poisonous plants (Ref. 9-6).

9-1.3 Exposure To Chemicals

The General Environment. It is common knowledge that air, water and food all contain some undesirable chemicals. Toxic chemicals in community air include oxides of sulfur, oxides of nitrogen, and ozone and other oxidants, as well as particulate matter. Chemical contaminants of water are too numerous to count. However, it is doubtful whether the traces of most of the organic chemicals that have as yet been detected in water cause adverse effects in man. Since diseases of the heart and blood vessels are responsible for approximately half the deaths in the United States, their relationship to water hardness should be of special interest. An inverse correlation between indices of water hardness and death rates from cardiovascular disease was noted in Japan, the United Kingdom, the United States, Sweden, and other countries (Ref. 9-7). The report by Crampton and Charlesworth (Ref. 9-5) lists several naturally occurring toxins in food. These include nitrites, nitrates, selenium, iodine, cyanide, solanine, and several others.

The Personal Environment. The following synthetic chemicals may be found on or in an average person: 4-methoxy-m-phenylenediamine; calcium thioglycolate; polyvinylpyrrolidine; (2, 5-dioxo-4-imidazolidinyl) urea, also known as allantoin; and boric acid. These chemicals, in the order mentioned, are constituents of hair dyes, depilator creams, hair sprays, creams for nipple care, and douche powders. However, these chemicals are not the most important in the personal environment with regard to their toxic effects. "Even in the workplace, it seems probable that the most important carcinogen is the cigarette" (Ref. 9-8). The relationship between lung cancer and cigarette smoking has long been established. Since inhalation is a common way of exposure to industrial chemicals, assessing the relationship between exposure to an industrial chemical and lung cancer would be difficult, and perhaps futile, unless the smoking habits of the people studied are known. Excessive consumption of alcoholic beverages is known to cause liver disease, including cancer. Certain chemicals, some of which are used in industry, similarly affect the liver.

Cigarette smoking has been implicated also in a variety of diseases in addition to lung cancer. Notable among these are chronic bronchitis and emphysema. Knowledge of the smoking habits of industrial workers is essential in determining whether these diseases are due to an industrial chemical or cigarette smoking.

One periodic claim has been that smoking may be linked to a higher incidence of spontaneous abortion (Ref. 9-9). The health of working women has recently become the focus of attention. The effects of most chemicals in the industrial and personal environments — and in air, water and food — on fetal health are unknown. There are chemicals in the industrial environment, e. g., lead, which are known to be toxic to the fetus. The effects of various chemicals on maternal and fetal health will probably be a major area of toxicological research in the near future.

Prescription and nonprescription (over-the-counter) drugs are important components of the personal environment. It has been recently reported that women given Dilantin[R] during pregnancy may have children who develop neuroblastoma, a tumor of nervous tissue (Ref. 9-10). Dilantin[R] is a prescription drug used in the treatment of epilepsy.

Para-aminobenzoic acid and certain of its esters are widely used in sunscreens. Proprietary sunscreens containing amyl paradimethyl-aminobenzoate were found to produce a skin reaction resembling sunburn. This "paradox" of a sunscreen that promotes "sunburn" is explained by the fact that, as affected users have concluded, the product is ineffective (Ref. 9-11).

The personal environment also includes chemicals used in hobbies and home repairs. Methylene chloride, an industrial solvent, is used also to remove paint. Strippers beware! It has been shown that the human body is capable of converting methylene chloride to carbon monoxide. Inhalation of high concentrations of methylene chloride, or carbon monoxide, could cause serious toxic effects, (Ref. 9-12). However, long-term low-level exposure to methylene chloride in industry has not been shown to cause adverse effects (Ref. 9-13).

A peculiar (and rare) example of harmful chemical exposure in the personal environment was reported in a patient who developed life-threatening leakage in one of his heart valves, following two successive nickel-containing valve prostheses. The patient was allergic to nickel (Ref. 9-14).

Chronic exposure to airborne lead in a poorly ventilated indoor firing range was reported to cause excessive lead absorption and mild lead poisoning in three instructors (Ref. 9-15). This is an occupational exposure that, under certain circumstances, could become a personal one.

The Occupational Environment, Industry and Agriculture. As previously mentioned, there are approximately 20,000 chemicals produced in the United States, 2000 of which are important starting materials. In agriculture, pesticides are perhaps the most important group of chemicals; their ingredients number approximately 1500. The modern farm in many industrial countries is not much different from the modern factory, with its tools and machinery and the various chemicals needed to operate such equipment.

9-2 INDUSTRIAL EXPOSURE TO CHEMICALS

9-2.1 Kinds of Exposure

Exposure to chemicals in industry varies considerably with the type of operation. Exposure to chemicals in a research laboratory is likely to be to small quantities of new chemicals, or old chemicals being tried for new uses. The employee frequently has a higher level of education than the average factory worker. Usually such education is associated with increased awareness of the hazards of chemicals, and knowledge of how to handle them safely.

Manufacturing chemicals in large quantities commonly occurs in closed systems. If they are adequately designed, there is little chance for exposure, except accidentally, as in the case of ruptured pipes or failing valves. Maintenance and repair personnel are more likely to be exposed to chemicals than operators of closed systems. Bulk manufacturing of chemicals is undertaken by large companies that can afford the services of health and safety professionals. Thus gross exposures and poor hygienic conditions in the workplace, contrary to popular belief, are not common. Many small manufactures and formulators are at a disadvantage; economic considerations in such operations not uncommonly lead to productivity taking precedence over safety.

9-2.2 Modes of Exposure

Inhalation and skin contact are the common modes of exposure to industrial chemicals, as well as exposure of the eye to chemicals in the air and accidental exposure to liquids. "Hand-to-mouth" exposure is uncommon, except when eating and smoking are permitted in the workplace and when personal hygiene is inadequate. The latter may be due to lack of washing facilities, or the lack of use of such facilities when available.

Inhalants may be particulate matter, vapors, or gases. Particulates include solid or liquid materials that are so finely divided that the size of the particles permits them to remain in the air for some time. If the particles are too large, they disappear by gravitational settling. If they are exceedingly small, they are cleared from the air by diffusion. Airborne particulate matter, aside from creating toxicity and explosion hazards, may be a nuisance by settling on the skin and clothing. Airborne particulates may also interfere with visibility and consequently with safety.

Vapors and gases include a wide variety of chemicals. Vapors arise from volatile liquids. A high vapor pressure at room temperature, a large surface area, and heat are all conducive to vaporization. Gases are commonly classified into the simple asphyxiants, the chemical asphyxiants, and the irritant gases. Hydrogen and nitrogen are examples of the simple asphyxiants; their harmful effects result from their displacing oxygen. Notorious among the chemical asphyxiants are carbon monoxide and hydrogen cyanide. Their toxic effects are the result of their ability to interfere with cell respiration. Carbon monoxide has been deservedly called "the silent killer"; it is not irritating and has no ordor, i. e., it has no warning properties. Chlorine, phosgene, and the oxides of sulfur and nitrogen are examples of the irritant gases. The concentration of a vapor or a gas at which it would be detectable to most people by its ordor may be considered a safety factor, provided that this odor threshold occurs at a level that is considered safe. For example, carbon tetrachloride, ozone, and chlorine can be detected by their odors at levels of approximately 80 ppm, 0.02-0.05 ppm, and 3-4 ppm, respectively. Their Threshold Limit Values (Ref. 9-16) are 10 ppm, 0.1 ppm, and 1 ppm, respectively. Ozone is thus the only one that can be detected by its odor at a concentration below its hygienic standard. However, it must be emphasized that odor cannot be relied on as a universal warning property. Olfactory fatigue (getting used to the odor) and the wide variation in the acuity of the sense of smell among individuals reduce the value of odor as a warning property.

Exposure of the skin to chemicals may be harmful at the site of contact, or in organs and systems far away from the site of contact if the chemical is absorbed through the skin. According to the Occupational Health and Safety Administration (OSHA) of the United States Department of Labor, skin disorders accounted for approximately 46% of the occupational illness reported in 1975 (Ref. 9-17). The principal effects on the skin are primary irritation, sensitization (allergy), and phototoxicity. The term "contact dermatitis" (inflammation of the skin) describes dermatitis that results from irritation, as well as that resulting from sensitization. Sometimes, contact dermatitis is used to denote only allergic contact dermatitis. Whether or not irritation of the skin will result from exposure to a chemical depends on the frequency, duration, and magnitude of exposure. Various acids and alkalies are examples of chemicals that can cause skin irritation. The degree of inflammation depends on, among other factors, the strength of the acid or the alkali, i. e., their pH. Liquids with a pH of 2.5 or less (very acidic), and a pH of 11.5 or greater (very alkaline) may cause serious inflammation of the skin. A chemical burn may result if contact with such liquids is prolonged. Acids burn the skin, alkalies dissolve it, solvents remove its natural greases, and oils block its pores. All four classes of chemicals, as well as others, are capable of causing irritant contact dermatitis. Bichromates, epoxy resins, formalin, and nickel and its compounds are examples of chemicals that can cause sensitization dermatitis (Ref. 9-18). Skin phototoxicity resembles an exaggerated sunburn. It is a form of irritation that is induced by light, and it occurs when photoactive chemicals reach the skin by direct contact or via the bloodstream, following ingestion or injection. "Bikini dermatitis" is an example of phototoxicity secondary to a dye in a bathing suit (Ref. 9-19). It is unlikely that "bikini dermatitis" is frequently noticed.

Some chemicals, such as carbon tetrachloride and other chlorinated aliphatic hydrocarbons, may cause primary irritation of the skin, as well as injury to internal organs such as the liver and kidneys. The effects on the liver and kidneys result from absorption of these chemicals through the skin into the bloodstream and their circulation to various body organs. Other chemicals are capable of causing toxic effects following contact with the skin, without causing skin irritation, e. g., the organo phosphates. These compounds, many of which are used as pesticides, include hundreds of chemicals. Parathion[R] and TEPP (Tetraethylpyrophosphate), are two of the most toxic. They stimulate certain parts of the nervous system, causing toxic effects in the gastrointestinal and respiratory systems, sweat and salivary glands, urinary bladder, and the heart (Ref. 9-20).

Eye injury may result from exposure to chemicals. Toxic effects in the internal organs following contact with chemicals is known, but is rare. Similar to the case with skin exposure, liquids with a pH less than 2.5 or greater than 11.5 may cause severe eye injury. Properties other than the pH also determine the seriousness of the injury. For example, sulfuric acid of the same strength as hydrochloric acid is more harmful to the eyes (and the skin) because of its ability to dehydrate living tissues, thus having a charring effect. Injury to the eyes by alkaline liquids is frequently more serious than injury by acidic liquids, because of the former's ability to penetrate tissues. If a liquid contains, a surfactant, e. g., a surface active agent such as a detergent, eye injury may be enhanced because the surfactant makes other chemicals in the liquid spread quickly.

Ingestion, as previously mentioned, is not a common way of exposure to chemicals in industry unless eating and smoking are permitted in the workplace, and when personal hygiene is inadequate. However, skin protection may be necessary when highly toxic chemicals are handled, since small amounts accidentally introduced into the mouth may have toxic effects.

Solvent vapors may be absorbed through disease-affected skin much more readily than through healthy skin. Absorption through undamaged skin of solvent vapors from the surrounding air in the work environment is likely to be insignificant (Ref. 9-21).

9-3 OCCUPATIONAL DISEASES CAUSED BY CHEMICALS

9-3.1 Distinguishing Occupational From Nonoccupational Diseases

An occupational disease is defined as an illness arising out of and in the course of employment, and in which a causal relationship has been established between the disease and the occupational exposure. Distinguishing an occuptional disease from a nonoccupational one is frequently difficult. There are several reasons. There is hardly an occupational disease that has distinctive or unique symptoms and signs. For example, shortness of breath and coughing, which result from exposure to cotton dust, are no different from the same symptoms that are caused by heavy cigarette smoking. Bronchial asthma caused by house dust in a sensitized individual is similar to the bronchial asthma that affects a worker who is allergic to toluene diisocyanate, flour, grain and wood dusts, or fungi and their decomposition products blown out from a ventilation system into workroom air. A chest x-ray of a foundry worker who has been inhaling free silica dust for several years is not different from the x-ray in numerous other diseases of the lungs. Not only do occupational diseases mimic nonoccupational ones, their earliest manifestations may not be clearly indicative of disease at all. For example, irritability, insomnia, and loss of appetite affect normal people every now and then. These symptoms may be also the earliest manifestations of poisoning with mercury and some of its compounds. Adequate training and experience in clinical toxicology, as well as clinical medicine, are essential in distinguishing occupational from nonoccupational diseases.

9-3.2 Occupational Diseases Caused By Chemicals

Systems and Organs. Almost any body organ or system, from the skin inwards, can be adversely affected by chemicals. Some chemicals affect more than one system. For example, arsenic and thallium predominately affect various parts of the nervous system, as well as the skin, nails and hair. A single substance, such as asbestos, may cause one or more diseases in the same person: asbestosis, pleural and peritoneal mesothelioma, lung cancer, benign pleural effusion, and pleural thickening and calcification.

Diagnosis of Occupational Diseases. Diagnosing an occupational disease, especially a serious one, frequently has important implications for the worker and his employer. The possibility of job change, compensation, concern of fellow workers, and measures that should be taken to prevent the occurrence of the disease in other employees are the principal implications. Recognition of occupational diseases requires thorough knowledge of the various occupations and the chemicals that may be associated with them. It also requires a clear understanding of the effects of chemicals on humans. Laboratory and other clinical studies, as in the case of nonoccupational diseases, are helpful in elucidating the cause of the illness.

The importance of the history and conditions of exposure will be illustrated by the following examples. Most of the illness caused by cadmium (lung and kidney disease) has occurred in workers cutting and welding cadmium-plated metals and in the manufacture of nickel-cadmium batteries, but not in metal-plating operations. Beryllium, a metal that can cause a serious, now-rare, lung disease, is a hazard in casting and machining alloys containing the metal, and formerly was a hazard in the fluorescent lamp industry. However, there have been no reported cases of beryllium disease in the miners of beryllium ore (beryl: beryllium aluminum silicate).

The duration of exposure to a substance is frequently important in determining whether or not harmful effects have been caused by it. For example, silicosis rarely develops before several years of exposure, usually ten or longer. Needless to say, a physician diagnosing an occupational disease caused by a chemical should be familiar with the effects of the chemical on people. For many chemicals there is more information on the effects on experimental animals than on humans. The undiscriminating use of such information in the diagnosis of occupational diseases could lead to erroneous conclusions.

It is unfortunate that the average physician receives very little, if any, training in Industrial Toxicology, although he is the one likely to encounter occupational diseases in their early stages. One patient who suffered from disabling anemia, the cause and treatment of which eluded a specialist in blood diseases, was once referred to this writer with the recommendation "no more exposure to chemicals!" The patient was a homemaker and was never employed by industry. A painstaking investigation of the chemicals in the patient's personal environment left this writer as puzzled as the specialist in blood diseases. None of the chemicals were known to cause anemia. However, the patient's anemia always improved during three hospital admissions, although she was not receiving any medicines for her ailment. It was suggested to the patient that she live with relatives for a few weeks and return for a fourth examination. No more anemia! It was boldly prescribed to the patient that she move out of her house, and preferably out of the whole neighborhood. Neither the referring physician nor this writer saw that patient again. Anecdotal as this may be, it is hoped to illustrate how difficult identifying the cause of an illness can be. It also illustrates "individual susceptibility" – all other members of the patient's family were in good health.

A clear understanding of the sequence of events that takes place following exposure to a chemical is helpful in establishing a diagnosis. An ancient metal and its inorganic compounds, lead, illustrate this point. Lead is abosrbed by inhalation, and to a lesser extent from the gastrointestinal tract if ingested. Lead and its compounds are excreted in urine and, if exposure is not excessive, it may not be possible to detect an increase in the blood levels. However, if absorption exceeds excretion, blood lead levels will rise. As exposure continues, constipation and abdominal cramps are likely to develop; then the blood and blood-forming organs begin to be affected. This results in anemia, and biochemical changes that are indicative of interference with heme synthesis. Massive continuous exposures may eventually cause disease in the central or peripheral nervous systems. It should be obvious from this sequence of events that constipation or anemia would be most unlikely to be due to lead poisoning when the blood lead levels are normal. Recent biochemical studies indicate that exposure to lead may be detectable before significant symptoms arise.

The history of exposure to a chemical should include the amounts used, the duration of exposure, the use of personal protective devices such as respirators and gloves, the perception of odors, hygienic conditions in the workplace, and the health of other workers.

As with nonoccupational diseases, textbook descriptions of occupational diseases are not commonly encountered. This writer has seen several cases of carbon monoxide poisoning and searched in vain for the cherry-red color of the skin and mucous membranes. Pallor or cyanosis are probably more frequent in carbon monoxide poisoning than pink lips and rosy cheeks. Upon consultation with a veteran occupational physician, it was pointed out that the cherry-red color of the skin requires an optimal combination of high carboxyhemoglobin, and low reduced hemoglobin, and is rare except in acute exposures to carbon monoxide at high concentrations. Pallor or cyanosis are common in prolonged exposure to small concentrations, such as may result from a malfunctioning heating furnace.

It is worth noting that a disease that follows exposure to a chemical may not have been caused by the chemical itself. It may have been caused by an impurity, a degradation product, or a reaction product. For example, commercial grades of toluene may contain appreciable amounts of benzene. Benzene is known to affect the blood forming organs, while pure toluene is not known to be a blood poison. Several chlorinated aliphatic hydrocarbons generate hydrogen chloride and phosgene when degraded by high temperatures. Hydrogen sulfide from decaying organic material, and hydrogen cyanide resulting from the addition of acids to cyanide salts, are examples of reaction products.

Laboratory and clinical studies are vital in the diagnosis of occupational diseases caused by chemicals. The average hopsital is equipped to conduct several studies that could be useful in diagnosing occupational diseases. These include x-ray examinations and biochemical and physiological studies to assess the function of various body organs and systems. Although most hospitals are equipped to analyze for common medications that are accidentally (or suicidally) ingested, such as aspirin and barbiturates, they are not usually equipped to analyze for trace metals and several other industrial chemicals. When such analyses are required, specialized laboratories should be consulted.

Prevention of Occupational Diseases. The prevention of occupational diseases entails the application of several sciences and arts — those of the design engineer, the safety specialist, the industrial hygienist, and the industrial physician. In summary, substitution, ventilation, enclosure, and segregation of the hazardous operation or chemical are the fundamentals of prevention. Personal protective devices are a last resort; they are cumbersome and disliked by most workers. Their main value is in temporary tasks of short duration, such as cleaning a reaction vessel, repairing a ruptured pipe, or replacing a malfunctioning valve.

9-4 CHEMICAL CARCINOGENS: PRACTITIONER'S APPROACH

"Cancer is one of the most frightening words in our language. And with good reason. This year 675,000 Americans will find out they have it. Some 370,000 will die of it — one every one-and-a-half minutes" (Ref. 9-22). This statement, though factual, gives the erroneous impression that cancer is an epidemic disease of modern times. The earliest references to what later became known to us as cancer are found in the Ebers Papyrus, which was written in Egypt in about 1500 B. C., and summarized medical knowledge as it was known at that time (Ref. 9-23).

9-4.1 Environmental Versus Occupational

It has been repeatedly stated that most cancers are caused by environmental factors. "Environmental" is frequently equated with "industrial," and exposure to chemicals is considered to be predominantly occupational. These are fallacies. Some examples of chemical exposures in the personal environment have already been mentioned. As previously quoted, it seems probable that the most important carcinogen in the workplace is the cigarette. The American Cancer Society estimates that smoking cigarettes may account for as much as 80% of all lung cancers — the leading cause of cancer deaths in males in the United States (Ref. 9-24). It has been estimated that the percent of total cancer related to diet is 40.9 in males and 60.1 in females (Ref. 9-25). Radiation, mostly solar, accounts for 5% to 8% of all cancers (Ref. 9-26). Indeed, the majority of cancers may be environmentally related. However, dietary, smoking, and drinking habits are much more important than previously thought in the causation of cancer, and the industrial work environment is but a small part of the total environment.

"The most hazardous industry in the United States, in terms of exposure of workers to carcinogens, may well be the manufacture of scientific and industrial instruments, according to a study prepared by John Hickey, James Kearney, and their associates at Research Triangle Institute for the National Institute for Occupational Safety and Health (NIOSH)." According to this report, the chemical industry, which many people would consider an odds-on choice to head the list, was ranked 12th. The investigators in this study combined potency, amount of exposure, and annual production to conclude that the ten most hazardous industrial chemicals are, respectively, asbestos, formaldehyde, benzene, lead, kerosene, nickel, chromium, coal tar pitch volatiles, carbon tetrachloride, and sulfuric acid. Their results differ from the conclusions of previous studies because in other studies only the volume of the carcinogens and not the amount of exposure were considered. The large quantities of materials manufactured by industry may actually be manufactured by only a very small number of people, so that consideration only of the volume of carcinogens grossly overestimates the potential hazard (Ref. 9-27).

"Among the many comparisons shown in the report, selected possible high-risk occupations are pointed out as examples of findings that are consistent with prior observations and others that warrant further follow-up. Among these are bladder cancer in leather workers and dairy farmers, lung cancer in brickmasons and smeltermen, and oral cancer in printing workers" (Ref. 9-28). Although several kinds of cancer have been caused by certain industrial chemicals, this recent report suggests that causation of cancer is complex, multifaceted, and not, in many instances, a simple effect of exposure to a single specific chemical. It seems that associating "occupational cancer" with "the chemical industry" has been an oversimplification, perhaps due to exaggerated publicity and the fact the chemical industry is more visible than dairy farming, bricklaying, and leather crafting. On the other hand, there is exposure to a multitude of chemicals in smelting, and to a few chemicals in printing. As is the case with retrospective surveys, this study (Ref. 9-28) identifies "association" rather than "causation." Definitive conclusions should await more rigorous and reproducible studies.

Asbestos, β-naphthylamine, vinyl chloride, and bis-chloromethyl ether are some of the chemicals for which there is substantial evidence of occupational cancer causation.

9-4.2 Has There Been a Cancer Epidemic?

Examination of the age-adjusted cancer death rates in males and females in the United States in the period 1930 to 1974 shows the following (Ref. 9-29): Lung cancer in men and women has risen steadily, more steeply in men. Cancer of the stomach in men and women, and cancer of the uterus have been declining. There has been no appreciable change in the death rates from other kinds of cancer over the 45-year period from 1930 to 1974. It is true that there are more cases of cancer now than 40 or 50 years ago. However, the population today is also greater — hence rates rather than numbers should be the basis of meaningful comparisons. Moreover, with increased life expectancies comes increased chances of developing cancer. Thus, rates of cancer incidence must be adjusted for age; otherwise the comparisons would be useless (and misleading) for determining whether or not cancer has become rampant over the past half century. Better methods of diagnosis may account for some of the apparent increase in any particular kind of cancer.

General estimates of the percentage of all human cancers related to occupational exposure range between 1% and 10% (Ref. 9-30). The percent of total deaths due to cancer in the United States (1974 vital statistics) is 18.6% (Ref. 9-29). Assuming that occupational cancer accounts for approximately 5% of total cancers, industry-related cancer causes $18.6 \times 0.05 = 0.9\%$ i. e., less than 1% of the total mortality. This is not an argument for ignoring occupational cancer; there is no question that some occupational exposures increase the risk for certain cancers. However, occupational cancer as a public health problem should be viewed in its proper perspective. The reason is simple. It is also trite. Resources to improve the health of any population are finite, and should be utilized in proportion to the anticipated benefit.

Because the essence of prevention of disease is the accurate identification of causes, the distinction between environmental and occupational is of paramount importance. Industrial workers are as likely to develop nonoccupational diseases as the general public, and preventive measures to improve their health should be commensurate with the magnitude of the hazards to which they are exposed in their occupations and in their personal and the general environment. Eradication of cancer is not the same as eradication of other diseases. Cancer was known long before industrial chemicals numbered in the thousands, and it affects children, long before they might be exposed to industrial chemicals. Cancer is a multitude of diseases; there are no less than one hundred, ranging from skin cancer, which has a survival rate of almost 100% when detected early, to cancers of the stomach and lung, with survival rates that are less than 20%, despite the best efforts for diagnosis and treatment. Public sentiment, fueled by well-intentioned misinformed pronouncements and regulatory zeal, have so far produced little more than confusion and controversy, and are unlikely to imporve the health of the people who work (and the people who do not), unless they are rationally directed at the major causes of illness, disability, and death, within and outside the workplace.

9-4.3 Evaluation of the Risk to Humans

Fewer than 40 chemicals and substances (the latter are not specific chemical entities) have been proved to be carcinogenic to man. In animal tests, where the quality of the data and rationality of conclusions varies widely, approximately 6000 chemicals have shown evidence of causing cancer (Ref. 9-4). Considering that there are approximately four million chemicals in existence, and a little more than 60,000 chemicals in use in the United States, it seems that the fear of chemically-induced cancer may have been exaggerated. In assessing the risk to humans, it would be useful to make a distinction between the ability of a chemical to cause cancer after it has been introduced into the body, its toxicity, and the likelihood of a chemical to cause cancer in man, the hazard.

Since the "evidence" of carcinogenicity of a chemical ranges from conjecture to convincing evidence, a system of categorization would be useful in ranking chemicals as to their potential for causing cancer *as a toxic effect*. The following is one example of such a classification:

I. Carcinogen
   A. Known human carcinogen as determined by sound epidemiological studies, or

B. Closely related in chemical structure to a known human carcinogen and satisfying either 1 or 2 below:
   1. Found to be carcinogenic in at least two mammalian species, in at least two laboratories. Testing procedures should be relevant to the potential for human exposure with regard to route of administration and dosage.
   2. Positive mutagenic activity in at least two short-term tests, one of which is a mammalian cell system.

II. Confirmed Animal Carcinogen
Positive mammalian test results in two or more species meeting criterion B1 above.

III. Possible Carcinogen
A chemical not meeting the criteria of categories I or II, but having positive mammalian test results of questionable significance.

A categorization system, the one here described or any other, should be used only as a frame of reference by a panel of experts to classify chemicals with regard to carcinogenicity. The fact that some chemicals may not exactly conform to the criteria of a particular category is unavoidable. Classification of some chemicals could be tentative, pending the acquisition of additional information. Ranking of a chemical based on its potential for causing cancer after it has been introduced into the body, as well as consideration of its physical properties, the quantities handled, and manner of usage, are all essential to the evaluation of the risk to humans. In other words, it is not only the carcinogenicity that matters, but also the probability of human exposure.

Several short-term tests for carcinogenicity are being developed. These include mutagenesis, DNA repair synthesis, *in vitro* cell transformation, reaction with nucleic acids, and cytological alterations. The Ames test is perhaps the best known. It tests the ability of a chemical to affect the heritable material of bacterial cells in a manner that is thought to be similar to the induction of cancer in man and laboratory animals. This test, and other short-term tests, provide circumstantial evidence. "In the course of time it is likely that a battery of such tests will be developed and appropriately validated to the point that they will provide a reliable indication of mutagenesis, with some guidance concerning carcinogenic potential. That stage has not yet been reached, and the most that can be said for the short-term tests at present is that positive results serve to raise the index of suspicion regarding the compound under study" (Ref. 9-31). In summary, "it remains to be seen whether the genetic bases of *E. coli* and *E. lephant* are the same and whether the response of their genetic apparatus to chemical mutagens may be equated" (Ref. 9-32). It also remains to be seen how predictive are mutations of cancer. If chemicals affect elephants in the same way they affect microbes, we are indeed at the threshold of a new era in toxicological testing. Understanding the mechanism of a toxic effect, be it cancer or irritation of the throat, is the crux of extrapolating test results among various creatures.

Finally, there is a modest body of research that tentatively links psychological stress and certain emotional characteristics to cancer. The Greek physician, Galen, "is said to have observed that women with melancholic dispositions seemed more inclined to breast cancer than those of a sanguine bent." Caroline Bedell Thomas of Johns Hopkins Medical School, in an epidemiological study of medical students that was begun in 1946, found strong similarities between the psychological profiles of those who developed cancer and those who committed suicide. The National Cancer Institute has recently funded several large epidemiological studies that are attempting to determine whether psychological factors are predictive of cancer (Ref. 9-33). The notion that emotions may influence the onset and course of cancer does not seem incredible, considering the research findings that correlated emotions, including abnormal ones, to biological-chemical events. Dr. D. M. Kissen of the University of Glasgow, from psychological studies of industrial workers (all of them smokers), concluded that, the more repressed the individual, the fewer cigarettes it took to induce cancer (Ref. 9-33). The possible interactions among chemicals and emotions in the causation of disease, including cancer, are yet to be explored.

9-5 TOXICITY TESTING

9-5.1 Testing in Animals

A "step-sequence" testing scheme for risk assessment of chemicals is described in Chapter 12. In this section, some fundamentals and shortcomings of toxicity testing are discussed. Regulatory agencies frequently approve or ban a chemical solely on the basis of animal test results. Eckardt and Scala (Ref. 9-34) classified the factors influencing toxicity into factors related to the toxic agent, factors related to the exposure situation, and factors related to the subject (the experimental animal). Among the factors related to the toxic agent they included chemical composition, physical characterisitcs, presence of impurities or contaminants, stability and storage characteristics of the toxic agent, solubility of the toxic agent in biologic fluids, and choice of the vehicle in which the toxic agent is to be administered to the test animals. The factors related to the exposure situation include the following: dose, concentration, and volume of administration; route, rate, and site of administration; duration and frequency of exposure; and time of administration (time of day, season of the year, etc.). The toxicity-influencing factors that are related to the subject may be inherent or environmental. The inherent factors include species and strain, genetic status, immunologic status, nutritional status, hormonal status (pregnancy, etc.), age, sex, body weight, maturity, central nervous system status (activity, crowding, handling, presence of other species, etc.), and the presence of disease or specific organ pathology. The environmental factors that may affect experimental animals include temperature and humidity, barometric pressure, ambient atmospheric composition, light and other forms of radiation, housing and caging effects, and noise and other geographic influences.

The following few examples will illustrate the importance of some of the factors that influence toxicity testing. If a chemical is unstable and is tested long after it has been synthesized, the toxicologist may be studying the toxic effects of a chemical other than the one that the chemist provided. As to the species of animal to be tested, it is a common misconception that nonhuman primates are the best animal model for extrapolating results to humans. The heart of a frog responds to electrolytes such as sodium and potassium salts in a way not much different from the human heart. Chickens are better predictors of the toxicity of organophosphates in man than are cows and monkeys.

Rats housed singly develop high blood pressure. This hypertension is often accompanied by increase in the activity of the adrenal glands, and these changes may affect kidney and electrolyte physiology, pituitary function, and the metabolism of chemicals. The exact nature and extent of these effects have not yet been established. The metabolism of some sex hormones is also affected by isolation in both male and female rats (Ref. 9-35).

Since the usual objective of testing chemicals in animals is to predict what might happen to humans, "the detection and measurement of risks to man would be facilitated if provision were made for the linking of records of exposure to industrial and other hazards with hospital discharge and mortality data" (Ref. 9-36). This report, recommends improved systems for monitoring the health of industrial workers exposed to new chemicals and indicates that experience with drugs shows "the folly of relying on results in animals."

It should be obvious that comparisons among animals, especially extrapolating animal test results to man, and comparisons among chemicals with regard to toxicity, is an intricate exercise in deductive and inductive thinking, that should be based on broad training and experience in this field. It is not an endeavor to be undertaken by amateurs.

9-5.2 Human Studies

Testing the Toxicity of Industrial Chemicals in Humans. Testing of chemicals with unknown properties using humans is obviously unethical. Only when the effects of a chemical on experimental animals have been extensively studied should the possibility of its administration to human volunteers be considered, in doses found to be relatively innocuous. Under such circumstances a safety factor of 100 would be employed, i.e., humans would be given one hundredth of the dose that was found to have minimal or no effects in animals. The rationale of this safety factor is based on the assumption that man may be ten times as susceptible to the chemical as experimental animals, and that certain individuals may be ten times more susceptible than others.

Biological Monitoring. Monitoring the exposure of industrial workers could yield useful information as to their exposure to chemicals in the work environment. Environmental monitoring, i.e., analysis of air, water, or food for chemicals, does not always correlate well with individual exposure. Zielhuis (Ref. 9-37), recently published a comprehensive report on biological monitoring. The following is predominately derived from Dr. Zielhuis' report. Biological monitoring is the measurement of internal exposure of an individual by analyzing a biological specimen such as blood or urine. Biological monitoring should be distinguished from health screening, e.g., for anemia, bladder cancer, and the like. Health screening is aimed at detecting individuals with impending or already evident adverse health effects. The objective of biological monitoring is to determine the exposure of groups of subjects, and of individual members of such a group.

Notable among the advantages of biological monitoring is its ability to pinpoint more adequately than environmental monitoring the groups that have the greatest risk. Also, the variability in biological parameters is usually smaller than the variability in air, water, and food concentrations. For example, while the concentration of a chemical in the air may remain the same, exposure of an individual to the chemical would be higher during physical activity than during rest. Since humans are exposed to many substances at the same time, biological monitoring is a more accurate indicator of total load of such chemicals as pesticides and metals to which there may be exposure in the occupational as well as the personal environment. For some chemicals, biological monitoring may be less expensive than environmental monitoring.

There are limitations and disadvantages to biological monitoring. Human beings are required as sampling units. This may cause inconvenience or anxiety. For most chemicals, there are no reliable biological permissible levels, i.e., no useful comparisons (to indicate harmful exposures) can be made. Moreover biological monitoring is not suitable for assessing the exposure to highly reactive chemicals such as chlorine and ozone, or chemicals with a very short biological half-life. Since biological monitoring uses human beings as sampling units, it should always be carried out under the supervision of medically qualified personnel. This may preclude its use in small workplaces where health professionals are not readily available.

Biological monitoring has been used to estimate the exposure of the general population to a certain chemical; polybrominated biphenyls (PBBs). By analyzing human breast milk for PBBs in a random-sample survey, Brilliant *et al.* (Ref. 9-38) were able to estimate that about 8 million of Michigan's 9.1 million residents have detectable body burdens of this chemical.

Epidemiological Studies.   Epidemiology started as the study of epidemics of infectious diseases.   At the present, epidemiological methods are also used to investigate the causes of noninfectious and chronic disease.   Epidemiological studies may be descriptive, that is, involving the distribution of disease in a defined population.   Analytical epidemiological studies, on the other hand, investigate the various factors that influence the incidence of disease.

Case-control and cohort studies are the two principal kinds of epidemiological investigations.   In case-control studies the researcher identifies a group of people who suffer from a certain disease and, for comparison, another group that is free from the disease, the control group.   Exposure to a suspect disease-causing agent (or agents) is then determined in both groups, in a quantitative or a semiquantitative manner.   If exposure among the people who have the disease is significantly higher than in the control group, this would suggest that exposure to the suspect agent caused the disease.   For example, such a comparison can be made between two populations, one suffering from anemia, and the other healthy (the control group).   In order to determine whether exposure to lead caused the anemia in the affected population, environmental or biological monitoring for lead is determined in both the study and control groups.   The presence or absence of anemia can be determined by the red blood cell count or by measuring the amount of hemoglobin in blood.   If exposure to lead in the anemic population is found to be significantly higher than in the healthy population, it may be concluded that exposure to lead is associated with anemia and may have caused it.

A cohort study starts with two groups of people, one of which is exposed to the disease-causing agent.   The two groups are followed for a certain period of time in order to determine the proportion in each that develops the disease being investigated.   This is a prospective cohort study.   A cohort study may also be retrospective, examining past records of two groups of people to assess both the exposure, the suspected cause, and the disease, the presumed effect.

Epidemiology has five main uses in occupational health.   It may be used to identify hazards, keep known hazards under control, help to find causes and to establish hygienic standards, correlate the prevalence of disease with levels of exposure, and to evaluate health services to determine how they are used, and their success in attaining certain standards (Ref. 9-39).

Epidemiological studies may also be classified into cross sectional and longitudinal.   In cross-sectional studies the state of health of an exposed group of people is determined at a certain point in time.   In these studies the prevalence of health effects in another group (the control group) not exposed to the same conditions must be known in order to compare the two groups. In a longitudinal study the changing health of each individual in the course of time is observed. Because various examinations take place at different points in time, the person is compared to himself. Because of the individuality of response, a relatively small change in health may become detectable; hence this method is preferable to the cross-sectional study (Ref. 9-40).

In human studies, it must be emphasized that the variability among humans is larger than the variability among species of experimental animals. The factors influencing variability (Ref. 9-40) have been classified into endogenous and exogenous. The endogenous factors include age, sex, and genetic pattern. For example, older people are more likely to show electrocardiographic abnormalities than younger subjects when they are exposed to carbon monoxide. However, considerable selection may have taken place among older workers, the more susceptible having already left because they experienced discomfort or illness. As to sex, women are probably more susceptible to lead toxicity than men. There are differences in the genetic pattern among individuals of the same ethnic background, as well as among various ethnic groups. For example, the normal total white cell count ranges from 4000 to 12,000 cells per cubic millimeter. It must be obvious that a white cell count of 4000 may be normal for some individuals, or indicate a toxic effect on the blood-forming organs in others. The ventilatory capacity of the lungs is one example of variability among ethnic groups. It has been found to be approximately 450 ml lower in African and Asian males than in Europeans (Ref. 9-40).

The exogenous factors influencing variability among humans include nutritional status, past or present disease, previous or concomitant exposure, and social conditions. Malnutrition may affect the metabolism of toxic agents and tolerance mechanisms. Past or present disease states, such as anemia or chronic bronchitis and emphysema, would make a person more susceptible to the effects of exposure to lead and cotton dust, respectively. Drugs may affect liver enzymes and alter the metabolism of toxic agents; conversely, toxic agents may affect the metabolism of drugs. As to previous or concomitant exposure, several chemicals could be present in the work as well as the personal environment. Pesticides, solvents, and carbon monoxide are examples. Social conditions include housing, crowding, nutrition, income, and the like. These have been shown to be related to the prevalence of disease, body weight, height, and even life expectancy (Ref. 9-40).

9-6 THE FUTURE

If rational decisions are to be made with respect to the effects of chemicals on human health, science must prevail over politics.

The following are only a few examples of scientific issues that need further study. It is a rare toxicological investigation that assesses the effects of a chemical on the functions of the heart and blood vessels, although cardiovascular disease acounts for almost half the mortality in the United States. Functions of the nervous system and kidneys are also rarely investigated in routine toxicological studies. Since extensive testing of chemicals is exceedingly expensive, the study of the relationship between chemical structure and biological activity, and the validation of such studies, could save time and money. Effects of chemicals on behavior, performance of various tasks, reproduction, and the fetus are unknown for many chemicals. Adequate knowledge of these effects is essential for the determination of safe levels of exposure, especially to low levels over a long period of time. Biological interactions among various chemicals (industrial, medicinal, and chemicals in the general and personal environments) is another area in which information is meager. For example, it has been recently reported that concurrent exposure to ethylene dibromide and disulfiram (Antabuse®) could be lethal. Ethylene dibromide is a synthetic intermediate and a fumigant-pesticide, among its uses. Disulfiram is widely used in alcoholism-control programs. In inhibits certain liver enzymes which presumably detoxify ethylene dibromide. Since ethylene dibromide was found to be carcinogenic in animals, a person treated with disulfiram, who is also exposed to ethylene dibromide, may be at a high risk of developing tumors (Ref. 9-41).

Although there is a body of knowledge of diseases caused by chemicals that has been developed over the years, starting with observations by Ramazzini on lead miners, and although much is known, the future holds promising possibilities of knowledge and understanding. Obviously not all chemicals need to be tested for their effects on the heart and blood vessels, behavior, performance, reproduction, or for toxic interactions. However, such testing, when relevant, could provide rational bases for decisions that would allow us to enjoy the benefits of chemicals and prevent their harmful effects.

REFERENCES

9-1. *Occupational Safety and Health Reporter,* 8, No. 9, p. 299, July 27, 1978.

9-2. Wallis, W. Allen, *An Overgoverned Society,* The Free Press, New York, 1976.

9-3. Maugh, T. H., II, "Chemicals: How Many are There?" *Science,* 199, p. 162, January, 1978.

9-4. Kraybill, H. F., "Chemical Selection Process and Federal Regulatory Interfaces," presented at the 1st Meeting of the Clearing House on Environmental Carcinogens, Bethesda, Maryland, November 8-9, 1976.

9-5. Charlesworth, F. A., and Crampton, R. F., "Occurrence of Natural Toxins in Food," *British Medical Bulletin,* 31, 3, pp. 209-13, 1975.

9-6. Arena, J. M., "The Peril in Plants," *Emergency Medicine*, 2, pp. 17-41, May 1970.

9-7. Higgins, I. T. T., "Importance of Epidemiological Studies Relating to Hazards of Food and Environment," *British Medical Bulletin*, 31, 3, pp. 230-35, 1975.

9-8. Montgomery, B. J., "Freedom of Choice Extends to Carcinogens," *JAMA*, 239, 22, pp. 2330-31, June 2, 1978.

9-9. Montgomery, B. J., "Are Spontaneous Abortion, Tobacco Linked?" *JAMA*, Vol. 239, No. 26, pp. 2749–53, June 30, 1978.

9-10. Gordon, R. S., (Ed.), "From the NIH, Research Findings of Potential Value to the Practitioner," *JAMA*, Vol. 239, No. 18, p. 1849, May 5, 1978.

9-11. Kaidbey, K. H., and Kligman, A. M., "Phototoxicity to a Sunscreen Ingredient," *Archives of Dermatology*, 114, pp. 547-49, 1978, cited in *JAMA*, 259, 18, pp. 18-19, 1978.

9-12. Stewart, R. D., and Hake, C. L., "Paint-Remover Hazard," *JAMA*, 235, 4, pp. 398-401, January 26, 1976, cited in *Information Bulletin B. I. B. R. A.*, 15, 4, p. 200, 1976.

9-13. Friedlander, B. R., Hearne, T., and Hall, S., "Epidemiologic Investigation of Employees Chronically Exposed to Methylene Chloride, Mortality Analysis," *J. Occup. Med.*, 20, 10, pp. 657-66, 1978.

9-14. Lyell, A., Bain, W. H., and Thomson, R. M., "Repeated Failure of Nickel-Containing Prosthetic Heart Valves in a Patient Allergic to Nickel," *The Lancet*, II, 8091, pp. 657-59, September 23, 1978.

9-15. Landrigan, P. J., *et al.*, "Chronic Lead Absorption, Result of Poor Ventilation in an Indoor Pistol Range," *JAMA*, 234, 4, pp. 394-97, October 27, 1975.

9-16. *Threshold Limit Values for Chemical Substances in Workroom Air*, American Conference of Governmental Industrial Hygienists, Cincinnati, 1978.

9-17. "Cutaneous and Eye Hazard," *Federal Register*, 43, 104, pp. 22999-23000, May 30, 1978.

9-18. Gardner, Ward, and Taylor, Peter, *Health at Work*, John Wiley and Sons, New York, 1975.

9-19. Maibach, H., "Cutaneous Pharmacology and Toxicology," *Annual Review of Pharmacology and Toxicology*, 16, pp. 401-11, 1976.

9-20. Drill, Victor A., and Lazar, Paul (Eds.), *Cutaneous Toxicity*, Academic Press, New York, 1977.

9-21. Riihimäki, V., and Pfäffli, P., "Percutaneous Absorption of Solvent Vapors in Man," *Scan. J. Work Environ, and Health*, 4, pp. 73-85, 1978.

9-22. Boland, Bill M., and Lehmann, Phyllis E. (Eds.), *Cancer and the Worker*, The New York Academy of Sciences, New York, 1977.

9-23. Braun, Armin C., *The Story of Cancer, on its Nature, Causes, and Control*, Addison-Wesley, Reading, MA, 1977.

9-24. *1977 Cancer Facts and Figures*, American Cancer Society, New York, 1977.

9-25. Statement by Gori, G. B., Deputy Director, Division of Cancer Cause and Prevention, National Cancer Institute. Presented before the Select Committee on Nutrition and Human Needs, United States Senate, July 28, 1976.

9-26. Newell, G. R., Acting Director, National Cancer Institute. Testimony to a Sub-Committee of the House Committee on Government Operations, June 15, 1977.

9-27. Maugh, T. H., II, "Carcinogens in the Workplace: Where to Start Cleaning Up," *Science, 197*, pp. 1258-69, 1977.

9-28. *A Retrospective Survey of Cancer in Relationship to Occupation*, DHEW (NIOSH) Publication No. 77-178, U. S. Government Printing Office, Washington, D. C., 1977.

9-29. Silverberg, E., "Cancer Statistics," *CA-A Cancer Journal for Clinicians, 27*, 1, pp. 30-31, 1977.

9-30. Wynder, E. L., and Gori, G. B., "Contribution of the Environment to Cancer Incidence: An Epidemiologic Exercise," *Journal of the National Cancer Institute, 58*, 4, pp. 825-32, April, 1977.

9-31. Golberg, L., Testimony before the United States Department of Labor, Assistant Secretary of Labor for Occupational Safety and Health Administration, Washington, D. C., OSHA Docket #113, April, 1977.

9-32. Grasso, P., "Ultra-Short-Term Tests," *Information Bulletin, B. I. B. R. A., 14*, 10, pp. 489-91, December, 1975.

9-33. Holden, C., "Cancer and the Mind: How Are They Connected?" *Science, 200*, pp. 1363-69, June 1978.

9-34. Eckardt, R. E., and Scala, R. A., "Toxicology: Assessing the Hazard," *J. Occup. Med., 20*, 7, pp. 490-93, 1978.

9-35. Editorial, "The Non-Science of Toxicology," *Information Bulletin B. I. B. R. A., 16*, 3, pp. 120-21, April, 1977.

9-36. "Long-Term Toxic Effects: A Study Group Report", London: Royal Society, cited in *The Lancet, II*, 8084, p. 328, 1978.

9-37. Zielhuis, R. L., "Biological Monitoring," *Scand. J. Work Environ. and Health, 4*, pp. 1-18, 1978.

9-38. Brilliant, L. B., *et al.*, "Breast-Milk Monitoring to Measure Michigan's Contamination with Polybrominated Biphenyls," *The Lancet,* Vol. II, pp. 643-46, September 23, 1978.

9-39. Schilling, R. S. F. (Ed.), *Occupational Health Practice,* Butterworths, London, 1973.

9-40. Report of WHO Study Group, *Early Detection of Health Impairment in Occupational Exposure to Health Hazards,* World Health Organization, Geneva, 1975.

9-41. Yodaiken, R. E., "Ethylene Dibromide and Disulfiram – A Lethal Combination," *JAMA, 239*, 26, p. 2783, June 30, 1978.

PART 2
CASE STUDIES

10

Syracuse Research Corporation's approach to chemical hazard assessment

---

**Philip H. Howard, Joseph Santodonato,
and Patrick R. Durkin**
Life and Material Sciences Division
Syracuse Research Corporation
Syracuse, New York

"... and suddenly evil mischief would be created and the harmony of The Chemicals would be lost forever..." (Ref. 10-1)

10-1 INTRODUCTION

As suggested by the above quote, chemicals designed to improve the quality of life occasionally run amuck and actually prove to be more a bane than a blessing. In the past, these chemicals were not identified until large quantities had been produced, used, released to the environment, and exposed to aquatic and terrestrial organisms, including man, in concentrations that in many cases were harmful. However, because of new regulations such as the Toxic Substances Control Act, attempts are being made to anticipate exposure to and adverse effects of chemicals before they are produced and used in large quantities. In either situation, methods must be developed to assess the risk or hazard involved in the commercial use of various chemicals. Such procedures provide priorities for research and testing, monitoring, and regulatory action.

Over the past five years, under support from the Office of Toxic Substances, EPA, the Syracuse Research Corporation has evaluated the hazards associated with the commercial use of many synthetic organic chemicals. So far, sixteen chemical groups have been completed and three more are in preparation (Ref. 10-2). In the following, we discuss the relevant information, its sources, approaches to its evaluation, and possible procedures for integrating the various forms in which it appears so that individual comparative risk assessments can be made. Examples from the various groups of chemicals that we have studied are provided in order to clarify the approaches that are recommended. Special emphasis is placed on the unique considerations necessary with synthetic organic chemicals.

There are two major factors that have an impact on the overall environmental hazard assessment: (1) exposure and (2) toxicity. In order for a chemical to have an adverse effect, it must come in contact with an organism in sufficient concentration. Estimating possible exposures is very difficult and requires evaluation of physical-chemical properties and information on production and use, current handling and control technology, monitoring, and environmental fate. It is also necessary to consider toxicology data on both acute and chronic effects to man, as well as organisms of ecological significance. Although there are many acceptable ways to organize these data, the following discussions are presented in much the same way as in the EPA reports mentioned above.

10-2   DATA TO BE CONSIDERED IN ENVIRONMENTAL HAZARD ASSESSMENT

10-2.1   Literature Search

A comprehensive environmental hazard assessment of a chemical or group of chemicals requires that all the available information—both published and unpublished—be used. Thus, the first step in a hazard assessment is a detailed literature search. Bibliographic data bases that provide citations to the published chemical and toxicological literature are a powerful tool for environmental hazard assessment. Recent developments in the computerized searching of bibliographic sources that allow for on-line interaction and "browsing" of data bases have aided immeasurably in the efficient retrieval of biomedical and chemical literature. However, inherent limitations in retrieval capabilities for computerized sources (Ref. 10-3) make it necessary to conduct manual searches as well. This is especially important in cases where an exhaustive literature search must be conducted for a particular substance, or when the historical literature must be searched (most on-line data bases go back only to 1965). We have found that a combination of TOXLINE, *Chemical Abstracts* (searched manually, by annual and cumulative indices, and by computer for newer issues), National Technical Information Service, and *Current Contents* are sufficient for identifying most relevant articles. In addition, specialized data bases are occasionally added: *Index Medicus* (when a specific biological effect has been attributed to the chemical or group of chemicals), *Science Citation Index* (when a very relevant paper is identified and may have been cited by other authors conducting similar studies), and Smithsonian Science Information Exchange (when current federally funded studies need to be identified) are examples. Specialized chemical marketing data bases are discussed in the section on Chemical Marketing Data and Estimates of Environmental Release.

However, these abstracting data bases are only a beginning of the literature search process. We have frequently found that anywhere from 20% to 50% of the most relevant articles are identified by examining the references cited in articles gathered by the abstract search or by personal contact with individuals publishing in the field. This fertile source of data is often overlooked. Once all the relevant articles are gathered, evaluation of the available data can begin.

10-2.2 Exposure Data

Estimates of potential exposure of man or other organisms to a commercial chemical may be released to the environment and (2) once released, the way in which it will behave (environmental fate and transport).

Chemical Marketing Data and Estimates of Environmental Release. The way a chemical is commercially produced and used and the quantities involved, as well as the physical-chemical properties of the chemical, can provide considerable insight into the amounts likely to be released. In fact, the amount of a chemical produced annually has frequently been used as a surrogate for environmental exposure because it has been assumed that the amount of a chemical produced and used has a direct relation to the likelihood of environmental release. However, this covers only one facet of an approach for environmental exposure, and its use by itself should be avoided if at all possible.

Production data can be obtained from a number of published sources such as the *SRI Chemical Economics Handbook (CEH), Census of Manufacturers, Minerals Yearbook,* and the U. S. International Trade Commission's (USITC) *Synthetic Organic Chemicals.* These data bases provide excellent information on chemicals that are produced in large volume (> 50 million lb per year) and are manufactured by several producers. However, for information on smaller volume chemicals, it may be necessary to obtain production figures from the producers. The amount of success in obtaining this information from the producer varies depending upon how proprietary the information is.

Other useful chemical marketing information includes (1) sites of production and use (release points), (2) type of use, (3) synthesis methods (contact with water, sealed reactor, etc.), (4) pollution control technology, and (5) waste disposal. Much of this information can be obtained from published sources such as the *CEH* or the *Kirk-Othmer Encyclopedia of Chemical Technology*; however, informatiön on control technology and waste disposal frequently can be obtained from the manufacturer. Information on the quantities used for different applications, combined with some common sense and information on past use-environmental release correlations, can be employed to provide some qualitative estimates of environmental release. For example, detergents and water emulsifiers will end up in water resources following passage, in many cases, through a treatment plant, whereas most aerosol propellants are so volatile that they will be released into the atmosphere and remain in a vaporous state. Physical-chemical properties are obviously useful in evaluating use-release relationships.

Effluent monitoring data are perhaps the best source of environmental release information, but unfortunately such information is rarely available. Instead of qualitative estimates based on chemical engineering considerations, actual quantitative data on concentrations would be available from monitoring. However, unless the monitoring sites, collection methods, and analytical procedures are appropriate, reliable estimates of the amount of chemical released are not possible.

Environmental Fate and Transport. The importance of environmental fate and transport is often underestimated in determining environmental exposure. It is quite conceivable that a highly toxic, readily degradable substance will cause less environmental damage than a less toxic, persistent chemical. For example, bis(chloromethyl)ether is a potent carcinogen but hydrolyzes in water to innocuous products in a matter of seconds. Evaluation of the behavior of a chemical in the environment requires the consideration of a number of parameters, including physical-chemical properties (water solubility, vapor pressure, octanol-water partition coefficient), soil adsorption and transport, volatilization, bioconcentration, and biological, chemical, and photochemical degradation in various media. Considerations for review and evaluation of environmental fate data have been discussed in detail elsewhere (Refs. 10-4, 10-5).

Due to the difficulty of simulating nature in the laboratory, these types of data, although quantitative in many instances (e. g., rates of hydrolysis), can only provide results that are indicative of quantitative results to be expected in any given microenvironment (e. g., rate of biodegradation by microorganisms can vary considerably). In order to provide quantitative data on exposure to man and the environment, ambient monitoring data are necessary.

Ambient Monitoring Data. Quantitative exposure estimates are best based on ambient monitoring data. Information on environmental release and fate provide only indications of potential exposure. The detection of a chemical in some environmental media indicates that the material is released and persists long enough that there is a high possibility of exposure to biological organisms. However, the lack of detection is not conclusive evidence since the collection and analytical methods may be inadequate. Information on collection efficiency, stability under collection and storage, recovery efficiency for various extraction and clean-up procedures, and specificity and sensitivity of the analysis method as well as the overall limit of detection need to be carefully considered.

Toxic Effects

The assessment of potential chemically induced human or ecological hazards is hardly an exact science and seldom leads to definitive or even clear answers. Both human and ecological risk assessment involve extrapolating the results of observable responses in test species to projected or acceptable risks in the species of concern (target species). Although the principles of extrapolation are similar in both types of risk analysis, certain practical differences between human and environmental risk assessment should be noted. In terms of the species involved, human risk assessment is directed to only a single species, *Homo sapiens*. Ecological risk assessment is concerned with the effect of xenobiotic stress on communities, groups of species interacting by mechanisms that are neither readily quantified nor clearly understood. In this respect, ecological risk assessment is much more complex than human risk assessment. However, the level of acceptable risk is much lower in human risk assessment than in ecological risk assessment. Consequently, the degree of high-to-low dose extrapolation is usually much greater in studies designed to detect human risk. Lastly, and perhaps most importantly, human risk assessment is usually based on studies using test species, such as rats or mice, that are not closely related to humans. In ecological risk assessment, data are sometimes available on test species that are closely related to the target species. Given the multiplicity of species involved in ecological risk assessment and the uncertainties of high-to-low dose and experimental-mammal-to-man extrapolations involved in human risk assessment, it often seems impossible to develop a realistic risk analysis. However, based on the types of data outlined below, a reasonable attempt can be made to identify the biological groups most likely to be affected and the nature of the adverse effect.

Pharmacology. This type of information generally includes data on absorption, tissue distribution, biotransformation, elimination, and physiological or biochemical effects. Although much of this information is difficult to apply directly to chemical risk assessment, it is sometimes useful in determining the potential for and mechanism(s) of chronic toxic action.

A compound's potential to exert chronic toxic effects is often related to the accumulation and persistence of the chemical within the organism. The concern for chronic toxicity can be reduced by demonstrations that the chemical is either not absorbed or is rapidly detoxified and eliminated. In human risk assessment, this type of information is generally obtained from metabolism studies on experimental mammals and occasionally from monitoring surveys of tissue levels in humans. In ecological risk assessment, parallel information is often restricted to aquatic systems and includes data on bioconcentration or biomagnification. For some chemicals, such as pesticides, tissue monitoring surveys may be available on both terrestrial and aquatic organisms. However, for many chemicals, no direct experimental information is available on bioconcentration or biomagnification. In such cases, it may be possible to estimate these parameters based on certain physicochemical properties such as octanol-water partition coefficients (Ref. 10-6) and water solubility (Ref. 10-7).

Within a class of chemicals, qualitative differences in chronic toxic effects can sometimes be related to differences in metabolites. For example, because polycyclic aromatic hydrocarbon (PAH)-induced carcinogenesis involves metabolic bioactivation to stereospecific arene oxide intermediates, Jerina and coworkers (Ref. 10-8) have been successful in using the "bay region" theory to predict the carcinogenic potential of various PAHs. Similarily, among the ketonic solvents, only methyl n-butyl ketone (MBK) appears to cause peripheral neuropathy. This toxic effect appears to be related to the metabolism of MBK to 2, 5-hexanedione, which is apparently not formed in the metabolism of other ketones.

Acute Toxicity. Acute toxicity information is usually expressed as time-specific LD_{50}s or LC_{50}s, i. e., doses or concentrations of a toxicant that result in 50% mortality of the test species during a specified exposure or observation period. Because acute toxicity studies are relatively inexpensive and are often required prior to initiating longer-term studies, acute toxicity information is the most commonly available type of data on the biological effects of chemicals. Such information can be directly applied to anticipating the potential short-term hazards of chemical exposure. In addition, within a class of chemicals, acute toxicity data can often be quantitatively related to structural features (e. g., lipophilic, electronic, and steric parameters). This type of analysis has been successfully applied in both mammalian and aquatic toxicology (Refs. 10-9, 10-10). Using this approach, it is possible to suggest how the addition or deletion of certain structural groups may influence biological activity. Thus, it becomes possible to estimate the acute toxicity of a chemical for which no experimental data are available.

Because the concern in environmental risk assessments primarily involves long-term low-level exposures, it would be beneficial if acute toxicity data could be used to estimate chronic no-effect levels. McNamara (Ref. 10-11) recently discussed the relationship between acute LD_{50}s and long-term no-effect levels in mammals. Using experimental data from a variety of sources, he found that the acute LD_{50} divided by 1000 resulted in a dose below that which produced no apparent adverse effects during repeated lifetime exposures for 95% of chemicals on which data were available. A similar approach, which has gained some popularity in aquatic toxicology, is based on the maximum concentration causing no apparent adverse effects in chronic exposures of a given species divided by the acute LC_{50} for that species. This ratio, called the application factor, is then used to estimate no-adverse-effect levels for other species by multiplying the acute LC_{50}s for the other species by the application factor (Ref. 10-12). Although both of these approaches have considerable practical appeal, both should be applied with caution. A major problem with them is that the mechanism(s) of acute and chronic toxicity may not be related. This is particularly important if the chemical is a carcinogen (see below). Further, the utility of application factors is limited by experimental variability and true differences in species susceptibility (Ref. 10-13).

Subacute and Chronic Toxicity. More direct measurements of environmentally meaningful no-effect levels can be made from subacute and chronic toxicity studies. Such studies are usually designed so that the test organism is repeatedly or continuously exposed to the test chemical for a significant portion of its lifespan.

For human risk assessment, subacute and chronic toxicity data from animal studies can be used to establish an acceptable daily intake (ADI), provided that the chemical under study is not a carcinogen. In practice, ADIs are derived by determining a no-effect dose in an experimental mammal and dividing this dose by an appropriate safety factor. The magnitude of the safety factor, which usually ranges from 100 to 1000, is dependent on the dose schedule used in the experimental study and is intended to account for the uncertainty encountered in extrapolating the results of experimental animal studies to man.

An analogous procedure, used in aquatic toxicology, is the estimation of a maximum acceptable toxicant concentration (MACT), defined as the maximum concentration having no adverse effects in the test species. Although safety factors are sometimes used to estimate "safe" concentrations for aquatic life (Ref. 10-14), the safety factors are usually substantially less than those used in human risk assessment because the test species is so closely related to the target species.

Epidemiological Studies. Strictly defined, epidemiology refers to the study of disease patterns in humans. When good epidemiological data are available, they can serve as the definitive estimate of potential human hazard. For example, in a review of the haloethers, epidemiological data on workers exposed to bis(chloromethyl)ether and/or chloromethyl methyl ether revealed that the risk of cancer was significantly increased. Occasionally, the unique nature of a chemically induced effect (e. g., liver angiosarcoma by vinyl chloride) will quickly lead to the recognition of a human risk. More often than not, however, epidemiologic studies in nonoccupational situations are seldom definitive enough to establish chemical cause-and-effect relationships with certainty. Although epidemiological studies will not be available on new chemicals, epidemiological data on related chemicals should be actively sought.

In ecological risk assessment, field studies, which directly measure the effects of a chemical on a complex ecosystem, provide a similar type of highly useful and directly applicable information. However, with the exception of certain pesticides and a few other widespread chemical contaminants, such information is rarely found in the literature. Because the measurement of ecosystem effects does directly address the major question of ecological risk assessment, the development of model ecosystems to predict ecological effects is an area of active research (e. g., Ref. 10-15).

Carcinogenicity. Although several different types of carcinogenicity bioassays are available, chronic feeding studies of rats and/or mice, such as those conducted by the National Cancer Institute, are the kind most commonly encountered in the literature. Carcinogenicity is an end point of great significance and controversy in human risk assessment. Because of the seriousness of the effect, very low levels of human risk are tolerated. One assessment model currently being used by the EPA adopts an acceptable increased lifetime risk of 1/100,000. An earlier model proposed by Mantel and Bryan (Ref. 10-16) used an acceptable risk of 1/100,000,000. Because practical studies cannot be conducted to directly detect such low risk levels, cancer bioassays in laboratory test species are conducted at elevated dose levels. Since, for practical and economic reasons, it is necessary to limit the numbers of animals used, doses must be elevated to establish the likelihood of low-level effects. This practical consideration for testing has led to a lively debate as to which mathematical models are best suited for the extrapolation from high to low dose levels (e. g., Refs. 10-17, 10-18, 10-19) and which assumptions are justified in extrapolating results from experimental mammals to man (Refs. 10-20, 10-21).

Because carcinogenicity bioassays are time consuming and expensive, several mutagenicity screening tests to detect potential carcinogenicity are becoming commonly used. These include tests with microorganisms (e. g., Ames assay), tests for genetic damage in cultured mammalian cells (e. g., unscheduled DNA repair synthesis, sister chromatid exchange, and point mutations), and tests for *in vitro* transformation of rodent cell lines. Although such tests have become extremely valuable in detecting chemicals that require further study (i. e., animal bioassay and/or epidemiology), they are not yet capable of detecting all potential carcinogens or indicating their relative potency in man.

Reproductive Effects. Chemically induced reproductive impairment is an important response parameter in both human and ecological risk assessment. In human risk assessment, teratogenicity, sterility, or decreased reproductive capacity can serve as end points in establishing no-effect levels from chronic exposures. Often, however, the threshold for adverse reproductive effects in mammals is above the threshold for more general toxic effects (e. g., decreased total body weight gain or altered organ weights). Further, many mammalian teratology studies involve single short-term exposures and are thus difficult to directly apply to estimating the risk from environmental exposure.

In ecological risk assessment, reproductive effects are often the critical factor in evaluating the environmental compatibility of a chemical. Because acceptable levels of risk are relatively high in ecological risk assessment and because test species closely resemble target species, it is possible to conduct tests that can reasonably approximate environmental exposure conditions and thus directly measure environmentally meaningful responses. Regrettably, relatively few chemicals have been tested for chronic ecological effects. Nonetheless, several tests have been developed: whole-life or egg and fry studies with fathead minnows, partial-lifespan studies with rainbow trout, several algal bioassays, and productivity studies using *Daphnia*. Since Maki (Ref. 10-22) has presented evidence that productivity studies in *Daphnia* may be predictive of whole-life studies in fish and since *Daphnia* productivity studies are much less expensive than chronic fish studies, the *Daphnia* test may become a popular screening tool in aquatic risk assessment.

Toxicant Interactions. One of the most formidable tasks confronting the discipline of environmental risk assessment is the problem of toxicant interactions. Even when the toxicology and environmental fate of a toxicant have been defined in great detail, statements of environmental compatibility must often be qualified because little information is available on how the toxicant will interact with other compounds (xenobiotic and natural) in the environment.

10-3 TYPICAL HAZARD ASSESSMENT RESULTS

10-3.1 Typical Report Outline

The exact outline to be used in presenting hazard assessment data is not as important as insuring that all pertinent pieces of information are considered. Our outline is presented in Table 10-1. Many of the individual sections cannot be written until other sections are complete. For example, Section II-C — Environmental Contamination — depends on Production (II-A), Uses (II-B), Physical Properties (I-A),

TABLE 10-1. Typical Outline of Potential Environmental Contaminant Studies for EPA's Office of Toxic Substances.

Executive Summary

Introduction
I. Physical and Chemical Data
    A. Structure and Properties
    B. Chemistry
        1. Reactions involved in uses
        2. Hydrolysis
        3. Oxidation
        4. Photochemistry
II. Environmental Exposure Factors
    A. Production, Consumption
    B. Uses
    C. Environmental Contamination Potential
    D. Current Handling Practices and Control Technology
    E. Monitoring and Analysis
        1. Analytical methods
        2. Monitoring studies
III. Health and Environmental Effects
    A. Environmental Effects
        1. Persistence
        2. Environmental transport
        3. Bioaccumulation
        4. Biomagnification
    B. Biology
        1. Absorption and elimination
        2. Transport and distribution
        3. Metabolism and excretion
        4. Metabolic and pharmacologic effects
    C. Toxicity – Humans
        1. Occupational studies
        2. Nonoccupational exposures
        3. Epidemiological and controlled human studies
    D. Toxicity – Birds and Mammals
        1. Acute animal toxicity
        2. Subacute and chronic toxicity
        3. Sensitization
        4. Mutagenicity
        5. Teratogenicity
        6. Carcinogenicity
        7. Possible synergisms
    E. Toxicity to Lower Animals
    F. Toxicity – Plants
    G. Toxicity – Microorganisms
IV. Regulations and Standards
V.   Summary and Conclusions
References

Handling and Control Technology (II-D), and Monitoring (II-E). Other sections are closely related to each other, such as Biological Degradation (III-A-1-a) and Toxicity — Microorganisms (III-G) and Biology (III-B) and Toxicity — Mammals (III-D). The following sections provide examples of the type of information available and how it was handled for several groups of chemicals.

10-3.2 Selected Data from Key Reports

Fluorocarbons. This report examined the one- and two-carbon chlorofluorocarbons that are used mostly as aerosol propellants and refrigerants. They are stable compounds that are released to the environment in significant quantities (Ref. 10-23) and cause cardiac arrhythmias at very high concentrations. From this data, we concluded in an early draft that although the levels of fluorocarbons in the atmosphere are increasing, they are well below the threshold of toxic response and, therefore, fluorocarbons could not be considered a significant environmental hazard. This was, of course, before the concept of stratospheric ozone destruction was proposed (Ref. 10-24) and illustrates the fact that hazard assessments must change as new data become available.

Silicone (Siloxane) Fluids. Silicone fluids are organic-inorganic, linear polymers that are produced in the United States in quantities of more than 40 million pounds a year (18,000,000 kg/yr) for such applications as car waxes and polishes, antifoams, cosmetics, food additives, and textile finishes. These compounds are both lipophobic and hydrophobic and extremely stable at neutral conditions. However, under acidic or basic conditions, they hydrolyze into shorter chains and cyclize to tetramers or trimers. Their uses result in considerable environmental release (e. g., 20% of car polish is silicone fluid). These materials exhibit a low degree of toxicity; therefore their environmental hazard seems low. Thus, silicones appeared to be another example of a group of chemicals that are produced and released to the environment in significant quantities, are persistent, but have no significant toxicologic effects at likely environmental concentrations.

As we were completing this study, Dow Corning provided us with information indicating that clays in soil would catalyze rearrangement of siloxanes to shorter straight chain polymers or cyclic polymers. These products have not been as well characterized toxicologically as the long straight chain polymers, but one, diphenylhexamethylcyclotetrasiloxane, has been shown to exert estrogenic effects. Thus, the soil-catalyzed rearrangement products may have considerably more significant toxicologic effects than the commercial siloxanes. This group of chemicals illustrates the importance of knowing any environmental metabolites that may be formed in order to provide a complete hazard assessment.

Mercaptobenzothiazole (MBT) Compounds. In some instances, only circumstantial evidence is available about the possible release of a group of chemicals to the environment. For example, over 90 million pounds (41,000,000 kg) of MBT compounds are used in the United States for rubber vulcanization accelerators (added at concentrations of 0.5 to 1.5%). Water effluent monitoring data were available, establishing significant releases during rubber production, but losses from rubber use (mostly automotive tires) are unknown. However, a Russian study demonstrated that MBT compounds are rapidly dissolved out of the matrix of rubber products with water. Also, it has been determined that approximately 1,200 million pounds (540,000,000 kg) of rubber dust are worn from automotive tires each year in the United States. Based on an average of 1% accelerator in the rubber, this would amount to 12 million pounds (5,000,000 kg) of MBT compounds being released into water systems each year. Little is known about the environmental fate of the MBT compounds, but they have been detected in raw and drinking water.

Even though the data are not that comprehensive, there appear to be at least three principal toxicologic effects of MBT: (1) the production of allergic contact dermatitis, (2) action on the central nervous system, and (3) inhibition of certain metalloenzymes that contain copper. Whether these or other effects will occur at likely environmental concentrations is unknown.

Monohalomethanes. There were four compounds (fluoromethane, chloro-methane, bromomethane, and iodomethane) that were considered in this report, only two of which (CH_3Cl, CH_3Br) were commercially significant. Chloro-methane, a colorless gas, is produced in quantities of 400 million pounds a year (180,000,000 kg/yr) and used mainly for the production of silicones and tetra-methyl lead (a gasoline additive). About 40 million pounds of bromomethane is produced in a year (18,000,000 kg/yr) and used principally as a fumigant for soil. A small amount (20,000 pounds or 9,000 kg) of iodomethane is produced annually for use as a laboratory and commercial alkylating agent. Although some of these compounds are produced in significant quantities and used in ways that would result in environmental release (e. g., fumigant application of bromo-methane), there are significant natural sources of these compounds that appear to contribute far greater amounts to the environment than the commercial sources.

All the halomethanes except fluoromethane have been detected in ocean wa-ters and in air above the ocean. It is believed that algae in the oceans are the major source of bromomethane and iodomethane, with chloromethane forming by the reaction of the chloride ion in seawater with iodomethane. Wofsy and coworkers (Ref. 10-25) have suggested that only 5% to 25% of the bromomethane in the atmosphere can be attributed to anthropogenic sources and Lovelock (Ref. 10-26) has calculated that almost 100% of the chloromethane and iodo-methane detected in the environment can be attributed to natural sources.

Chloro-, bromo-, and iodomethane all produce central nervous system effects with acute exposure with a probable mechanism of action being methylation of essential enzymes, cofactors, and intracellular proteins. Chloromethane and iodomethane have both been reported to be mutagenic in the Ames microbial assay; iodomethane has increased the incidence of tumors in rats and mice. Bromomethane has not been tested yet, but because of the alkylating properties of these compounds they are all potential carcinogens.

The overall hazard assessment on these groups of chemicals was extremely difficult to make because, although their biological effects were significant, the major sources did not result from commercial activities. We concluded that chloromethane and iodomethane were not high environmental risks because they were either used in closed systems (both) or in small amounts (iodomethane) and the background sources are much more significant than natural sources. With bromomethane, there is a considerable possibility of local exposures to high concentrations that far exceed background. All of the compounds are occupational health hazards.

Haloethers. Chloroalkyl ethers are the more commercially significant haloethers. In terms of environmental hazard, the chloroethers can be divided into two categories – (1) α-chloroethers and (2) non-α-chloroethers – because of the drastic difference between these compounds in terms of uses, stability, and toxicity.

Only two α-chloroethers are commercially significant. Chloromethyl methyl ether (CMME), containing a minor amount (1% to 7%) of bis(chloromethyl)ether (BCME), is used as a chemical intermediate in the production of strong base anion exchange resins. The exact production volume is unknown. Both compounds are unstable in water (half-life of <1 sec and <1 min for CMME and BCME, respectively), although they appear to be more stable in the vapor phase (half-life >390 min and >25 hr for CMME and BCME, respectively). Because of this instability, these compounds are not likely to be widespread environmental contaminants and are definitely not water contaminants. However, we were surprised to note that some early water monitoring studies reported detection of these two α-chloroethers. Because of the hydrolysis rates of these compounds, we decided that these reports were erroneous and should be discounted in the hazard assessment evaluation. Toxicologically, there is convincing evidence available from both laboratory animals and epidemiological investigations of worker populations to suggest that BCME is a potent carcinogen.

In contrast to the α-chloroethers, β-chloroethers are (1) not produced commercially but formed as a byproduct from the chlorohydrin synthesis for making epoxides from olefins, (2) very stable and, apparently, widespread contaminants based upon available monitoring data, and (3) only weak carcinogens if they are carcinogenic at all. The detection of bis(2-chloroethyl)ether and bis(2-chloroisopropyl)ether in waters receiving effluent from ethylene and propylene glycol/oxide plants is well documented and these compounds have been detected in the Kanawha, Mississippi, Ohio, and Delaware Rivers. From the chlorohydrin byproducts in the synthesis of ethylene and propylene chlorohydrins, we were able to also suggest that major amounts of bis(2, 2'-dichloroisopropyl)ether would be formed during the chlorohydrin synthesis of epichlorohydrin from allyl chloride. Bis(2-chloroethyl)ether has induced liver tumors in mice, but bis(2-chloroisopropyl)ether produced no increased incidences of tumors.

We concluded that α-chloroethers were not significant environmental hazards, because for occupational protection they are now produced in elaborately controlled closed systems. Their instability also indicates that the possibility of any environmental exposure is remote. In contrast, the β-chloroethers are of considerable concern because (1) they are produced as byproducts in large quantities, (2) they have been detected in numerous rivers, (3) they appear to persist in the environment, and (4) they may be carcinogenic.

10-4   ASSIGNING PRIORITIES FROM HAZARD ASSESSMENT

10-4.1 General Priority Setting

Perhaps the most difficult task in hazard assessment is making comparative assessments of hazard for various chemicals. This makes it necessary to combine all the relevant parameters discussed previously in some rational fashion. The number of chemicals that have to be considered often has considerable impact on the prioritizing scheme chosen. When only a hundred or so chemicals are being evaluated, qualitative processes may be used. When tens of thousands of compounds are involved, some type of quantitative number must be assigned to individual compounds. Ideally the process of prioritizing chemicals should be as objective as possible. However, because of the many factors involved (production-use, environmental fate, toxicity),the weighting assigned will reflect the bias of the individuals developing the prioritization scheme and, therefore, the final result will be somewhat subjective. It is also important to realize that many of the parameters that need to be considered have nonquantitative results (e. g., test results on biodegradation are quantitative but are only qualitatively related to what will occur in nature) and will result in priorities that may be assigned a quantitative number while, in reality, remaining only qualitative.

Developing priorities for totally dissimilar chemicals is extremely difficult because all of the factors to be considered are usually different. It is extremely difficult to decide, for example, whether a persistent chemical that has no toxicity data is more important than a chemical that is toxic to fish but for which little other information is available. Chemically related compounds are somewhat easier because they frequently have similar uses, physical-chemical properties, and toxicity. We conducted such an exercise on nitroaromatic compounds and the approach used is described in the next section.

10-4.2 Priority Setting for Nitroaromatic Compounds

Approximately 250 to 300 compounds are listed as commercial nitroaromatic compounds (nitro substituent on an aromatic ring) but most of these compounds are produced in such small quantities that they are little more than laboratory curiosities. However, approximately 40 compounds are consumed annually in quantities greater than 0.5 million pounds (200,000 kg). In many instances, the production figures had to be obtained from the manufacturer.

The first step in priority setting was to list all the compounds that were (1) listed as a commercial chemical [1000 pounds (500 kg) or $1000 annually], and (2) had some environmental fate or biological effects information. Commercial compounds having no data on fate or effects were included if available information indicated that they were produced or imported in over 100,000 pounds (45,000 kg) annually. The total number listed was 92.

Nitroaromatic compounds find applications as pesticides, perfumes, explosives, and chemical intermediates. Each of these uses results in a different potential for release of the compound to the environment. In order to convert the production figure into a number related to the quantities likely to be released to the environment, a release factor was developed for each application. Pesticides and perfumes were assigned a factor of 1.00, since they would be used in such a way that all the material would be released to the environment or come into contact with humans. Chemical intermediates were given a small factor (0.05, which is probably the upper limit) because most of the chemical is converted to another material. Explosives were also given a low factor (0.05) because most of them are destroyed during use.

An attempt was made to include the biodegradation data in the contamination factor calculation. However, after examining the available information, it did not appear that a quantitative value should be assigned, as much of the biodegradation work was conducted with pure cultures of microorganisms and correlation of these results to nature is difficult. Other environmental fate data were very sparse, but bioconcentration factors were estimated when octanol-water partition coefficients were available. In addition, any ambient or effluent monitoring data were included in the table. No air-monitoring data were available but considerable water-monitoring data were found.

Nitroaromatic compounds exhibit several distinct and important biological effects. Nitrobenzene and its hydrocarbon derivatives (dinitrobenzene, nitrotoluenes) primarily affect the hematologic system through the production of methemoglobinemia, sulfhemoglobinemia, Heinz bodies, and red cell destruction. 2, 4-Dinitrophenol and related structures (2-*sec*-butyl-4, 6-dinitro-phenol, 4, 6-dinitro-*o*-cresol) are unique in their ability to "uncouple" oxidative phosphorylation by suppressing the coupling of electron flow to synthesis of adenosine triphosphate (ATP). This uncoupling effect produces a profound disturbance of metabolic function.

In most instances of acute exposure, the resulting effects on the hematologic system or metabolic function can rapidly be reversed after removal from exposure. In only a few cases (e. g., 4, 6-dinitro-o-cresol; m-nitrobenzene) does it appear that cumulative toxicity may occur from long-term, low-level exposures similar to the exposures that would result from environmental contamination. In addition, a number of nitroaromatic chemicals (most of which are not produced in large commercial quantities) are active tumor-producing agents in animals. Because of similar metabolic products (believed to be nitroso or hydroxylamino derivatives), it has been possible to demonstrate that most nitro analogs of aromatic amine carcinogens are likewise active tumor-producing agents.

Of the 92 compounds that were initially listed, the following numbers of compounds had mammalian toxicity information available: acute oral LD_{50} in rats $-$ 41; tested for metabolic effects $-$ 9 positive, 25 negative; tested for hematogenic effects $-$ 22 positive, 11 negative; tested for tumorigenic effects $-$ 7 positive and 9 negative. The only other toxicity information available were acute bioassays on fish. Only nine chemicals were tested in this way. Given the lack of very complete toxicity data, it was concluded that toxicity could not be used in any systematic way to provide an ordering of the hazard from various nitroaromatics.

After completion of the first table, another table was prepared to order compounds in a descending contamination index (production times release factor). Thirty-seven compounds were included and the first two pages of the table are presented in Table 10-2. Because we were focusing on nonpesticide compounds, no pesticides were included in Table 10-2, even though many had high contamination indices. Although the compounds seem to be ordered in decreasing environmental exposure, as somewhat confirmed by the incidence of detection in water, the order is not necessarily appropriate for overall environmental hazard. For example, 1-nitronaphthalene may be a significant environmental contaminant because it is related by reduction to α-naphthylamine, a suspected carcinogen. It is produced in only one plant, where it is subsequently used to produce α-naphthylamine. Whether any nitronaphthalene is lost can be determined only by effluent monitoring combined with a plant survey. The chloronitrobenzenes are all produced in large quantities, released to the environment, and persistent. However, none have been tested for carcinogenicity. In fact, only three of the compounds in Table 10-2 have been tested for carcinogenicity, which makes it very difficult to assign hazard when extensive contamination is indicated.

p-Nitrophenol is also an interesting compound whose hazard assessment needs qualification. Its major use is in the synthesis of Methyl and Ethyl Parathion, which are widely used pesticides. During Parathion use, the organophosphate is hydrolyzed and p-nitrophenol is a major product. Thus, the amount of p-nitrophenol entering the environment from commercial production and use is

TABLE 10-2. Nitroaromatic Compounds Which Have a High Potential for Being Environmental Pollutants.

Chemical	Contamination Factor	Biodegradable	Monitored in Water Effluents or River Water	Monitored in Drinking Water	Oral LD$_{50}$ (rats) mg/kg	Fish Toxicity (96 TL$_M$; mg/liter)	Tested for Carcinogencity
Nitrobenzene	32750	?	Yes	Yes	640-664	–	No
Dinitrotoluene	23561	Yes	Yes	Yes	268-707	16	Yes – Neg.
2,4,6-Trinitrotoluene (TNT)	21600	Yes	Yes	No	–	2.6	Yes – Neg.
1-Chloro-4-nitrobenzene	5500	No	Yes	No	420	–	No
p-Nitrophenol	5000	Yes	Yes	No	350-467	–	Yes – Neg.
1-Chloro-2-nitrobenzene	3000	No	Yes	No	288	–	No
4-Nitrotoluene	888	?	Yes	No	2144	20-50 (6 hr LC$_{min}$)	No
o-Nitrophenol	750	?	Yes	No	2828	46.3-51.6 (48 hr TL$_M$)	Yes – Neg.
p-Nitroaniline	700	?	No	No	1410-3249	–	No
2-Nitrotoluene	600	No	Yes	No	891	18-40 (LC$_{min}$)	No
1,3-Dinitrobenzene	600	No	Yes	Yes	–	–	No
4,4'-Dinitrostilbene-2,2'-disulfonic acid	493	–	No	No	–	–	No
5-Nitro-o-toluenesulfonic acid	398	–	No	No	–	–	No
1-Chloro-3-nitrobenzene	395	No	Yes	Yes	555	–	No
1-Chloro-2,4-dinitrobenzene	331	No	Yes	No	500-1593	–	No
1-Nitronaphthalene	315	–	No	No	120	–	No
o-Nitroaniline	300	?	No	No	535-3520	–	No
m-Nitrobenzenesulfonic acid	183	–	No	No	–	–	No

miniscule in comparison to the amount resulting from pesticide use. Thus, from the above, it can be concluded that it is no easy task to assign relative priorities as a result of environmental hazard assessment, even for a closely related group of chemicals.

10-5 SUMMARY

It is possible to assess environmental hazard from commercial chemicals by reviewing and integrating available information on production and use, monitoring, chemistry and environmental fate, and toxicity. This chapter specifies information sources and tests to be considered. A format is presented for organizing and interpreting results. Example data from key chemical classes are included.

However, environmental risk assessment is an inexact process, subject to error, and difficult with chemically and commercially related compounds and more difficult with less related compounds. A major problem is the lack of adequate data in the many categories that must be considered. However, properly conducted hazard assessment reviews are extremely helpful in setting research and data-gathering priorities and in identifying high-risk compounds. Such assessments should be continuous in order to incorporate new data as it becomes available.

REFERENCES

10-1.   Brautigan, R., *The Hawkline Monster,* New York: Simon and Schuster, 1974.

10-2.   The following chemical groups have been assessed (the number which follows the chemical group is the report number for ordering from the Nat'l Technical Information Service, Springfield, VA 22161, USA): Chlorinated naphthalenes, silicones, fluorocarbons, benzene polycarboxylate, and chlorophenols — PB 238 074; One and two carbon fluorocarbons — PB 246 419; Liquid siloxanes — PB 247 778; Benzene — PB 244 139; Haloethers — PB 246 356; Chlorinated paraffins — PB 248 634; Ketonic solvents — PB 252 970; Acrylamides — PB 257 704; Mercaptobenzotriazoles — PB 252 662; Nitroaromatic compounds — PB 275 078; Haloalkyl phosphates — PB 257 910; Benzotriazoles — PB 266 366; Monohalomethanes — PB 276 483; Styrene and Ethylbenzene — in draft; Epichlorohydrin — in draft; Ethylene, propylene, and butylene oxide — in preparation; and Haloalcohols — in preparation.

10-3.   Santodonato, J., "A Comparison of On-Line and Manual Modes in Searching Chemical Abstracts for Specific Compounds," *J. Chem. Inform. Comput. Sci.,* Vol. 16, August 1976, p. 135-137.

10-4.   Howard, P. H., Saxena, J., and Sikka, H., "Determining the Fate of Chemicals," *Environ. Sci. Technol.,* 12, April 1978, p. 398-407.

10-5. Howard, P. H., Saxena, J., Durkin, P. R., and Ou, L. T., "Review and Evaluation of Available Techniques for Determining Persistence and Routes of Degradation of Chemical Substances in the Environment," EPA-560/5-75-006, U. S. Nat. Tech. Inform. Serv., PB 243 825, 1975.

10-6. Neely, W. B., Branson, D. R., and Blau, G. E., "Partition Coefficients to Measure Bio-concentration Potential of Organic Chemicals in Fish," *Environ. Sci. Technol.,* 8, December 1974, p. 1113-1115.

10-7. Metcalf, R. L. and Lu, P. Y., "Environmental Distribution and Metabolic Fate of Key Industrial Pollutants and Pesticides in a Model Ecosystem," University of Illinois at Urbana-Champaign, Water Resources Center, UILV-WRC-0069, U. S. Nat. Tech. Inform. Serv., PB 225 479, 1973.

10-8. Jerina, D. M., Yagi, H., Levin, W., and Conney, A. H., "Carcinogenicity of Benzo-[a]pyrene," *Drug Design and Adverse Reactions,* Alfred Benzon Symposium X, 1977, p. 261-277.

10-9. Gould, R. F. (Ed.), "Biological Correlations – The Hansch Approach," *Advances in Chemistry Series,* 114, American Chemical Society, Washington, D. C., 1972, 304 pp.

10-10. Veith, G. C., and Konasewich, D. E. (Eds.), "Symposium on Structure-Activity Correlations in Studies on Toxicity and Bioconcentration with Aquatic Organisms," Great Lakes Research Advisory Board, Windsor, Ontario, Canada, 1975.

10-11. McNamara, B. P., "Concepts in Health Evaluation of Commercial and Industrial Chemicals," *Advances in Modern Toxicology,* John Wiley and Sons, New York, Vol. 1, Part 1, pp. 61-140, 1976.

10-12. Macek, K. J., Burton, K. S., Derr, S. K., Dean, J. W., and Sauter, S., "Chronic Toxicity of Lindane to Selected Aquatic Invertebrates and Fishes," EPA-600/3-76-046, May 1976.

10-13. Mount, D. I., "An Assessment of Application Factors in Aquatic Toxicology," EPA-600/3-77-085, U. S. Nat. Tech. Inform. Serv., PB 273 500, 1977.

10-14. International Joint Commission, United States and Canada, *New and Revised Great Lakes Water Quality Objectives,* International Joint Commission Great Lakes Regional Office, Ontario, Canada, 1977.

10-15. Giddings, J. M., and Eddleman, G. K. "The Effects of Microcosm Size and Substrate Type on Aquatic Microcosm Behavior and Arsenic Transport," *Arch. Environ. Contam. Toxicol.,* 6, pp. 491-505, 1977.

10-16. Mantel, N., and Bryan, W. R. " 'Safety' Testing of Carcinogenic Agents," *J. Nat. Cancer Inst.,* 27, pp. 455-470, 1961.

10-17. Guess, H. A., and Crump, K. S., "Best-estimate Low-dose Extrapolation of Carcinogenic Data," *Environmental Health Perspectives,* 22, pp. 149-152, February 1978.

10-18. Hartley, H. O., and Sielken, R. L., Jr., "Estimation of 'Safe Doses' in Carcinogenic Experiments," *Biometrics,* 33, pp. 1-30, March 1977.

10-19. Safe Drinking Water Committee of the National Research Council, *Drinking Water and Health,* National Academy of Science, Washington, D. C., 1977.

10-20. Ehrenberg, L. and Holmberg, B., "Extrapolation of Carcinogenic Risk from Animal Experiments to Man," *Environmental Health Perspectives,* 22, pp. 33-36, February 1978.

10-21. Saffiotti, V., "Experimental Identification of Chemical Carcinogens, Risk Evaluation, and Animal-to-Human Correlations," *Environmental Health Perspectives*, **22**, pp. 107-114. February 1978.

10-22. Maki, A. W., "Predictions of the Chronic Fish Toxicity of Test Materials Using *Daphnia magna,*" *Proceedings of the 8th Annual Conference in Environmental Toxicology*, U. S. Nat. Techn. Inform. Serv., AD 4051334, 1977, pp. 306-326.

10-23. Howard, P. H., and Hanchett, A., "Chlorofluorocarbon Sources of Environmental Contamination," *Science*, **189**, pp. 217-219, July 18, 1975.

10-24. Molina, M. J., and Rowland, F. S., "Stratospheric Sink for Chlorofluoromethanes: Chlorine Atom-Catalyzed Destruction of Ozone," *Nature*, **249**, pp. 810-812, June 28, 1974.

10-25. Wofsy, S. C., McElroy, M. B., and Yung, Y. L., "Chemistry of Atomspheric Bromine," *Geophys. Res. Lett.*, **2**, pp. 215-218, 1975.

10-26. Lovelock, J. E., "Natural Halocarbons in the Air and in the Sea," *Nature*, **256**, pp. 193-194, 1975.

11

Environmental risk analyses of wastewaters produced by synthetic fuels technologies[1]

B. R. Parkhurst

Environmental Sciences Division
Oak Ridge National Laboratory
Oak Ridge, Tennessee

11-1 INTRODUCTION

Increased utilization of the large coal reserves of the United States is being promoted as one means of reducing rapid consumption of dwindling petroleum and natural gas reserves, thus reducing our dependence on imported fossil fuels. However, there are two disadvantages to using coal as a fuel: (1) its unwieldy solid form is much less versatile than liquid or gaseous fuels, and (2) there are serious environmental and health hazards associated with its burning. One way of overcoming these problems is to convert coal to a liquid or gaseous product, which would also be relatively free of sulfur, and thus could be burned as a cleaner fuel.

[1] Research sponsored by the Office of Health and Environmental Research, U. S. Department of Energy, under contract W-7405-eng-26 with Union Carbide Corporation. Publication No. 1497, Environmental Sciences Division, ORNL.

Several coal conversion technologies are now under development. It is anticipated that by 1990, as much as 5% of the gas needs of the U.S. and 2% of the petroleum needs will be met by products produced from coal (Ref. 11-1). Currently, an
intensive effort is underway to identify potential environmental or health hazards
that may result from the development of a large coal conversion industry.

This chapter outlines the scientific approach, the chemical and biological
methods used, and some of the results of the Advanced Fossil Energy Program
(AFEP) at Oak Ridge National Laboratory (ORNL) to analyze the effects of
wastewaters from synthetic fuel technologies on the aquatic environment.

11-2 ORNL SCIENTIFIC APPROACH

To convert coal to a liquid or gaseous fuel, the hydrogen-to-carbon ratio of the
coal must be increased. This is accomplished by pyrolysis, which involves high
temperatures and pressures and a reducing atmosphere. These extreme conditions with the complex chemical composition of coal result in production of
products and wastewaters containing hundreds of organic compounds whose
environmental behavior is often poorly known (Ref. 11-2).

The chemical complexity of coal conversion products and wastewaters
necessitates a different approach for environmental risk analyses than would be
used for a single chemical compound; therefore, the approach used in the AFEP
has two components (Ref. 11-3). One involves the testing of individual chemical
compounds; the other involves the testing of the wastewaters produced by individual coal conversion processes. The objective of the first is to analyze the environmental fate and effects on aquatic organisms of individual compounds present
in most coal conversion wastewaters and products. Because of the many compounds present, it is impossible to analyze adequately the potential environmental
risks of each individual compound. Therefore, based on results from an intensive
analysis of information available in the literature, organic components of coal
conversion wastewaters were divided into five classes of compounds for environmental study (Ref. 11-2). These classes are shown in Table 11-1: (1) phenols, (2)
azaarenes, (3) monoaromatic hydrocarbons (HC), (4) polycyclic hydrocarbons,
and (5) sulfur-containing compounds. An analysis of the data available on these
classes of compounds led to our selection of the polycyclic aromatic hydrocarbons
(PAHs) as the class deserving the highest research priority (Ref. 11-2). This selection was based primarily on the known carcinogenic and mutagenic properties of
many PAHs, the lack of information on their behavior in the aquatic environment,
and their expected presence in significant quantities in synthetic fuel products
and wastewaters. Several individual PAHs were then selected whose chemical

TABLE 11-1. Concentrations and Wastewater Treatment Efficiencies of Major Constituents of Coal Conversion Effluents.

Effluent Constituent	Anticipated Effluent Concentration Range (mg/liter)	Wastewater Treatment Removal Efficiency (%)	Expected Levels in Final Effluent, (mg/liter)
Phenols	10^4	99.9+	1
Azaarenes	$10^2 - 10^3$	30 – 50	50 – 700
Monoaromatic HC	$10^1 - 10^3$	90+	1 – 10
Thiophenes	10^1	?	1 – 10
Polycyclic HC	$10^{-1} - 10^0$	30 – 80	0.02 – 0.7

Source: Adapted from Herbes *et al.*, Ref. 11-2.

structure represented the PAH as a class. These representitive PAHs were subjected to an analysis of their environmental fate and toxicity to some aquatic organisms. It was then possible to use the information gained from these studies of individual compounds to predict the behavior of the PAHs as a class.

The second component of our environmental risk analyses of synthetic fuel wastewaters involves the testing of wastewaters from individual processes. These materials are complex mixtures of many chemicals. The goals of this research are the following: (1) The determination of relative acute toxicities of wastewaters from individual processes to selected species of aquatic organisms. These data permit comparisons of the relative toxicity of the various types of synthetic fuel process under development to aquatic organisms. (2) The determination of the relative acute toxicities to aquatic organisms of wastewaters from individual processes operated under different conditions such as various types of coal, operating modes (liquid or gaseous), or operating conditions (temperature and pressure). (3) The identification and quantification of the toxicity of the chemical components of the wastewater. This permits the identification of chemicals that may be hazardous to aquatic organisms and may lead to changes in operating or wastewater treatment practices to reduce the quantities of these hazardous chemicals in wastewaters released to the aquatic environment.

The following is a summary of the chemical and biological methods and the results of some of the research done in the AFEP at Oak Ridge National Laboratory. Only our work on the effects on aquatic organisms of wastewaters from synthetic fuel processes is discussed. A large portion of the research conducted in the AFEP is devoted to studies on the transport and transformation of organic chemicals present in synthetic fuel wastewaters and products. Those interested in a discussion of the approach used in the transport and transformation work should refer to Herbes *et al.* (Ref. 11-4) and Southworth (Ref. 11-5).

11-3 STUDIES WITH SINGLE COMPOUNDS:
TOXICITY OF POLYCYCLIC AROMATIC COMPOUND
TO SEVERAL AQUATIC SPECIES

Our studies on the toxicities of PAHs to aquatic organisms are designed to determine the potential hazards to aquatic organisms of PAHs that are likely to be present in synthetic fuel products and wastes and to determine the relative toxicities to several aquatic species of PAHs.

We have conducted acute and some chronic tests with several species including the cladoceran, *Daphnia magna,* the fathead minnow, *Pimephales promelas,* the green alga, *Selanastrum capricornutum,* and the bluegreen alga, *Microcystis aeruginosa.* Some of our results are presented in Table 11-2.

From the results of these tests, several conclusions can be drawn concerning the toxicity of PAHs to aquatic organisms. Saturated solutions of anthracene, 2-methylanthracene, 9-methylanthracene, and 1-methylphenanthrene were not acutely toxic to any of the species tested except *S. capricornutum.* Phenanthrene was generally the most toxic of the PAHs tested while naphthalene and 1-methyl-napthalene were generally of intermediate toxicity. The PAHs of intermediate molecular weight and water solubility, 1-methylnaphthalene and phenanthrene, were the most toxic while the PAHs of the highest molecular weight and lowest solubility (the anthracenes) were the least toxic. It appears that only the more soluble intermediate and low-molecular-weight PAHs (naphthalenes and phenanthrenes) pose a potential acute toxicity hazard to aquatic biota, although we have not tested sufficiently to say that this relationship applies to all the PAHs.

The addition of a methyl group to the parent compound does not appear to have a consistent effect on PAH toxicity. Methylnaphthalene was more toxic to *D. magna* and *S. capricornutum* than naphthalene; however, methylnaphthalene was less toxic than naphthalene to *M. aeruginosa.* Methylphenanthrene was generally less toxic than phenanthrene whereas for only one species (*S. capricornutum*) methylanthracene was more toxic than anthracene.

None of the test species was consistently the least or most sensitive to the PAHs tested. Although the test conditions and/or exposure periods used with each species differed, their sensitivities to PAH toxicity, as indicated by the respective EC_{50} estimates, were quite similar.

The chronic toxicity of PAHs to aquatic organisms has been studied for only two of the representative PAH compounds. The effects of a 28-day exposure to naphthalene and phenanthrene were studied on reproduction of *D. magna.* The highest test concentration that produced no significant observed effects on *D. magna* reproduction (the NOEC) was 3.0 mg/liter for naphthalene and 0.3 mg/liter for phenanthrene (Table 11-2). These concentrations are 1/8 and 1/4, respectively, of the 48-hr EC_{50}s. The 14-day NOEC of *S. capricornutum* exposed to phenanthrene was 1.1 mg/liter (Table 11-2). This value was actually greater

TABLE 11-2. Toxicity of some Polycyclic Aromatic Hydrocarbons to Daphnia magna, Pimephales promelas, Selanastrum capricornutum, and Microcystis aeruginosa.

Polycyclic Aromatic Hydrocarbon	Daphnia magna		Pimephales promelas	Selanastrum capricornutum		Microcystis aeruginosa
	48-hr EC_{50} (mg/liter)[a]	28-day NOEC (mg/liter)[b]	48-hr EC_{50} (mg/liter)	4-hr EC_{20} (mg/liter)[c]	14-day NOEC (mg/liter)	4-hr EC_{20}
Naphthalene	24.1	3.0	8.5	5.7	n.d.	0.9
1-methylnaphthalene	1.2	n.d.[d]	n.d.	3.8	n.d.	3.6
Phenanthrene	1.1	0.3	no effect[e]	0.7	1.1	0.2
1-methylphenanthrene	no effect[e]	n.d.	n.d.	(100%)[f]	n.d.	n.d.
Anthracene	no effect[e]	n.d.	n.d.	no effect[e]	n.d.	no effect[e]
2-methylanthracene	no effect[e]	n.d.	n.d.	no effect[e]	n.d.	n.d.
9-methylanthracene	no effect[e]	n.d.	n.d.	(100%)[f]	n.d.	n.d.

[a]The 48-hr EC_{50} is the estimated concentration of the chemical that would immobilize 50% of the test organisms during 48 hr of exposure.

[b]The NOEC is the highest test concentration which had no significant observed effects on the test organisms during 28 days of exposure.

[c]The 4-hr EC_{20} is the estimated concentration of the chemical which would reduce photosynthesis of the test cultures by 20% during 4 hr of exposure.

[d]n.d. = not determined.

[e]No effects were observed in a saturated solution of the chemical.

[f]Number in parentheses is the lowest concentration of the chemical, measured as the percent of saturation of the chemical in water, that had a significant effect on photosynthesis.

Sources: Parkhurst, unpublished data, and Giddings, Ref. 11-6.

than the estimated 4-hr EC_{50}. The apparent disparity is probably a result of the different end points measured in the 4-hr and 14-day tests. The 4-hr test measured effects on photosynthesis, while the 14-day test measured effects on algal growth; thus photosynthesis appears to be more sensitive to phenanthrene toxicity than algal growth.

11-4 TESTS WITH SPECIFIC SYNTHETIC FUEL WASTEWATERS

Our methods with specific synthetic fuel wastewaters consist of acute toxicity tests in which large numbers of materials can be rapidly screened to determine their toxicity to aquatic organisms. The objectives of these tests are (1) to compare the relative toxicities of wastewaters from several synthetic fuel processes to several aquatic species, and (2) to identify and evaluate the toxicity of some of the fractions and chemical constituents of synthetic fuel wastewaters. The screening is generally done with static exposure systems because often only small quantities of the wastewaters are available for testing, and larger quantities would be required to perform continuous flow-through exposure tests.

11-4.1 Comparative Toxicity of Several Synthetic Fuel Wastewaters

The acute toxicity to fathead minnows, bluegills, and zooplankton was investigated for untreated wastewaters from four synthetic fuel processes: (1) two coal liquefaction processes, the hydrocarbonization process and the COED process, (2) a coal gasification process, Synthane, and (3) an oil shale retorting process. The acute toxicities of the wastewaters were calculated as both the concentration of the wastewater and the total organic carbon (TOC) concentration of the wastewater estimated to kill 50% of the test organisms in 48 hours, i.e., the 48-hr LC_{50}.

All of the wastewaters were found to have very high acute toxicities to the three test species (Table 11-3). The highest toxicity, a 96-hr LC_{50} of 0.004% and 1 mg/liter TOC, was measured for the hydrocarbonization wastewater to *D. pulex*. The lowest toxicity in percent concentration of the wastewater, 0.305%, was estimated for the hydrocarbonization wastewater to bluegills.

Considerable interspecific variability in sensitivity to the four wastewaters was apparent. *Daphnia pulex* was generally the most sensitive species to all the wastewaters. Bluegills were the least sensitive of the three species to all of the wastewaters except the shale oil product, to which the fathead minnows were least sensitive.

There does not appear to be any direct relationship between the TOC concentration of the wastewaters and their acute toxicity to the three test species. The hydrocarbonization wastewater had the highest TOC content. While it also had the

TABLE 11-3. Acute Toxicities of Untreated Wastewaters from Synthetic Fuel Processes to some Aquatic Animals.

Process Effluent	Fathead Minnow (96-hr LC_{50})		Bluegill (96-hr LC_{50})		Daphnia pulex (96-hr LC_{50})	
	(%)	(mg/liter TOC[a])	(%)	(mg/liter TOC)	(%)	(mg/liter TOC)
Hydrocarbonization scrubber water (20,000 mg/liter TOC)	0.093	19	0.305	26	0.004	1
Synthane condensate (12,780 mg/liter TOC)	0.040	5	0.233	30	0.043	5
Shale oil product water (4740 mg/liter TOC)	0.091	4	0.070	3	0.043	2
COED product separator liquor (9550 mg/liter TOC)	0.040	4	0.127	12	0.024	2

[a]Total organic carbon content at the LC_{50} of the wastewater.

Source: Parkhurst, unpublished data.

highest toxicity to *D. pulex,* it had the lowest toxicity to fathead minnows. The Synthane wastewater, with the second highest TOC content, had a lower toxicity to *D. pulex* than either the shale oil wastewater or the COED wastewater, both of which had lower TOC contents. We have found (Parkhurst *et al.,* Ref. 11-7) that the major toxic components of the hydrocarbonization wastewater are phenol, *o-, p-,* and *m*-cresol and ammonia. The individual proportions of these materials in the wastewaters would largely determine the toxicity of the wastewaters and, then, toxicity would not be expected to be directly related to TOC concentration.

11-4.2 Identification and Evaluation of the Toxic Components of Synthetic Fuel Wastewaters

The next step in the screening of synthetic fuel wastewaters is to identify the toxic components of the wastewaters. The sequential procedure used (Ref. 11-8) is shown in Fig. 11-1. Acutely toxic mixtures are separated by solvent extraction into their organic and inorganic fractions and these fractions are then tested for acute toxicity. If the organic fraction is toxic, the fraction is further separated into acid, base, and neutral fractions and the relative amount of each fraction is determined. The toxicity of each fraction is then determined and its contribution to the toxicity of the original mixture is calculated from the relative toxicity of the fraction and its concentration in the original mixture. If more specific identification of the toxic components is desired, further fractionation and testing can be done. For example, up to 14 fractions have been separated from synthetic fuel process wastewaters and tested for their mutagenicity (Rubin *et al.,* Ref. 11-9). Finally, tests with specific compounds can be performed to evaluate the toxicity of individual chemical components of the wastewater.

A different approach from the above is used to identify the toxic components of the inorganic fraction of the wastewater. If found to be acutely toxic, this fraction is characterized chemically. The toxicities of the individual components are then determined, and the contribution of each component to the toxicity of the original wastewater is calculated as described above for the organic fraction.

In the last step of the screening program, known components of the original wastewater are combined to produce a "reconstituted" wastewater. As a check, the acute toxicity of the reconstituted wastewater is then tested and compared with the toxicity of the original wastewater to insure that the toxicity of the latter has been identified.

The procedure outlined above has been used to identify the toxic components of several types of synthetic fuel wastewaters. The following is a discussion of the test results with two of these wastewaters. The first material tested was an untreated process-water effluent from the Solvent Refined Coal (SRC) pilot plant in Ft. Lewis, Washington. The second material, Hydrocarbonization (HCZ) process-water from the process development unit at the Oak Ridge National

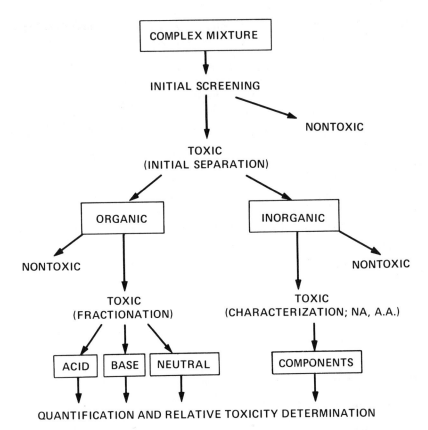

Fig. 11-1. Sequential testing procedure for identifying the toxic components of complex aqueous mixtures. *Source:* Gehrs *et al.,* Ref. 11-8.

Laboratory, consisted of two effluents, an untreated and a biologically treated sample.

The complete SRC effluent was acutely toxic to *D. magna,* with the 48-hr LC_{50} estimated to be at a dilution of 15.7% (Parkhurst *et al.,* Ref. 11-10). The inorganic portion of the effluent was relatively nontoxic. Of the three organic fractions, the neutral fraction was the most toxic with a 48-hr LC_{50} of 9 mg/liter (Table 11-4). The 48-hr LC_{50}s of the acid and base fractions were 29.5 and 45.8 mg/liter, respectively. The contribution of each fraction to the toxicity of the whole SRC wastewater ranged from 51.3% for the acid fraction to 3.5% for the base fraction (Table 11-4).

Testing the whole wastewater as well as its fractions allows detection and measurement of the toxic interactions among fractions, which may affect toxicity. Comparisons of results between whole wastewater before fractionation and "re-

TABLE 11-4. Contributions of Fractions from a Solvent Refined Coal Process Wastewater to the Toxicity of the Wastewater to *Daphnia magna.*

Fraction	48-hr LC_{50} (mg/liter)	Fraction Concentration (mg/liter) at 48-hr LC Concentration		Contribution to Whole Wastewater Toxicity (%)[b]	
		Reconstituted Wastewater	SRC Wastewater	Reconstituted Wastewater	SRC Wastewater
Acid	29.5	15.2	15.6	51.5	51.3
Base	45.8	1.6	1.7	3.5	3.5
Neutral	9.0	3.8	3.9	42.2	41.9
Inorganic	1000[a]	32.7	33.6	3.0	3.3

[a]No mortality occurred at highest concentration in test series. It is assumed that 1000 mg/liter is a conservative estimate of the 48-hr LC_{50}.

[b]The contribution of each fraction to the toxicity of the whole wastewater is given by the expression,

$$\frac{C_i\,(LC_{50_i})^{-1}}{\sum\limits_{i=1}^{n} C_i\,(LC_{50_i})^{-1}}$$

where i is the component, LC_{50_i} is the toxicity (48-hr LC_{50}) of the individual component, and C_i is the component concentration at the 48-hr LC_{50} of the wastewater.

Source: Parkhurst *et al.,* Ref. 11-10.

constituted" wastewaters may show whether the fractionation procedure affected the toxicities of the fractions. In the tests with the SRC wastewater, toxic interactions were not found. The 48-hr LC_{50}s of the whole wastewater and the reconstituted wastewater both calculated as a percent concentration of the wastewater were 15.7% and 15.5%, respectively. These values were not significantly different (t-test, $p = 0.05$) and indicated that (1) the chemical fractionation procedure did not alter the toxicity of the fractions, and (2) all of the toxicity of the effluents was accounted for in the four fractions. The additivity index (Marking and Dawson, Ref. 11-11) was 0.03 for the whole SRC wastewater and 0.00 for the reconstituted wastewater. These values were not significantly different from each other nor from zero (t-test, $p = 0.05$), which indicates that all of the major toxic components of the effluents were included in the fractions and that the toxicities of the individual fractions of the wastewaters were additive.

In the second example of the AFEP environmental screening approach wastewaters from the Oak Ridge National Laboratory Hydrocarbonization Unit (HCZ) were tested. First, the toxicities of the untreated and treated wastewaters were compared.   Treatment reduced the toxicity of the wastewater to *D. magna* by 99%.   Further testing revealed that 99.5% of the wastewater toxicity was due to inorganic components (Table 11-5) and the remainder to organics (principally phenols).   Of the inorganic constitutents, ammonia contributed 77% of the toxicity.   Ammonia can be efficiently removed from wastewaters by currently available technology.   Removal of the ammonia from the hydrocarbonization wastewater in an additional treatment step would be expected to significantly reduce the acute toxicity of the wastewater.   The HCZ wastewater components demonstrated an additive toxicity.

11-5   DISCUSSION OF APPROACH AND RESULTS

The application of our approach to analyzing the acute toxicity of wastewaters from various synthetic fuel processes has demonstrated that screening tests can rapidly provide valuable information for environmental risk analyses.   By using a combination of tests on individual constituents of synthetic fuel wastewaters with fractionation and identification of the toxic components of the wastewaters, we have been able to analyze some of the risks to aquatic biota of several synthetic fuel process wastewaters.   It was demonstrated that the untreated aqueous wastewaters produced by synthetic fuel processes have high acute toxicities.   The toxicity of one of the wastewaters was found to be derived primarily from their acidic and neutral fractions.   The acidic fraction contains the phenolic compound, which can be efficiently removed by biological oxidation. This was demonstrated in our study of the untreated and treated hydrocarbonization wastewaters.

The neutral fraction will largely contain the PAHs.   Our studies on the toxicity of PAHs have demonstrated that their toxicity will primarily be a result of the more soluble low- and intermediate-molecular-weight PAHs, such as naphthalene and phenanthrene.

TABLE 11-5.   Concentration and Acute Toxicities (48-hr LC_{50}s) to *Daphnia magna* of Fractions from a Hydrocarbonization Wastewater.

Parameter	Fraction	
	Organic	Inorganic
Concentration in wastewater (mg/liter)	110	2846
Relative quantity (%)	3.7	96.3
D. magna 48-hr LC_{50} (mg/liter)	774	31.7
Relative contribution of fraction to wastewater toxicity (%)	0.5	99.5

Source: Gehrs *et al.,* Ref. 11-8.

The low solubilities of PAHs in water appears to limit their acute toxicity to aquatic organisms. The higher-molecular-weight and more insoluble PAHs such as anthracene are nontoxic at water saturation, while the lower-molecular-weight and more soluble PAHs such as naphthalene and phenanthrene are only toxic at concentrations near water saturation. Algae, fish, and daphnids appear to have similar sensitivities to PAH toxicity.

We have also demonstrated that inorganic chemicals may cause toxicity problems in some synthetic fuel wastewaters and will not be a problem in others. Ammonia was found to be the major toxic compound of the treated hydrocarbonization wastewater; however, ammonia can be efficiently removed by presently available wastewater treatment techniques.

Areas needing further research include (1) the effects of chronic exposure to synthetic fuel process wastewaters and their components, and (2) extrapolation of laboratory single species effects studies to more complex ecological systems.

REFERENCES

11-1. Exxon Company, *Energy Outlook, 1978-1990.* Public Affairs Dept., Exxon Company, USA, Houston, TX, 1978.

11-2. Herbes, S. E., Southworth, G. R., and Gehrs, C. W., "Organic Contaminants in Aqueous Coal Conversion Effluents: Environmental Consequences and Research Priorities," in *Trace Substances in Environmental Health,* D. D. Hemphill (Ed.), University of Missouri, Columbia, pp. 295-303, 1977.

11-3. Gehrs, C. W., "Environmental Implications of Coal-Conversion Technologies: Organic Contaminants," in *Energy and Environmental Stress in Aquatic Systems,* J. H. Thorp and J. W. Gibbons (Eds.), CONF-771114, Technical Information Center, U.S. Department of Energy, pp. 157-175, 1978.

11-4. Herbes, S. E., Southworth, G. R., Shaeffer, D. L., Griest, W. H., and Maskarinec, M. P., "Critical Pathways of Polycyclic Aromatic Hydrocarbons in Aquatic Environments," in *Proc., Symposium on the Scientific Bases for Toxicity Assessment,* Gatlinburg, Tennessee, April 17-19, 1979, Oak Ridge National Laboratory, Oak Ridge, Tenn.

11-5. Southworth, G. R., "Transport and Transformation of Anthracene in Natural Waters," in *Aquatic Toxicology,* ASTM, STP 667, L. L. Marking and R. A. Kimerle (Eds.), American Society for Testing and Materials, Philadelphia, PA, pp. 359-380, 1979.

11-6. Giddings, J. M., "Acute Toxicity to *Selanastrum capicornutum* of Aromatic Compounds from Coal Conversion, *Bull. Environ. Contam. Toxicol.* **23**, pp. 360-364, 1979.

11-7. Parkhurst, B. R., Bradshaw, A. S., Forte, J. L., and Wright, G. P., "An Evaluation of the Acute Toxicity to Aquatic Biota of a Coal Conversion Effluent and Its Major Components," *Bull. Environ. Contam. Toxicol.* **23**, pp. 349-356, 1979.

11-8. Gehrs, C. W., Parkhurst, B. R., and Shriner, D. S., "Environmental Screening of Complex Chemical Mixtures," in *Application of Short-Term Bioassays in the Fractionation and Analysis of Complex Environmental Mixtures,* EPA-600/9-78-027. U.S. Environmental Protection Agency, Research Triangle Park, NC, pp. 317–330, 1978.

11-9. Rubin, I. B., Guerin, M. R., Hardigree, A. A., and Epler, J. L., "Fractionation of Synthetic Crude Oils from Coal for Biological Testing," *Environ. Res.* **12**, pp. 358–365, 1976.

11-10. Parkhurst, B. R., Gehrs, C. W., and Rubin, I. B., "Value of Chemical Fractionation for Identifying the Toxic Components of Complex Aqueous Effluents," in *Aquatic Toxicology,* ASTM STP 667, L. L. Marking and R. A. Kimerle (Eds.), American Society for Testing and Materials, Philadelphia, PA, pp. 122–130, 1979.

11-11. Marking, L. L., and Dawson, V. K., "Method for Assessment of Toxicity of Efficacy of Mixtures of Chemicals," in *Investigations in Fish Control No. 67.* U.S. Department of the Interior, Fish and Wildlife Service, Washington, D.C., 1975.

12

Sequential testing for chemical risk assessment

Bernard D. Astill, Ph.D.,
Haines B. Lockhart, Jr., Ph.D.,
Jay B. Moses, M.D.,
Ahmed Nasr, M.D., Ph.D.,
Robert L. Raleigh, M.D., and Clarence J. Terhaar, Ph. D.

Health, Safety, and Human Factors Laboratory
Eastman Kodak Company, Rochester, New York

12.1 INTRODUCTION

The professional engaged in making assessments of hazards to health or the environment is faced with a steady growth in the number and type of tests that may be selected to define the hazard. There has also been a rapid growth in recent years in the areas of concern associated with exposure to chemicals. Perhaps the most direct approach possible would be to assemble a list of toxic effects, along with a list of tests that establish the presence or absence of those effects. In an entirely arbitrary manner, and at great expense, one could then define the toxicological and environmental properties of a chemical as completely as is currently possible. Such an exhaustive approach has been commonly required for food additives, pesticides, new human drugs, and animal drugs.

Testing on such a scale may be justifiable where large populations are exposed to a chemical in an involuntary manner. It is unnecessary and inpracticable for most of the chemicals used in industry and commerce. While such an approach may define the biological effects due to a substance, it is necessarily incomplete, since such information is only a part of what we need to know to define the actual hazard to health or the environment. The real need is to obtain the information and select the tests most appropriate to define the hazards involved in production, use, and disposal of a chemical. One also must be certain that the information and tests are adequate for that purpose. This needs to be done in a consistent manner from chemical to chemical, and in a manner that relates to a body of established practice acceptable to professionals in the field.

There is, therefore, a clear and pressing need for schemes that, in a systematic manner, organize, relate, and evaluate information concerning chemical hazards to health and the environment. Such schemes should identify the types of information needed to evaluate the hazards involved, and should enable the user to select the types of testing needed to define the hazard. They should relate intended uses to testing requirements and should provide some means of deciding when information or testing is sufficient. In particular, such schemes should be free of mandatory requirements; rather should they allow for judgment in the selection of appropriate tests. Any scheme evolved to meet these needs should also allow likely testing requirements to be predicted from the inspection of the available information, even though that may be incomplete or fragmentary.

A scheme for the evaluation of chemical hazard to health and the environment that attempts to meet these requirements is presented in this article. To date, it has been in use for the past two years, and in its present form reflects the experience gained in evaluating over 500 chemicals. The scheme sets up four basic categories for which information is required. Within each category criteria are established to define a category. Each criterion is evaluated numerically. By summing numerical values within a category, scores for the four basic categories are obtained. These are then combined to yield final scores for Health Hazard, and for Environmental Hazard. These scores are used, in turn, as decision points or triggers in a hierarchical assembly of health and environmental tests. A final score can be used as an aid in determining whether the existing information is adequate to define the hazard, or whether more testing is required. It also indicates what the extent of that testing might be. Inherent in the scheme are requirements for information that make it possible to judge the most appropriate tests in a level of testing. On completion of any testing that may be indicated, the scheme provides for a reevaluation to confirm the original score or obtain a new one. Thus a decision can be made to test further, or whether the information is sufficient to establish the health or environmental hazard.

12-2 CATEGORIES AND CRITERIA

The Toxic Substances Control Act provides a useful guide to the categories of importance in making risk assessments. Section 5 (B) (4) (A) (ii) states that the following factors are to be considered in compiling a list of chemical substances that may present an unreasonable risk of injury to health or the environment:

1. The effects of the chemical substance on health and the magnitude of human exposure to such a substance.
2. The effects of the chemical substance on the environment and the magnitude of environmental exposure to such substance.

This suggests that there are four categories of information. These are

M_H — The magnitude of health exposure.
M_E — The magnitude of environmental exposure.
H — The effects on human health of an exposure.
E — The effects on the environment of an exposure.

Categories M_H and M_E may be regarded as approximations to the dose, intake level, or exposure concentration. Criteria for these categories should, therefore, be chosen so that these entities may be defined in a usable manner. It is desirable to introduce quantification for criteria, such that a rating of intensity or severity can be made. Summation of such ratings provides a total score for a category. One useful numerical rating system relies upon two boundaries, the upper boundary being chosen to define the lower maximum intensity limit for a criterion, and the lower boundary to define the upper minimum limit. These boundaries establish a rating system of three levels of "intensity," in ascending order 1, 2, and 3.

12-2.1 Magnitude of Health Exposure

The criteria chosen to define M_H (Table 12-1) are (a) quantities manufactured annually, (b) number of people exposed, (c) duration of exposure, and (d) population groups exposed.

The quantity of a chemical manufactured annually is significant because it represents the maximum possible exposure. Exact definition of exposure level may not always be possible, because of the wide diversity of uses of many chemicals, and the wide range of possible exposure conditions. The boundaries selected for quantities manufactured are realistic in terms of annual production volumes in the chemical industry.

The second M_H criterion uses a lower boundary of 1,000 and an upper boundary of 1,000,000 for people exposed. These values approximate the range of population size between a moderate-sized industrial plant and a large city. The lower score for a lower number of people exposed does not mean there is less concern for the individual who is a member of a small population. In a large population, there is greater opportunity for variations in susceptibility, and in

TABLE 12-1. Criteria for Magnitude of
Health Exposure (M_H).

Criterion	Value	Rating
Production in lb[a] yr	$<1 \times 10^6$	1
	$1 - 100 \times 10^6$	2
	$>100 \times 10^6$	3
Number of people exposed	<1000	1
	$1000 - 1 \times 10^6$	2
	$>1 \times 10^6$	3
Duration of exposure	<200 hr/yr	1
	$200 - 2000$	2
	>2000	3
Populations exposed:		
In-plant or manufacturing; consumer; general public		
	Single population	1
	Two populations	2
	Three populations	3

[a]lb $\times$ 0.45 = kg.

some cases it is not possible to control exposures to a chemical as well as for a small population. The level of information needed to determine the hazard is thus greater for a larger population.

The lower boundary for duration of exposure reflects the situation of an individual in the workplace; 200 hr would be the period for four consecutive weeks of working day exposure. The upper boundary for duration of exposure reflects the difference between an individual in the workplace and a consumer or member of the general population. Thus 2,000 hr is the limit to occupational exposure in a year; exposures in excess of this amount would be more likely to be encountered in the general population. Finally, three distinct populations can usually be discerned in considering the manner of exposure and the extent to which it can be controlled. These are the working population, the consumer population, and the general public. The preceding criteria can be applied to each of these population groups, and indeed each could be treated as a separate category. The evaluation and scoring procedure is, however, complicated greatly without any clear advantage.   It appears adequate to rate for the number of population groups, and include this value in the category score, rather than to score for each group separately. The boundaries for this criterion are thus one and three population groups exposed, respectively.

12-2.2 Magnitude of Environmental Exposure

The criteria chosen to define M_E (Table 12-2) are (a) amount of chemical discharged, (b) number of sites of discharge, (c) frequency of discharge, and (d) duration or dwell time of the potential pollutant in the environment. The boundaries chosen for amounts discharged are at least an order of magnitude below those chosen for annual production volume. This reflects the consideration that, in general, chemicals are produced to be used rather than to be wasted. The lower boundary for the number of sites of discharge reflects the situation of the manufacturer with a number of plants handling a given chemical, or the discharge situation in a small city, or that of a limited number of manufacturers using a chemical. The upper boundary reflects the discharge situation for a chemical used over a fairly wide geographical area. The boundaries for frequency of discharge distinguish between an infrequent or sporadic occurrence, the lower boundary, and a fairly common occurrence, the upper boundary. In addition to criteria that establish quantity, extent, and frequency of a discharge, an important additional concern is the length of time a chemical may remain in the environment. A criterion of duration or dwell time, expressed as a half-life in days, is included to meet this concern. Boundaries are chosen to reflect the distinction between readily degraded or transformable substances and those that show a considerable environmental stability or persistence.

TABLE 12-2. Criteria for Magnitude of Environmental Exposure (M_E).

Criterion	Value	Rating
Amount discharged in lb[a] yr	$<1 \times 10^5$	1
	$1 \times 10^5 - 5 \times 10^6$	2
	$>5 \times 10^6$	3
Number of sites of discharge	<25	1
	$25 - 2500$	2
	>2500	3
Frequence of discharge in times/year	<5	1
	$5 - 25$	2
	>25	3
Dwell time after discharge (half-life in days)	<5	1
	$1 - 7$	2
	>7	3

[a] lb $\times$ 0.45 = kg.

12-2.3 Health Effects

As has been indicated, Categories H and E refer to effects on health and environment, respectively. In selecting criteria for health effects, it became clear that the number and variety of biological effects would make a scoring system based on assessments of specific effects, such as hepatotoxicity, mutagenicity, and neurotoxicity, too complicated to be readily usable. It was felt that criteria should be as general as possible. Those selected are therefore (a) the acute oral LD_{50}, (b) immediate effects, and (c) prolonged or delayed effects (Table 12-3). The rat oral LD_{50} was selected since it is clearly desirable to know the lethality of a chemical, and the oral LD_{50} is the basis of some commonly used classifications of relative toxicities (Gleason *et al.,* Ref. 12-1). The lower boundary for the LD_{50} defines a "highly toxic" substance (Federal Hazardous Substances Act, 1967, Ref. 12-2), and the upper boundary is a commonly accepted value for substances of moderate or little lethality.

It is convenient to distinguish between immediate and prolonged effects when assessing the health effects of chemicals. Immediate effects are those that follow an exposure within a short time. The amount of exposure may range from a low dose or concentration to an acute nonlethal dose. Prolonged or delayed effects may result from single or a few exposures of short duration, or may result from continuous or prolonged intermittent exposures. A further distinction can be made between reversible and irreversible effects. Effects criteria range from no effects through reversible effects to irreversible effects. A chemical would receive a rating of 1 in the absence of any effects. If the effect produced was a moderate skin (or eye) irritation, the rating for immediate effects would be 2, since irritation is usually reversible. If destructive skin or eye damage occurred, the rating

TABLE 12-3. Criteria for Health Effects (H).

Criterion	Value	Rating
LD_{50} (P. O., rats), mg/kg	>500	1
	50 - 500	2
	<50	3
Immediate effects	None	1
	Reversible	2
	Irreversible	3
Prolonged effects	None	1
	Reversible	2
	Irreversible	3

for immediate effects would be 3. Similar considerations apply in rating prolonged systemic effects.

12-2.4 Environmental Effects

There is a large body of practice that is helpful in establishing criteria for health effects. Such a body of practice does not as yet exist for environmental effects. In choosing criteria for the effects of chemicals on the environment it was, therefore, felt best to define areas that have, by consensus, become of the most concern. These are the persistence of a chemical in the environment, its compatibility with secondary waste treatment facilities, its toxicity to aquatic organisms, and its ability to bioconcentrate and thus have potential to ascend the food chain. Persistence has already been identified as dwell time in the exposure criteria. The effects criteria chosen are thus (a) compatibility with secondary waste treatment, (b) toxicity to fish, and (c) bioconcentration potential (Table 12-4).

The compatibility of a chemical with secondary waste treatment may be considered as its effect on the performance of secondary waste treatment microorganisms. A quantitative measure of this effect, referred to as the 5-hr IC_{50}, is the concentration of test chemical that inhibits the metabolic rate of such microorganisms by 50% in 5 hr (Lockhart et al., Ref. 12-3). Experience with this test has provided an upper boundary of 5,000 mg/liter for chemicals that produce essentially no effect, and a lower boundary of 250 mg/liter for those that produce a marked effect on waste treatment microorganism metabolism.

The criterion for aquatic toxicity is derived from the aquatic concentration that is lethal to 50% of the test species in 96 hr (96-hr LC_{50}) using fathead minnows (Terhaar et al., Ref. 12-4). The boundaries chosen are 100 mg/liter for

TABLE 12-4. Criteria for Environmental Effects (E).

Criterion	Value	Rating
Compatibility with waste treatment, 5-hr IC_{50} (mg/liter)	>5000	1
	250 - 5000	2
	<250	3
Aquatic toxicity (fish) LC_{50} (mg/liter)	>100	1
	10 - 100	2
	<10	3
Bioconcentration potential (octanol-water distribution coefficient)	<100	1
	100 - 10,000	2
	>10,000	3

substances not likely to be lethal to fish, and 10 mg/liter for lethal substances. The potential of a chemical that is discharged to the environment to be accumulated by aquatic and other species is usually derived from its ability to concentrate in the fatty tissues of an organism. For most chemicals there are no actual data of this type available. An adequate indicator of bioconcentration potential may be provided by the octanol-water partition coefficient of the chemical (Neely *et al.*, Ref. 12-5). Boundaries for this criterion are 100 for chemicals of little or no bioconcentration potential, and 10,000 for chemicals of considerable bioconcentration potential.

12-2.5 Worksheet for Categories and Criterion

The collection and assessment of the above information and the assignment of scores is facilitated by the use of a worksheet such as that presented in Fig. 12-1. The worksheet lists the various criteria and provides for their tabulation and rating. Summation of ratings then gives the final health and environmental scores. The worksheet permits estimated ratings regardless of the quality or availability of information in a category. The value of a final score is considerably enhanced when precise data are available on the use of a chemical and its biological effects. Predictions from structure, experience, and intuition allow the worksheet to be used for screening purposes, and for guidance as to the type of testing that may then be needed. The worksheet is also a useful checklist for required information. The use of final health and environmental scores as decision points in a testing scheme is described in Section 12.4.

12-3 HIERARCHICAL TESTING

We have found it helpful in this scheme to identify several levels of information and testing for health and environmental hazard assessments. In addition, we have recognized an initial level, referred to as baseline information, for which data are obtained before embarking upon any program of testing (Table 12-5).

12-3.1 Base-Line Information

Base-line information can usually be obtained before carrying out any chemical or biological testing. It includes information compiled on quantities manufactured and disposed, product application and function, and the type, extent, and duration of exposure in scoring for M_H and M_E. The completeness of

CHEMICAL OR SUBSTANCE		Q = Quantities or Information R = Rating	HEALTH INFORMATION AND RATINGS							ENVIRONMENTAL INFORMATION AND RATINGS							TOTAL SCORES	
			EXPOSURE				EFFECTS			DISCHARGE					EFFECTS			
NAME, FORMULA OR STRUCTURE	IDENTIFICATION		PRODUCTION VOLUME (lbs/yr)	PEOPLE (No.)	DURATION (hrs/yr)	POPULATION GROUPS	ACUTE ORAL LD$_{50}$ (mg/kg)	IMMEDIATE	PROLONGED	QUANTITY (lbs/yr)	SITES (No.)	FREQUENCY (Times/yr)	DWELL TIME (T½ days)	PARTITION COEFFICIENT	COMPATIBILITY IC$_{50}$ (mg/l)	96 hr-LC$_{50}$ (mg/l)	HEALTH	ENVIRONMENTAL
		Q																
		R																
		Q																
		R																
		Q																
		R																

Fig. 12-1. Worksheet for risk assessment.

TABLE 12-5. Information and Testing Levels.

Base-Line Information
Quantities manufactured and disposed of
Exposure estimates
Product function and application
Structure activity correlation
Literature search
Cancer hazard evaluation

Level I
A. Physicochemical information
B. Health and environmental screening

Level II
Basic toxicology and environmental tests

Level III
Subacute exposures and intermediate environmental testing

Level IV
Long-term health and environmental effects

such information usually depends on the stage of development or use of the product.

Base-line information includes estimates of biological and environmental activity gained from inspection of the chemical structure. Although quantitative structure activity relationships are presently restricted to a few areas of pharmacological activity, a number of general relationships are accepted. These include the neurotoxicity of several organophosphates and carbamates, the mutagenic and carcinogenic potential of alkylating agents, the irritant properties of strong acids and acid chlorides, the mutagenic and carcinogenic potential of unsaturated and strained ring lactones, the potential for skin sensitization and aquatic toxicity of substituted phenols, and the potential antithyroid activity of certain thiazoles. Base-line information includes the results of literature searches for health, biological, and environmental effects. It may include an evaluation of the likely cancer hazard of the chemical, based on structure activity correlations and literature searches.

Base-line information may be complete enough for hazard assessment, particularly where previous toxicity studies are available from the literature. Where biological testing is indicated, tests are most conveniently arranged in levels.

12-3.2 Level I: Physicochemical Tests and Health and Environmental Screening

This level of testing is conveniently divided into physicochemical and biological testing. Level IA (Table 12-6) consists of a number of physicochemical tests, in-

TABLE 12-6. Level I-A: Physicochemical Tests.

Molecular weight
Molecular formula
pH
Physical state, color
Particle size
Density
Flammability
Flash point
Explosivity
Corrosivity
Purity
Gross solubilities (H_2O, Acetone, Corn Oil, Dimethyl Sulfoxide)
Solubility in H_2O (to 1 ppb)
Melting point
Boiling point
Vapor pressure
pK_a

formation from which is valuable in making hazard assessments. Level IA includes descriptions of physical state, fundamental properties of the molecule such as molecular weight and melting point, solubilities and reactivity as expressed by pH and pK_a and corrosivity. Knowledge of purity is important since trace amounts of contaminating reactive substances may affect the results of biological and environmental testing. It is prudent to perform physicochemical and biological tests with properly characterized materials that are closely similar to those intended for use in commerce. The tests in Level IA are neither mandatory nor comprehensive; the selection of tests should depend on the test chemical and its proposed uses.

Level IB (Table 12-7) consists of a group of tests for health and environmental screening purposes. Selection from these tests will be dictated by the structure, properties, and the proposed use of the test chemical. Commonly selected tests are the range finding LD_{50}, skin irritation and fish LC_{50} tests, and BOD_5/COD ratio. Brief descriptions of the tests follow; more complete descriptions of health effects tests may be found in publications by the National Academy of Sciences (Ref. 12-6) and the Food Safety Council (Ref. 12-7).

Range Finding LD_{50}. This is usually the initial measurement in toxicity testing, using lethality as the response. It is commonly performed by orally dosing rats, and calculating the dose that will kill 50% of the exposed animals.

Skin Irritation. This may be estimated in rabbits, guinea pigs, or other small animals, and is intended to predict the likely damage to human skin from a single

TABLE 12-7. Level I-B: Health and
Environmental Screening.

Acute studies – select from
LC_{50} (range finding)
Skin Irritation
Eye Irritation
Skin sensitization
Acute inhalation

In vitro mutagenesis battery

Environmental tests
LC_{50} fish
BOD_5/COD

exposure to a chemical. The test chemical may be held in contact with the skin under a cuff.

Eye Irritation. This is usually determined by placing the test chemical into the conjunctival sac of the rabbit eye. The test estimates the likely damage to the human eye from the single exposure such as a splash.

Skin Sensitization. A test for allergic response resulting from skin exposures is usually performed on guinea pigs. Repeated exposures are followed by a challenging exposure several days later. The severity of the response is estimated from the degree of skin inflammation.

Acute Inhalation. Where exposures are likely to be to dusts, vapors, or gases, it is important to have an estimate of the potential for harmful effects. Using lethality as an index, acute inhalation studies will establish the calculated airborne concentration which is lethal to 50% of the test animals.

In Vitro Mutagenesis Battery. Preliminary estimates of mutagenic and, by implication, carcinogenic potential may be indicated from structural considerations. A useful guide to such activity may be obtained from the so-called short-term tests; a battery of these should be run, preferably consisting of a bacterial mutagenesis test, a mammalian mutagenesis test, and a mammalian cell transformation test, with and without metabolic activation (Food Safety Council, Ref. 12-7).

Ninety-Six Hour LC_{50}. A variety of species could be suggested in testing for the effects of a chemical after discharge to receiving bodies of water. For screening purposes a useful index is the amount that can be tolerated by fish in a short time, using lethality as an end point. Fathead minnows or rainbow trout may be used (Terhaar *et al.*, Ref. 12-4).

BOD_5/COD. The BOD_5 (five day biochemical oxygen demand) test measures the uptake of oxygen by a standard microbiological population in the presence of a test compound. It thus can indicate the extent to which a chemical is being metabolized by the biological population. The COD (chemical oxygen demand) test measures the extent to which a test chemical may be oxidized in solution by a strong oxidant (Taras *et al.*, Ref. 12-8). Preliminary estimates of the biodegradability of a chemical can be obtained from the ratio of the BOD_5 and COD values.

The base-line information and the health and environmental screening level may provide adequate data for assessments of the hazard due to relatively small volume chemicals, and to chemical intermediates that are not isolated. In particular, where a chemical used in commerce is completely enclosed for its use or function, information obtained up to and including Level I may be sufficient for assessment.

12-3.3 Level II: Basic Toxicology and Environmental Tests

This level of testing is intermediate between acute tests and subchronic or 90-day feeding studies; it also provides more precise environmental information (Table 12-8). Animal tests employing repeated feedings or exposures lasting up to 2 weeks can provide preliminary no-effect levels and information about target organs. Environmental tests will provide information about the compatibility of

TABLE 12-8. Level II: Basic Toxicology and Environmental Tests.

Basic toxicology testing
    (Level I plus 2-week feeding or inhalation with clinical chemistry
    and pathology)
    Repeated skin application test

In vitro mutagenesis battery

Environmental tests
    Activated sludge effects
    Photodegradation $T_{\frac{1}{2}}$
    Biodegradation $T_{\frac{1}{2}}$
    Plant growth and germination effects
    Octanol-H_2O and Soil-H_2O partition coefficients

---

the chemical with secondary waste treatment facilities, its dwell time in the environment, and its toxicity to plants. Capacity for bioconcentration and retention in the soil may be indicated by appropriate partition coefficients.

Short-Term Feeding, Inhalation, and Skin Application Studies. Selection of one or more of these studies should be based on the type of exposure to be anticipated. Information on procedures to be followed can be derived from the references cited in Level I-B testing. Feeding studies can last as long as 2 weeks. They may consist of a control and two dose levels in rodents, accompanied by hematology and clinical chemistry and followed by gross- and micro-pathology. A two-week inhalation study could similarly be performed in rodents at two exposure levels plus controls, with hematology, clinical chemistry and pathology. Short-term experimental exposures of this type will frequently provide information adequate to assess the risks of handling candidate chemicals, particularly in an occupational setting where control measures can be devised.

Ten-Day Skin Irritation. Repeated skin exposures may exacerbate an initial reaction or produce other skin damage. Repeated skin exposures similar to the acute skin test in animals will indicate the extent of any nonallergenic trauma that may be produced.

In Vitro Mutagenesis Tests. If these are omitted from Level I, and if structure activity considerations combined with a sufficient volume of production or extent of discharge suggest that information on mutagenic potential may be useful, these tests may be included in Level II.

Effects on Activated Sludge. The compatiiblty of a chemical with secondary waste treatment may be estimated by selecting suitable microorganisms from secondary waste treatment plants, which are then exposed to the test chemical. A suitable technique of estimating such effects is a radiorespirometry procedure that shows the effect of the chemical on the rate of production of $^{14}CO_2$ from ^{14}C glucose as a substrate. As indicated above, the concentration that inhibits 50% of the metabolism in 5 hr (5-hr IC_{50}) is a useful index of compatibility (Lockhart et al., Ref 12-3).

Photodegradation and Biodegradation. The rate of disappearance or breakdown of the chemical in aqueous solution either photolytically, or in the presence of activated sludge microorganisms, provides a useful estimate of the persistence of the chemical after discharge in an effluent (Lockhart and Blakely, Ref. 12-9). The increasing availability of gas-liquid and high-pressure liquid chromatographic procedures makes such measurements more feasible than hitherto. The half-life of the parent compound $(T_{1/2})$ appears to be a reasonable index of biodegradation or photodegradation.

Plant Tests. Estimates of effects on plant germination and growth due to test chemicals are of value, where the use or discharge pattern may bring the chemical into contact with plant life. Suitable germination tests may be performed with radish or lettuce seeds, and growth tests with growing radish, lettuce, sweet corn and marigold plants. The test chemical may be applied in aqueous solution.

Partition Coefficients: Octanol-Water, Soil-Water. The potential of a chemical to accumulate in aquatic species, so that its uptake exceeds its depuration and it ascends the food chain, can be ascertained from the bioconcentration factor (see Section 12-3.4). This is a biochemical measurement, and at the level of testing under discussion it is adequate to calculate or measure the octanol-water partition coefficient. Good correlation has been obtained between this value and the tendency of a substance to bioconcentrate (Neely et al., Ref. 12-5). Similar considerations apply to the use of the soil-water partition in determining the likelihood of a substance being retained by or leaching through soil.

The information on a test chemical obtained up to and including Level II serves as a useful guide to what further testing may be needed. Alternatively, for chemical intermediates and low-volume consumer chemicals with low exposure and limited discharge potential, there may be sufficient information at this level to make an adequate risk assessment.

12-3.4 Level III: Subacute Exposures and Intermediate Environmental Testing

The so-called subacute feeding or exposure has been used as the basis of Level III testing, together with environmental studies of a similar level of complexity (Table 12-9). A subacute study usually involves a 90-day exposure of an experimental species at three dose levels with a control. The rat is the most frequently animal chosen. The study is usually accompanied by hematology and clinical chemistry measurements on all animals. Gross and histologic pathology of selected tissues and organs of some animals is usual. The feeding study may be replaced by an inhalation study, depending on the properties of the chemical and the type of exposure envisaged in manufacture or use. This level of testing is also appropriate for a reproductive performance study. This could take the form of a one-generation reproduction study in rodents accompanied by a teratology study. Descriptions of such tests are given in publications by the World Health Organization (Ref. 12-10) and in references for Level IB health testing.

Preliminary pharmacokinetic data may be useful in this level of testing. Thus, it may be possible to obtain the half-life of the test chemical in rodents, as well

**TABLE 12-9. Level III: Subacute Exposures and
Intermediate Environmental Testing.**

A. Health Effects
  Ninety-day rodent feeding or inhalation study
  Fertility studies
  Teratology
  $T\frac{1}{2}$ in rodents; metabolites

B. Environmental Effects
  Nitrification inhibition
  Algal toxicity
  Fourteen to 21-day biodegradation
  Short term simulated fate study
  Egg-fry (larval) fish studies
  Bioconcentration factor

as readily identifiable metabolites, from analysis of serum from animals in feeding or inhalation studies (Food Safety Council Ref. 12-7). The tests for health effects in Level III may thus provide us with information on the effects of a chemical following the prolonged exposure of a growing animal, its effects on fertility, and preliminary metabolic information.

The environmental effects selected for Level III provide more information on the compatibility of a chemical with secondary waste treatment, its fate and persistence, and its ability to bioconcentrate. This level extends the range of aquatic toxicity testing to other life forms.

Nitrification Inhibition. Important constituent species of secondary waste treatment microorganisms are the so-called nitrifying bacteria, which convert ammonia and amines to nitrate. Measurement of this activity is an additional index of compatibility of the test chemical with secondary waste treatment (Hockenburg and Grady, Ref. 12-11).

Fourteen to Twenty-One Day Biodegradation. A more complete account of the biodegradability of a test chemical can be obtained with 14 to 21-day incubation periods, by using mixed or pure microbial cultures, and by estimating major chemical breakdown products (Pitter, Ref. 12-12).

Short-Term Simulated Environmental Fate Studies. A useful estimate of the persistence or degradation of a chemical in a receiving body of water can be obtained, using the volume of water in a large open container as a simulated natural environment. A suitable container would be a 55-gallon drum (Robillard and Lockhart, Ref. 12-13). In effect, the study is a combined biodegradation and photodegradation study, since the drum is exposed to a variety of environmental conditions. Analytical measurements are performed to determine disappearance rates for the compound and for its readily identifiable metabolites.

Bioconcentration Factor. This is defined as the ratio of the uptake rate of a chemical to its depuration rate in selected aquatic species, under equilibrium conditions. It is thus the measure of a chemical's capacity for bioconcentration, and can be compared with values for other chemicals tested under comparable conditions (Neely *et al.*, Ref. 12-5).

Algal Toxicity. As a further index of aquatic toxicity, selected algal species may be studied for the effect of a test chemical on growth or metabolic activity. Useful parameters are the measurement of biomass, i. e., growth and development, and of adenosine triphosphate formation i. e., viability (Patterson *et al.*, Ref. 12-14).

Egg-Fry Larval Fish Toxicity. A useful guide to the effects of prolonged exposure to a chemical on aquatic species may be obtained from studies on the response of egg to fry growth in fish to a test chemical (McKim, Ref. 12-15).

12-3.5 Level IV: Long-Term Health and Environmental Tests

The final level of testing includes biological and environmental tests that are recommended for evaluation of food additives and pesticides (Table 12-10). They are clearly more appropriate for chemicals that are widely disseminated as a result of their use, and to which the general population or considerable segments of the environment are likely to be exposed on a continuous basis. Health effects testing, as described in Refs. 12-6, 12-7, and 12-10, may include lifetime feedings and oncogenicity studies in one or more rodent species. Multigeneration reproduc-

TABLE 12-10. Level IV: Long-Term Health and Environmental Effects.

A. Health Effects
  Two-year rodent study
  Three-generation reproductive performance
  Teratology
  Pharmacokinetics in test and target species

B. Environmental Effects
  Biodegration products identification
  Soil interaction tests
  Long term aquatic studies

---

tive performance tests may be included, together with mutagenicity and terato-genicity studies. Biochemical information may include pharmacokinetics in both experimental animals and humans, and elucidation of metabolic profiles. Studies on environmental effects may include the elucidation of the breakdown products of the chemical in the environment. There may be some concern that biodegradation should not give rise to materials of increased or altered potential for harm. It may be necessary to perform lifetime feeding studies in selected aquatic organisms, as well as to assess the stability of the test chemical in contact with soils and sediments.

It is clear that tests selected from Level IV may require a substantial commitment of resources and time. A sequential scheme of the type described in this article provides information that could be of great assistance in deciding whether to perform Level IV tests. It would be unrealistic to embark on Level IV testing without a reasonable certainty beforehand that such testing would only reveal effects of little consequence. Level IV testing is presumably most appropriate when the use of a material in commerce is so widespread that there is a need to be certain that such use is not a matter of concern to health or the environment, or that such testing will enable conditions of use to be properly defined.

12-4 DECISION MAKING

This scheme divides the process of hazard assessment into two parts, i. e., tabulation and organization of information and data, and hierarchical ordering of tests for environmental and health effects. The key requirement is, therefore, the relation between the numerical scoring of the data and the four levels of testing. The following relationship between scores and levels of testing has been reached by a process of trial and error.

Level of Testing	Health or Environmental Score
I	Less than or equal to 9
II	10-13
III	14-17
IV	Equal to or greater than 18

The health or environmental score may be used independently to arrive at a level of testing. Thus it is possible to have a health score of between 10 and 13 and an environmental score of greater than 18. Under such circumstances the level of testing for health effects would be II, and for environmental effects would be IV.

The apparent discrepancy in boundaries for each level of testing, e. g., that for Level I is 9, and the lower level for Level II is 10, is deliberate. The discrepancies were introduced to permit judgment in selecting the test level, since in fact there may not be clearcut distinctions between levels of testing. Assignment to a level by a score does not imply that all the tests in that level would be required, but that they should be selected as appropriate to the use or discharge of the chemical in question.

The scheme is employed for hazard assessment in the following manner:

1. Assemble information.
2. Tabulate and rate information using the worksheet.
3. Obtain scores for health or environmental categories by summation of ratings.
4. Apply scores to obtain testing levels for health and environment.
5. Select and perform appropriate test and evaluate the results.
6. Tabulate results of testing and obtain revised ratings.
7. Obtain new scores.
8. Reassess level of testing.

The flowchart (Fig. 12-2) indicates a suggested use of the scheme for premanufacturing notification under Section 5 of the Toxic Substances Control Act. For example, steps 1 and 2 may place the chemical in Level II. Examination of Level II may indicate that information from certain tests may be needed to complete the assessment. If on completing the tests, the score still remains within the boundaries for Level II, a judgment may be made that the information on health and environmental effects would be adequate for premanufacturing notification purposes. Alternatively, the results of testing may take the new score outside the upper boundary of Level II, and in that event, additional testing may be needed to be selected, this time from Level III. The flowchart can be used independently for health or environmental scoring.

12-5 COMMENTS

Experience with a very wide range of industrial chemicals has shown that a scheme of this type possesses a number of useful attributes. It is first and foremost an organized and systematic way to look at exposure and effects data and information. In addition, it provides a quantitative assessment of the criteria involved, which is not encumbered by complicated systems of weighting for various factors.

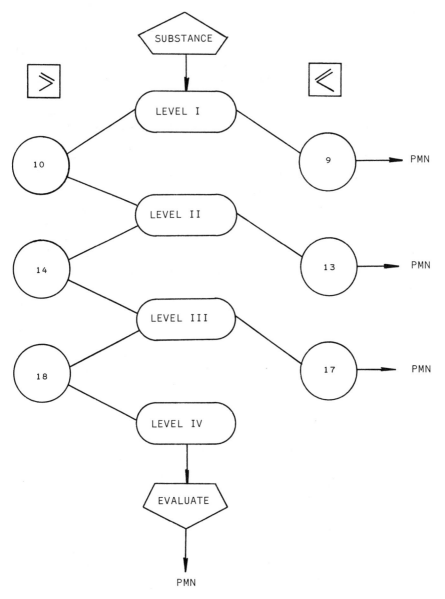

Fig. 12-2. Decision making: use of test scores to ascertain the appropriate level of testing, e.g., for premanufacturing notification (PMN).

In this sense a score for the health or environment is readily broken down into identifiable components.

The scheme is readily adaptable to information retrieval systems and, correspondingly, modifications or changes in criteria ratings are easily interpolated. The scheme relates the type and scope of testing to conditions of use, exposure, and disposal, and permits flexibility in judgment in the selection of testing. While the scheme permits the separation of health and environmental testing, it requires that health and environmental concerns be examined at the same time. The scheme is a valuable adjunct in integrating testing requirements with research and development needs because it permits early identification of testing needs. It may be of particular value in both cost and time control, as it can be used to prevent unnecessary or inappropriate testing.

In permitting early assessments of toxicity the scheme is valuable for chemicals under investigation for research and development. Chemicals that may pose a high degree of potential hazard may be identified early, allowing for the selection of feasible substitutes. As indicated, the scheme may be used to respond to changes in production volume or discharge pattern, so that reassessment of testing levels under changed conditions can be made. It also distinguishes between unisolated intermediates, isolated intermediates, final compounds, and manufacturing processes in the selection of appropriate tests.

REFERENCES

12-1. Gleason, M. N., Gosselin, R. E., Hodge, H. C., and Smith, R. P., *Clinical Toxicology of Commercial Products,* 3rd ed., Williams and Wilkins, Baltimore, 1969.

12-2. Federal Hazardous Substances Act, 1967, SEC. 2 (h) (1) (a).

12-3. Lockhart, H. B., Jr., Blakely, R. V., Cunningham, S. L., and Astill, B. D., "The Compatibility of Process Chemicals and Effluents with Secondary Waste Treatment as Determined by Radiorespirometry," 51st Annual Waste Pollution Federal Conference. Anaheim, California, 1978, unpublished.

12-4. Tehaar, C. J., Ewell, W. S., Dziuba, S. P., and Fassett, D. W., "Toxicity of Photographic Processing Effluents to Fish," *Photographic Sci, and Engineering,* 16, pp. 370-376, 1972.

12-5. Neely, W. B., Branson, D. R., and Blou, G. E., "Partition Coefficients to Measure Bioconcentration of Organic Chemicals in Fish," *Environ. Sci. Technol.,* 8, pp. 1113-1115, 1974.

12-6. National Academy of Sciences, *Principles and Procedures for Evaluating Household Substances,* Committee for the revision of NAS Publication 1138, Committee on Toxicology National Research Council, Washington, D. C., 1977.

12-7. Food Safety Council Scientific Committee, "Proposed System for Food Safety Assessment," *Fd. and Cosmet. Toxicol.* 16 Suppl. 2, pp. 29-108, 1978.

12-8.   Taras, Michael T., Greenberg, Arnold E., Hoak, R. D., and Rand, M. C., *Standard Methods for the Examination of Water and Waste Water.* Washington, D. C., American Public Health Association, 1970.

12-9.   Lockhart, H. B., Jr. and Blakely, R. V., "Aerobic Photodegradation of Fe (III) Ethylenedinitrilotetraacetrate, Ferric EDTA," *Env. Sci. and Technol.* **9**, pp. 1035-1038, 1975.

12-10. World Health Organization, *Principles and Methods for Evaluating the Toxicity of Chemicals,* Part I, Geneva, 1978.

12-11. Hockenburg, M. R. and Grady, C. P. L., Jr., "Inhibition of Nitrification Effects of Selected Organic Compounds." *J. Water Pollut. Cont. Fed.,* **49**, pp. 768-777, 1977.

12-12. Pitter, P., "Determination of Biological Degradability of Organic Substances," *Water Res.,* **10**, pp. 231-235, 1976.

12-13. Robillard, K. R., and Lockhart, H. B., Jr., Unpublished Data, Health, Safety and Human Factors Laboratory, Eastman Kodak Company, Rochester, New York.

12-14. Patterson, J. W., Bezonils, P. L., and Putnam, M. D., "Measurement and Significance of Adenosine Triphosphate in Activated Sludge," *Environ. Sci, Technol.,* **4**, pp. 569-575, 1979.

12-15. McKim, J. M., "Evaluating of Tests With Early Life Stages of Fish for Predicting Long-Term Toxicity," *J. Fisheries Res. Board of Canada,* **34**, pp. 1148-1154, 1977.

13

Chemical pollution dossiers for environmental decisions*

Stephen L. Brown, Ph. D., Director

Center for Resource and Environmental Systems Studies
SRI International, Menlo Park, California

13-1 PURPOSE AND SCOPE OF THE DOSSIER PROGRAM

13-1.1 Background and Objectives

For several years, the Central Unit on Environmental Pollution of the United Kingdom Department of the Environment has been developing a concept for a network of data on environmentally significant chemicals (DESCNET). The concept includes (1) a directory function, which would allow a user to find sources of information on chemicals, (2) an alert function, which would provide a network of interested parties with news of chemical incidents, laboratory findings, environmental regulations, and so on, and (3) a national pollution

---

*The research for this case study was supported by the United Kingdom Department of the Environment, Central Unit on Environmental Pollution, 2 Marsham Street, London SWIP 3EB. The guidance of Dr. G. N. J. Port is gratefully acknowledged.

dossier function, which would gather together the information needed for environmental appraisal of specified pollutants in a systematic fashion.

Although the categories of information desired in a National Pollution Dossier had been laid out previously to some extent in terms of a "CHEMORECORD," until recently little attention has been given to the question of how much information is needed at the different stages of appraisal versus the cost of acquiring and displaying the different types of information. Furthermore, the order and format of information display that would most effectively support environmental decisions had not been definitely specified.

Consequently, a need was evident for designing and evaluating alternative dossier formats as the first step to implementing the dossier feature of the DESCNET system. SRI International (formerly Stanford Research Institute) was asked to contribute to the design and evaluation exercise.

The objectives of the dossier program were the following:

- To construct prototype dossier formats for discussion, review, and comments.
- To refine the dossier format and define an assessment procedure that could operate on the data in the dossiers.
- To design and test procedures for gathering information on chemicals for the dossiers, with a summary of the sources of information and their use.

13-1.2 Environmental Protection Against Chemical Pollution in the United Kingdom

It is difficult to steer a path between too much and too little government regulation of chemicals in the environment. On one hand, too much attention to chemicals not only is wasteful of government resources, but also tends to place unduly severe economic penalties on the producers and users of chemicals, especially in situations where the assimilative capacity of receiving environments is large enough to reduce risks to an acceptable level. On the other hand, chemical risks to man and his environment must be kept within bounds, and particular care must be given to ensuring that insidious long-term effects do not eventually lead to consequences of disastrous proportions, as are feared by some experts with respect to ozone depletion or polychlorinated biphenyls (PCBs) in the environment.

To carry out its responsibilities, the Department of the Environment needs to undertake a number of activities, ranging from first recognizing a chemical pollutant as a potential environmental concern to arranging the eventual legal or voluntary arrangement that controls pollution of the environment by the chemical. Those activities are almost totally the responsibility of the Department of the Environment in the early stages, but gradually require more and more the contributions of and concurrence by other government ministries and eventually by industry, labor, and the general public.

The following are some of the most important activities:

- *Monitoring* of all indicators that a chemical pollutant may be of environmental concern. Significant indicators include reports of incidents linked to the pollutant, alerts by the scientific community of new findings, notification by industry of new chemical uses, or rising concentrations in environmental media.
- *Preliminary assessment* of chemical pollutants to determine those most likely to be of environmental significance. This activity includes both a systematic aspect, in which candidate chemicals are assessed for potential hazards, and a reactive aspect, in which known environmental problems advance the candidacy of a chemical.
- *Designation* of a chemical pollutant as a significant environmental issue. As the step beyond the assessment process, designation constitutes formal recognition that the chemical may be worthy of governmental attention, but does not prejudge the level of concern or actions to be taken.
- *Systematic appraisal* of the hazards of chemical pollutants that are environmental issues. Here, most of the easily available information about the pollutant is appraised to determine whether there is good reason for the government to take action.
- *Decision* on governmental regulatory actions. After a decision to act has been made, the nature of the response is developed by a thorough examination of all relevant information on the chemical's behavior in the environment.
- *Implementation* of regulatory intent. Several activities are embraced by this rubric, including the following:
  Monitoring – determining the extent of chemical pollution and of the effects due to it.
  Enforcement – determining if regulatory guidelines are being met and, if not, taking appropriate sanctions against violators.
  Emergency response – developing and carrying out plans to deal with incidents (beyond routine discharges) related to the chemical pollutant.

Although the activities have been listed roughly chronologically (with respect to a given chemical pollutant), there are connections among the activities that deviate from this linear development pattern. The more important linkages are diagrammed in Fig. 13-1. Notice that chemicals that are not assessed or appraised as being of highest concern during one cycle of assessment and appraisal can become so at some succeeding cycle. Notice also that monitoring can begin when a pollutant becomes an environmental issue, and that such information feeds the

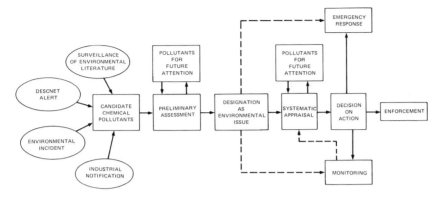

Fig. 13-1. Activities on chemical pollutants.

assessment process. Moreover, monitoring also provides a feedback loop to the decision on action, which can be modified if it proves to be ineffective.

13-1.3 Information Needs

Each of the above activities is either supported by or generates information on the chemical pollutants. As the intensity and importance of the decision activities increase as time passes, there is needed an increasing amount of information in order to support further decisions. However, because resources are limited, not every pollutant can proceed through all the stages of activity, and only a few can be afforded the luxury of a full development of information. Thus, there is a hierarchy of information packages, as shown in Figure 13-2:

- The list of candidate chemicals serves simply as a vehicle for storing very limited information related to identifying the chemicals that come to the attention of the Department of the Environment.
- The information file is a mechanism for passively receiving information about chemicals. A chemical file is established whenever a pollutant is determined to be of potential environmental concern.
- When an information file contains a significant cross section of information, or when an incident or alert draws attention to a chemical, a chemical dossier is prepared; this dossier contains a basic set of information on the chemical (as available) in a relatively standardized format.
- Periodically, the set of chemical dossiers undergo a preliminary assessment to determine pollutants to be designated as environmental issues. For these, a more comprehenisve national pollution dossier is prepared, also in a reasonably standardized pattern.
- The national pollution dossiers are then periodically subjected to a system- atic appraisal that identifies those chemicals requiring governmental regula-

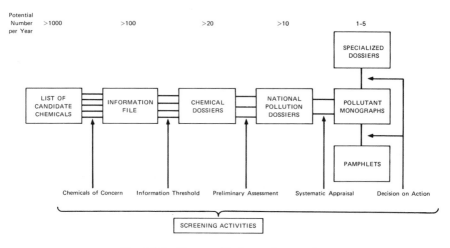

Fig. 13-2. Stages of information needs.

tory action. To support the action, a complete pollutant monograph is prepared, which contains all available information relevant to the decision.

• When the actions have been taken, various specialized dossiers and pamphlets may be produced to support a variety of implementation activities, such as a handbook of chemical analysis techniques for monitoring purposes or a guide to therapeutics for emergency response.

13-2 DESIGN OF THE DOSSIER HIERACHY

13-2.1 Considerations in the Design

The term *dossier* refers in this section to any of the information steps described in the preceding section. In designing a dossier format, the most important consideration is the use to which the dossier will be put. Although in some cases dossiers may serve several purposes, in every case their utility depends on two key properties:

• The information contained in the dossiers must be sufficient for the purposes for which they are designed.
• The information must be presented in a way that will facilitate their use by the intended audience.

For brevity, these two properties will be called *content* and *format,* respectively.

Design that examines purpose exclusively, however, will fail because it does not give due consideration to the availability of information to meet the design and to the costs and other constraints that limit both content and format. Furthermore, harmony with other attempts to design similar information systems is a benefit that must be considered. The relationships between criteria for content and format, availability of information and tools for its presentation, and the desire for harmonization are displayed in Fig. 13-3.

The three considerations of possible content, criteria for desirable content, and available information are shown as determining the selection of a set of information items to be included in the dossier. Similarly, questions of format are combined to set the tone for the order and form of information presentation. Subsidiary considerations in design include provisions for correcting and updating the information in the dossiers to keep it as current as needed, and procedures for qualifying and referencing the material so that it can be used with confidence. Together with an important, but not overriding, consideration of harmonization, these selections complete the dossier design.

13-2.2 Availability of Information

For dossiers to support environmental decisions about chemical pollutants, an excellent listing of potentially useful information items is contained in the

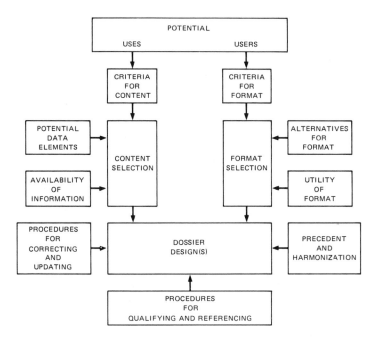

Fig. 13-3. Considerations in design of National Pollution Dossiers.

CHEMORECORD. They have been classified into ten categories, each addressing a facet of environmental significance. However, there is considerable overlap among categories, and the assignment of a specific data type to a category may be largely arbitrary. For example, statistics on exports and imports are classified under "Handling, Transportation, and Disposal" because they require transportation and handling to leave or enter the country, but they also are closely related to statistics under the category of "Production and Use." The ten categories are the following:

    I. Compound identification.
   II. Chemical structure.
  III. Physical and chemical properties.
  IV. Chemical analysis data and methods.
   V. Production and use.
  VI. Handling, transportation, and disposal.
 VII. Environmental involvement.
VIII. Toxicology.
  IX. Treatment.
   X. Environmental protection standards and legal requirements.

To determine the availability (and to some extent the utility) of information in these categories, the items of the CHEMORECORD were compared with information types in a number of chemical information projects carried out by SRI and other expert groups. The analysis indicated that many of the CHEMORECORD items are either regularly sought or generally available, and are thus true candidates for the dossier design. The following were exceptions: elaborate chemical structure data, which are neither easily available nor particularly relevant to environmental decisions; chemical analysis methods, which are not needed for the earlier stages of pollutant appraisal; details of production and use, which are difficult to obtain; environmental involvement, which, though valuable, is not easily generated; and treatment, which is valuable only in the later stages of dossier development. The information items that appeared to be generally available *or* exceedingly useful were retained as candidates for the dossier design.

13-2.3 Levels of Dossier Development

Only limited resources can be applied to develop dossiers at each level of detail. Accordingly, the information types for each level are limited by the resources necessary to search for and present the specific information for a given pollutant. Estimates of time required to retrieve information items in the CHEMORECORD were developed and used to help determine at which level of dossier each item should be added. In doing so, groups of items were not subdivided when information on all items is most efficiently gathered simultaneously.

Descriptions of the six levels of development are presented in Table 13-1 and include sections on content, selection criteria, storage and retrieval, and estimates of the number and effort required on an annual basis. The latter two estimates assumed that the number of pollutants treated had to decline sharply as the information requirements rose in the more advanced stages. Overall, more effort was assigned to those stages than to the earlier ones. Effort was estimated both for additions to the files and for maintenance of existing files. Although additions might be expected to decrease with time, the effort for maintenance increases with the total number of chemicals in each file. The total effort devoted to the dossier program should grow from about 7000 hr per year initially to perhaps as much as 20,000 hr per year for long-term maintenance.

13-3 ASSESSMENT FOR SETTING ENVIRONMENTAL PRIORITIES

13-3.1 Rationale for Assessment

As indicated earlier, one of the most difficult problems in attempting to deal with chemicals of environmental significance is selecting which few chemicals to treat with the limited resources available. These screening decisions begin with the selection of chemicals to be listed and are still necessary when final regulatory actions are being contemplated.

Obviously, the most environmentally important pollutants should be treated first, but the definition of environmental significance is far from obvious. Should one concentrate on the pollutants that have been demonstrated to cause acute effects in humans (sulfur dioxide, mercury, carbon monoxide)? Should one be on the lookout for potentially insidious, long-term problems (vinyl chloride, DDT, PCBs, chlorofluorocarbons)? Should highly visible problems (dioxins at Seveso) be given the most attention? There are no easy answers, and none are supplied here.

Consequently, the emphasis here is on preliminary assessment of candidate chemicals for designation as environmental issues (see Figs. 13-1 and 13-2). At that stage of screening, it is necessary to select among (say) 25 candidates a year to nominate (say) 10 for systematic appraisal. Both subjective and objective elements are involved. The subjective part includes selection of those chemicals for which a combination of scientific and political opinion holds that thorough appraisal is necessary. The objective element is designed to operate on the data contained in the chemical dossiers in order to construct an index of environmental significance upon which the chemicals can be ranked for further attention. If both the subjective and objective methods are working well, there should be a high degree of correlation between the two sets of priorties. However, a good objective system should occasionally point out the potential significance of a chemical that has not come to the attention of decision makers; conversely, the

TABLE 13-1. Levels of Dossier Development

Level 1: List of Candidate Chemicals
Content
  Chemical Abstracts System (CAS) No. (if available)
  Systematic Name (if available)
  Trivial Name (if different – one only)
  Code for reason on list
  Status of information development (level reached)
Selection
  Passive – brought to attention through incident, alert, notification, and the like.
  Active – systematic addition of chemicals in high usage, of high toxicity, with high con-
    centration in the environment, and the like.
Storage and retrieval
  Computerized; searchable on all categories shown under "Content"; hard copy generated
    quarterly with several sortings.
Annual Additions
  1,000 to start, declining to a few hundred.
Effort
  Additions – 1 hr per chemical.
  Maintenance – 0.1 hr per chemical per year.

Level 2: Information File
Content
  Identification (from Level 1).
  Photocopies from standard sources.
  Articles, newspaper clippings, excerpts, as available.
  Contractor supplied information(?)
Selection
  Any suggestion of significant environmental hazards. Content includes passive receipt of
    material marked for file plus active search in standard sources and, possibly, contractor
    support.
Storage and Retrieval
  File cabinets, alphabetical by trivial name. Retrieval assisted by quarterly Level 1 lists.
    Possible subcategories: Production and Use; Environmental Involvement; Toxicology;
    Control; Miscellaneous. File by date.
Annual Additions
  500 to start, dwindling to under 100.
Effort
  Additions – 1 hr per chemical, average.
  Maintenance – 2 hr per chemical per year.

Level 3: Chemical Dossiers
Content
  Identification (number, name, synonyms, formula, structure).
  Structure (structural formula).
  Properties (molecular weight, melting/boiling points, vapor pressures, solubilities, density).
  Production and use (production quantities, exports and imports, use pattern).
  Handling, etc. (flammability, explosiveness, corrosiveness, disposal techniques, known
    accidents).

(Continued)

TABLE 13-1. (Continued.)

Environmental Involvement (environmental flow chart, degradation half-lives, release rates).

Toxicology (LD_{50}'s, carcinogenesis, mutagenesis, clinical effects).

Standards, etc. (principal known guidelines, criteria, standards).

Special considerations specific to the chemical.

Selection

Information files reaching an initial size (say 50 sheets with some in every category).

Chemicals of concern.

Systematic sampling of information file.

Storage and retrieval

Paper file appended to information file; emphasizes numerical information and supporting text. Computer storage of selected data possible.

Annual additions

25.

Effort

Additions — 40 hr per chemical.

Maintenance — 4 hr per chemical per year.

Level 4: National Pollution Dossiers

Content

Chemical dossier content.

Structure (line notations).

Properties (remainder).

Analysis (spectra, summary of methods).

Production and use (remainder).

Handling, etc. (remainder).

Environmental Involvement (remainder).

Toxicology (remainder).

Treatment (summary of therapeutics).

Standards, etc. (exhaustive search for guidelines, etc., presentation of limited assessments).

Additional special considerations specific to the pollutant.

Selection

Preliminary assessment of chemical dossier list.

Pollutants of concern to DOE.

Storage and Retrieval

Paper file appended to information file; more elaborate text, selected graphical displays.

Annual additions:

10.

Effort

Additions — 120 hr per pollutant.

Maintenance — 8 hr per pollutant per year.

Level 5: Pollution Monograph

Content

Everything available of potential environmental significance, including all information on National Pollution Dossier. Organized according to characteristics of pollutant under consideration, but including, at a minimum, sections on identification, production and

(Continued)

TABLE 13-1. (Continued.)

use, environmental involvement, toxicology and other effects, therapeutics and other controls, transport and handling, and overall assessment.

Selection
  Systematic appraisal of National Pollution Dossier list.
Storage and Retrieval
  Published volumes; emphasize text with supporting tabular and graphical material.
Annual Additions:
  1 to 3.
Effort
  Additions – 1,000-2,500 hours per monograph.
  Maintenance – 40 hours per monograph per year.

Level 6: Specialized Dossiers and Pamphlets
Content:
  As required for specialized purpose. For example, a dossier might focus on analytic techniques for use by monitoring agencies, or a pamphlet might describe the dangers of improper disposal of certain materials.
Selection:
  Needs revealed by monograph and decisions on action.
Storage and retrieval
  Published documents.
Annual Additions
  1-5 (starting in second year).
Effort
  Additions – 100-500 hr per document.
  Maintenance – essentially none.

---

objective system should sometimes indicate that a subjective interpretation has resulted in excessive concern because a critical mitigating circumstance has been overlooked. For example, the key factor might be the rate at which the chemical degrades in real environments.

13-3.2 A Prototype System for Preliminary Assessment

The objective of the preliminary assessment is thus to examine systematically the data in chemical dossiers and to recommend selected chemicals (25%-40% of those examined) for designation as environmental issues. The selected chemicals would be subject to systematic appraisal of data in an expanded Natonal Pollution Dossier.

Chemicals (or classes thereof) are to be ranked on an index of environmental significance, and selections for further examinaton are to be made on the basis of the ranking. The index is designed to relate – at least qualitatively – to the degrees of risk to which humans and other species are exposed and to the total societal hazards posed by the chemical.

Because the Department of Environment has few resources for either dossier construction or assessment, the system emphasizes data that are comparatively easy to obtain, methods that are relatively simple, and provisions for operation even when data are missing. The system is only a prototype and should undergo considerable further development as experience is gained with its use. Until then, its outputs should be examined closely to see if they conform with scientific opinion, and if not, to identify the reasons.

The system's basic guiding principles are the following:

- The chemical pollutant must be capable of significantly damaging a sensitive organism (or, occasionally, property).
- A significant population of the sensitive organisms (man and other species) must be exposed to the chemical pollutant.
- The exposures of the sensitive populations must reach or surpass the levels required for significant damage. Large exposures can be inferred from the following:

  Evidence that chemicals are accumulating in organisms *or*
  Evidence that environmentally significant concentrations are found in the media of exposure (air, water, etc.) *or*
  Evidence that chemicals persist and should be transported to populations at risk, *and*
  Evidence that significant quantities of the chemical are being discharged into the environment

These principles are illustrated on Fig. 13-4, where it is shown how the various pieces of information combine to suggest the degree of hazard that may be attributable to the chemical pollutant. The shading on the hazard "meter" is intended to reflect the idea that there is a "danger zone" of estimated environmental hazard; when that zone is reached, the pollutant should be designated as an environmental issue.

In translating the guiding principles into a preliminary assessment procedure, one must picture a model of the entire pollution process. The model should recognize that most of the components are connected in a multiplicative way: if the release to the environment is trebled, then the damages (hazards) are probably also *at least* trebled. On the other hand, if persistence of the chemical in the environment is near zero, then so will be the hazard, even when toxicity is high or the quantities released are large. The multiplicative property is reflected (not very precisely) in the prototype system by addition of the logarithms of many of the quantitatively expressed pieces of information.

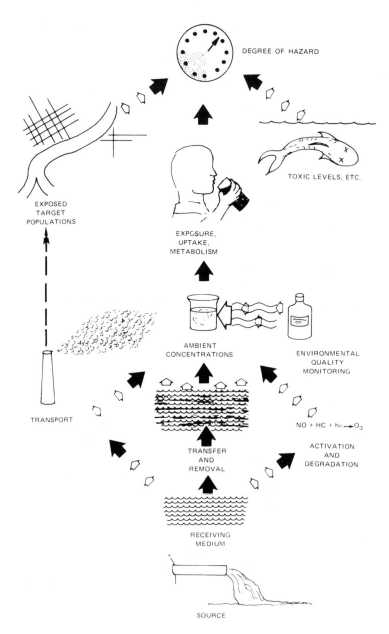

DEGREE OF HAZARD

TOXIC LEVELS, ETC.

EXPOSED
TARGET
POPULATIONS

EXPOSURE,
UPTAKE,
METABOLISM

AMBIENT
CONCENTRATIONS

ENVIRONMENTAL
QUALITY
MONITORING

TRANSPORT

$NO + HC + h\nu \longrightarrow O_3$

ACTIVATION
AND
DEGRADATION

TRANSFER
AND
REMOVAL

RECEIVING
MEDIUM

SOURCE

Fig. 13-4. Assessment information needs.

13-3.3 Procedures for Preliminary Assessment

The system for preliminary assessment always assigns values to the following indices:

A. Consumption in the United Kingdom: Add imports to production and subtract exports (metric tons per year). If data are unavailable, exports and imports are assumed negligible, production is assumed equal to 100. Index is the logarithm (base 10) of consumption, to two significant figures.

B. Release factor: Estimate fraction used in closed systems or for production of intermediates. Subtract from 1.0. If data are unavailable, use release factor of 0.1 (default value). Release factors below 0.001 are probably unduly optimistic. Take logarithm (base 10) of release factor. This number will not be positive.

C. Transport: Select an index value from the table below, adjusting for any special conditions. Default value is 1.

Populations Are	Air	Transport Through Water	Groundwater
Nearby (~1 km)	2	2	0
Intermediate (~10 km)	1	1.5	-2
Distant (~100 km)	0	1	-5

D. Transfer and removal: Determine whether transfer to another medium (e.g., by evaporation) or removal (e.g., by sedimentation) is likely to be rapid (order of days), moderate (order of weeks, months), or slow (order of years). Assign indices of -2, -1, 0, respectively. Default value is -1.

E. Degradation and activation: If chemical degradation is rapid, moderate, or slow, assign -2, -1, or 0, respectively. If activation of compound is possible, add 1 to this index. Default value is –1.

F. Ambient concentrations: Determine typical concentrations in mg/liter for drinking water or $\mu g/m^3$ for urban air. Take logarithm (base 10) and add 6. Add 1 for suggestion of significant levels in biological tissues. Compare with sum of A through E and use larger quantity in subsequent calculations. If no concentrations are available, use total of A through E.

G. Exposed populations: Estimate the number of humans exposed to the pollutant. Take logarithm (base 10). If any nonhuman population is thought to be significantly exposed, add 1. Default value is 100,000 (index = 5).

H. Exposure, uptake, metabolism: If there are any mitigating circumstances on exposure (no absorption through gastrointestinal tract; rapid detoxification in body), set index to -1. Otherwise 0.

I. Toxic levels: Take logarithms (base 10) of LD_{50}s in mg/kg body weight. Divide by 5 for nonhuman species. Add these together and then make result negative. Add 2 for positive carcinogenicity; add 1 for positive mutagenic assay if no carcinogenic data are indicated. Add 0.2 for each nonfatal effect of importance. Default value is 0.

J. Controllability: Use index of -2 if already well controlled, -1 if partly controlled, 0 if essentially uncontrolled. Default is 0.

K. Other considerations: Use small positive or negative indices for additional factors like potential for synergistic action, irreversibility, transportation accident potential, etc.

The indices A through E, or F, and G through I are combined additively; chemical pollutants are then ranked on the total index value.

Only limited testing of the system has been conducted, using carbon tetrachloride, chromium, and nitrates as test chemicals. A sample dossier for carbon tetrachloride, utilizing a graphic presentation format, is shown on Fig. 13-5. The test revealed that although dossier data were objective, a subjective interpretation of the data was necessary for preparing the rankings. That subjectivity appears to be necessary to perform the preliminary assessment within available resources. If the prototype scheme is to be used in practice, it should be undertaken under the following rules:

- At least three knowledgeable scientists should attempt ratings independently.
- The ratings should then be compared and argued in committee.
- The consensus should be biased toward the opinion of the rater with the greatest expertise on a particular index.
- The final rankings should again be reviewed for credibility before using the results for designating environmental issues.
- Procedures and default values should be revised as experience accumulates.

If batches of ten or more chemicals are examined at once, the total professional effort expended per chemical should average less than 8 hr, presuming the chemical dossiers have already been prepared.

13-4 ACQUISITION OF INFORMATION FOR THE DOSSIERS

13-4.1 Two Aspects of Information Retrieval

The discussion of information and assessment activities would be incomplete without a concrete guide to information acquisition. The following sections

(Text continued on p. 453)

SAMPLE DOSSIER—GRAPHICAL STYLE

NUMBER: 56235   NAME: CARBON TETRACHLORIDE

IDENTIFICATION AND STRUCTURE

    CAS No: 56-23-5   CAS NAME: TETRACHLOROMETHANE

    COMMON NAME(S): CARBON TETRACHLORIDE   TRADE NAME(S): FREON 10

    MOLECULAR FORMULA: CCl_4 STRUCTURAL FORMULA:

$$Cl - \underset{\underset{Cl}{|}}{\overset{\overset{Cl}{|}}{C}} - Cl$$

    DESCRIPTION: COLOURLESS; HEAVY, ETHEREAL ODOR

PHYSICAL AND CHEMICAL PROPERTIES

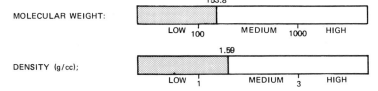

MOLECULAR WEIGHT:

DENSITY (g/cc);

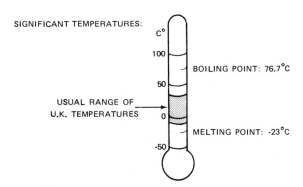

SIGNIFICANT TEMPERATURES:

BOILING POINT: 76.7°C

USUAL RANGE OF U.K. TEMPERATURES

MELTING POINT: -23°C

VAPOUR PRESSURE:

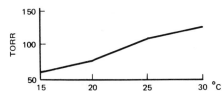

Fig. 13-5. Sample Dossier—Graphical Style.

(Continued)

NUMBER: 56235   NAME: CARBON TETRACHLORIDE

PRODUCTION AND USE

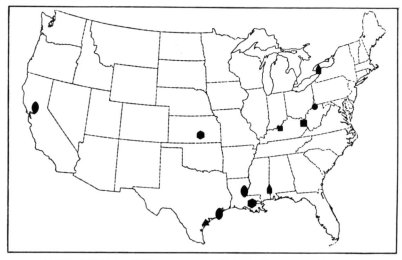

● ALLIED CHEMICAL CORP   ■ FMC CORP.
◖ DOW CHEMICAL   U.S.A.   ▲ STAUFFER CHEMICAL CO.
▲ EI DU PONT DE NEMOURS   ⬣ VULCAN MATERIALS CO.

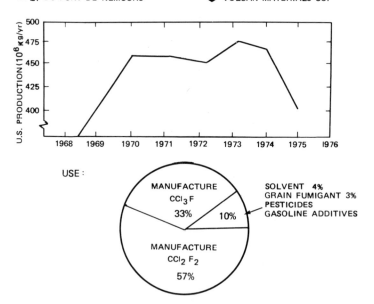

USE:

Fig. 13-5. (Continued)

NUMBER: 56235   NAME: CARBON TETRACHLORIDE

ENVIRONMENTAL INVOLVEMENT

PATHWAYS:

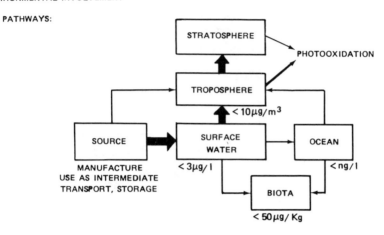

CONCENTRATION:

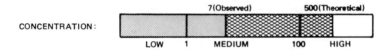

PARTITION COEFFICIENT

PERSISTENCE:

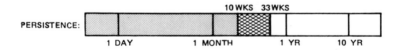

DEGRADATION: $CCl_4 + O_2 + h\gamma \longrightarrow CO_2 + 4Cl$

METABOLIC
BREAKDOWN: $CCl_4 \longrightarrow CHCl \dashrightarrow CO_2 + Cl^-$

CONTAMINANTS: $HCl, CS_2, Br_2, CHCl_3$

EPIDEMIOLOGY: LIVER, KIDNEY, LUNG AND HEART DAMAGE; POSSIBLE LIVER CANCER

Fig. 13-5. (Continued)

NUMBER: 56235   NAME: CARBON TETRACHLORIDE

TOXICOLOGY

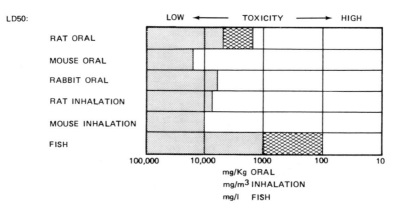

LD50:

IRRITATION/ALLERGY: NONE CITED

TERATOGENESIS: NONE CITED

MUTAGENESIS: NONE CITED

CARCINOGENESIS: CASE STUDIES OF HUMAN HEPATOMA ARE OF DOUBTFUL SIGNIFICANCE
                      LIVER CANCER IN EXPERIMENTAL ANIMALS HAS BEEN REPORTED

METABOLISM:

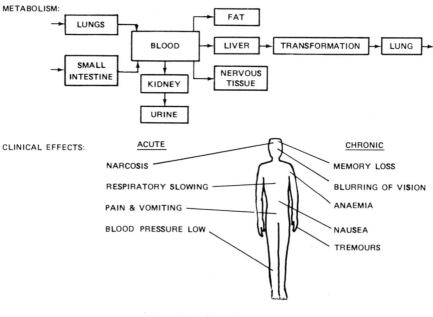

Fig. 13-5. (Continued)

NUMBER: 56235   NAME: CARBON TETRACHLORIDE

HANDLING TRANSPORTATION AND DISPOSAL

HAZARDS:

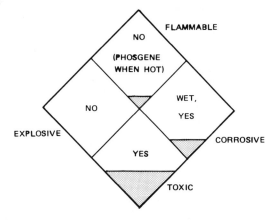

EXPORT/IMPORT:

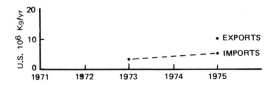

DISPOSAL OF WASTES:

SOLVENT RECLAMATION

LANDFILL OF RESIDUES AND SLUDGES

Fig. 13-5. (Continued)

are based primarily on SRI experience in gathering and processing information for other chemical information projects. It is essential that the following two aspects of information acquisition are treated: identification of the sources of information, and suggestions for information analysis and synthesis for dossier purposes. As with previous recommendations, the emphasis is on conserving scientific resources. The information sources are therefore principally secondary literature compilations rather than the primary literature, and the analysis procedures are as simple as can be justified for each category of information.

NUMBER: **56235**   NAME: CARBON TETRACHLORIDE

ENVIRONMENTAL PROTECTION STANDARDS AND LEGAL REQUIREMENTS

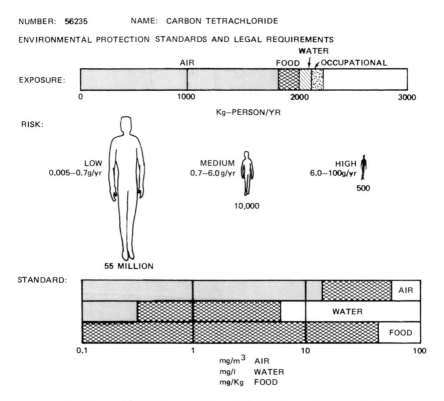

RISK:

STANDARD:

REBUTTABLE PRESUMPTION AGAINST REREGISTRATION AS PESTICIDE (U.S.) APRIL 1977

NO EMISSION/EFFLUENT STANDARDS

ENVIRONMENTAL CONTROL TECHNOLOGY:

    CLOSED SYSTEM DESIGN

    VENTILATION OF WORKPLACE

    STEAM-STRIPPING OF SLUDGE

    INCINERATION WITH ACID SCRUBBING

TRANSPORTATION:

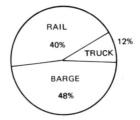

Fig. 13-5. (Concluded)

13-4.2 Sources of Information

A listing of published documents and standard information, mostly from the United States, is shown by CHEMORECORD category in Table 13-2. Both a short code and the complete citation are shown; the complete citation is omitted where the source is relevant to a second or succeeding category. Category V, Production and Use, is divided into six subcategories, as this type of information is much less likely to be compiled in single volumes.

TABLE 13-2. Published Documents and Standard Information Systems, by CHEMORECORD Category

I. *Chemical Identification*

Biosis	*Abstracts on Health Effects of Environmental Pollutants,* Biosciences Information Service, Philadelphia, Pennsylvania, 1975.
Chemline	*Chemline,* National Library of Medicine, Bethesda, MD.
Chem. Abs.	*Chemical Abstracts,* Chemical Abstracts Service, American Chemical Society, Columbus, OH, 1907-present (weekly publication with cumulative index).
DAT	*Desktop Analysis Tool for the Common Data Base,* Chemical Abstracts Service, American Chemical Society, 1968.
Product bulletins	Obtained from chemical producers.

II. *Chemical Structure*

Chemline	See above.
Merck	Windholz, M. (Ed.), *The Merck Index,* 9th ed., Merck and Company, Inc., Rahway, NJ, 1976.
Toxic Effects	Christensen, H. E., *Registry of Toxic Effects of Chemical Substances,* 1976 Edition, U. S. Department of Health, Education, and Welfare, Public Health Service, National Institute for Occupational Safety and Health, Rockville, MD, June 1976.
SOCMA	*SOCMA Handbook: Commercial Organic Chemical Names,* Chemical Abstracts Service, American Chemical Society, Washington, D. C., 1965.
PHS 149	*Survey of Compounds Which Have Been Tested for Carcinogenic Activity,* U. S. Department of Health, Education, and Welfare, National Institutes of Health, Bethesda, MD, 1951-1973.

III. *Physical and Chemical Properties*

CRC	*Handbook of Chemistry and Physics,* 51st ed., The Chemical Rubber Company, Cleveland, Ohio, 1970-1971.
Lange's	Lange, N. A., Handbook of Chemistry, 10th ed., McGraw-Hill Book Company, Inc., New York, 1961.
CCD	*Condensed Chemical Dictionary,* 8th ed., Van Nostrand Reinhold Publishing Company, New York, 1971.
Aldrich	*Aldrich Catalog/Handbook of Organic and Biochemicals,* Aldrich Chemical Company, Inc., Milwaukee, WI, 1974.
SAX	Sax, N. I., *Dangerous Properties of Industrial Materials,* 4th ed., Van Nostrand Reinhold Publishing Company, New York, 1975.

(Continued)

TABLE 13-2. (Continued.)

IV. *Chemical Analysis Data and Methods*

IARC — *IARC Monographs on the Evaluation of Carcinogenic Risk of Chemicals to Man.* Vols. 1-12, World Health Organization, International Agency for Research on Cancer, Lyon, France, 1971-1976.

Chem. Abs. — See above.

CRC Atlas — Grasselle, J. G. (Ed.), *Atlas of Spectral Data and Physical Constants for Organic Compounds,* The Chemical Rubber Company Press, Cleveland, OH, 1973.

Official Methods — Horwitz, W. (Ed.), *Official Methods of Analysis of the Association of Official Analytical Chemists,* Twelfth Edition, Association of Official Analytical Chemists, Washington, D. C., 1975.

V. *Production and Use*

A. *General*

SOC — *Synthetic Organic Chemicals, United States Production and Sales,* United States Trade Commission, U. S. Government Printing Office, Washington, D. C., 1968-1975.

CEH — *Chemical Economics Handbook,* Stanford Research Institute, Menlo Park, California (updated continuously).

IARC — See above.

Merck — See above.

K. O. — *Kirk-Othmer Encyclopedia of Chemical Technology,* 2nd ed., Vols. 1-22, Interscience Publishers, New York, 1963-1970.

F, K & C — Faith, W. L., Keyes, D. B., and Clark, R. L., *Industrial Chemicals,* John Wiley and Sons, Inc., New York, 1966

B. *Inorganic Chemicals*

Min. Year — *Minerals Yearbook,* Bureau of Mines, U. S. Department of the Interior, U. S. Government Printing Office, Washington, D. C. (published annually).

Min. Facts — *Mineral Facts and Problems,* U. S. Department of the Interior, Bureau of Mines, 1975.

C. *Pharmaceuticals*

Facts — Kastrup, E. K. (Ed.), *Facts and Comparisons,* Facts and Comparisons, Inc., St. Louis, MO, 1947-1977 (updated monthly).

Remingtons — Hoover, J. E. (Ed.), *Remington's Pharmaceutical Sciences,* 15th ed., Mack Publishing Company, Easton, PA, 1975.

Pharmacopeia — *The United States Pharmacopeia,* 19th Revision, United States Pharmacopeial Convention, Inc., Rockville, MD, 1974.

D. *Surface-Active Agents*

McCutcheon's — *McCutcheon's Detergents and Emulsifiers,* North American Edition, McCutcheon Division, MC Publishing Company, Glen Rock, NJ (published annually).

Encyclopedia of SA Agents — Sisley, J. P., *Encyclopedia of Surface-Active Agents,* Chemical Publishing Company, Inc., New York, Vol. 2, 1964.

E. *Flavors and Fragrances*

Furia — Furia, T. E., and Bellanca, N., *Fenaroli's Handbook of Flavor Ingredients,* The Chemical Rubber Company, Cleveland, OH, 1971.

TABLE 13-2. (Continued.)

Study of F. & F. Fragrance	Jones, M. C., *A Study of Flavors, Frangrances and Essential Oils,* Delphi Marketing Services, Inc., New York, 1976.
Monographs	"Monographs on Fragrance Raw Materials," in *Food and Cosmetics Toxicology,* Research Institute for Fragrance Materials, Inc., Pergamon Press, New York, (updated continuously).
Cosmetic Dict.	Estrin, N. F., *CTFA Cosmetic Ingredient Dictionary,* First Edition, The Cosmetic, Toiletry, and Fragrance Association, Inc., Washington, D. C., 1973.
Food Codex	*Food Chemicals Codex,* 2nd ed., Committee on Specifications, Food Chemicals Codex of the Committee on Food Protection, National Research Council, National Academy of Sciences, Washington, D. C., 1972.

F. Pesticides

Farm Chems	Berg, G. L., (Ed.), *Farm Chemicals Handbook, 1976,* Meister Publishing Company, Willoughby, OH, 1976.
Spencer	Spencer, E. Y., *Guide to the Chemicals Used in Crop Protection,* Publication 1093, Sixth Edition, Research Institute, University of Western Ontario, Ontario, Canada, 1973.
Chem. Week	Johnson, O., "Pesticides '72," *Chemical Week,* McGraw-Hill Publications, Inc., New York, June 21, 1972, Part I; July 26, 1972, Part II.
Herbicide Handbook	*Herbicide Handbook of the Weed Science Society of America,* Third Edition, Weed Science Society of America Herbicide Handbook Committee, Champaign, IL, 1974.
Compendium	*Environmental Protection Agency Compendium of Registered Pesticides,* Pesticide Regulation Division E238, Office of Pesticides Program, U. S. Environmental Protection Agency, Washington, D. C. (updated continuously).
Prodn. Processes	Sittig, Marshall, Pesticide Production Processes, Chemical Process Review No. 5, Noyes Development Corporation, Park Ridge, NJ, 1967.

VI. *Handling, Transportation and Disposal*

Toxic Effects	Christensen. H; E., *Registry of Toxic Effects of Chemical Substances,* U. S. Department of Health, Education, and Welfare, Public Health Service, National Institute for Occupational Safety and Health, Rockville, MD, June 1976.
Handbook of Poisoning	Dreisbach, R. H., *Handbook of Poisoning: Diagnosis Treatment,* Lange Medical Publications, Los Altos, CA, 1974.
SAX	See above.
Doc. of TLV's	*Documentation of Threshold Limit Values,* Third Printing, Committee on Threshold Limit Values, American Conference of Governmental Industrial Hygienists, Cincinnati, OH, 1976.
IARC	See above.
TADS	Environmental Protection Agency Data Base, *Oil and Hazardous Materials, Technical Assistance Data System* (guide to original literature).
Toxline	*Toxline,* National Library of Medicine, Rockville, MD, (guide to original literature).

TABLE 13-2. (Continued.)

Medline	*Medline,* National Library of Medicine, Rockville, MD (guide to original literature).

VII. *Environmental Involvement*

Toxline	See above.
Medline	See above.
IARC	See above.
Marine Pollutants	Windom, H. L., and Duce, R. A., *Marine Pollutant Transfer,* Lexington Books, D. C. Heath and Company, Lexington, Massachusetts, 1976.
Soil and Water	Guenzi, W. D. (Ed.), *Pesticides in Soil and Water,* Soil Science Society of America, Inc., Madison, WI, 1974.

VIII. *Toxicology*

SAX	See above.
Toxic Effects	See above.
PHS 149	See above.
IARC	See above.
Doc. of TLV's	See above.
Toxline	See above.
Medline	See above.
EMIC	*Environmental Mutagen Information Center,* Oak Ridge National Laboratory, Oak Ridge, Tennessee (guide to original literature).
ETIC	*Environmental Teratogen Information Center,* Oak Ridge National Laboratory, Oak Ridge, Tennessee (guide to original literature).
GHSG	Gosselin, Hodge, Smith, and Gleason, *Clinical Toxicology of Commercial Products,* Fourth Edition, The Williams and Wilkins Company, Baltimore, MD, 1976.

IX. *Treatment*

IARC	See above.
Handbook of Poisoning	See above.
SAX	See above.
Hazardous Materials	*Fire Protection Guide on Hazardous Materials,* 4th Ed., National Fire Protection Association, Boston, MA, 1972.
Desk Reference	*Physicians Desk Reference,* 30th ed., Medical Economics Company, Oradell, NJ, 1976.

X. *Environmental Protection Standards and Legal Requirements*

Doc. of TLVs	See above.
IARC	See above.
Food additives	Furia, *Handbook of Food Additives,* The Chemical Rubber Company, Cleveland, OH, 1968.
CFR	*U. S. Code of Federal Regulations,* Titles 1-50, Office of the Federal Register, Washington, D. C. (continuously updated).
ELC	Environmental Law Centre, International Union for Conservation of Nature and Natural Resources, Bonn, West Germany.

13-4.3 Techniques of Analysis

In analyzing chemical information acquired for any level of presentation in dossiers, the purposes for which the information is being compiled should be kept clearly in mind. The depth of analysis required for preliminary assessment is clearly less than for systematic appraisal, which in turn is less demanding than for governmental regulatory actions. Because the preliminary assessment procedure is relatively crude, there is little advantage to expending great effort in obtaining precise information at that stage. In this spirit, order of magnitude estimates are considerably better than no estimates at all.

However, care must be exercised to ensure that crude estimates for one purpose are not misinterpreted as authoritative for another. One way of dealing with that problem is to identify the quality of information along with each estimate. Such identification can be accomplished quantitatively (e.g., confidence limits) or qualitatively (codes or descriptive terms).

At the other extreme, attention to detail is very important in a pollutant monograph because it may be used as the basis for sensitive negotiations over voluntary arrangements or for litigation regarding environmental responsibility. Not only must the information be as accurate as possible, but it also must be very carefully qualified whenever the scientific evidence is less than conclusive.

13-5 POTENTIAL APPLICATION BY OTHER USERS

The chemical dossiers and priority assessment system described in the preceding sections were specifically designed to be used in the framework of the United Kingdom's environmental management program. However, modifications or extentions of the concepts should be applicable to the problems of industrial producers or users of chemicals as well as those of other governmental agencies concerned with chemical hazards, such as the Environmental Protection Agency in the United States.

Most organizations that must deal with environmental aspects of chemicals have limited resources to apply to their investigations. Thus, it is important to decide what information should be gathered, in what format, and for what chemicals. In making such decisions, the key is to recongize what items of information are important for evaluating chemical hazards. The dossier concept not only indicates the most useful data elements for such evaluations but shows how a screening procedure can be applied using a few elements to decide which chemicals should be subject to further data gathering and analysis. The screening criteria can be adjusted to the resources of the organization and the total number of chemicals with which it must deal.

Environmental regulatory agencies should be able to apply the dossier/ assessment system in the following manner:

- Set priorities for research and development on chemical hazards.
- Decide which chemicals to establish as candidates for regulations.
- Determine appropriate regulatory strategies.
- Monitor results of regulatory policies.

Industries dealing with chemicals could use the system in the following manner:

- Identify chemicals that might pose environmental hazards.
- Anticipate regulatory attention to the chemicals.
- Allocate resources for acquiring information about various chemicals.
- Determine appropriate environmental control strategies for selected chemicals.

The data sources presented in Table 13-2 are generally applicable to the assessment of chemicals in the environment and should serve as useful references for both classes of potential users.

14

Models for predicting bioaccumulation and ecosystem effects of Kepone and other materials

Lowell H. Bahner
Statistical Consultants, Inc.
198 East Nine Mile Road
Pensacola, Florida

Jerry L. Oglesby
Faculty of Mathematics and Statistics
University of West Florida
Pensacola, Florida

14-1 PURPOSE AND SCOPE OF MODEL STUDY

Procedures for testing aquatic organisms with toxic chemicals often require that discrete testing and sampling schemes be chosen to represent continuous variables such as toxicant concentration or time. The most effective method of approximating these discrete data is to design a continuous model that can interpolate over the ranges of the continuous variables tested. Choice of an appropriate linear or nonlinear model is often difficult; however, if such a model can be chosen or designed, it can be used to make predictions not possible with other methods of analysis. Regression techniques can effectively be used to solve model equations that involve few (less than five) variables; if intrinsically nonlinear, these models can become quite complex to solve and interpret (Ref. 14-1).

In biological and ecological systems, so many factors affect each other and the dependent variable of interest that regression models may not be feasible. It is unlikely that enough data could be collected for a regression ecosystem model of more than five variables. Deterministic models using kinetic differential equations are more commonly used for problems of this magnitude. These large models are deterministic because rates and parameter estimates (the "unknowns" in regression models) are considered to be known constants; these constants are then used to solve the desired model equations. Differences between model solutions and data are reduced by adjusting the "constants" within their ranges of variance until an acceptable model is established.

In this report, regression and kinetic ecosystem models are discussed. Non-linear regression models are applied to laboratory data concerning (1) uptake of the insecticide, Kepone, from water, food, or sediments by estuarine fishes and invertebrates, (2) stimulation or inhibition of growth of several algal species that were exposed to textile industry effluents or Kepone in flask assays, and (3) cumulative spawning by grass shrimp exposed to several concentrations of the pesticide, endrin. Also briefly discussed is a complex ecosystem model using mass–balance kinetic equations to compute the distribution of Kepone in the James River.

14-2 BIOACCUMULATION MODEL DEVELOPMENT AND USE

14-2.1 Kepone Uptake Model Development

A statistical model developed to describe uptake and depuration of Kepone and other toxic materials by aquatic organisms has been previously reported (Refs. 14-2, 14-3). This model was applied to a limited data base consisting of 25 data sets and was found to effectively describe uptake of toxic materials by estuarine biota. The model was initially developed to allow computation of Bioconcentration Factors [BCF, the ratio of the chemical concentration in biota at chemical equilibrium (ng/g, ppb) to the concentration in exposure water (μg/liter, ppb)]. The model

$$\ln \text{RESIDUE} = 1/[\text{Parameter } A + 10^{(-\text{Parameter } C \times \text{DAY})}]$$

and

in FORTRAN notation ($*$ = multiply, $**$ = exponentation)

$$\ln \text{RESIDUE} = 1/[A + 10 ** (-C * \text{DAY})] \qquad \text{(Eq. 14-1)}$$

was computationally stable and required estimation of only two model parameters A and C, with DAY ranging from beginning to end of exposure. There are two inadequacies of this model: (1) calculation of parameter estimates require that DAY be used as a weighting factor, thus necessitating more complex weighted nonlinear regression computation; and (2) data near the "knee" region of the uptake curve are sometimes overestimated so that the maximum response parameter, A, can be estimated accurately; this phenomenon was most apparent when the model was applied to data of uptake of Kepone by oysters (Ref. 14-2). Since an uptake model should be appropriate for all toxic material uptake data sets, a similar but more refined statistical model was developed and has proven to be of greater general utility.

Uptake Model Equation. Using TIME of sample (days) and the natural logarithm of the measured chemical RESIDUE ($\mu g/kg$, ppb) as input data, a nonlinear regression estimation computer program can quickly and accurately calculate chemical uptake by aquatic organisms using the following equation:

$$Y = \ln \text{RESIDUE} = P1/[1 + P2 ** (\text{TIME} - P3)] \qquad \text{(Eq. 14-2)}$$

where Y is the dependent variable expressed as the natural log of RESIDUE, TIME (days) ranges from beginning to end of exposure, parameter $P1$ is the natural log of the maximum predicted residue concentration (in $\mu g/kg$ or ppb), parameter $P2$ is the rising slope ($0 < P2 < 1$), and parameter $P3$ is the TIME (days) at which $Y = 0.5 * P1$. This model (Eq. 14-2) was derived from a logistics equation developed for grain crop forecasting (Ref. 14-4) and is graphically illustrated (Fig. 14-1a). Calculation of the Bioconcentration Factor for any one data set merely requires that the antilog of $P1$ be divided by the mean chemical concentration in the exposure medium ($\mu g/liter$ or ppb):

$$\text{BCF} = \frac{\text{antilog } P1 \text{ (ng/g)}}{\text{average exposure concentration } (\mu g/liter)} \qquad \text{(Eq. 14-3)}$$

The statistical uptake model (Eq. 14-2) utilizes the natural logarithm of the dependent variable, RESIDUE, in order to stabilize the regression error variance over the range of TIME tested.

14-2.2 Application of Uptake Model

This statistical uptake model (Eq. 14-2) has proven to be valuable for describing toxic chemical uptake by estuarine organisms exposed to contaminated water, food, or sediment. Application of this model to data for the uptake of Kepone from water (Ref. 14-5), food (Ref. 14-6), and estuarine sediments (Ref. 14-7) is graphically illustrated (Figs. 14-1b, 14-1c, and 14-1d). For all these data, the regression curves appear to accurately describe the uptake of Kepone by the test organisms.

Replication of test samples for at least one sample period make it possible to use generalized lack-of-fit analyses that are sensitive indicators of the applicability

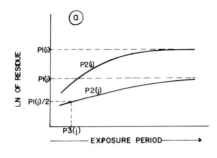

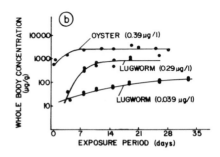

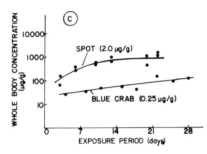

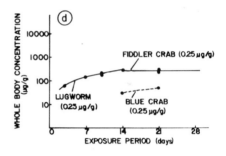

Fig. 14-1. (a) Plot of hypothetical uptake curves. Model equations:

$$Y(i) = P1(i)/(1 + P2(i)**(TIME-P3(i)))$$
$$= 7.5/(1 + .8**(TIME-3))$$
$$Y(j) = P1(j)/(1 + P2(j)**(TIME-P3(j)))$$
$$= 5.0/(1 + .9**(TIME-1.7))$$

(b) Plots of Kepone uptake from flowing water by:
Oysters (exposed to 0.39 μg/l):
$Y = 7.88/(1 + 0.59**(TIME-1.81))$
Lugworms (exposed to 0.29 μg/l):
$Y = 6.72/(1 + 0.63**(TIME-3.98))$
Lugworms (exposed to 0.039 μg/l):
$Y = 5.01/(1 + 0.90**(TIME-0.45))$

(c) Plots of Kepone uptake from food by:
Spot (exposed to 2.0 μg/g):
$Y = 6.81/(1 + 0.79**(TIME-(-0.47)))$
Blue crabs (exposed to 0.25 μg/g):
$Y = 5.97/(1 + 0.95**(TIME-0.0))$

(d) Plots of Kepone uptake from sediments by:
Fiddler crabs (exposed to 0.25 μg/g):
$Y = 5.5/(1 + 0.27**(TIME-10.6))$
Lugworms (exposed to 0.25 μg/g):
$Y = 5.5/(1 + 0.73**(TIME-(-0.46)))$
Blue crabs (exposed to 0.25 μg/g):
Too few data; analysis not accomplished.

of a model to experimental data. Although no lack-of-fit analytical procedures have been developed specifically for nonlinear regression models, the lack-of-fit test methods for linear models (Ref. 14-8) can be effectively used with nonlinear models. For most practical applications, these tests are acceptable even though the desired alpha-level cannot be certain. The regression curve for uptake by Spot fed Kepone (Fig. 14-1c) showed no lack of fit by these methods.

Often it is desirable to test the statistical difference of one data set from another to determine the effect of some variable such as exposure concentration, time, exposure method, or medium. As it has been determined that the statistical model (Eq. 14-2) gives a reasonably good estimation of uptake of Kepone by Spot fed contaminated food (Fig. 14-1c), differences between calculated regression curves can be tested. For instance, to determine whether the amount of food consumed significantly affects the final concentration of Kepone in predatory fish, two or more regression curves of interest can be compared with a calculated F statistic (Ref. 14-8). If there is no difference, this fact can be used advantageously by allowing the data sets to be pooled for further analyses. If the regression curves differ, determining which parameter estimate(s) of the model caused the difference is considerably more difficult for a nonlinear model than for a linear model.

14-3 ALGAL EFFECTS

14-3.1 Algal Stimulation and Inhibition Statistical Model

Analyses of estuarine algae grown in flask cultures under controlled conditions indicate that additions of pollutants or toxic chemicals can cause increased growth (stimulation), decreased growth (inhibition or toxicity), or "no effect" when compared to control treatments. Historically, straight-line interpolation or probit analysis has been applied to these data for estimation of the EC_{50} (concentration of toxicant that reduced growth to 50% of the observed control value); normally, stimulation might be noted, but not treated statistically. Since increases of minor nutrients can cause quite dramatic eutrophication, it would be beneficial to analyze stimulation data so that projections of environmental hazard can be implied. The following statistical model was designed to analyze algal stimulation and/or inhibition data:

$$Y = \text{Growth Index}$$
$$= P6/[1 + P7 ** (\text{CONC} - P8)] - P6/[1 + P9 ** (\text{CONC} - P10)] \quad \text{(Eq. 14-4)}$$

where Y (Growth Index) is the dependent variable expressed as culture density, percent of control, or other measure of growth; parameter $P6$ is the maximum predicted growth (units are those of Y); parameter $P7$ is the stimulation function slope; CONC is the toxicant concentration of the test cultures; parameter $P8$ is the concentration of toxicant that is predicted to cause growth equal to $0.5 * P6$; parameter $P9$ is the inhibition function slope; and parameter $P10$ is the concentration of toxicant that is predicted to cause growth inhibition equal to $0.5 * P6$.

Application of this model (Eq. 14-4) to algal bioassay data from selected industrial effluent tests is graphically illustrated to show the versatility of the model (Fig. 14-2) for describing stimulation and/or inhibition compared to control growth (Ref. 14-9).

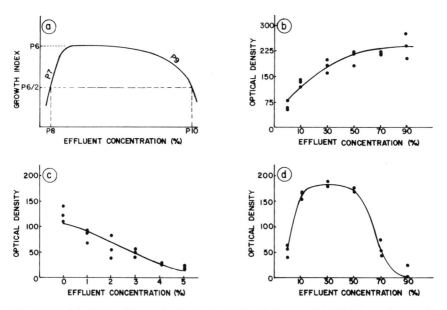

Fig. 14-2. (a) Plot of hypothetical algal stimulation and inhibition of growth curve.

Model equation: $Y = P6/(1 + P7**(CONC-P8))$
$- P6/(1 + P9**(CONC-P10))$.

(b) Plot of stimulated growth of *Skeletonema costatum* exposed to a textile industry effluent.

$Y = 238./(1 + 0.94**(CONC-11.42))$.

(c) Plot of inhibited growth of *Skeletonema costatum* exposed to a second textile industry effluent.

$Y = 125.-125./(1 + 0.47**(CONC.-2.33))$.

(d) Plot of stimulated and inhibited growth of *Skeletonema costatum* exposed to a third textile industry effluent.

$Y = 182./(1+0.78**(CONC-3.49)) - 182./$
$(1 + 0.84**(CONC-65.77))$.

In practice, Eq. 14-4 can be simplified from five to four parameters, which allows the model to be used for analyzing a minimum (5) number of data points. However, as with all parametric statistical methods, increased numbers of data points with replication within and between tests give increased accuracy in statistical prediction. The simplified equation is:

Y = Growth Index
= $MEAN/[1 + P7 ** (CONC - P8)] - MEAN/[1 + P9 ** (CONC - P10)]$

(Eq. 14-5)

where MEAN is the mean growth (units are those of the dependent variable Y) of the test treatment concentration that caused maximum growth. The MEAN is entered as a constant since it is the best unbiased estimate of parameter $P6$ (Eq. 14-4), the maximum growth estimator.

Special cases of the above models (Eqs. 14-4 and 14-5) are:

1. Stimulation only.
2. Inhibition (toxicity) only.

Stimulation-only data can easily be treated with the first portion of Eq. 14-4:

$$Y = \text{Growth Index} = P6/[1 + P7 ** (CONC - P8)] .$$ (Eq. 14-6)

As before, if parameter $P6$ does not need to be estimated, $P6$ can be replaced by the experimental mean growth for the highest concentration tested; only two parameters are then estimated, making a comprehensive analysis quite rapid. However, for many applications of analyzing stimulation data, calculation of maximum growth (parameter $P6$) is desired.

Inhibition (toxicity) data can easily be treated with the following equation:

$$Y = \text{Growth Index} = P6 - P6/[1 + P9 ** (CONC - P10)] .$$ (Eq. 14-7)

Analysis of this equation shows that it is assumed that control growth is maximum and all other treatments cause equal or decreased growth. Therefore, parameter $P6$, which can be replaced by MEAN, the mean control growth, is the initial value of the function at concentration equal to control, and the right-hand portion of the equation (Eq. 14-7) is subtracted from $P6$ (or MEAN) to describe increasing toxicity with increased concentration. The EC_{50} is therefore represented by parameter $P10$, because $P10$ is the concentration that equals $0.5 * P6$ (50% of control growth).

14-3.2 Application of Algal Inhibition Model

Kepone was toxic to three species of algae grown in enriched media (Ref. 14-10). In no assay was algal growth stimulated by the addition of Kepone; therefore analysis of the assay data was best performed using the algal toxicity model (Eq. 14-7). Parameter $P6$ was replaced with the mean control growth index (optical density) for each data set, so that only two parameters ($P9$, $P10$) were estimated. Minimization of the number of parameters in the model assured that the variances of the parameter estimates (and, therefore, the confidence limits on the estimates) were also minimized. Further analyses using these reduced variance estimates (such as lack-of-fit or comparisons of parameter estimates) were more discriminatory than were models with increased numbers of parameters.

Application of Eq. 14-7 to the algal assay data is graphically illustrated (Figs. 14-3a, 14-3b, and 14-3c). Parameter $P10$ is the EC_{50} estimate for this toxicity model; confidence limits on the EC_{50} (i.e., parameter $P10$) were also computed by the nonlinear regression estimation program.

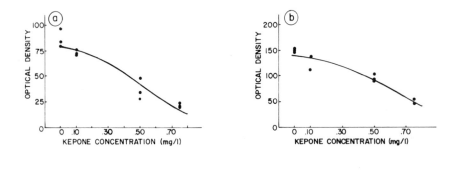

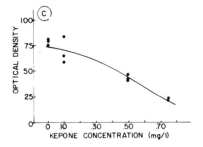

Fig. 14-3. (a) Plot of Kepone toxicity to the alga, *Chlorococcum* sp.
$$Y = 87/(1 + 0.95**(CONC-0.47))$$
EC50 = 0.47   95% C.I. = .408 – .536

(b) Plot of Kepone toxicity to the alga, *Dunaliella tertiolecta*.
$$Y = 149/(1 + .95**(CONC-0.61))$$
EC50 = 0.61   95% C.I. = .562–.658

(c) Plot of Kepone toxicity to the alga, *Nitzschia* sp.
$$Y = 79/(1 + 0.95**(CONC-0.55))$$
EC50 = 0.55   95% C.I. = .486 – .609

14-4 CUMULATIVE RESPONSES

14-4.1 Cumulative Response Model Development

Several biological responses are most effectively analyzed when the response is accumulated over a period of time. Cumulative responses, such as percent of metamorphosis or total production of young or offspring for test populations, can be statistically analyzed using a modified form of the regression model for algal stimulation (Eq. 14-6). The model appropriate for cumulative response analysis is

$$Y = \text{Cumulative Response} = 100\% / (1 + P11 ** (\text{TIME} - P12)] \quad \text{(Eq. 14-8)}$$

where the maximum cumulative response is 100%, parameter $P11$ is the response slope, TIME is the independent variable at which the responses Y were summed, and parameter $P12$ is the TIME at which the estimated cumulative response equals 50%.

14-4.2 Application of Cumulative Response Model

The cumulative response model was applied to data on the cumulative number of female grass shrimp that had been chronically exposed to the pesticide Endrin and spawned (Ref. 14-11). The plot of these data (Fig. 14-4) represents an analog (continuous) approximation to the data derived from discrete sampling periods. Differences of time of spawning can be tested; all female shrimp exposed to endrin required more time to spawn then control shrimp. The time to 50% spawning by endrin-exposed shrimp was significantly greatly than for that of controls since the confidence intervals on parameter $P12$ for endrin-exposed shrimp did not include $P12$ of the regression for control shrimp. Whereas spawning was slowed by 5.1 days during exposure to 30 parts-per-trillion endrin, spawning was slowed by 55.4 days during exposure to 500 parts-per-trillion endrin in seawater.

14-5 INTEGRATION OF MODELS

14-5.1 James River Model

Kepone contamination of the James River caused the U. S. Environmental Protection Agency to investigate the movement of Kepone within the James River and estuary and their biota. Concern over the contamination has not been confined to the James River, because it flows into the Chesapeake Bay, one of the largest in North America. Extensive laboratory and field-analyses of the hydrodynamics of the James River system have been completed, as hydrody-

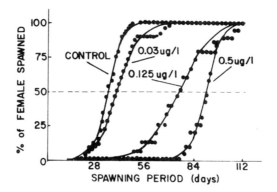

Fig. 14-4. Plots of cumulative deposition of eggs from grass shrimp exposed to no endrin, 0.03 μg endrin/l, 0.125 μg endrin/l, and 0.5 μg endrin/l.

Y(control) = $100/(1 + 0.78^{**}(TIME-35.2))$
Time to 50% deposition = 35.2 days
95% C.I. = 35.0 – 35.4 days

Y(.03 μg/l) = $100/(1 + 0.85^{**}(TIME-40.3))$
Time to 50% deposition = 40.3 days
95% C.I. = 40.0 – 40.6 days

Y(.125 μg/l) = $100/(1 + 0.9^{**}(TIME-75.0))$
Time to 50% deposition = 75.0 days
95% C.I. = 74.6–75.4 days.

Y(.5 μg/l) = $100/(1 + 0.8^{**}(TIME-90.6))$
Time to 50% deposition = 90.6 days
95% C.I. = 90.3–90.9 days.

namics play a critical role in the distribution of persistent compounds such as Kepone. The extensive hydrodynamics data base for the James River coupled with the Water Quality Analysis Simulation Program[1] (WASP) are being used as a research project[2] by the U. S. EPA to determine the distribution of Kepone in the James River.

The WASP model equations are founded on the principle of conservation of mass. WASP, conserving mass both in time and space, accounts for and traces water quality constituents from their point of spatial and temporal input to their

[1] WASP programs purchased from Hydroscience, Incorporated, 363 Old Hook Road, Westwood, NJ.
[2] Mathematical Models of Fates of Pollutants in Estuaries, Environmental Engineering and Science Division, Manhattan College, Bronx, NY. U. S. Environmental Protection Agency Project: R 804563-02.
[3] Hydroscience, Incorporated. 363 Old Hook Road, Westwood, NJ.

final points of export. For these computations, the model segmentation, advective and dispersive transport fields, boundary conditions, waste loads, segment parameters, kinetic model constants and time variable functions, and the kinetic model equations must be determined from laboratory or field-derived information. Once these data and model equations are available, WASP numerically integrates the mass–balance equations with time. The Model Verification Program (MVP), an auxilliary part of the WASP program package, can then be used to compare the theoretical computations with observed data from the river; discrepancies can then be adjusted by altering values for the model constants.

Complete documentation on the model development and use of WASP has been described in detail[3].

14-5.2 Regression Models

Modification of a logistics model equation has resulted in model equations that can be used to analyze uptake of toxic chemicals by fishes and invertebrates, and stimulation or inhibition of algal growth by industrial effluents and pesticides, as well as to compare cumulative response curves.

Uptake Model. Kepone uptake from water, food, and sediments by estuarine invertebrates and fishes were analyzed using the nonlinear equation, $Y = P1/[1 + P2 ** (TIME - P3)]$, which requires estimation of three parameters ($P1$, $P2$, and $P3$). The maximum predicted residue concentration (ng/g, ppb) in the exposed organism is equal to e^{P1}; the bioconcentration factor is, therefore, e^{P1}/exposure concentration. The benefits of using such a model are (1) more realistic BCFs can be calculated from the data because variability due to exposure and analytical techniques are smoothed and (2) differences due to exposure method or media, concentration, or time can be statistically analyzed.

Algal Toxicity Model. The effects of industrial effluents and pesticides on growth of marine algae were analyzed with the aid of a model equation that initially increases in value so that stimulated algal growth can be represented, and then decreases in value so that inhibitory (toxic) effects can be modeled. Selected portions of the total model can be fixed at constant values or deleted, as necessary, so that stimulation-only or inhibition-only data are readily analyzed. Kepone was toxic to all species of algae tested; therefore, analyses for stimulated growth were not performed. One parameter of the algal growth inhibition model estimates the EC_{50}, and 95% confidence intervals on the EC_{50} are also available. Generally, there is no lack-of-fit for this model.

The full model for algal growth was applied to data from 14 assays of textile plant effluents. Application of the model allowed interpolation between data so that the concentrations of effluents that caused 20% stimulation or 50%

inhibition of growth could be calculated for possible enforcement actions. Thirteen of the fourteen data sets were easily modeled; one data set was more difficult since considerable stimulation of growth occurred at 0.1% effluent and nearly complete toxicity was found at 5% effluent concentration. The calculated EC_{50} was reasonably close to the data; however, confidence intervals on the parameter estimates were wide in comparison to other tests analyzed using the model.

Cumulative Response Model. The model equation for chemical uptake and algal growth stimulation was also used to analyze cumulative spawning by female grass shrimp exposed to endrin. Application of this model to such cumulative-over-time data allows comparisons of slope of accumulation or time to 50% of total. For the grass shrimp data, statistically significant delays in spawning were judged when the 95% confidence interval of spawning time for endrin-exposed shrimp did not include the estimated mean spawning time for non-exposed control shrimp.

14-5.3 James River Ecosystem Model

WASP has been calibrated against salinity and has been utilized to model the transport and spatial distribution of suspended solids in the James River estuary.

Field monitoring and laboratory data indicate that the majority of the Kepone in the James River is associated with particulates; therefore, if the suspended and bed solids are accurately modeled, the Kepone distribution will be known. Particular attention is being paid to the rates of diffusion of Kepone within the bed sediments because it appears that these rates will play a major role in computing the long-term washout rate of Kepone from the River. The WASP programs compute the concentration of Kepone in water and sediments of each estuarine model segment; these concentrations will be used to project concentrations of Kepone in biota.

REFERENCES

14-1. Daniel, C., and Wood, F. S., *Fitting Equations to Data*, Wiley, New York, 1971.

14-2. Bahner, L. H., and Oglesby, J. L., "Test of Model for Predicting Kepone Accumulation in Selected Estuarine Species," in *Aquatic Toxicology and Hazard Evaluation*, L. L. Marking and R. A. Kimerle (Eds.), American Society for Testing and Materials, Philadelphia, PA, pp. 221–231, 1979.

14-3. Oglesby, J. L., and Bahner, L. H., "Nonlinear Model for Pesticide Accumulation in Selected Estuarine Species," *Proc. Third Annual Conference of the SAS Users Group International*, R. H. Strand and M. Farrell (Eds.), SAS Institute, Inc., Raleigh, NC, pp. 288–292, 1978.

14-4. Larsen, G. A., *Forecasting 1977 Kansas Wheat Growth*, Research and Development Branch, Statistical Research Division, ESCS, United States Department of Agriculture, Washington, D. C., 1978.

14-5. Bahner, L. H., Wilson, A. J., Jr., Sheppard, J. M., Patrick, J. M., Jr., Goodman, L. R., and Walsh, G. E., "Kepone Bioconcentration, Accumulation, Loss, and Transfer through Estuarine Foodchains," *Chesapeake Sci.*, 18, pp. 299–308, 1977.

14-6. Schimmel, S. C., Patrick, J. M., Jr., Faas, L. F., Oglesby, J. L., and Wilson, A. J., Jr., "Kepone: Toxicity and Bioaccumulation in Blue Crabs," *Estuaries*. 2, pp. 9–15, 1979.

14-7. Bahner, L. H., Rigby, R. A., and Faas, L. F., "Bioavailability of Kepone from Sediments to Several Estuarine Species," Environmental Protection Agency's Gulf Breeze Research Laboratory, Contribution No. 365. In preparation.

14-8. Neeter, J., and Wasserman, W., *Applied Linear Statistical Models*, Irwin, Homewood, IL, 1974.

14-9. Walsh, G. E. and Bahner, L. H., "Assessment of Textile Waste Toxicity with Marine Bioassays," Environmental Protection Agency's Gulf Breeze Research Laboratory, Contribution No. 379, 1979.

14-10. Walsh, G. E., Ainsworth, K., and Wilson, A. J., Jr., "Toxicity and Uptake of Kepone in Marine Unicellular Algae," *Chesapeake Sci.* 18, pp. 222–223, 1977.

14-11. Tyler-Schroeder, D. B., "Use of the Grass Shrimp, *Palaemonetes pugio*, in a Life-cycle Toxicity Test," in *Aquatic Toxicology and Hazard Evaluation*, L. L. Marking and R. A. Kimerle (Eds.), American Society for Testing and Materials, Philadelphia, PA, pp. 160–170, 1979.

15

An environmental
fate model leading to
preliminary pollutant
limit values for human
health effects

David H. Rosenblatt, Ph.D. and Jack C. Dacre, Ph.D.
Environmental Protection Research Division
U. S. Army Medical Bioengineering Research and Development
Laboratory
Ft. Detrick, Frederick, Maryland

David R. Cogley, Ph.D.
Technology Division, GCA Corporation
Burlington Road
Bedford, Massachusetts

15-1 INTRODUCTION

15-1.1 Historical Perspective

Guidelines, criteria, and standards for single-path, single-transfer exposures to water and air pollutants have been appearing regularly for over a decade. The chief problem in their formulation has been to develop and interpret chronic toxicity data and to apply these data to predict effects on large human populations at low exposure levels. Our society has recently become concerned with the need to deal similarly with potentially significant concentrations of toxic pollutants in soil, whence multiple transfers represent the chief means of exposure. Moreover, a given toxicant (or group of such substances) may reach its target by several routes, originating in soil, water, and possibly gaseous emissions; clearly guidance is also needed for this. The regulatory process has not yet addressed these requirements in detail.

The consequences of chemical waste dumping discovered at Love Canal in Niagara Falls, NY (Ref. 15-1), West Point, KY (Ref. 15-2), and southwest Philadelphia, PA (Ref. 15-2) emphasize the need for regulation. Responsible industries would like to anticipate the standards that will eventually be set so that they could plan, develop, and institute operational controls over a reasonable span of time. It would appear that the U. S. Army was among the first Federal organizations to recognize a responsibility in this field.

Early in 1975 the Army's attention was focused on Rocky Mountain Arsenal (RMA), a 25-square mile reservation near Denver, Colorado. RMA had been the site of military and civilian-lessee chemical processing operations for over a third of a century; as a result, certain sections had become heavily contaminated with waste materials (Ref. 15-3). Consideration was given in 1975 to restoring the land to a near-pristine condition. The Army Medical Research and Development Command was tasked to develop a scientific data-base, involving 16 potential pollutants (Ref. 15-4), and to generate a rational approach to guidelines for removal of these substances from the soil (Ref. 15-3). This was the origin of the concept of *soil pollutant limit values* (Ref. 15-5). The immensity of the proposed task led the Army to adopt the more realistic goal of stopping additional pollution and of treating contaminated groundwater to prevent its spread beyond RMA's borders. The need was therefore shifted to *water pollutant limit values*. The approach to the limit values, as outlined below, was essentially the same, whether for soil or water. To emphasize the tentative, transitory, and nonregulatory nature of these numbers, the term *preliminary pollutant limit values* (PPLVs) was adopted for them.

A clear distinction must be drawn between the concept of PPLVs and widely recognized criteria or legally sanctioned health or environmental standards. The latter are usually promulgated shortly after public hearings on recently announced agency intentions. They then require speedy compliance. The interval from first disclosure of planned pollution standards to mandated implementation is exceedingly short. It cannot encompass the design and construction of water-treatment facilities or the necessary preparations for restoration of contaminated land areas. Setting such efforts in motion is especially lengthy for Federal agencies, because construction or renovation plans must be submitted for Congressional approval and funding. Thus, it was necessary to develop a methodology to forecast probable environmental limits for use in preliminary design criteria for pollution control systems; the methodology had to make do with such information as might be available. Such an approach would permit the planning process to start as soon as a pollutant was recognized, before all the information on which to base health or environmental standards had been gathered.

15-1.2 Basic Approach to PPLV Derivation

PPLVs should be calculated separately for human health effects and for effects on aquatic organisms, provided both types of effect are relevant. The lower of the two values should then be the accepted final value. In this chapter, we address only human health effects PPLVs; the approach to PPLVs for aquatic biota would resemble that shown here in many respects. Because human beings are potentially exposed by several routes, shown conceptually for soil via water in Fig. 15-1, the separate *single-pathway PPLVs* (SPPPLVs) for human health effects are "normalized" to the final PPLVs for water and soil (so that the lowest

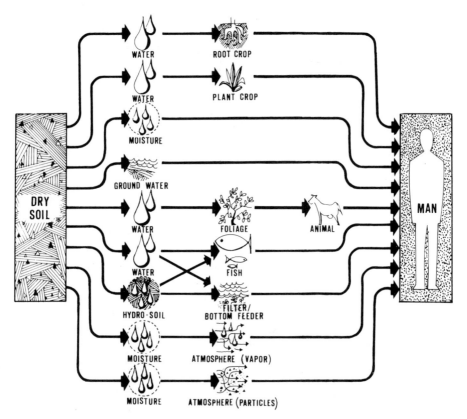

Fig. 15-1. Pollutant pathways from soil to man.

SPPPLV value is diminished to lead to the correct overall exposure level); this is done without any attempt to weight the included pathways according to their relative importance.

The principal merit of the PPLV procedure developed for human health effects is that all factors may be separately identified and quantified even in the absence of complete data, thus allowing ready examination of the following:

1. The sufficiency of the number of pathways considered.
2. Site-specific determinants at appropriate places in the procedure.
3. The ranking of pathways with respect to the pollutant dose, in order to identify the more critical pathways.
4. Sensitivity to errors in parameter values.

The completeness of the calculations for a given pollutant is assured by listing all pathways concerned, together with relevant partition (intermedia transfer) coefficients. With this structure, qualified reviewers can scrutinize all assumptions regarding the pathways taken by a pollutant. Consideration of each possible pathway should effectively deter a reviewer from preselecting favorite pathways. Eventually, though, those pathways critical to the situation under review must be identified. This requires that site-specific elements be brought into consideration. When this has been accomplished, every pathway can be ranked quantitatively with respect to the pollutant flux reaching the receptor organism, which is man in these examples. If data gaps are sufficient to warrant experimental research, one can properly allocate available resources by specifying research only for key parameters.

15-1.3 Alternatives to PPLVs

The establishment of criteria and standards for toxic pollutants in drinking water or food, when these media are considered the sole sources of the toxicants of interest, is relatively straightforward. It requires toxicological experiments by the ingestion route, whereby maximum acceptable daily intakes may be adduced. In the United States, nevertheless, national interim primary drinking water standards, water quality criteria, and toxic pollutant effluent standards related to human health effects have so far been established for only a relatively small number of compounds (Refs. 15-6, 15-7, 15-8). The situation is essentially similar for water quality criteria to protect aquatic biota (Refs. 15-8, 15-9). It should be noted that carcinogens represent a special subset of toxic substances for which standard-setting is more difficult and that decisions are based on a case-by-case hazard assessment.

It is more difficult to formulate standards for soil than for drinking water and food. Although soil can contain very high pollutant concentrations, the pollutants, with very rare exceptions, are available to human beings and higher animals only indirectly. However, the interactions with organisms that live in the soil may be rather important and should be addressed. Thus far, it would appear that standard-setting activities for soil pollutants have been limited to the Soviet Union. Most notably, Perelygin (Refs. 15-10, 15-11), citing previous Russian work, has discussed the basis for "hygienic standardization" of toxic substances introduced into soil by industrial effluents used for agricultural irrigation. Citing experimental studies and using previously recognized maximum permissible concentrations in food plants and in drinking water, he has proposed soil standards to avoid the harmful effects of selected pesticides on agricultural soil (microorganisms), subsurface water, growing plants, agricultural workers, and consumers of food; the compounds, with their standards, were DDT (1 ppm), hexachlorocyclohexane (1 ppm), γ-hexachlorocyclohexane (1 ppm), polychloropinene (0.5 ppm), polychlorocamphene (0.5 ppm), and carbaryl (0.05 ppm). These standards were apparently adopted by the USSR Ministry of Public Health.

Such experimental approaches as the Russian work are to be preferred to the large number of assumptions involved in the PPLV concept. The results of any known experiments should certainly be integrated as fully as possible into the PPLV-setting process.

15-2 HUMAN HEALTH-EFFECT PPLVs

15-2.1 Derivation of PPLVs—the Basic Approach

There are six steps in the human health-effect PPLV calculation procedure as practiced herein. These are (1) specification of the systems to be studied—pollutants and pathways, (2) listing of required data—acceptable daily doses (D_T) and partition (or intermedia transfer) coefficients, (3) establishment of a data base, and determination or estimation of data values, (4) calculations of single-pathway PPLVs (SPPPLVs) for all pathways (not site-specific), (5) selection of the most critical pathways for each pollutant (often site-specific), and (6) normalization of PPLVs.

The first step involves recognizing that a problem exists, undertaking research and interviews to discover the probable whereabouts of waste chemicals, and instituting a sampling and analysis program to define the actual distribution of known or suspect pollutants. Although known or suspect compounds are of paramount importance, the analysis should also attempt to identify any unsuspected contaminants present in the soil or water samples. Careful examination of local conditions—especially geohydrology, demography, and the local economy—aids in defining the more likely pathways from the sources to the human receptors.

Step two is the listing of required data, as they pertain to the more likely pathways.

The third step, establishment and interpretation of a data base, requires the services of a multidisciplinary team to acquire and sift the literature, summarize it, and interpret it. Especially important for PPLV calculations are toxicity and health/environmental standards, on which to base acceptable daily doses, and any information from which to calculate partition (intermedia transfer) coefficients. If data are unavailable for a particular compound, parameter values may be estimated from knowledge of the same parameter for a related class of compounds, or a default value may be chosen.

Fourth is the calculation of SPPPLVs for all specified pathways; the pathways with the lowest SPPPLVs are most carefully studied. Input parameters and SPPPLVs are checked and reviewed for appropriateness and completeness.

In the fifth phase of the calculation procedure, site-specific "critical" SPPPLVs are chosen. Subjective judgments are made of the most likely among the significant pathways for the site under consideration.

The final phase, normalization, consists of deriving PPLVs from the SPPPLVs. Where pathways have their origin in the same medium or have a common point of intersection, a simple calculation is used to adjust the concentration at the origin or the intersection such that SPPPLVs taken together provide the target organism, man, with an exact value for D_T (acceptable daily dose of toxicant). The calculation is similar to the addition of electrical resistances in parallel DC circuits. Thus, for SPPPLVs via three pathways from water:

$$PPLV = [1/(SPPPLV)_1 + 1/(SPPPLV)_2 + 1/(SPPPLV)_3]^{-1} \qquad \text{(Eq. 15-1)}$$

If several independent sources of the pollutant have to be considered, the individually calculated PPLVs must be reduced to some arbitrary combination consistent with the D_T. All notation is defined in Appendix 15-1.

15-2.2 Pathways and Factors for Pollutant Access to Man

Major transport pathways of pollutants from source to human receptor are described in this chapter for a generalized site. Physical transport pathways, pathways through biological intermediary sinks, loss mechanisms, and physical, chemical, and biological factors affecting transport along these pathways are also considered.

The history of pollutants is categorized according to original sources (soil surface, soil subsurface, ponds, or surface water) and intermediate sinks (air, groundwater, surface water, or living organisms such as natural flora, crops, wildlife, or livestock). Toxic material can move from one site to another by a variety of physical and biological processes. Ultimately, the material finds its way to human beings by the three major exposure mechanisms: surface contact, respiration, and ingestion. However, surface contact (percutaneous) doses required for a given biological effect are usually so much higher than doses required by either of the other exposure routes that the percutaneous route can nearly always be eliminated from consideration. Furthermore, for practical purposes, we choose to consider D_T the same for inhalation as for ingestion with but few, usually apparent, exceptions (for instance, hydrogen chloride).

Chemical losses can occur by transformation of the original compound into another compound of different toxicological properties. The four most common loss mechanisms are photochemical, hydrolytic, oxidative, and biodegradative. Because of uncertainty as to the degree of exposure of a pollutant to such mechanisms, however, losses of this type are only estimated when the particular circumstances warrant such consideration.

The model accounts for a number of factors affecting pollutant transport along the various pathways, in particular through surface water, groundwater, vapor, and soil movement.

Where the human receptors are distant from the site of contamination of a surface water, it may well be feasible to introduce factors that take dilution, hydrolysis, biodegradation, and phototransformation into account.

The groundwater system usually provides the major pathway for movement from deposits in soil of moderately or extremely water-soluble pollutants. Chemically stable pollutants can follow the groundwater's general flow, appearing at considerable distances from their sites of deposition. In practice, it is not possible to quantitatively predict the effects of factors governing the subsurface travel of pollutants with groundwater. Nevertheless, the general direction (though not the elapsed time) of a subsurface pollutant plume at a particular site can be predicted by means of geohydrological analysis.

Movement of a pollutant from the soil into and through the atmosphere starts either with volatilization of the pollutant from the soil or surface water, or with entrainment of dust particles containing adsorbed material or material dissolved in free water in the particles. Volatilization depends on the vapor pressure and solubility of the species, the wind structure of the surface boundary (mass transfer), the soil porosity and moisture, and the vertical distribution of material. Vapor pressure, in turn, depends to some extent on the degree to which adsorptive forces of the particles have lowered the activity relative to the pure substance. Strongly adsorbed species such as the insecticides aldrin or dieldrin, exhibit significantly lowered vapor pressure over dry organic-containing soil, although not over wet soil in which water competes for the adsorption sites. Vaporization from the soil may suddenly assume importance; for example, when excavation exposes contaminated particles to the atmosphere. The vapor route may also be signficant for enclosed spaces, such as the crawl spaces under houses.

Windblown particles can convey nonvolatile adsorbed pollutants over considerable distances. The concentration of particulates in the air is generally very low, however, rarely exceeding 0.06 mg m^{-3} in nonurban areas (Ref. 15-12). Thus, industrial exposure to chemical dusts would normally be the only important type of hazard from organic toxicants associated with airborne particles.

Biological processes can lead to the movement of compounds within the ecosystem through plant uptake and translocation through the food chain. Plants assimilate a variety of substances from the soil during the growth cycle. Little is known of the factors governing the accumulation of exogenous chemicals in plants.

Transport through the food chain, beginning with microorganisms or plankton in the aquatic environment, can progress from innocuous concentrations of some substances to toxic concentration levels. This process occurs, for example, with the fat-soluble chlorinated hydrocarbon pesticides that are neither readily metabolized nor readily excreted. As a rule, however, transport processes tend to *reduce* the concentration of pollutants in the food chain.

15-2.3 Numerical Methods for Calculating SPPPLVs and PPLVs

Our approach to calculating SPPPLVs assumes that contaminants are in stepwise equilibrium (steady state) from source to receptor. Although we have here applied the model to humans, it is equally applicable to plants and wildlife. It is expressed in such a manner that any part can be modified as required, without necessarily invalidating the other parts. For example, pollutant residues in fish are considered to exist in equilibrium with pollutant concentrations in water. However, bottom feeders, such as catfish, may actually be in equilibrium with insecticide residues in sediments. The calculation procedure can be tailored to accommodate such a situation.

SPPPLVs, and thus PPLVs, are based on parameter values for chronic human exposures (Table 15-1). In exceptional cases, the calculations may take into account sensitive subgroups within the population, such as embryos, infants, and aged individuals.

Consider a particular pathway from source to human receptor (Fig. 15-2). Each compartment represents a medium in which the toxic material can reside. Maximum concentration values and pseudosteady-state conditions between media are presumed; for instance, it is assumed in the example that pollutants in cattle eating contaminated forage crops reach some steady-state level that can be readily calculated by considering the cattle feed and water uptake regime. Each double arrow represents a process by which that equilibrium is achieved. The last step involves no assumed equilibrium, but depends on the ingestion rate. With reference to the notation in Appendix 15-1, the relationship between the acceptable daily dose, D_T, and the limiting pollutant level in soil, C_s, is given by an equation of the form

$$D_T = \frac{f \times DFI \times C_a}{BW} = K_{pa} \times K_{wp} \times K_{sw} \times C_s \times f \times DFI/BW, \quad \text{(Eq. 15-2)}$$

TABLE 15-1. Parameter Values for Human Chronic Exposure.

Parameter	Value
Body weight (BW)[a]	70 kg
24-hr breathing rate (RB)[b]	18.5 m³/24 hr
Workday breathing rate (RB')[b]	12.1 m³/8 hr
Daily food intake (DFI), dry weight basis[c]	0.63 kg day^{-1}
Daily water intake (W_w)[a]	2 kg day^{-1} (2 liter day^{-1})
Temperature (T), unless otherwise stated	25°C (77°F)

[a]The Safe Drinking Water Committee of the National Research Council (Ref. 15-13) used the 70-kg person and 2 liter day^{-1} as average values, which are therefore used here. Possibly 60 kg and 2.5 liter day^{-1} would have been more representative, according to Tables 27, 29, and 31 in Albritton (Ref. 15-14).

[b]Average breathing rates were obtained from Cleland and Kingsbury (Ref. 15-15):

Light work, adult male, 68.5 kg body weight, 28.6 liter min^{-1}.
Light work, adult female, 54 kg body weight, 16.3 liter min^{-1}.
Resting, adult male, 68.5 kg body weight, 7.43 liter min^{-1}.
Resting, adult female, 54 kg body weight, 4.5 liter min^{-1}.

For each condition (light work and resting), the breathing rate divided by body weight was obtained and averaged (male and female). This gave values of 0.360 liter min^{-1} kg^{-1} for light work and 0.096 liter min^{-1} kg^{-1} for the resting condition. Converted to an 8-hr day, the light work breathing volume for a 70-kg person is 12.1 m³ (RB'); the resting breathing volume over 16 hr for a 70-kg person is 6.4 m³. Thus, $RB = RB' + 6.4 = 18.5$ m³.

[c]According to Albritton (Ref. 15-14), daily food intake figures vary somewhat from one information source to another (Tables 27, 29, and 31), as well as between males and females. This is due, in part, to the representation by different categories in each table. Thus, the United States (Table 31) divides the adult population into 25-yr, 45-yr, 65-yr, pregnancy, and lactation subgroups; Canada (Table 27) divides it into sedentary, moderate-activity, heavy-activity, very-heavy-activity, pregnancy, and lactation subgroups; the United Kingdom (Table 29) classifies it into over-65, over-60, sedentary, moderately-active, active, very-active, pregnancy, and lactation subgroups. A reasonable value in the total range of 7.0 to 13.7 g/kg is 9.0 g/kg, which works out to 630 g per 70-kg person. According to the Canadian and United Kingdom data (Tables 27 and 29), this figure lies between sedentary and moderate activity for both men and women. By U.S. statistics (Table 31), it lies between the male and female figures for 45-year-old persons.

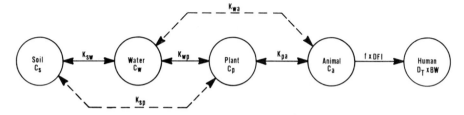

Fig. 15-2. Pollutant pathway from soil to man via water, plant, and animal compartments.

where the right-hand part of the equation is derived from the definitions:

$$C_w = K_{sw}C_s \qquad \text{(Eq. 15-3)}$$
$$C_p = K_{wp}C_w \qquad \text{(Eq. 15-4)}$$
$$C_a = K_{pa}C_p \qquad \text{(Eq. 15-5)}$$

and f refers to the fraction of meat in the diet. We suggest that the values of f listed in Table 15-2 might be used in lieu of better sources of information (e.g., Ref. 15-16); these are our own estimates and should not be otherwise construed.

By inversion of Eq. 15-2, the SPPPLV in soil by the pathway shown in Fig. 15-2 is

$$C_s = (BW \times D_T)/(K_{pa} \times K_{wp} \times K_{sw} \times f \times DFI) \qquad \text{(Eq. 15-6)}$$

Establishment of a D_T. The acceptable daily dose of a toxicant, D_T, relative to human health effects, is central to the calculations discussed in this chapter. Yet, it is exceedingly difficult to propose a value that everyone would agree is really safe and not unreasonably low. The economic impact of a needlessly stringent limitation on human exposure could be enormous; "safer than safe" should not be the objective in establishing PPLVs. Moreover, excessively conservative assumptions might result in PPLVs lower than background levels and, in some cases, lower than the required concentrations of "pollutants" that are also essential trace-level nutrients (for example, selenium or zinc).

Carcinogens pose a special challenge; there has been no consensus reached among scientists regarding a suitable mathematical model for carcinogenesis; neither is there an accepted "safe" level for a carcinogen. At the same time, the Environmental Protection Agency, using very simple mathematical models, for example, the "one-hit" model, appears to find a hazard level of one cancer death per hundred thousand lifetime exposures to be an acceptable criterion for most carcinogenic pollutants (Ref. 15-17).

TABLE 15-2. Assumed Fractions, f, of Diet Represented by Given Types of Foods (Dry Weight Basis).

Type of Food	Fraction, f, of Total Diet
One crop (e.g., root, grain, or fruit)	0.1
Meat	0.2
Fish	0.05

Faced with the need to decide about a carcinogen, the engineer may do well to accept a concentration that represents (a) the level of detectability of the more easily detected nonradioactive toxic substances or (b) the level at which a variety of potent, but ubiquitous carcinogens may be found in drinking water. If such a level is used, and we accept the figure of 2 liters day^{-1} for a 70-kg adult, the value of D_T (in mg kg^{-1}) will be $\frac{1}{35}$ the concentration permitted in the drinking water (in ppm or mg liter^{-1}). A case in point, at least for consideration, might be 2, 3, 7, 8-tetrachlorodibenzodioxin (TCDD), which is a highly toxic substance to some animals and may be a carcinogen (Ref. 15-18). The limit of detection for this compound, for which sensitive analytical methods are known, is about 5×10^{-6} mg liter^{-1} (Ref. 15-18), which indicates that such a figure might represent a practical lower limit for drinking water PPLVs. This implies that D_T has a lower limit of about 10^{-7} mg kg^{-1}. Alternatively, one might adopt the lowest water quality criterion (though based on toxicity to aquatic organisms) promulgated by the Environmental Protection Agency (Refs. 15-8, 15-9). For example, that adopted for the insecticides mirex, heptachlor, polychlorobiphenyls, and DDT gives a PPLV for water of 10^{-6} mg liter^{-1}. Even though current equipment is not able to detect this concentration, such a level is not far from the state-of-the-art.

Table 15-3 lists six sources of information from which D_T values may be drawn, listed in order of preference. From this, it is seen that if there is available an ADI (Acceptable Daily Intake) value, originating with the World Health Organization (Ref. 15-19), then that figure should be used as D_T.

A second excellent, but limited, source of information is the list in the National Interim Primary Drinking Water Regulations; its MCL (Maximum Concentration Level) values (Ref. 15-6) are directly convertible to D_T values through the use of the figures in Table 15-1, i.e.,

$$D_T = (MCL \times W_w)/BW = MCL/35 \qquad \text{(Eq. 15-7)}$$

A third generally accepted source of values is the collection of TLVs (Threshold Limit Values) published by the American Conference of Governmental Industrial Hygienists (Ref. 15-20), which is utilized by the Occupational Safety

TABLE 15-3. Information Sources from Which to Derive Values of Acceptable Daily Doses (D_T's) of Toxic Pollutants for Human Beings (in Order of Priority).

Input Information	Calculation Required	Reference
Existing standards		
1. Acceptable daily intake (*ADI*)	None	WHO (Ref. 15-19)
2. Maximum concentration level (*MCL*) in drinking water	Adjust for water consumption level	EPA (Ref. 15-6)
3. Threshold limit value (*TLV*) for occupational exposures	Use factors for breathing rate, exposure time, safety factor of 10^2	ACGIH (Ref. 15-20)
Experimental results in laboratory animal studies		
4. Lifetime no-effect level (*NEL$_L$*)	Use safety factor of 10^2	(Ref. 15-21)
5. Ninety-day no-effect level (*NEL$_{90}$*)	Use safety factor of 10^3	(Ref. 15-21)
6. Acute toxicity (*LD$_{50}$*)	Use safety factor of $\sim 10^5$	See text

and Health Administration. Conversion of these to D_T (Ref. 15-15) involves three factors: The first is division by $7/5 = 1.4$ to convert from a normal 5-day, work-week to a 7-day exposure week. The second is division by 100; this allows for exceptionally sensitive individuals who would not normally be part of the work force, and takes into consideration the completely involuntary and unsuspecting nature of the exposure (Ref. 15-15). The third factor converts from TLV (expressed in mg m^{-3}) to a total dose; the breathing rate, RB', for a 70-kg person doing light work is taken as 12.1 m^3/8-hr day (see Table 15-1). Thus,

$$D_T = (TLV \times RB')/(140 \times BW) = TLV/810 \qquad \text{(Eq. 15-8)}$$

There is one important caveat for using converted TLV values. When TLV values were chosen, they were not permitted to exceed 1000 ppm except for carbon dioxide (Ref. 15-20). Thus, if the TLV is 1000 ppm (or its equivalent in mg/m^3), the actual safe level of exposure, and therefore the D_T, may be higher. It is also necessary to consider whether the TLV is strongly related to a degree of lung irritancy not likely to be translated, after ingestion, to systemic toxicity. A flagrant example is hydrogen chloride, where the formula shown above would lead to the ridiculously low D_T of 0.009 mg kg^{-1} day^{-1}; the figure for chlorine is half this value.

The remaining three sources from which D_T might be derived involve animal experiments. Although similar experiments were the ultimate source of the first three methods for calculating D_T, the toxicological experiments referred to here have not gone through the process of extrapolation, evaluation, and consensus. Thus, they are used only in the absence of better data. The no-effect level from a chronic or lifetime study in a laboratory animal is diminished by a factor of 100 (Ref. 15-21), i.e.,

$$D_T = NEL_L \times 10^{-2} \qquad \text{(Eq. 15-9)}$$

This takes into account interspecies differences and especially susceptible individuals or groups within the population.

The no-effect level from a subchronic (90-day) study requires an additional safety factor of 10 because of the shorter period of exposure (Ref. 15-21). Hence

$$D_T = NEL_{90} \times 10^{-3} \qquad \text{(Eq. 15-10)}$$

The most likely toxicity value to be found in the literature is the LD_{50} (dose lethal to 50% of the animals) for some laboratory species, usually rat or mouse. A fine source of such information is the *Registry of Toxic Effects of Chemical Substances (Ref. 15-22)*. This value is generally obtained by plotting on probit paper the fraction of experimental animals killed against the acute dosage. There is seldom enough information to permit extrapolation to a dosage at which only a very small (e.g., 1%) fraction of the animals would be killed, much less to an acceptable risk level. Handy and Schindler (Ref. 15-23), however, assume a safe limit for the maximum body concentration of a toxic substance to be $5 \times 10^{-4} \times LD_{50}$. Based on experimental studies, they also assume a biological half-life of 30 days, which implies a disappearance rate of 2.31% per day. If the daily intake of the toxic substance is made equal to the daily disappearance rate at the safe concentration limit, then that safe concentration is maintained. Thus,

$$D_T = 2.31 \times 10^{-2} \times 5 \times 10^{-4} \times LD_{50} = 1.155 \times 10^{-5} \times LD_{50} \quad \text{(Eq. 15-11)}$$

This is the least desirable method of estimating D_T.

Establishment of Partition Coefficients. Our model assumes that between any two adjacent media (such as soil and water or water and crops) the pollutant is partitioned in a perfectly constant manner, e.g., Eqs. 15-3 to 15-5. The assumption is seldom true, of course, because equilibrium is rarely achieved and because the equilibrium ratio need not be constant for all concentration levels. Moreover, it is difficult to find the needed information, and one must often accept a single literature value as typical of a given intermedia transfer; for example, one might use the concentration of dieldrin in soil and in carrots grown in that soil to provide a value of K_{sp}. The per-acre application of a pesticide could be converted to a pseudoconcentration by assuming pesticide penetration to, say, 15 cm; this, in conjunction with residue levels in root crops grown in that soil, would provide a rough value of K_{sp}. As another example, the concentration (or range of concentrations) of a pollutant is sometimes given for a body of water and for the associated sediment. The ratio of these two concentrations would be taken as the partition coefficient, that is, $K_{sw} = C_w/C_s$. It may very well be that K_{sw} would change for a different soil, perhaps one with less organic matter and more clay or sand. These cases illustrate how attentiveness and ingenuity may be required to develop even poor information. When no pertinent information can be found, one may use algorithms that relate soil–water partition coefficients (Ref. 15-24) and water–fish partition coefficients (Ref. 15-9) to octanol–water partition coefficients; the latter can be estimated for most organic compounds from structural fragment constants and algorithms available in the literature (Refs. 15-25, 15-26). In the event that there is no source whatsoever on the basis of which to calculate a given partition coefficient, let $K = 1$. It should be observed that a partition coefficient is not required for the transfer of pollutant between the final medium and the receptor organism, man. Here one considers the total (assumed) quantity of water or food that might be ingested.

SPPPLVs for Pollutants Originating in Surface Water or Groundwater. It is obvious that a calculated PPLV (for an aqueous solution of the pollutant) has little meaning if it is greater than the solubility of the pollutant. However, an SPPPLV above the solubility should be kept in the calculations until the PPLV is obtained by normalization, since the latter might still be less than the solubility value.

Ingestion of water:

$$C_w = \frac{BW \times D_T}{W_w} = 35\, D_T \qquad \text{(Eq. 15-12)}$$

Use of this equation entails the assumption that the water has undergone filtration or settling to the extent that particulates are not likely to be ingested. Alternatively, $C_w = MCL$.

Fish ingestion:

$$C_w = \frac{BW \times D_T}{f \times DFI \times K_{wf}} \qquad \text{(Eq. 15-13)}$$

For example, assume $f = 0.05$ (Table 15-2), in which case, $C_w = 2222 D_T/K_{wf}$.

If bioaccumulation in aquatic organisms occurs almost entirely through water contact, rather than food, it is usually true that the estimated value of C_w is not affected by the position of the fish in the food chain (trophic level).

Water used for irrigation or pond culture of plants for human consumption:

$$C_w = \frac{BW \times D_T}{f \times DFI \times K_{wp}} \qquad \text{(Eq. 15-14)}$$

The sources of food in the United States are normally quite diverse, so that the value of f shown in Table 15-2 (0.1) would not be inappropriate to represent a food crop grown in water contaminated with a unique (not widespread) pollutant. Values of K_{wp} are typically in the vicinity of unity. Thus, a calculation embodying the above assumptions along with the parameters listed in Table 15-1 would give $C_w = 1111 D_T$.

K_{wp} may tend to be higher for a root crop than for a grain or leafy vegetable.

Water used to irrigate crops for consumption by food animals (the crops assumed to be 100% of the animal diet):

$$C_w = \frac{BW \times D_T}{f \times DFI \times K_{wp} \times K_{pa}} \qquad \text{(Eq. 15-15)}$$

If food animals consume polluted water and the value of K_{wp} is low, it may be better to consider the direct water-animal equilibrium involving K_{wa} and ignore the plant compartment. Since data for K_{wa} and K_{wp} are not often found, one might be tempted to use the equivalents $K_{wa} = K_{wf}$ and $K_{pa} = K_{wf}/K_{wp}$. This may be inappropriate. For example, consider the case of a dairy cow (Ref. 15-27) consuming a pollutant that is totally accumulated—without either excretion or metabolism. If the animal is slaughtered after 10 years at a typical weight, 455 kg, and has consumed a generous daily water ration of 55 kg, its lifetime water consumption will have been 440 kg/kg body weight; thus, the highest possible value of K_{wa} is 440. Similarly, for K_{pa}, the animal is well fed on 10 kg of feed per day (90% solids), and will have eaten 36,500 kg in its 10 years, or 80 kg/kg body weight; thus, the highest possible value of K_{pa} is 80. In the case of cattle, if $K_{wp} < 1$, the route soil-water-plant-animal-man can be ignored; if $K_{wp} > 10$, the route soil-water-animal-man can be ignored. Consideration of the two routes together might be appropriate when one is dealing with $1 < K_{wp} < 10$.

Water to respired air. We do not believe that this pathway is ordinarily significant. If treated, it would require complex calculations beyond the scope of this chapter. Still, it might demand consideration in some special contexts, such as that of a relatively volatile, and not overly soluble, material of high toxicity in a large waste-disposal lagoon.

SPPPLVs for Pollutants Originating in Soil and Transferred Through Water. The calculation of an SPPPLV or a PPLV for soil through water implies a certain concentration of pollutant in the water phase, namely $C_w = K_{sw} C_s$. Should C_w

be greater than the solubility of the pollutant (*after normalization*) then no concentration in the soil would be high enough to provide a toxic level to human beings solely by the soil–water routes.

Soil leached by eventually ingested groundwater:

$$C_s = \frac{BW \times D_T}{K_{sw} \times W_w} = 35 D_T / K_{sw} \qquad \text{(Eq. 15-16)}$$

Soil used to grow crops for human consumption:

$$C_s = \frac{BW \times D_T}{K_{sw} \times K_{wp} \times f \times DFI} \qquad \text{(Eq. 15-17)}$$

or

$$C_s = \frac{BW \times D_T}{K_{sp} \times f \times DFI} \qquad \text{(Eq. 15-18)}$$

Soil leached by water used by food animals:

$$C_s = \frac{BW \times D_T}{K_{sw} \times K_{wa} \times f \times DFI} \qquad \text{(Eq. 15-19)}$$

Soil used to grow crops fed to or grazed by food animals (the crops assumed to be 100% of the animal diet):

$$C_s = \frac{BW \times D_T}{K_{sw} \times K_{wp} \times K_{pa} \times f \times DFI} \qquad \text{(Eq. 15-20)}$$

or

$$C_s = \frac{BW \times D_T}{K_{sp} \times K_{pa} \times f \times DFI} \qquad \text{(Eq. 15-21)}$$

The earlier discussion of K_{wa} and K_{pa} applies here. Note that sheep in pasture may consume a considerable amount of soil, since they graze close to the ground. An estimate would have to be made of soil ingested and this would be used for a new calculation.

River sediment via water and fish flesh:

$$C_s = \frac{BW \times D_T}{K_{sw} \times K_{wf} \times f \times DFI} \qquad \text{(Eq. 15-22)}$$

SPPPLVs for Pollutants Originating in the Upper Layers of Soil and Transferred by Respiration.

Soil from which pollutant vapors are inhaled by man. The worst situation would involve inhalation of air saturated with the vapors. Vapor buildup to any level near such a concentration is exceedingly unlikely, except in an inclosed space over a period of time. Even then, the vapor pressure over adsorbent soil-particulate surfaces could be considerably less than the vapor pressure, P_0, of the pure chemical. Note that if P_0 is not known, it can sometimes be estimated (Ref. 15-28). The main problem with the worst-case model is that it predicts a saturation vapor pressure in the air at any pollutant concentration in the soil, no matter how small. Furthermore, the model does not account for dispersion with increasing distance from the source, which occurs under most conditions. The chief value of the calculation is to determine whether the calculated saturated air concentration comes anywhere near the health-effect limiting concentration. In the latter case, the search for a more sophisticated model would be in order (e.g., Ref. 15-29), following which a determination might be made of the likely hazard. The saturation vapor density is a function of the vapor pressure (P_0 expressed in units of atmospheres at the temperature of interest), molecular weight and temperature:

$$VD_0 = P_0 \times MW \times 10^6 / RT \qquad \text{(Eq. 15-23)}$$

However, if P_0 is expressed in mm of mercury (Torr) and $T = 25°C$ ($298°K$), then

$$VD_0 \text{ (mg m}^{-3}) = 54 \times P_0 \times MW \qquad \text{(Eq. 15-24)}$$

This must be compared with the limiting concentration of the pollutant vapor in air. That value would preferably be

$$C_{air} = TLV \times RB'/140\ RB = TLV/214 \qquad \text{(Eq. 15-25)}$$
(See discussion under Establishment of a D_T.)

If there is no TLV, then it is necessary to use the breathing rate over a 24-hr period, R_B (see Table 15-1), in the following equation:

$$C_{air} = BW \times D_T/R_B = 70\ D_T/18.5 = 3.78\ D_T (\text{mg m}^{-3}) \qquad \text{(Eq. 15-26)}$$

If $C_{air} \gg VD_0$, then inhalation is an unimportant route of exposure. If the two values are close, some judgment may be necessary. Thus, for an assumed vapor pressure of the pollutant, P (P_0 or less, in Torr), we may calculate a daily dose via the vapor route, D_{air}. For $25°C$:

$$D_{air} = R_B \times VD/BW = 0.264\ VD = 14.3 \times P \times MW \qquad \text{(Eq. 15-27)}$$

This value can be used in normalization, as discussed below.

Soil dispersed as particulates (dust) inhaled by man. Unlike the relationship expressed for the previous case, the calculated exposure is dependent on the concentration of pollutant in the soil. However, as in the previous case, no account is taken of dispersion with increasing distance. Nor does the calculation consider settling out of contaminated particles. The new symbol, C_{ss}, refers to the concentration of suspended particles (suspended solids) in the air. The highest normally measured value of C_{ss} for nonurban air (Ref. 15-12), 6×10^{-2} mg m^{-3}, is chosen as the value of C_{ss} for calculation:

$$C_s = (BW \times D_T \times 10^6)/(RB \times C_{ss}) = 6.3 \times 10^7 D_T \qquad \text{(Eq. 15-28)}$$

Except for abnormally high concentrations of dust, a pollutant of very high toxicity, and very high soil concentrations acting together, the particulate inhalation route is not apt to be significant.

Normalization of SPPPLVs to PPLVs. (*Note*: It is only here that we use the symbols C_{wi}, C_{wf}, C_{si}, C_{sf} defined in Appendix 15-1.) Most likely, the pollutant will be found in soil and associated groundwater. It does not matter, then, whether each value for soil is converted to an SPPPLV value for water,

$$C_{wi} = K_{sw} C_{si} \qquad \text{(Eq. 15-29)}$$

or each value for water is converted to an SPPPLV value for soil,

$$C_{si} = C_{wi}/K_{sw} \qquad \text{(Eq. 15-30)}$$

All that matters is that each pathway be represented once and only once. For example, if ingested water is considered both as polluted groundwater, and as leachate from soil, the total water ingested is still 2 liters day^{-1}, and the water ingestion pathway is only represented once. A PPLV value is calculated (with C_{wi} values being SPPPLVs C_{w_1}, C_{w_2}, etc.) as:

$$C_{wf} = \left(\sum_{i=1}^{n} 1/C_{wi} \right)^{-1} \qquad \text{(Eq. 15-31)}$$

Thus, if the SPPPLV values, C_{wi}, were 10 ppm by water ingestion, 5 ppm by fish ingestion, and 20 ppm by crop ingestion, then the PPLV would be

$$C_{wf} = (1/5 + 1/10 + 1/20)^{-1} = 2.86 \text{ ppm} \qquad \text{(Eq. 15-32)}$$

If C_{wf}, by such a calculation or by

$$C_{wf} = K_{sw} C_{sf} \qquad \text{(Eq. 15-33)}$$

were higher than the solubility of the pollutant in water, and all probable pathways had been adequately considered, it would be unnecessary to set a PPLV. This assumes that the ingested water would be free of suspended particulates.

If inhalation of suspended particles looked like a significant route of exposure, one would preferably use the soil-concentration equation.

$$C_{sf} = \left(\sum_{i=1}^{n} 1/C_{si} \right)^{-1} \qquad \text{(Eq. 15-34)}$$

in order to include this route of exposure. Here, however, it is possible that the calculated C_{wf} would exceed the solubility of the pollutant in water. Whereupon the maximum exposure dose, D_i, would be calculated for each water-borne route, starting with the solubility limit, C_{sol}, in water. Consider the pollutant pathway through water, crops, and animals to man (letting subscript i be 4, arbitrarily). The dose reaching the receptor, man, by this route is

$$D_4 = C_{sol} \times f \times DFI \times K_{wp} \times K_{pa}/BW. \qquad \text{(Eq. 15-35)}$$

Since the value of C_{sf} can be arbitrarily large (i.e., exposure is solubility limited), as far as routes involving water are concerned, the acceptable daily dose via the suspended particulate route may be written as

$$D_T' = D_T - \sum_{i=1}^{n} D_i \qquad \text{(Eq. 15-36)}$$

whereupon

$$C_{sf} = 6.3 \times 10^7 D_T' \qquad \text{(Eq. 15-37)}$$

Here we have obtained the acceptable daily dose by the water-borne route and subtracted it from D_T to leave the daily dose D_T', allowed for the inhalation route. We have then used D_T' to obtain the PPLV for soil.

Finally, consider the case of a fairly volatile pollutant. Since D_{air} (Eq. 15-27) is not dependent on the concentration of the pollutant in the soil, this value, if significant, should be subtracted from the originally calculated value of D_T and the reduced D_T value (D_T'') used in place of D_T in calculations by all other routes:

$$D_T'' = D_T - D_{air} \qquad \text{(Eq. 15-38)}$$

15-2.4 Sample Calculations of SPPPLVs and PPLVs

The methodology discussed in Section 15-2.3 can be reduced to a practical case. At the outset, it must be decided which pathways can be eliminated completely and which should be emphasized. As a case in point, consider hexachlorobutadiene (HCBD) at the Geismar, LA site of the Vulcan Materials Company (Ref. 15-30). Being discharged into a river, the pollutant is most likely to reach man via ingestion of water, root crops, fish, and domestic animals. We should also calculate a vapor pressure to see how near the value of D_{air} calculated from P_0 comes to D_T, because the chemical might be released by dried sediment. Table 15-4 provides input data for these calculations.

TABLE 15-4. Compound-Specific Data for SPPPLV Calculations Relating to Hexachloro-1, 3-butadiene at 25°C

Property	Data as Found in Literature	Data as Required for Calculation	Reference
Solubility	4.0 mg liter^{-1} @ 20°C 7.9 mg liter^{-1} @ 40°C	4.76 mg liter^{-1} by inter-polation of $S = 4 \times 2^{(t-20)/20}$	15-31
Vapor Pressure at 25°	1.5 mm Hg @ 40°C	1 mm Hg @ 25°C by Clausius-Clapeyron equation, assuming $\Delta H_v = 6000$ cal	15-32
K_{wf}	~ 740 ppb in fish ~ 0.6 ppb in water	1.2×10^3	15-30
K_{sw}	938 ppb in sediment equilibrated with 3.6 ppb in water	3.84×10^{-3}	15-30
K_{wp}	20 ppb in emergent plants, 0.6 ppb in water	33	15-30
K_{pa}	Assume this is K_{wf}/K_{wp}	36	—
$D_T{}^a$	$NEL_{180} = 5 \times 10^{-4}$ mg kg^{-1}	$D_T = 5 \times 10^{-7}$ mg kg^{-1}	15-33
MW	——	260.7	—

[a]The value shown for D_T is considerably lower than D_T derived from other information sources. In particular, a LD$_{50}$ of about 200 mg/kg in rats (Ref. 15-33) would give a D_T of 2.3×10^{-3}, i.e., about 4600 times higher than the one we are using.

Table 15-5 is a worksheet for making the calculations. Although the calculations look straightforward, there are, in reality, many judgment factors. Thus, one could use a Russian MCL as a standard (Ref. 15-34), i.e., 10^{-2} mg liter^{-1} in drinking water, and a corresponding 2.6 mg kg^{-1} in soil. The probability of accumulation in meat might be considered very low, according to the nearly total clearance of HCBD in rats within two weeks after dosing (Ref. 15-35). If similar clearance occurs in cattle, the water and feed consumption on which to base accumulation would provide $K_{wa} < 1.7$ and $K_{pa} < 0.3$, according to the general lines of discussion following Eq. 15-15 (i.e., $14 \times 55 \div 455$, etc.). In fact, if one assumes that nearly total clearance means exponential reduction to one-tenth the original concentration in a period of 2 weeks, one can make a steady-state calculation for the ratio of HCBD concentration in the animal to that in the ingested water:

$$K_{wa} = (55 \times 14)/(455 \times \ln 10) = 0.73.$$

TABLE 15-5. PPLV Worksheet (with the Calculated Values for Hexachlorobutadiene Originating in Soil).

Formulas and Calculations	Equation
1. Water Ingestion	
$C_s = 35\,D_T/K_{sw} = 35 \times 5 \times 10^{-7}/(3.84 \times 10^{-3})$	(15-16)
$= 4.6 \times 10^{-3}\,\text{mg kg}^{-1}$ (SPPPLV)	
2. Root Crops	
$C_s = BW \times D_T/(K_{sw} \times K_{wp} \times f \times DFI)$	
$= 70 \times 5 \times 10^{-7}/(3.84 \times 10^{-3} \times 33 \times 0.1 \times 0.63)$	(15-17)
$= 4.4 \times 10^{-3}\,\text{mg kg}^{-1}$ (SPPPLV)	
3. Grain Crops	
$C_s = BW \times D_T/(K_{sw} \times K_{wp} \times f \times DFI)$	(15-17)
(Not Considered)	
4. Meat (by lifetime accumulation, but see text of Section 15-2.4)	
$C_s = BW \times D_T/(K_{sw} \times K_{wp} \times K_{pa} \times f \times DFI)$	
$= 70 \times 5 \times 10^{-7}/(3.84 \times 10^{-3} \times 33 \times 36 \times 0.2 \times 0.63)$	(15-20)
$= 6.09 \times 10^{-5}\,\text{mg kg}^{-1}$ (SPPPLV)	
5. Fish Ingestion	
$C_s = BW \times D_T/(K_{sw} \times K_{wf} \times f \times DFI)$	
$= 70 \times 5 \times 10^{-7}/(3.84 \times 10^{-3} \times 1.2 \times 10^3 \times 0.05 \times 0.63)$	(15-22)
$= 2.41 \times 10^{-4}\,\text{mg kg}^{-1}$ (SPPPLV)	
6. $C_{\text{air}} = 3.78\,D_T$	(15-26)
$= 3.78 \times 5 \times 10^{-7}$	
$= 1.89 \times 10^{-6}\,\text{mg m}^{-3}$	
$VD_0 = 54\,P_0$ (mm Hg) $\times$ MW $= 1.4 \times 10^4\,\text{mg m}^{-3}$	(15-24)
The large ratio of VD_0/C_{air}, i.e., 7.4×10^9, indicates the need for concern over exposure by the vapor route under some conditions.	
7. Normalization	
Only routes 1, 2, 4 and 5 are normalized.	
$C_{sf} = (1/0.0046 + 1/0.0044 + 1/0.0000609 + 1/0.000241)^{-1}$	(15-34)
$= 4.8 \times 10^{-5}\,\text{mg kg}^{-1}$	
8. Comparison of C_{wf} with C_{sol}	
$C_{wf} = K_{sw}C_{sf} = 3.84 \times 10^{-3} \times 4.8 \times 10^{-5}\,\text{mg liter}^{-1}$	(15-33)
$= 1.84 \times 10^{-7}\,\text{mg liter}^{-1}$	
$C_{\text{sol}} = 4.76\,\text{mg liter}^{-1}$	
$\therefore C_{\text{sol}} \gg C_{wf}$, so that solubility is not a limiting consideration.	

Similarly,

$$K_{pa} = 0.13.$$

15-2.5 Human Health Effect SPPPLVs for Some Compounds of Interest

The purpose of this section is not to provide SPPPLVs (or PPLVs) to be used indiscriminately for the cited compounds at all locations, but to demonstrate the individualized approach that must be used with compounds that may differ widely in their physicochemical and toxicological properties. To this end, we have chosen seven illustrative compounds, listed in Table 15-6, with some suitable inputs for D_T.

The first item, chlorate salts, presents a special problem. Potassium chlorate is regarded as too toxic to be used as a food additive (Ref. 15-39). However, we can make the assumption that the toxicity of chlorates is equal to or less than that of the corresponding bromates; the symptoms associated with toxic exposure to these ions are similar (Ref. 15-40). Potassium bromate is permitted in enriched bromated flour up to a maximum of 50 ppm (Ref. 15-41). Conversion of the bromate figure to chlorate, on an equimolar basis, gives a value of 32 mg kg^{-1} as chlorate. Assumption of a maximum daily consumption of 0.5 kg of bread and other flour products containing about 50% enriched bromated flour provides

$$0.5 \text{ kg day}^{-1} \times 0.5 \times 32 \text{ mg kg}^{-1} = 8.0 \text{ mg chlorate day}^{-1}$$

which is equivalent to a D_T of 0.114 mg kg^{-1} day^{-1} of chlorate.

Chlorate's low potential for bioconcentration is reflected in the fact that the lowest SPPPLV is by the water ingestion route.

Arsenic levels, especially in soil, are highly variable because arsenical agricultural pesticides have been used in some areas and are often quite persistent. Background levels were above 10 ppm in about 20% of presumably untreated soil types examined by Williams and Whetstone (Ref. 15-42). The concentration range in lakes and rivers is 0.01 to 1.1 ppm (Ref. 15-43).

Mercury in lake and stream sediments presents a greater environmental hazard than mercury in soils because mercury is more readily transmitted from sediments than from soils to man. In effect, lake and stream sediments should have lower permissible mercury levels than soils if there is a signficant chance that fish or other lake- and stream-dwelling animals will be ingested. The same remarks apply to lipid-soluble, bioaccumulated pesticides such as dieldrin.

The pesticide (dieldrin, aldrin, and endrin) SPPPLVs are quite low for the fish ingestion pathway; this is due to the importance of bioaccumulation. It should be noted that aldrin is readily converted to dieldrin in the biosphere (Ref. 15-4), so that it has been convenient to treat these two compounds together for certain purposes (cf. Table 15-8).

TABLE 15-6. Some Pollutant Standards and D_T Values for Seven Compounds.

	ADI (mg kg^{-1} day^{-1})	MCL (ppm)	TLV (mg m^{-3})	D_T (mg kg^{-1} day^{-1})
Chlorate salts[a]	None	—	—	1.14×10^{-1}
Arsenic compounds	5×10^{-2} (Ref. 15-36)	5×10^{-2}	5×10^{-1}	5×10^{-2}
Mercury compounds	4.7×10^{-4} (Ref. 15-37) (methyl-Hg)	2×10^{-3} [b]	1×10^{-2} (alkyl-Hg)	4.7×10^{-4} (alkyl-Hg)
Aldrin	1×10^{-4} (Ref. 15-38)	1×10^{-3}	2.5×10^{-1}	1×10^{-4}
Dieldrin	1×10^{-4} (Ref. 15-38)	1×10^{-3}	2.5×10^{-1}	1×10^{-4}
Endrin	2×10^{-4} (Ref. 15-38)	2×10^{-4}	1×10^{-1}	2×10^{-4}
Toluene	—	—	375	4.6×10^{-1}

[a] See text.
[b] As inorganic mercury.

For toluene, the most hazardous route of exposure may be by inhalation. Nevertheless, soil loading would have to be heavy, continuous, and widespread for this to constitute a chronic exposure threat to public health.

Individual pollutant data required for SPPPLV calculations appear in Table 15-7. The unreferenced distribution coefficients were estimated. For highly soluble, weakly adsorbed pollutants, soil–water distribution coefficients were set at unity to allow for the amount of soluble material that could readily precipitate from solution and redissolve when wetted. SPPPLVs (soil only) were calculated for various routes of exposure (Table 15-8).

15-3 COMMON SENSE IN THE DERIVATION AND USE OF PPLVs

There is a great temptation to develop PPLVs and to institutionalize them as quasistandards. This should be avoided. Perhaps each PPLV should be given a mandatory date for reconsideration. Moreover, the data base used in its formulation should be continually upgraded and examined for impact on the PPLV. One of the products of the PPLV-setting exercise should be the initiation of research to answer questions, especially where there is an indicated weakness in a critical pathway. Some of this research might be relatively easy to perform and would greatly diminish the initial uncertainty regarding suitable limit values.

PPLVs should always be considered in connection with the site-specific situation. Is the groundwater likely to undergo dilution before it reaches the nearest well? Is the exposed area of a disposal site sufficient to generate hazardous quantities of vapors? Is the background concentration of the pollutant close to the proposed PPLV (for example, the present calculations for arsenic)? The answers to such questions can be used to modify the model presented here, which must be regarded as a framework.

A word is in order about the choice of human-effects versus aquatic PPLVs. If it is unlikely that any fish population will be exposed to the pollutant, then neither the SPPPLV via fish ingestion nor aquatic PPLVs would be applicable. It is entirely possible that the latter would be a controlling factor while the former would not need to be considered (but not the converse), since the exposed organisms could be game-fish in numbers too small to constitute a significant part of anyone's diet.

It must be emphasized that the PPLV-calculation exercise should never be mechanical. It requires expertise and judgment. Each factor should be weighed carefully. The environmental data-base should be scrutinized for clues that would at least be better than wild guesses.

TABLE 15-7. Individual Pollutant Data for SPPPLV Calculations Relating to Seven Pollutants.

Property	Chlorates	Arsenic	Mercury	Aldrin	Dieldrin	Endrin	Toluene
Molecular Weight (g/mole, M)	84	139	200	365	381	381	92
Water Solubility (mg/kg, C_{sol})	2×10^5 (Ref. 15-4)	a	7×10^3 b (Ref. 15-4)	2.0×10^{-1} (Ref. 15-4)	2.5×10^{-1} (Ref. 15-4)	2.3×10^{-1} (Ref. 15-4)	570 (Ref. 15-44)
Equilibrium Vapor Pressure at 298°K (mm Hg, P_0)	—	1×10^{-5} c	1.2×10^{-3} d (Ref. 15-4)	1.4×10^{-4} (Ref. 15-4)	5.4×10^{-6} (Ref. 15-4)	2×10^{-7} (Ref. 15-4)	28 (Ref. 15-44)
Distribution Coefficients:							
Soil-Water (K_{sw})	1	1 (Refs. 15-45, 15-46)	0.2 (Refs. 15-47, 15-48)	1	1 (Refs. 15-49, 15-50)	1	0.1
Water-Root Plant (K_{wp})	1	1 (Refs. 15-51, 15-52)	1 (Ref. 15-53)	0.5 (Ref. 15-51)	0.5 (Ref. 15-51)	0.5 (Ref. 15-51)	1
Water-Grain Plant (K_{wp})	1	0.3 (Refs. 15-51, 15-52, 15-54)	1 (Refs. 15-53, 15-55)	0.5 (Ref. 15-51)	0.5 (Ref. 15-51)	0.5 (Ref. 15-51)	1
Plant-Grazing Animal[e] (K_{pa})	1×10^{-2}	5	5	5	5	5	5
Water-Fish (K_{wf})	1×10^{-2}	30 f	3×10^3 (Ref. 15-56)	3×10^3 (Ref. 15-4)	3×10^3 (Ref. 15-4)	10^3 (Ref. 15-4)	0.8 (Ref. 15-44)

[a] Highly variable.
[b] $HgCl_2$.
[c] Estimated for As_2O_3.
[d] As elemental Hg.
[e] Estimated.
[f] Highest value reported for freshwater fish (Ref. 15-57).

TABLE 15-8. SPPPLVs for Seven Pollutants Originating in Soil (mg kg^{-1})[a].

Routes of Exposure	Chlorates	Arsenic	Mercury	Aldrin/ Dieldrin	Endrin	Toluene
Ingestion of water[b]	4.0	1.75	8.2×10^{-2}	3.5×10^{-3}	7×10^{-3}	160
Ingestion of food plants[c]	1.3×10^2	56	2.6	0.22	0.44	5.1×10^3
Ingestion of foraging animals[d]	6.3×10^3	19	0.26	2.2×10^{-2}	4.4×10^{-2}	5.1×10^2
Ingestion of fish[e]	2.5×10^4	3.7	1.74×10^{-3}	7.4×10^{-5}	4.4×10^{-4}	1.3×10^4
Inhalation of vapors[f] (VD_0/C_{air})	0	32	2.8×10^5	2.4×10^{3}g 9.5×10^{1}h	8.8	7.9×10^4
Inhalation of particulates	∞	∞	3.0×10^4	6.3×10^3	1.3×10^4	i

a Except for inhalation of vapors.

b In the case of SPPPLVs derived for compounds originating in water, use MCL values.

c Root crop K_{wp} used, $f = 0.1$.

d Grain crop K_{wp} used, $f = 0.2$.

e Freshwater fish, $f = 0.05$.

f VD_0/C_{air} compares the saturation vapor pressure with the permissible concentration in air (which in all these cases is $TLV/214$).

g Value for aldrin.

h Value for dieldrin.

i Compound too volatile to adhere to particles.

Anticipation of all questions is impossible in any treatment of the type presented here. The authors hope, instead, that this chapter will provide the stimulation to seek and apply answers.

REFERENCES

15-1. Elliott, J., "Lessons from Love Canal," *J. Am. Med. Assoc.*, **240**, pp. 2033-2034, p. 2040, 1978.

15-2. Anon., "Chemical Industry Warned of Waste Dumps," *Chem. Eng. News,* **57**, No. 8, p. 6, February 19, 1979.

15-3. Rosenblatt, D. H., Miller, T. A., Dacre, J. C., Muul, I., and Cogley, D. R. (Eds.), *Problem Definition Studies on Potential Environmental Pollutants. I. Toxicology and Ecological Hazards of 16 Substances at Rocky Mountain Arsenal,* Technical Report 7508, U. S. Army Medical Bioengineering Research & Development Laboratory, Ft. Detrick, Federick, MD, December, 1975 (AD B039661L).

15-4. Rosenblatt, D. H., Miller, T. A., Dacre, J. C., Muul, I., and Cogley, D. R. (Eds.) *Problem Definition Studies on Potential Environmental Pollutants. II. Physical, Chemical, Toxicological, and Biological Properties of 16 Substances,* Technical Report 7509, U. S. Army Medical Bioengineering Research & Development Laboratory, Ft. Detrick, Frederick, MD, December, 1975 (AD AO30428).

15-5. Dacre, J. C., Rosenblatt, D. H., Woodard, G., and Cogley, D. R., "Toxic Potential Evaluation of Soil Pollutant Chemicals," *Toxicol. Appl. Pharmacol.,* **37**, p. 104, 1976.

15-6. U. S. Environmental Protection Agency, "National Interim Primary Drinking Water Regulations," *Fed. Regist.,* **40**, pp. 59565-59588, December 24, 1975.

15-7. U. S. Environmental Protection Agency, "Toxic Pollutant Effluent Standards," *Fed. Regist.,* **42**, pp. 2588-2621, January 12, 1977.

15-8. U. S. Environmental Protection Agency, "Water Quality Criteria," *Fed. Regist.,* **44**, pp. 15926-15981, March 15, 1979.

15-9. U. S. Environmental Protection Agency, "Water Quality Criteria," *Fed. Regist.,* **43**, pp. 21506-21518, May 18, 1978.

15-10. Perelygin, V. M., "The Theoretical Bases of Hygienic Standardization of Noxious Substances in the Soil," *Gig. Sanit.,* **40**, No. 1, pp. 29-33, 1975.

15-11. Sidorenko, G. I., and Perelygin, V. M., "The Problems of Hygienic Standards of Chemical Substances in the Soil," *Cesk, Hyg.,* **20**, pp. 103-107, 1975.

15-12. McCormick, R. A., "Air Pollution Climatology," *Air Pollution*, A. C. Stern (Ed.), Academic Press, New York, Vol. I, p. 316, 1968.

15-13. National Research Council, *Drinking Water and Health*, National Academy of Sciences, Washington, DC, pp. 11, 498, 1977.

15-14. Albritton, E. C. (Ed.), *Standard Values in Nutrition and Metabolism*, W. B. Saunders Company, Philadelphia and London, pp. 48-57, 1954.

15-15. Cleland, J. G., and Kingsbury, G. L., *Multimedia Environmental Goals for Environmental Assessment, Vol. I.* EPA 600/7-77-136a, Environmental Protection Agency, Washington, DC, pp. 60-61, November 1977.

15-16. Albritton, E. C. (Ed.), *Standard Values in Nutrition and Metabolism*, W. B. Saunders Company, Philadelphia and London, pp. 75-76, 1954.

15-17. U. S. Environmental Protection Agency, "Water Quality Criteria," *Fed. Regist.,* **44,** pp. 15977-15981, March 15, 1979.

15-18. Rawls, R. L., "Dow Finds Support, Doubt for Dioxin Ideas," *Chem. Eng. News,* **57,** No. 7, pp. 23-25, 28-29, February 12, 1979.

15-19. World Health Organization, *Evaluation of the Toxicity of a Number of Antimicrobials and Antioxidants*, Sixth Report of the Joint FAO/WHO Expert Committee on Food Additives, WHO Tech. Rep. Ser. No. 228, pp. 9-11, 1962.

15-20. American Conference of Governmental Industrial Hygienists, *Documentation of the Threshold Limit Values* (for Substances in Workroom Air), 3rd ed., 2nd printing, American Conference of Governmental Industrial Hygienists, Cincinnati, OH, 1974.

15-21. Vettorazzi, G., "Safety Factors and Their Application in the Toxicological Evaluation." in *The Evaluation of Toxicological Data for the Protection of Public Health*, Proc. Int. Colloq. Commission of the European Communities, Luxembourg, 1976, W. J. Hunter, and J. G. P. M. Smeets (Eds.), Pergamon Press, Oxford/New York, pp. 207-223, 1976.

15-22. Lewis, R. J., and Tatken, R. L., (Eds.), *Registry of Toxic Effects of Chemical Substances*, 1978 Edition, DHEW (NIOSH) Publication No. 79-100, National Institute for Occupational Safety and Health, Cincinnati, OH, January 1979.

15-23. Handy, R. and Schindler, A., *Estimation of Permissible Concentrations of Pollutants for Continuous Exposure*, EPA 600/2-76-155, Environmental Protection Agency, Washington, DC, p. 61, June 1976.

15-24. Kenega, E. E., and Goring, C. A. I., "Relationship Between Water Solubility, Soil Sorption, Octanol-Water Partitioning, and Concentration of Chemicals in Biota," In *Aquatic Toxicology*, ASTM STP 707, J. G. Eaton, P. R. Parrish, and A. C. Hendricks (Eds.), American Society for Testing and Materials, Philadelphia, PA, pp. 78-115, 1980.

15-25. Leo, A., Hansch, C., and Elkins, D., "Partition Coefficients and Their Uses," *Chem. Rev.,* **71,** pp. 525-616, 1971.

15-26. Rekker, R. F., *The Hydrophobic Fragmental Constant*, Elsevier Scientific Publishing Company, Amsterdam/Oxford/New York, 1977.

15-27. Siegmund, O. H. (Ed.), *The Merck Veterinary Manual*, Merck & Co., Inc., Rahway, NJ, 1973.

15-28. Reid, R. C., Prausnitz, J. M., and Sherwood, T. K., *The Properties of Gases and Liquids*, 3rd ed., McGraw-Hill Book Company, New York, 1977.

15-29. Turner, D. B. *Workbook of Atmospheric Dispersion Estimates*, Publ. No. AP-26, Environmental Protection Agency, Office of Air Programs, Research Triangle Park, NC, 1970.

15-30. Laseter, J. L., Bartell, C. K., Laska, A. L., Holmquist, D. G., Condie, D. B., Brown, J. W., and Evans, R. L., *An Ecological Study of Hexachlorobutadiene (HCBD)*, EPA 560/6-76-010, Environmental Protection Agency, Washington, DC, April 1976.

15-31. Antropov, L. I., Pogulyai, V.E., Simonov, V. D., and Shamsutdinov, T. M., "Solubility in Chlorocarbon–Water Systems," *Russ. J. Phy. Chem.,* **46**, pp. 311–312, 1972.

15-32. Mumma, C. E., and Lawless, E. W., *Survey of Industrial Processing Data: Task I - Hexachlorobenzene and Hexachlorobutadiene Pollution from Chlorocarbon Processing*, EPA 560/3-75-003, Environmental Protection Agency, Washington, DC, June 1975.

15-33. Berkowitz, J. B., Harris, J. C., Lyman, W. L., Horne, R. A., Nelken, L. H., Harrison, J. E., and Rosenblatt, D. H., *Literature Review–Problem Definition Studies on Selected Chemicals, Volume II. Chemistry, Toxicology and Potential Environmental Effects of Selected Organic Pollutants* (Appendix G, Chemistry, Toxicology and Potential Environmental Effects of Hexachloro-1, 3-butadiene), Arthur D. Little, Inc., Cambridge, MA, June 1978.

15-34. Murzakaev, F. G., "Maximum Permissible Concentrations of Hexachlorobutadiene and Polychlorobutanes in the Water of Reservoirs," *Gig. Sanit.*, **28**, No. 2, pp. 9–14, 1963.

15-35. Gul'ko, A. G., and Dranovskaya, L. M., "Distribution and Excretion of Hexachlorobutadiene from Rats," *Aktual. Vopr. Gig. Epidemiol.*, pp. 58–60, 1972; *Chem. Abstr.*, **81**, p. 346q, 1974.

15-36. World Health Organization, *Specifications for the Identity and Purity of Food Additives and Their Toxicological Evaluation: Some Emulsifiers and Stabilizers and Certain Other Substances*, Tenth Report of the Joint FAO/WHO Expert Committee on Food Additives, WHO Tech. Rep. Ser. No. 373, pp. 14–15, 1967.

15-37. World Health Organization, *Evaluation of Certain Food Additives and the Contaminants Mercury, Lead, and Cadmium*, Sixteenth Report of the Joint FAO/WHO Expert Committee on Food Additives, WHO Tech. Rep. Ser. No. 505, FAO Nutrition Meetings Report Series No. 51, 1972.

15-38. World Health Organization, *Pesticide Residues in Food*, Report of the 1974 Joint Meeting of the FAO Working Party of Experts on Pesticide Residues and the WHO Expert Committee on Pesticide Residues. WHO Tech. Rep. Ser. No. 574, FAO Agric Studies No. 97, Annex 1, 1975.

15-39. World Health Organization. *Specifications for the Identity and Purity of Food Additives and Their Toxicological Evaluation: Some Food Colors, Emulsifiers, Stabilizers, Anticaking Agents, and Certain Other Substances*, Thirteenth Report of the Joint FAO/WHO Expert Committee on Food Additives, WHO Tech. Rep. Ser. No. 445, 1970.

15-40. Gleason, M. N., Gosselin, R. E., Hodge, H. C., and Smith, R. P., *Clinical Toxicology of Commercial Products. Acute Poisoning,* 3rd ed., The Williams & Wilkins Co., Baltimore, MD, Section III, pp. 48–51, 1969.

15-41. Food and Drug Administration, *Code of Federal Regulations, Title 21,* Part 15, Cereal Flour Standards, Superintendent of Documents, Washington, DC, April 1978.

15-42. Williams, K. T., and Whetstone, R. R., "Arsenic Distribution in Soils and Its Presence in Certain Plants," U. S. Dept. of Agric. Tech. Bull. 732, pp. 1–20, 1940.

15-43. Durum, W. H., Hem, J. D., Heidel, S. G., "Reconnaissance of Selected Minor Elements in Surface Waters of the United States," U. S. Geol. Surv. Circ. 643, 1971.

15-44. Miller, T. A., Rosenblatt, D. H., Dacre, J. C., Pearson, J. G., Kulkarni, R. K., Welch, J. L., Cogley, D. R., and Woodard, G., (Eds.), *"Problem Definition Studies on Potential Environmental Pollutants: IV. Physical, Chemical, Toxicological, and Biological Properties of Benzene; Toluene; Xylenes; and p-Chlorophenyl Methyl Sulfide, Sulfoxide, and Sulfone,"* Technical Report 7605, U. S. Army Medical Bioengineering Research & Development Laboratory, Fort Detrick, Frederick, MD, June, 1976 (AD A040435).

15-45. Sundd, D. K., and Bansal, O. P., "Studies on the Adsorption of Arsenites by a Few Typical Indian Soils," *Indian J. Appl. Chem.* 29, pp. 23–26, 1966.

15-46. Jacobs, L. W., Syers, J. K., and Keeney, D. R., "Arsenic Sorption by Soils," *Soil Sci. Soc. Amer. Proc.*, 34, pp. 750–754, 1970.

15-47. Andersson, A. A., "Mercury in the Soil," *Grundforbattring,* 20, pp. 95–105, 1967.

15-48. Jenne, E. A., "Atmospheric and Fluvial Transport of Mercury," in *Mercury in the Environment*, W. T. Pecora (Ed.), U. S. Geol. Surv. Prof. Pap. No. 713, pp. 40–45, 1970.

15-49. Eye, J. D., "Aqueous Transport of Dieldrin Residues in Soils," *Diss. Abstr. B.,* 27, pp. 3548–3549, 1967.

15-50. Boucher, F. R., and Lee, G. F., "Adsorption of Lindane and Dieldrin Pesticides on Unconsolidated Aquifer Sands," *Environ. Sci. Technol.,* 6, pp. 538–543, 1972.

15-51. Nash, R. G., "Plant Uptake of Insecticides, Fungicides, and Fumigants from Soils," in *Pesticides in Soil and Water*, W. D. Guenzi, (Ed.), Soil Science Society of America, Inc., Madison, WI, pp. 257–313, 1974.

15-52. Woolson, E. A., "Arsenic Phytotoxicity and Uptake in Six Vegetable Crops," *Weed Sci.,* 21, pp. 524–527, 1973.

15-53. Gracey, H. I., and Stewart, J. W. B., "Distribution of Mercury in Saskatchewan Soils and Crops," *Can. J. Soil Sci.,* 54, pp. 105–108, 1974.

15-54. Moore, L., Fleischer, M., and Woolson, E. A., "Distribution of Arsenic in the Environment–Natural Sources," in *Arsenic*, NRC Committee on Medical and Biological Effects of Environmental Pollutants, National Academy of Sciences, Washington, DC, pp. 16–26, 1977.

15-55. Van Loon, J. C., "Mercury Contamination of Vegetation Due to the Application of Sewage Sludge as a Fertilizer," *Environ. Lett.,* 6, pp. 211–218, 1974.

15-56. Lagerwerff, J. V., "Lead, Mercury, and Cadmium as Environmental Contaminants," in *Micronutrients in Agriculture*, Soil Science Society of America, Inc., Madison, WI, pp. 593–636, 1972.

15-57. Pratt, D. R., Bradshaw, J. S., and West, B., "Arsenic and Selenium Analyses in Fish," *Proc. Utah Acad. Sci., Arts Lett.,* 49, No. 1, pp. 23–26, 1972.

APPENDIX 15-1. Symbols and Abbreviations.

ADI	=	acceptable daily intake, mg kg^{-1} day^{-1}
BW	=	body weight (human), kg
C_a	=	limiting concentration of pollutant in a food animal, typically cattle, mg kg^{-1}
C_{air}	=	limiting concentration of pollutant as vapor in air, mg m^{-3} (or ppm)
C_p	=	limiting concentration of pollutant in a crop used for human or meat animal food, mg kg^{-1}
C_s	=	limiting pollutant level in soil for human health effects, mg kg^{-1}
C_{sf}	=	final C_s value (a PPLV), mg kg^{-1}
C_{si}	=	initial C_s value (an SPPPLV), mg kg^{-1}
C_{sol}	=	solubility limit for a pollutant in water at the temperature under consideration, mg liter^{-1} = mg kg^{-1}
C_{ss}	=	assumed maximum probable (nonurban) particle concentration in air, mg m^{-3}
C_w	=	limiting pollutant level in water for human health effects, mg liter^{-1} = mg kg^{-1}
C_{wf}	=	final C_w value (a PPLV), mg liter^{-1} = mg kg^{-1}
C_{wi}	=	initial C_w value (an SPPPLV), mg liter^{-1} = mg kg^{-1}
D_{air}	=	acceptable daily dose via the vapor route, mg kg^{-1} day^{-1}
DFI	=	daily food intake, kg day^{-1}
D_i	=	some daily dose less than D_T by any applicable route, mg kg^{-1} day^{-1}
D_T	=	acceptable daily dose of a toxic substance, mg kg^{-1} day^{-1}
D_T'	=	acceptable daily dose of a toxicant by the inhalation of soil particles, mg kg^{-1} day^{-1}
D_T''	=	acceptable daily dose in addition to that accounted for by the inhaled vapor route, mg kg^{-1} day^{-1}
f	=	fraction of total diet represented by food of a given type
HCBD	=	hexachloro-1, 3-butadiene
K_{pa}	=	C_a/C_p
K_{sp}	=	C_p/C_s
K_{sw}	=	C_w/C_s
K_{wa}	=	C_a/C_w
K_{wf}	=	Concentration in fish/C_w, i.e., bioconcentration factor
K_{wp}	=	C_p/C_w
LD_{50}	=	single dose level that kills 50% of a group of test animals of a given species (and preferably strain and sex), with 14-day observation of the animals, mg kg^{-1}
MCL	=	maximum contaminant level (provisional water limit) in water, mg liter^{-1} = mg kg^{-1}
MW	=	molecular weight, g mol^{-1}
NEL_L	=	lifetime (chronic) no-effect level, mg kg^{-1} day^{-1}
NEL_{90}	=	90-day (subchronic) no-effect level, mg kg^{-1} day^{-1}
P	=	partial (less-than-saturation) pressure of a pollutant vapor, atmospheres = Torr/760 (Torr = mm of mercury)
P_0	=	saturation vapor pressure of a pollutant, atmospheres or Torrs
PPLV	=	preliminary pollutant limit value, mg liter^{-1} or mg kg^{-1}
ppm	=	parts per million (mg liter^{-1} *in water*; mg kg^{-1} *in soil, food or living organisms*; molecules per million molecules *in air*)

APPENDIX 15.1. (Continued.)

ppb	=	parts per billion = ppm $\times$ 1000
R	=	gas constant, 0.082054 liter atm deg^{-1} mol^{-1}
RB	=	breathing rate, volume of air breathed by a 70-kg human being in 24 hours, m^3 day^{-1} (see text)
RB'	=	breathing rate, volume of air breathed by a 70-kg human being in an 8-hr workday, m^3 day^{-1} (see text)
SPPPLV	=	single-pathway preliminary pollutant limit value, mg liter^{-1} or mg kg^{-1}
T	=	temperature, degrees Kelvin
TL	=	tolerance level in food of a given type, mg kg^{-1}
TLV	=	threshold limit value for occupational exposure to toxic airborne pollutants over a 40-hour week, mg m^{-3} (or ppm)*
VD	=	less-than-saturation vapor concentration, mg m^{-3}
VD_0	=	saturation vapor density, mg m^{-3}
W_w	=	daily water intake, liter day^{-1} = kg day^{-1}

*mg m^{-3} = ppm $\times$ MW/24 at 25°C.

16

The development of testing requirements under the Toxic Substances Control Act

A. Karim Ahmed, Ph.D.
Senior Staff Scientist
Natural Resources Defense Council, Inc.
New York, New York

George S. Dominguez
President
Springborn Management Consultants, Inc.
Enfield, Connecticut

The Toxic Substances Control Act (TSCA) was signed into law on October 11, 1976 and became effective January 1, 1977. In simplest terms, the objective of the law is to protect human health and the environment against unreasonable risk from toxic substances. The term "unreasonable risk" is one of the key statutory guidelines of TSCA, and a recognition of this concept is essential because it embodies the legal and scientific standard that must be met in order to implement the law or impose regulation.

Inherent in the law is authority given to the Environmental Protection Agency (EPA) over both new and existing chemicals. Various legal mechanisms are established, enabling EPA to require testing and impose regulations. In the case of new chemicals or new uses for existing chemicals. Premarketing

Notification (PMN) requires that manufacturers report their intention to begin production of a new chemical substance.

With new chemicals, greater reliance must be placed on predictive testing. This in fact is the basis for establishing the PMN requirements because it is at this stage that the manufacturer or processor of the chemical must submit information on a new chemical substance, including results of environmental and health effects testing. PMN thus helps to provide EPA with sufficient information to make its decision based on the "unreasonable risk" criterion.

A group convened under the auspices of the Conservation Foundation recognized the need for communication among the public interest and the industrial community. The following individuals participated in this project:

Sam Gusman, Chairman
Conservation Foundation Environmental Issues Dialogue Group*

A. Karmin Ahmed
Senior Staff Scientist
Natural Resources Defense
    Council

Linda M. Billings
Washington Representative
Sierra Club

Terry Davies
Executive Vice President
The Conservation Foundation

George S. Dominguez
Director of Government Relations
Safety, Health & Ecology
CIBA-GEIGY Corp.

A. Blakeman Early
Legislative Director
Environmental Action

K. Warren Easley
Director of Regulatory Affairs
Monsanto Company

Fred Hoerger
Director, Regulatory and
    Legislative Issues
Health & Environmental Research
Dow Chemical

Glenn Paulson
Assistant Commissioner for Science
New Jersey Department of
    Environmental Protection

James I. Reilly
Director of Environmental Affairs
E.I. du Pont de Nemours Company

Richard P. Nalesnik
Vice President
National Association of Manufacturer

John G. Tritsch
Assistant Technical Director
Manufacturing Chemists Association

Jacqueline M. Warren
Staff Attorney
Environmental Defense Fund

---

*These persons, while affiliated with the organizations indicated, were participating as individuals and not as institutional representatives.

This project, which was one of a number of programs initiated by the Conservation Foundation, was structured in order to provide an opportunity for discussion on some of the environmental and health issues pertaining to toxic substances. This led to the identification of some critical issues, such as testing under TSCA, where it was felt that preparation of a report could lead to specific recommendations to the Agency.

The problems inherent in testing and the determination of risk were recognized by the group as being among the most important areas of concern. This led to the formation of a panel under the joint chairmanship of Dr. A. Karim Ahmed, Senior Staff Scientist, Natural Resources Defense Council, and George S. Dominguez, President, Springborn Management Consultants. Panel members are listed in Appendix 16-1. In the following sections of this chapter, the results of the joint effort are presented.

As with any attempt at such an ambitious project, it was clearly recognized that a consensus would not be possible in all instances, and so there are commentaries available on several specific areas (Appendix 16-2). However, considering this was the first effort directed towards developing a conceptual approach to testing under TSCA by major representatives of the public interest, academic, professional testing laboratories, and the chemical industry communities, and further considering that there was consensus in overall principles and guidelines, this project stands as an example of the success that can be achieved through cooperative effort.

The report was presented at several seminars and comments were solicited. The report, together with these comments, was submitted to EPA on July 20, 1978, with the recommendation that it be considered as the basis for development of a hierarchical testing program under TSCA. EPA acknowledged the proposal and indicated that it would be taken into consideration in the development of Agency testing guidelines and policy.

16-1 PURPOSE AND SCOPE OF TSCA TESTING GUIDELINE STUDY

We present in this chapter an approach for the development of a comprehensive and sound testing program for determining the potential effects of chemical substances on health and the environment, as required under the Toxic Substances Control Act (TSCA). This effort is a product of over nine months of extensive deliberations between members of a panel representing industry, academic institutions, and public interest organizations. The chapter outlines a basic conceptual framework of a testing program for adoption as rules or guidelines under the premarket notification provisions of TSCA. We also believe that the same guidelines should be applied to potentially toxic chemical substances placed on the Priority List by the Interagency Committee on Testing and may be used for the testing of other existing chemical substances.

The conceptual nature of this study should be stressed: no attempt has been made to provide more than a brief outline to define decision rules. More precise definition of specific criteria for the selection of tests at the various tier levels obviously represents a major aspect of any testing guideline program and may be the subject of future consideration by the panel. For the most part, general types of tests are recommended, we have not attempted to define actual testing procedures or detailed protocols here.

16-2   BASIC APPROACH FOR DEVELOPING SUGGESTED TSCA TESTING GUIDELINES

The approach for developing the suggested guidance was to attempt consensus among the panelists, recognizing that it would not always be possible to achieve complete agreement. The resultant recommendations, however, reflect consensus in major areas. Minority positions were accepted and transmitted with the report; these are listed in Appendix 16-1.

The principal features and recommendations are as follows:

1. That a tiered or hierarchical approach using a logically consistent series of tests be employed in the testing of chemical substances or mixtures. We believe that this is a more cost-effective and scientifically acceptable approach than any requirement not providing for a sequential assessment of potential risks. The requisite number of tiers of testing for a particular chemical substance will be determined by factors such as the type and extent of potential exposure of humans and the environment to the chemical substance. This minimum number of tiers will not be decreased by negative results in the lower tiers. This number may range from Tiers 0 and I minimally required for almost all substances to complete testing in all tiers.
2. That all relevant aspects of the potential impacts on human and animal health and the natural environment be tested and analyzed, beginning with relatively simple and inexpensive tests, which are followed by increasingly complex and lengthy testing, as required. In general, we have recommended those types of tests for which adequate protocols or standard methodologies already exist. We have also indicated separately further areas of research for the development of tests for which there are no generally recognized protocols or procedures. We stress that the guidelines are open to revision as new information and technologies become available.
3. That tests within each tier level should be required on a selective and necessary basis, i.e., the decision to test at a higher tier level should be

made based on relevant information gathered at a lower tier level and on considerations of production/use/exposure, nature of the hazard, physical-chemical properties and on Structure-Activity Relationships (SAR). For example, if a substance that has low exposure potential to the environment is shown in early tier tests to be readily soluble in water, to have a low octanol-water partition coefficient, and to be biodegradable, there may be little reason to conduct elaborate bioaccumulation tests at the latter tier levels.   On the other hand, if early information indicates that adverse, chronic effects are likely at environmental concentrations, or if water solubility and the partition coefficient data make bioconcentration likely, both higher tier health and bioaccumulation testing would be required.

The tier system is constructed so that successive tests confirm, clarify, or expand information from earlier tiers.   It should be stressed that the ultimate decision on the safety or risk of a chemical substance would be based on total information developed in all tiers.

4. That tests in Tier 0 and I should be viewed as being minimally required of almost all substances (exceptions may include substances used strictly for limited experimental purposes only).   Some flexibility may also be appropriate at the Tier 0 and I stage of testing.   For example, a liquid substance with a high boiling point and low vapor pressure may not need to be subjected to acute inhalation tests at the Tier I level.   Thus, we stress the need to use good professional judgment when it comes to selecting tests in these tiers in order to avoid overliteral application of the testing guidelines.   On the other hand, where there is any question of possible harmful effect—even a "border-line" case—the testing guidelines should be followed as closely as possible.

16-3 PROPOSED TIERS OF TESTING

The following is an outline of the proposed tier testing guidelines which are presented on the following pages:

TIER 0

- Physical/chemical properties
- Elementary mass balance analysis
- Preliminary analytical methods

TIER I

- Acute toxicity tests: Mammalian
- Acute toxicity tests: Aquatic and avian species
- Screening tests for toxicity to plants
- Screening tests for environmental transformation and degradation

- Screening tests for chronic health effects
- Refinement and application of analytical procedures

TIER II

- Subacute toxicity tests: Mammalian
- Reproduction and teratogenicity tests
- Further chronic health effects tests
- Chronic toxicity tests: Aquatic species
- Chronic toxicity tests: Plants
- Transformation and degradation tests
- Bioaccumulation tests: Aquatic flora and fauna
- Sludge toxicity tests
- Further refinement and application of analytical methods

TIER III

- Chronic toxicity tests: Mammalian long-term study
- Photochemical degradation tests
- Soil transfer and degradation tests
- Chronic toxicity tests: Terrestrial and aquatic organisms
- Metabolism studies
- Further refinement and application of analytical methods

16-3.1 Tier 0

Tier 0 requires basic information on the physical and chemical properties of the chemical substance. This information, combined with the "elementary mass balance analysis" also required at this stage, allows a preliminary assessment of chemical properties, the probable environmental fate of the substance, and the degree to which humans and wildlife might be exposed.

Physical Chemical Properties. The following tests are designed to provide the most elementary information about a chemical substance. Standard procedures and methodology exist to conduct them:

- Molecular weight
- Empirical formula
- Chemical structure
- Product assay for purity of the chemical substance
- Physical properties of solid materials
- Solubility in polar and nonpolar solvents
- Partition coefficient (water–octanol): Calculated or measured
- Vapor pressure
- Melting point/boiling point

- Flash point
- Stability/reactivity (includes storage stability)
- Physical state
- Acid/base properties including determination of ionization constants
- Density
- Absorption/desorption properties

Elementary Mass Balance Analysis. This includes an initial analysis of potential production levels or capacity, release rates, exposure potential and distribution (air, water, land, products), and storage and disposal data. Where there are significant changes in production or use, it would be necessary to again analyze the mass balance of the chemical substance.

Preliminary Identification and Development of Analytical Methods for Purity. A preliminary assessment of known analytical techniques (such as gas–liquid chromotographic detection capabilities, mass spectrometry, etc.) and their reliability is required at this stage. If sound methods do not exist, development of such techniques should be initiated.

16-3.2 Tier I

Tier I testing expands the physical/chemical data on the chemical substance and begins a preliminary but critical evaluation of biological activity. Results of Tier 0 will provide the basis for many decisions regarding a chemical substance's safety.

Acute Toxicity Tests: Mammalian. Tests used to characterize immediate effects in mammals, by a variety of routes of exposure, are well-defined. The actual routes of exposure used in acute toxicity tests should anticipate the ways in which mammals might be exposed to them. If the exposure route is multiple or cannot be anticipated, then we suggest that all the following tests be conducted:

- Oral toxicity (LD_{50})
- Inhalation toxicity (LC_{50})
- Dermal toxicity (LD_{50})
- Dermal irritation
- Eye irritation

The statistical calculation of LD_{50} and LC_{50} may not always be required for materials in low production and use if preliminary tests indicate very low toxicity to a test species; in these cases, "range finding" data may be sufficient.

Acute Toxicity Tests: Aquatic and Avian Species. The sensitivity of aquatic and avian species to toxic chemical substances is a well-known phenomenon.

Appropriate species of fish (such as rainbow trout, bluegill, or fathead minnows) and birds should be chosen as a predictive organism using a prescribed protocol for LC_{50}/LD_{50} determination. An LC_{50} and EC_{50} test with *Daphnia magna* is also recommended (Ref. 16-1).

Screening Tests for Toxicity to Plants. Screening tests for effects on plants include the algal bottle test (EPA method) and tests with duckweed that measure altered growth potential and other effects (Ref. 16-2). Short-term tests for effects on seed germination (ASTM) are also included.

Screening Tests for Environmental Transformation and Degradation. We believe that Tier I also represents the stage of analysis where the chemical substance under scrutiny must be considered in conjunction with other chemical and biochemical events. A number of transformation and degradation screens are suggested:

- Biochemical oxygen demand (BOD tests — 5, 10, and 20 days) using enriched mixed culture
- Reactivity with air and water (a qualitative assessment of oxidation, reduction and hydrolysis potential with an emphasis on preliminary determination of reaction rates)
- Preliminary assessment and/or discussion of probable degradation products

Mutagenicity/Carcinogenicity Screens. Mutagenicity and carcinogenicity short-term testing is a complex area and has been the subject of investigation by a number of expert committees (Ref. 16-3). Increasing recognition that a chemical carcinogen may operate in the same way that a mutagen induces hereditary changes—by interacting with cellular DNA—has led to the increasing use of short-term *in vitro* tests capable of detecting mutagenicity as indicators of carcinogenicity. A number of studies bear out this relationship and show very good (85%-96%) correlation between known mammalian carcinogens and positive results in various *in vitro* mutagenicity test systems (Refs. 16-4 to 16-8).

Transformation of cells in culture (as demonstrated, for example, by altered growth potential) may also be the result of a mutation (Refs. 16-9, 16-10). Therefore, cell transformation tests are being considered by the National Cancer Institute (NCI) as potential indicators of mutagenicity/carcinogenicity (Ref. 16-10). One of these systems—the Syrian hamster embryo cell transformation test—shows a high correlation both with results from animal testing and with the Ames/*Salmonella* system (Refs. 16-7 to 16-11).

Advantages of *in vitro* screening tests are low cost (on the order of $350-$5,000), the speed with which results are obtained (generally 2 weeks to several months), and the fact that these tests offer the means of studying the mechanisms of chemical carcinogenesis.

At the Tier I Level, *we recommend not less than three well validated, complementary screening tests.* Three well-validated short-term tests are recommended as a goal, although at present three such tests may not necessarily be available. In choosing an appropriate group of tests, one should consider the degree of validation and the reproducibility and reliability of a test system (i.e., whether it can be performed with consistent results by a number of laboratories). There are a number of tests which meet these criteria sufficiently well to be recommended.

These tests should be appropriate to the substance tested. For example, if a test system does not detect mutagenic properties in certain classes of substances this system would not be appropriate for substances in these categories. A number of tests are recommended because a single, simple test system is unable to detect all potential types of genetic changes (e.g., gene chromosomal mutations or primary DNA damage) with sufficient sensitivity. Further, the use of complementary tests will insure that one test system will compensate for weaknesses inherent in another.

By "well-validated" we mean reproducible tests that have been shown to have at least 85% correlation between a positive response and evidence of mammalian carcinogenicity for a number of representatives of different chemical classes of carcinogens. In other words, well-validated tests are those that generate no more than 15% false-positives or false-negatives in comparison to appropriate *in vitro* tests.

Several short-term methods have been used to test a number of chemical substances, and validation studies showing high correlation with animal carcinogenicity tests results have been published (Refs. 16-4 to 16-13). The study by Purchase *et al.,* (Ref. 16-4) is a summary of results; one should await the complete publication of data before making final judgments on the test systems described.

These methods include the Ames/*Salmonella* test (Ref. 16-12), the Syrian hamster (embryo fibroblast) cell transformation test validated by Pienta (Ref. 16-7), DNA repair in mammalian cells (Ref. 16-13), and the Drosophila sex-linked recessive lethal test (Ref. 16-14). Several other methods are widely used and appear reliable, although formal validation studies have not yet been published. Other tests include the following:

- Point mutation in mammalian cells in culture e.g., the mouse lymphoma (L5178Y) cell system (Ref. 16-15).
- Chinese hamster ovary (CHO/HGPRT) test (Ref. 16-16).
- DNA repair in bacteria; for example, the *E. coli* (Pol A-1) assay (Ref. 16-17).*
- Ames urine assay test (Ref. 16-19).
- Mitotic recombination and/or gene conversion in yeast (Ref. 16-23).*
- Sister chromatid exchange in mammalian cells in culture (Ref. 16-20).

- Mouse cell transformation tests such as BALB/c 3T3:** and C3H 10T 1/2 tests (Refs. 16-21, 16-22).
- Fisher rat embryo cells infected with Rauscher leukemia virus (Ref. 16-10).**
- Rat liver epithelial cell transformation test (Ref. 16-10)
- Human DNA synthesis tests (i.e., W138) (Ref. 16-11).
- V79 Chinese hamster cell transformation (Ref. 16-24).
- *In vivo* and *in vitro* cytogenetic tests for chromosomal breaks and aberrations (Refs. 16-25, 16-26, 16-27) (for example, combined with mammalian culture tests, or sister chromatid exchange).

Since chemical substances are often nonmutagenic unless converted to an active mutagen by metabolic processes, metabolic activation should be incorporated into any *in vitro* test system.

Subacute 10 or 30 Day Whole Animal Tests on a Single Species. These tests are well defined and indicate possible neuropathic and other toxic effects (e.g., organ site specificity).

Dermal Sensitization and Irritation Tests on Mammals. Sensitization tests are required to observe potential systemic effects of immune system activation. Subacute doses are generally administered by intradermal injection or by topical application (patch test).

Refinement and Application of Analytical Procedures as Required. Reliable and sensitive analytical techniques are to be developed and applied at this stage in order to reasonably characterize all trace contaminants and impurities. Methods and procedures should be developed for detecting and monitoring the chemical substance and its degradation products in the environment in concentrations as low as 1-10 ppm.

16-3.3 Tier II

Tests in Tier II are designed to clarify and expand upon the results of Tier I health and environmental tests. These include tests to evaluate subacute, chronic, and teratogenic effects in greater depth through both *in vitro* and *in vivo* systems.

---

*DNA repair in bacteria and mitotic recombination in yeast have been evaluated recently by the National Cancer Institute (NCI). While these tests were found to be less sensitive than several other short-term tests, they did not generate many false-positives and can, therefore, be considered useful tests (Ref. 16-18).

**This initial cell culture system has been selected for evaluation by NCI (Ref. 16-10).

Tier II also calls for further tests to determine the environmental fate of the chemical substance and the degradation products from oxidation, reduction, and hydrolysis. There are also included tests that allow judgments to be made on the degree to which the substance bioconcentrates—and therefore can contaminate the food chain or web—and on the effects it may have on bacteria and other microorganisms in wastewater treatment systems.

Subacute Toxicity Tests: Mammalian. These are 90-day subacute tests, with complete histopathology, using both sexes of one or two species. Standard test procedures to characterize subacute effects, by various routes of exposure, are well defined. As in the selection of acute testing protocols, the most relevant and appropriate route(s) of exposure should be selected.

Teratogenicity Tests: Full Scale Reproduction/Teratology Study. Standard protocols for teratogenic effects in mammalian species exist: two species are dosed during gestation; the fetus is examined before birth; there is a gross organ examination; skeletal malformation is sought; histopathological tests are performed; and so on. Similarly, methodologies exist for determining effects on reproduction in mammalian test species.

Further Chronic Health Effects Tests. For any substance that shows potential mutagenic/carcinogenic effects in Tier I tests, or where considerations of Structure/ Activity Relationship or production/use/exposure dictate, Tier II chronic tests will be necessary. The purpose of these Tier II tests is to clarify earlier test results when necessary, especially when Tier I tests give uncertain or equivocal results. The systems used would include both *in vitro* systems and *in vivo* tests for detecting chromosomal aberrations (e.g., mammalian cytogenetics). It is understood that these tests must be validated systems appropriate to the chemical substance being tested. The Ames urine assay test is an example of an *in vitro* test which might be included in this tier.

They might also include tests that are known to be complementary with Tier I tests: e.g., DNA repair in mammalian cells (Ref. 16-13), which would be complementary to a Tier I sister chromatid exchange test. Another such complementary pair would be the Chinese hamster ovary (CHO/HGPRT) test and the mouse lymphoma cell system. Tier II could include both *in vitro* and *in vivo* cytogenetics to measure changes in the morphological structure of chromosomes (Refs. 16-25, 16-26, 16-27).

Chronic Toxicity Tests: Aquatic Species. Further testing with aquatic organisms will be required if indicated by Structure/Activity relationships, levels of production, use, and exposure, and/or if the substance is persistent, nonbiodegradable, and will not be stripped from water rapidly. Tests with fish, egg larvae, and shellfish (shrimp and oyster) using appropriate species and protocols are required. A multigeneration *Daphnia magna* test (either renewal or flowthrough) is recommended to give information on reproductive effects, behavior, delayed neurotoxicity, and lethality. The American Society for Testing and Materials (ASTM) has proposed a number of standard practice procedures for testing with these and other aquatic organisms (Ref. 16-28).

Chronic Toxicity Tests: Plants. Where production, use, exposure, and so on, indicate the need, tests will be required to measure seed germination and leaf toxicity in terrestrial macrophytes (which may include major crop species) or representative aquatic plants. Further testing to measure changes in growth and reproduction (partical life-cycle test) using aquatic or terrestrial macrophytes will be included in this tier.

Transformation and Degradation Tests. At this stage, further information on chemical and biochemical degradation and transformation is developed. The toxicological effect of the substance on microbes and potential conversion by microbes to more or less harmful substances are investigated as the following are performed:

- Further biodegradability tests, including bacterial screens to determine the effect on specific bacterial functions such as carbon reduction, nitrogen fixation, and ammonification.
- Chemical degradation tests with soil and air mixtures.
- Characterization of major degradation products from oxidation and hydrolysis.

Bioaccumulation Tests with Aquatic Flora and Fauna. Accumulation of residues in aquatic organisms is an indication of contamination of the food web. Flowthrough test methods would be used unless it is demonstrated that a static test is appropriate to the substance being tested.

Sludge Toxicity Tests. If toxic to microorganisms a chemical substance may disrupt the treatment process by killing the microorganisms responsible for normal wastewater treatment. Methods of testing for this effect have been developed and may be employed at this stage.

Further Refinement and Application of Analytical Methods. Fully developed analytical procedures should be defined and applied at this stage to characterize trace contaminants, impurities, and degradation products. Environmental tests and field monitoring methods should also be further developed at this stage.

16-3.4 Tier III

Tests in this tier are aimed at further expanding the basis on which to judge the safety and environmental fate of a chemical. Long-term tests with live animals are undertaken to further clarify results from Tier I and II health tests. Tier III mutagenicity testing is performed to confirm earlier test results and to expand the data base where appropriate by more complex tests. The ultimate decision on the substance will be founded on the total information base from all tiers. Wherever necessary, elucidation of the major biochemical mechanisms and identification of the products of metabolism will be undertaken. In-depth degradation studies with characterization of degradation products in soil and air may be required, as well as relevant evaluation of the chemical's effect on terrestrial plants and animals.

Chronic Toxicity Tests: Mammalian Long-Term Study. Chronic tests are conducted with two appropriate mammalian species of both sexes (over lifetime with complete histopathological examination and analysis of data) unless there are compelling reasons for selecting only one species. Mutagenicity tests in mammalian species are also conducted to confirm or further expand the data from mutagenic tests at lower tiers.

Photochemical Degradation Tests. At this stage, photodegradation studies should be performed, with an emphasis on reaction rates. In addition, characterization of degradation products should be undertaken where meaningful results can be obtained.

Soil Transfer (Leachability) and Degradation Tests with Characterization of Degradation Products. The movement of the chemical substance or its degradation product(s) through the soil may result in contamination of the food chain and entry into water supplies, with serious ecological and health effects. Soil transfer (or leachability) data are generally obtained by laboratory studies using radioisotopic and nonradioisotopic analytical techniques with various types of soil. Degradation in the soil is determined by microbial metabolism studies using radiolabeling or nonradioisotopic techniques in both aerobic and anaerobic conditions.

Chronic Toxicity Tests: Terrestrial and Aquatic Organisms (Plants, Fish, Birds and Other Wildlife). Chronic testing of terrestrial and aquatic organisms is required for substances released to the environment in large quantities and/or for highly persistent substances that biotransfer or bioaccumulate. Examples of tests recommended for further evaluation of environmental hazard are life-cycle plant growth and reproduction studies (biomass, pollination, and seed or propagule production) using appropriate plant species, reproductive and neurological tests for birds and mammals, the brook trout life-cycle test, and the midge test (ASTM-draft). The fathead minnow may be used if test results can be related to results from tests in more sensitive species. Relevant monitoring, trophic level system modeling, or restricted field studies are recommended at Tier III to measure effects of metabolites or degradation products.

Metabolic Studies. These include pharmacological studies to determine the biochemical mechanism responsible for the activity of the chemical along with identification, wherever possible, of the products of metabolism. Absorption and excretion tests are also included.

Further Refinement and Application of Analytical Methods. Further refinement and application of analytical techniques is required in order to give as complete a qualitative and quantitative description as possible of the trace contaminants, impurities, and degradation products of the chemical substance. Detection and monitoring methods and procedures should be well developed in order to characterize as fully as possible the environmental fate of the chemical substance.

16-4 SUGGESTED RULES FOR TESTING DECISIONS

No consideration of a tier testing approach would be complete without assessment of the decision-making process involved in moving from tier to tier in order to fully determine human health or environmental impacts of the chemical substance under consideration. As mentioned earlier, no attempt has been made to do more than develop general principles for such decision-making. These two underlying principles should be as follows:

1. The decision to move to a successive tier or tiers should be predicated upon assessment of each or a combination of the following five criteria and/or upon the results obtained in the preceding tier or tiers:

    a. Volume of production.
    b. Anticipated or actual use, and environmental or human exposure.
    c. Nature of the potential hazard.
    d. Physical/chemical properties.
    e. Structural activity relationships.

As an example, for a chemical substance anticipated to have a very high exposure level or potential for bioaccumulation, a decision to move to higher tier(s) of testing (II, III) could be made even if there have been negative or inconclusive results in earlier tier tests.

2. Tests should be selected within each tier based on any one of the above five criteria and/or upon results obtained in the preceding tier or tiers.

The system in its entirety can be applied to both new and existing chemical substances. The application will vary in detail, however, and it is suggested that testing decisions should be related to the category of application (new or existing) as follows.

New Chemical Substances or Significant New Uses of Chemical Substances. Where new chemical substances or significant new uses of existing chemical substances are involved.

1. All substances should be subject to Tier 0 and Tier I testing. The only exception would be for chemical substances used for preliminary research or experimental purposes.
2. Tier II testing is to be undertaken for specific environmental or health effects depending upon (a) the five criteria indicated above, e.g., high anticipated production or exposure, and/or (b) the results of Tier 0 and Tier I tests.
3. Tier II tests are to be undertaken for environmental or health effects depending upon the five criteria indicated above and/or the results of Tier I and Tier II tests.

Existing Chemical Substances. The large numbers of existing chemical substances now in use make it necessary to subject these substances to ongoing priority review for purposes of selecting those which require further testing. Guidelines for priority review are being developed by the Environmental Protection Agency. The Conservation Foundation has prepared a separate document outlining priority review guidelines. Following selection, individual substances would be subject to the testing guidelines as follows:

1. a. All chemical substances placed on a priority list by the Interagency Committee on Testing should be subject to Tier 0, I, and II Tests, as appropriate and necessary.
   b. For these substances Tier III tests will also be required where earlier tier tests indicate the need. Tier III tests may also be warranted based on the five criteria at the beginning of Section 16-4.
2. a. For other selected existing chemical substances, Tier 0 and I tests should be minimally required where information provided by the guidelines does not exist.
   b. Tier II and III tests should be required when earlier tests indicate a need or based on the five criteria previously mentioned.

16-5 ADDITIONAL RECOMMENDATIONS FOR TSCA TESTING

During the development of this hierarchical testing system, a number of areas emerged as worthy of additional consideration for possible inclusion in the guidelines. All of these areas should be subject of ongoing discussion, particularly since they represent fields of rapid change and advancement (Refs. 16-29 to 16-36).

16-5.1 Areas of Agreement Requiring More Detailed Research

This category includes those tests generally believed by the panel to belong in the system, but where the panel was unable to agree on methods and procedures that were acceptable to all.

Areas of Agreement Requiring More Detailed Research. Photochemical oxidation tests range from simple differentiation of persistent and degradable chemicals through exposure to high and low wavelength light (Activity and Sensitivity Testing) to elaborate tests that simulate as far as possible the photochemical environment in which the substance would be found. Some members of the panel believe that many of these more complicated tests, used in pesticides testing, are too costly and complex to use in early tiers. Increased research and development work on more accurate and reliable photochemical screening procedures is therefore recommended.

Further Chronic Health Effects Testing. As mentioned in the discussion of Tier I and II chronic health effects tests, some screening tests have been generally accepted as validated and a number of others show promise in the near future. The panel strongly believes that more work is urgently needed to validate these additional screening tests. Similarly, members of the panel concluded that there was a critical need for validated screening procedures for teratogenicity, behavioral, and other chronic effects.

Aquatic Species Testing. Tests to determine effects in aquatic species (toxicity, bioaccumulation, and so on) have been included in the proposed guidelines. There remains to be more carefully defined the actual criteria for species selection (fish, shellfish, and larvae), as well as specific techniques to be employed for related static and dynamic bioassay.

16-5.2 Areas Requiring Additional Consideration

The second category includes aspects for which consensus was not achieved, in some cases because members of the panel were not sufficiently familiar with the types of test systems available. These aspects, however, were considered to be of interest and have potential applicability in the future.

Ecological Effects Testing. It is recognized that testing of the effects of chemical substances on aquatic and terrestrial plants and animals is an important part of the guidelines. There is a need for an integrated approach to ecological testing — one which is aimed at elucidating the processes involved rather than isolated events. There is a need to develop methods that will measure not only biotransfer and trophic level accumulation (like the Metcalf microcosm test) but also the biological significance of these processes. The effect of body-burden residues and metabolites should be studied.

For plant testing, there exist a number of screening procedures, such as the Callus Bioassay (Bednar test) for detection of forward mutations and developmental changes in the plant cell line. Other plant screening tests include the lipid droplet bioaccumulation test and mitochondrial respiratory pathway tests to determine effects on oxidation.

A number of innovative screening tests have been mentioned, including tests with honeybees to determine effects on pollination or to indicate potential toxic, neurotoxic, and behavioral effects of chemical substances. The leaf-litter test, which detects toxicity to reducer organisms, may also have potential future application.

Health Effects Testing. Karyotypic testing for screening for substances that cause chromosomal mutations undeniably has potential application; these tests require further assessment by the panel before any precise application can be recommended.

The question of synergistic as well as inhibitory health effects is important. However, the state of the art does not now readily permit quantifiable determinations in these areas. Consideration must be given to the development of research techniques in this area.

Testing for the ability to cause point mutations which result in metabolic disorders in an area to be considered further. There is some indication that a number of these diseases is increasing.

Avian Chronic Testing. Methods for testing with birds for acute toxicity, delayed neurotoxicity, biochemical changes (e.g., effects on hormone production, mixed function oxidase system, and so on), reproductive and teratogenic effects have been developed largely for use in pesticides testing (Ref. 16-37). The FDA also routinely employs tests with poultry and hen eggs to determine potential human hazards from drugs or food additives. Results of avian testing have therefore been important in determining potential hazard both to birds and to humans (Ref. 16-38).

In this document, avian testing is required in cases where birds will be directly exposed to the chemical substance, its metabolites, or degradation products, if the substance is accumulated in plant or animal tissues, or if it is otherwise persistent in the environment. Acute avian tests (LD_{50}) for substances in these categories are included in Tier I. In Tier III, avian tests are required to determine whether the substance may cause neurological and reproductive effects.

The use of bird tests as screens for chronic effects such as neurotoxicity or teratogenicity should be given consideration for future inclusion either in Tier I or Tier II because these tests are potentially valuable indicators of health effects in humans. Subacute avian testing should also be considered under Tier I or Tier II for chemical substances to which birds may be generally exposed through soil residues (tests on seed-eating birds such as quail, pheasants, and chickens) or which are either released into the aquatic environment or biotransfer by the predator food chain (tests on mallard ducks and barn owls).

Metabolic Studies. It is possible to determine the metabolites of a chemical substance through the use of radiolabeled functional groups in an enzyme system followed by analysis with chromatographic techniques. Those tests may be considered for future inclusion in Tier II.

REFERENCES

16-1. Peltier, W., *Methods for Measuring the Acute Toxicity of Effluents to Aquatic Organisms,* Ecological Monitoring Series, EPA-600/4-78/012, U. S. Environmental Protection Agency, Washington, D. C., January 1978.

16-2. Walbridge, C. T., "A Flowthrough Testing Procedure with Duckweed (*Lemma minor L.*)," EPA-600/3-77/108 Ecological Research Series, U. S. Environmental Protection Agency, Washington, D. C., September 1977.

16-3. Consumer Products Safety Commission, *Mutagenesis,* Pubn. No. 1138, Washington, D. C., National Academy of Sciences; Department of Health, Education and Welfare, Committee to Coordinate Toxicology and Related Programs, "Approaches to Determining the Mutagenic Properties of Chemicals: Risk to Future Generations," in preparation; Environmental Protection Agency, "FIFRA, Section 3 Guidelines on Hazard Evaluation of Humans and Domestic Animals," in preparation; and Food and Drug Administration, *Criteria for Evaluation of the Health Aspects of Using Flavoring Substances as Food Ingredients,* Bethesda, Md., Federation of American Societies for Experimental Biology, 1976.

16-4. Purchase, I. F. H, *et al.*, "Evaluation of Six Short-Term Tests for Detecting Organic Chemical Carcinogens and Recommendations for Their Use," *Nature,* **264**, p. 624, 1976.

16-5. McCann, J., Choi, E., Yamasaki, E., and Ames, B., "Detection of Carcinogens as Mutations in the *Salmonella* Microsome Test: Assay of 300 Chemicals," *Proc. Natl. Acad. Sci., USA,* **72,** p. 5135, 1975.

16-6. McCann, J., and Ames, B., "Detection of Carcinogens as Mutagens in the *Salmonella* Microsome Test: Assay of 300 Chemicals: Discussion," *Proc. Natl. Acad. Sci., USA,* **73,** p. 970, 1976.

16-7. Pienta, R. J., Poiley, J. A., and Lebherz, W. B., "Morphological Transformation of Early Passage Golden Syrian Hamster Embryo Cells Derived from Cryopreserved Primary Cultures as a Reliable *In Vitro* Bioassay for Identifying Diverse Carcinogens," *Int. J. Cancer,* **19,** p. 642, 1977.

16-8. Sugimura, F., *et al.*, "Overlapping of Carcinogens and Mutagens," in *Fundamentals in Cancer Prevention,* P. M. Magee (Ed.), Univ. Park Press, Baltimore, MD, 1976.

16-9. Huberman, E., Mager, R., and Sachs, L., "Mutagenesis and Transformation of Normal Cells by Chemical Carcinogens," *Nature,* **264,** p. 360, November 25, 1976.

16-10. Dunkel, V. C., "*In Vitro* Carcinogenesis: A National Cancer Institute Coordinated Programmer, Screening Tests in Chemical Carcinogenesis," R. Montesano, H. Bartsch, and L. Tomatis (Eds.), IARC Scientific Publication No. 12, Lyon, 1976.

16-11. Dunkel, V. C., Personal Communication, National Cancer Institute, Director of *In Vitro* Assay Program, November 5, 1977.

16-12. Ames, B. N., McCann, J., and Yamasaki, E., "Methods for Detecting Carcinogens and Mutagens with the *Salmonella*/Mammalian-Microsome Mutagenicity Test," *Mutat. Res.,* **31,** p. 347, 1975.

16-13. San, R. H. C., and Stich, H. F., "DNA Repair Synthesis of Cultured Human Cells as a Rapid Bioassay for Chemical Carcinogens," *Int. J. Cancer,* **16,** p. 284, 1975.

16-14. Vogel, E., "Identification of Carcinogens by Mutagen Testing in Drosophila: The Relative Reliability for the Kinds of Genetic Damage Measured," *Origins of Human Cancer,* **4,** Cold Spring Harbor Conferences on Cell Proliferation, 1977.

16-15. Clive, D., Flamm, W. G., and Patterson, J. B., "Specific Locus Mutational Assay Systems for Mouse Lymphoma Cells," in *Chemical Mutagens: Principles and Methods for Their Detection,* A. Hollaender (Ed.), Plenum Press, NY, Vol. 6, p. 79, 1973.

16-16. O'Neill, J. P., *et al.*, "A Quantitative Assay of Mutation Induction at the Hypoxanthine-Quanine Phosphoribosl Transferase Locus in Chinese Hamster Ovary Cells (CHO/HGPRT System): Development and Definition of the System," *Mutat. Res.,* Vol. 55, pp. 91-101, 1977.

16-17. Rosenkranz, H. S., Gutter, B., and Speck, W. T., "Mutagenicity and DNA – Modifying Acitivity: A Comparison of Two Microbial Assays," *Mutat. Res.,* **1,** p. 61, 1976.

16-18. Simmon, "*In Vitro* Mutagenicity Assays with *Saccharamyces cerevisae*," *J. Natl. Cancer Inst.*, in press.

16-19. Yamasaki, E., and Ames, B., "Concentration of Mutagens from Urine by Absorption with the Nonpolar Resin XAD-2: Cigarette Smokers have Mutagenic Urine," *Proc. Natl. Acad. Sci.,* **74,** p. 3555, 1977.

16-20. Stetka, D. G., and Wolff, S., "Sister Chromatid Exchange as an Assay for Genetic damage Induced by Mutagen-Carcinogens, II, *In Vitro* Test for Compounds Requiring Metabolic Activation," *Mutat. Res.,* **41,** p. 343, 1976.

16-21. Kakunage, T., "A Quantitative System for Assay of Malignant Transformation by Chemical Carcinogens Using a Clone Derived from BALB/3T3," *Int. J. Cancer,* **12,** p. 463, 1973.

16-22. Reznikoff, C. A., Bertram, J. S., Brankow, G. W., and Heidelberger, C., "Quantitative and Qualitative Studies of Chemical Transformation of Cloned C3H Mouse Embryo Cells Sensitive to Postconfluence Inhibition of Cell Division," *Cancer Res.,* **33,** p. 3239, 1973.

16-23. Zimmermann, F. K., "Procedures Used in the Induction of Mitotic Recombination and Mutation in the Yeast *Saccharamyces cerevisae,*" *Mutat. Res.,* **31,** p. 71, 1975.

16-24. Hubberman, E., and Sachs, L., "Mutability of Different Genetic Loci in Mammalian Cells by Metabolically Activated Carcinogenic Polycyclic Hydrocarbons," *Proc. Natl. Acad. Sci.,* **73,** p. 188, 1976.

16-25. Legator, M. S., Palmer, K. A., and Adler, I., "A Collaborative Study of *in vivo* Cytogenetic Analysis I. Interpretation of Slice Preparations," *Toxicol. Appl. Pharmacol.,* **24,** p. 337, 1973.

16-26. Evans, H. J., and O'Riordan, M. L., "Human Peripheral Blood Lymphocytes for the Analysis of Chromosome Aberrations in Mutagen Tests," *Mutat. Res.,* **31,** p. 135, 1975.

16-27. Nichols, W. W., *et al.*, "Chromosome Methodologies in Mutation Testing," *Toxicol. Appl. Pharmacol.,* **22,** 269, 1972.

16-28. *Proposed Standard Practice for Conducting Renewal Life Cycle Toxicity Tests with the Daphnid, Daphnia magna,* Draft No. 2, Am. Soc. Testing and Materials, Philadelphia, PA, Oct. 12, 1977.

16-29. *Reports of the ACMRR/IABO Working Party on Biological Effects of Pollutants,* First, session, Rome, Italy, 27-31 October 1975; Second session, Dubrovnik, Yugoslavia 22-25 November 1976; and *Report of the ACMRR/IABO Expert Consultation on Bioassays with Aquatic Organisms in Relation to Pollution Problems,* Dubrovnik, Yugoslavia 16-19, November 1976, Food and Agriculture Organization of the U. N., Rome, 1977.

16-30. "Attendees at an ACS Symposium Heard about Systems Designed to Detect Biological Effects of Environmental Pollutants on Man or Surrogate Hosts," *Environ. Sci. Technol.,* **11,** p. 1050, 1977.

16-31. *Fates of Pollutants, Research and Development Needs,* Committee on Natural Resources, National Research Council, National Academy of Sciences, Washington, D. C., 1977: *Principles for Evaluating Chemicals in the Environment,* National Academy of Sciences, Washington, D. C., 1975.

16-32. *Screening Program for the Identification of Potential Chemical Mutagens and Carcinogens,* Litton-Bionetics, Inc., Nov. 1, 1977.

16-33. Mayer, F. L., and Harzelink, J. L., (Eds.), *Aquatic Toxicology and Hazard Evaluation,* Proceedings of the First Annual Symposium on Aquatic Toxicology, American Society for Testing and Materials, Philadelphia, PA, 1977.

16-34. Lee, S. D., *Biochemical Effects of Environmental Pollutants,* Ann Arbor Science Publishers, Ann Arbor, MI, 1977.

16-35. *The Testing of Chemicals for Carcinogenicity, Mutagenicity, and Teratogenicity,* published by authority of the Honorable Marc LaLonde, Minister of Health and Welfare, Canada, 1975.

16-36. *Pathology and Toxicology: Modern Concepts and Techniques for Assessment of Long-term Low-dose Studies of Toxicity,* Universities Associated for Research and Education in Pathology, Inc., Prepared for Food and Drug Admin., Dec. 1972.

16-37. *Hazard Evaluation: Wildlife and Aquatic Organisms Subpart E.,* Working Group, U. S. Environmental Protection Agency Draft, April 27, 1977.

16-38. Walker, R., "Pre-1972 Knowledge of Nonhuman Effects of Polychlorinated Biphenyls," in Proceedings of the National Conference on PCBs, November 19-21, 1975, Chicago, Illinois, U. S. Environmental Protection Agency, March 1976.

APPENDIX 16-1

PANEL MEMBERS*

Dr. A. Karim Ahmed (Co-Chairman)
Natural Resources Defense Council, Inc.
122 East 42nd Street
New York, NY 10017

Eileen Choffnes
Citizens for a Better Environment
Suite 2610
59 E. Van Buren
Chicago, IL 60605

Dr. J. Clarence Davies, III
The Conservation Foundation
1717 Massachusetts Ave., N. W.
Washington, D. C. 20036

Dr. Paul F. Deisler, Jr.
Shell Chemical Company
1 Shell Plaza
Houston, TX 77001

George Dominguez (Co-Chairman)
Springborn Management Consultants, Inc.
Enfield, CT 06082

Marcia Fine
Environmental Defense Fund
1525 18th Street, N. W.
Washington, D. C. 20036

*These persons, while affiliated with the organizations indicated, were participating as individuals and not as institutional representatives.

Dr. William Gorham
Union Carbide Corporation
River Rd.
Bound Brook, NJ 08805

Dr. Sam Gusman
The Conservation Foundation
1717 Massachusetts Ave. N. W.
Washington, D. C. 20036

Dr. Joseph Highland
Environmental Defense Fund
1525 18th St. N. W.
Washington D. C. 20036

Dr. George Levinskas
Monsanto Company
800 North Lindburgh Blvd.
St. Louis, MO 63166

Dr. Kaye Kilburn
Pulmonary Division
Mt. Sinai School of Medicine
100th St. & 5th Ave.
New York, NY 10029

Dr. Paul Kotin
Johns-Manville Co.
P. O. Box 5108
Greenwood Plaza
Denver, CO 80217

Dr. Joyce McCann
Dept. of Biochemistry
University of California
Berkeley, CA 94702

Dr. Blaine McKusick
Haskell Laboratories
Du Pont de Nemours Co.
1007 Market St.
Wilmington, DE 19898

Dr. Robert Moolenaar
Environmental Sciences Research
Dow Chemical Company
1702 Building
Midland, MI 48640

Dr. William Nicholson
Environmental Sciences Lab.

Mt. Sinai School of Medicine
100th St. & 5th Ave.
New York, NY 10029

Dr. Glenn Paulson
N. J. Dept. of Environmental Protection
P. O. Box 1390
Trenton, NJ 08625

Frederica P. Perera
Natural Resources Defense Council
122 East 42nd St.
New York, NY 10017

Dr. Robert Weir
Litton Bionetics
5516 Nicleson
Kensington, MD 20795

APPENDIX 16-2 List of Organizations Which Submitted Minority Opinions on the Testing
Requirements

CIBA-GEIGY Corporation
State of New Jersey, Department of Environmental Protection
The Johns Hopkins University
Chemical Industry Institute of Toxicology
Dow Chemical U. S. A.
The Procter & Gamble Company
The University of Texas Medical Branch
Union Carbide Corporation
Shell Oil Company
University of California, Berkeley
Manufacturing Chemists Association
Litton Bionetics
Monsanto Company
Celanese
Environmental Defense Fund
Citizens for a Better Environment
Veterans Administration Hospital, Bronx, New York
National Research Council – Commission on Natural Resources

17

A risk assessment approach for evaluating the environmental significance of chemical contaminants in solid wastes

---

G. Fred Lee and R. Anne Jones

Colorado State University
Department of Civil Engineering
Environmental Engineering Program
Fort Collins, Colorado

17-1 INTRODUCTION

Increasing attention is being given to the environmental significance of chemical contaminants in solid wastes.* This is the result of an increasing awareness that improper disposal of liquid and solid wastes on land can result in significant

---

*For the purposes of this discussion, "solid wastes" are defined as municipal and industrial solid wastes, as well as the concentrated liquid wastes from certain manufacturing operations, which are typically disposed of by land application or burial. The approaches discussed in this chapter are equally applicable to assessing the environmental significance of land disposal of domestic wastewater sludges.

environmental degradation. The "Love Canal" situation in Niagara Falls, NY, is just one of many examples across the United States in which land disposal of wastes has caused or has the potential to cause significant environmental degradation. As recently as two years ago, based on a survey by Green *et al.* (Ref. 17-1), there were no federal, and, with a few exceptions, no state regulations governing the disposal of hazardous chemical wastes on land. In an effort to remedy this situation the U.S. Congress and many State Legislatures have passed laws enabling pollution control agencies to develop regulations governing solid waste disposal. The U.S. EPA, in accord with the provisions of the Resource Conservation and Recovery Act (RCRA), is attempting to develop regulations that would minimize environmental contamination from land disposal of solid wastes.

One of the key requirements of RCRA is the classification of solid wastes with respect to their potential environmental hazards. There is considerable controversy, however, about the validity of the U.S. EPA's planned approach. The basic approach advocated by the U.S. EPA for classifying solid wastes by toxicity involves a single leaching test with arbitrarily established concentration limits for contaminants in the leachate. As discussed by Lee (Ref. 17-2), this approach is not technically valid for assessing the environmental hazard associated with solid wastes primarily because the transport and transformation (environmental chemistry) of the solids-associated contaminants can be markedly different for each specific environment, and usually play a dominant role in determining the hazard associated with the introduction of the solid into the environment. The environmental chemistry of contaminants is influenced by many factors, as discussed later in this chapter, and must therefore be evaluated on a site-specific and solid waste-specific basis. No single leaching test will properly simulate chemical behavior for all cases. In addition to causing municipalities, industry, and others to spend unnecessarily large amounts of money for controlling wastes that would not cause environmental degradation, implementation of such an approach into public policy may result in instances of inadequate protection being provided as a result of improper classification of the environmental hazard associated with contaminants present in solid wastes (Lee, Ref. 17-2). This chapter presents general guidance that is useful not only for the development of appropriate leaching test and hazard assessment procedures for solid waste disposal evaluation, but also in the site-specific application of these approaches.

Some groups within the U.S. EPA and Corps of Engineers have attempted to make site-specific assessments of the hazard associated with solids (dredged sediment) -associated contaminants by determining the bulk content of potentially hazardous chemicals in the solid. It has been well documented, however, that the release of solids-associated contaminants when exposed to water, or the impact of the contaminants on water quality, cannot be predicted based on the total concentration of the contaminants in the solid (Refs. 17-3, 17-4). In order to develop technically valid, cost-effective, environmentally protective solid waste disposal regulations, an "environmental hazard assessment approach" must be used to classify solid wastes on a site-specific basis. This chapter describes the elements of such a hazard assessment approach.

17-2 THE ENVIRONMENTAL RISK ASSESSMENT APPROACH — AN OVERVIEW

The environmental hazard assessment approach is an outgrowth of the work that has been done by the chemical industry and more recently by governmental agencies such as the U.S. EPA Office of Toxic Substances, in prescreening new chemicals for their potential environmental hazard prior to their large-scale manufacture. There are two basic factors motivating work in this area. One is the series of chemically-caused environmental crises that have resulted in widespread environmental contamination by potentially highly toxic, persistent contaminants such as PCBs, DDT, mercury, and, most recently, kepone. There is no doubt that it is necessary to prescreen all chemicals in order to try to prevent chemical crises of this type. The other factor is the need to establish technically valid chemical screening requirements so that environmental protection can be ensured and thus governmental agencies can be prevented from making the kinds of arbitrary decisions with respect to the chemical safety as occurred with nitrilotriacetic acid (NTA), a replacement for phosphate in household laundry detergent formulations. It is now well known that large amounts of money were lost by the chemical industry and, hence, consumers because the U.S. Department of Health, Education, and Welfare (HEW) at the last minute prevented the use of NTA, based on controversial studies conducted by HEW staff. A comprehensive review of the suitability of the use of NTA as a replacement for phosphate in detergents has since been conducted by an International Joint Commission for Great Lakes Water Quality task force that has recommended its use in the Great Lakes Basin. The development of an appropriate hazard assessment would, if properly conducted, prevent the next PCB-type crisis and protect the chemical industry and consumers from inappropriate governmental decisions with respect to new chemical - new product development.

While there is general agreement concerning the need for an environmental hazard assessment approach, there is some controversy regarding the detailed procedures to be included in it. It is generally agreed, however, that the overall approach that is being formulated by the American Society for Testing and Materials (ASTM) Subcommittee E-35.21 "Safety to Man and the Environment," is a technically valid, cost-effective, environmentally protective approach that should be followed for prescreening new chemicals for their potential environmental impact. A discussion of the principles of a hazard assessment for prescreening new chemicals is presented in the ASTM publication edited by Cairns *et al.* (Ref. 17-5) and, for existing chemicals, in a publication by Lee *et al.* (Ref. 17-6).

As designed for aquatic systems, hazard assessment is developed from two types of information: aquatic toxicology and environmental chemistry–fate. Aquatic toxicology defines the response of selected aquatic organisms to given concentrations of particular forms of contaminants or given dilutions of a particular effluent, when exposed for a given duration. Environmental chemistry–fate defines the transport and transformation of the chemical or effluent and defines the expected concentration of the chemical and its potentially significant transformation products in the aquatic compartments of concern, the water, the sediment, and so on. An assessment of aquatic environmental hazard is then made by comparing the expected concentrations to those that have been found to be harmful to aquatic organisms, to man when he eats the aquatic organisms, or to other beneficial uses of the water. As discussed by Cairns *et al.* (Ref. 17-5) and Lee *et al.* (Ref. 17-6), a hazard assessment should be conducted in a series of levels or tiers of increasing sophistication so that highly innocuous and highly hazardous materials can be readily identified with limited testing and so that it is possible to make situation-specific judgments regarding necessary testing.

The basic philosophy that the authors feel should apply to any situation of potentially hazardous chemical use or discharge is to allow the source of the contaminant of concern (the municipality, industry, operator of a disposal site, and so on) the option of conducting a detailed evaluation to determine the potential environmental hazard associated with the manufacture and use and/or disposal in the desired manner. If the source is unwilling to make such an evaluation, then worst-case assumptions should be made to ensure that unacceptable environmental degradation does not result from these activities.

17-3 HAZARD ASSESSMENT APPROACH FOR SOLID WASTES

A key component in defining the fate and behavior of contaminants associated with solid wastes is a leaching procedure, since it is this process that mobilizes the solids-associated contaminants in disposal sites so that they may affect water quality. As discussed previously, however, this leaching should not be mechanically performed by a single rigid procedure, but should be conducted to take into consideration the characteristics of the material and the disposal site. While there does not yet exist sufficient information to develop a detailed hazard assessment program or a widely applicable leaching test to evaluate the solid waste-associated contaminants, the RCRA guidelines should describe a general leaching procedure to be used, indicating the factors to be considered in developing site-specific leaching tests, and giving guidance on how these factors are to be incorporated into such leaching tests.

The authors have conducted a comprehensive study of the U.S. EPA and Corps of Engineers' elutriate test, a leaching test designed to evaluate the release of dredged sediment-associated contaminants upon their open water disposal. This test was developed because the previously used bulk sediment criteria lacked validity; but when the elutriate test was developed, essentially nothing was known about appropriateness of the test conditions specified (although they were chosen on the basis of what was believed appropriate for hydraulic dredging-hopper disposal operations) or its predictive capabilities. The Lee et al. (Ref. 17-3) and Jones and Lee (Ref. 17-4) studies, undertaken to resolve these questions, included the evaluation of over 50 sediments from about 20 river and harbor areas in the United States and in situ monitoring of about 20 disposal operations. Their studies provide insight into the factors that should be considered in developing a leaching test for assessing the hazard associated with solid-waste-associated contaminants. The following section is based largely on the findings of these studies. It is important, as discussed above, that each of these parameters be carefully evaluated for solid waste disposal sites.

17-3.1 Factors to be Considered in Designing Leaching Test for Solid Waste

The first important factor to be considered in developing a leaching test for solid waste is the leaching solution to be used. For some situations, the only liquid with which the solids will come in contact at the disposal site is water. In other instances, liquid wastes may be disposed of at the same site. There is no point in leaching solid wastes with acetic acid — "garbage juice" — such as is proposed by the U.S. EPA, if these wastes will not be in contact with a dilute acetic acid solution or its equivalent at the disposal site. The leaching solution used in the testing must bear some relationship to that acutally encountered at the disposal site, because markedly different chemical release from solids can be promoted by different leaching solutions.

Another important factor that must be considered in establishing chemical leaching tests for solid wastes is the control of the oxidation–reduction conditions present during leaching. If the solid waste disposal site is likely to be anoxic (i.e., no oxygen present), then the leaching test should be conducted under an anoxic condition. Lee et al. (Ref. 17-3) and Jones and Lee (Ref. 17-4) evaluated the effects of redox conditions of elutriate test results and found that, for some contaminants, markedly different release occurred under oxic as opposed to anoxic conditions, with generally greater release occurring under anoxic conditions. They found that, unless this parameter is controlled (i.e., assurring oxic or anoxic conditions as desired), the results were variable and largely uninterpretable because the oxygen demand of some sediments was sufficient to deplete the oxygen in the test vessel if no supplemental oxygen was provided, a depletion that would not be encountered at most dredged sediment disposal sites. For most dredging cases, unless the disposal site or chemical characteristics dictate otherwise, the elutriate test should be run under oxic conditions, as the sediment is generally dumped into oxic waters. The redox conditions must be examined for each solid waste disposal site and appropriate conditions incorporated into the leaching test for solid wastes.

Specification of leaching test parameters also should consider the importance of the liquid-to-solid ratio, i.e., the ratio of the leaching solution to the solid waste, in the disposal area. The elutriate test liquid-to-solid ratio is usually 80% by volume, based on the optimum dredge pumping ratio. It was found, in evaluating the impact of this factor on elutriate test results, that the liquid-to-solid ratio affected the amount of release of some contaminants (Ref. 17-4, 17-3). This effect was, however, highly sediment- and contaminant-specific and was not predictable. It underlines the importance of mimicking the liquid-to-solid ratio appropriate for the system under investigation. Because it is difficult to define the actual liquid-to-solid ratio at each particular solid waste disposal site, the leaching test procedure should be conducted with several liquid-to-solid ratios that cover the range likely to be encountered and define the sensitivity of the contaminant release to this condition. In general the liquid-to-solid ratio associated with water percolating through the pores of the solid material is a more appropriate guide to the leaching test ratio than an arbitrarily selected value. Studies need to be conducted on a variety of solid wastes and at a variety of solid waste disposal areas in order to determine what range of liquid-to-solid ratios should be used and to determine the influence this parameter will have on the release of contaminants in question.

Another factor that needs more detailed investigation before a general procedure can be developed for formulating site-specific leaching tests is the contact time between the leaching solution and the solid waste sample. Lee *et al.* (Ref. 17-3) and Jones and Lee (Ref. 17-4) found that, for some contaminants, the sediment–water contact time could have a marked effect on the release of the contaminant in the elutriate test, depending on sediment charactersitics. Because the retention time in solid waste disposal sites may be highly variable, it is necessary to choose this condition on a site-specific basis, at least until its impact on leaching test results has been defined.

Similarly the degree and method of agitation used during the leaching procedure must be chosen so that leaching test results properly simulate environmental conditions. Alternatively, a completely mixed system could be used, which would likely represent worst-case release; data interpretation would then have to take that into account.

Other factors that will need to be evaluated in establishing a widely applicable leaching test are the impact of leaching solution pH, temperature, and the organic content of the solids. The method of separation of the leachate from the solid must also be evaluated for each contaminant of interest. For some contaminants, there can be considerable sorption on membrane filters typically used for this purpose. Centrifugation can become quite cumbersome if a large number of different analyses need to be run, and may result in more of the contaminant being separated out than would be retained by the disposal area. Settling alone does not generally adequately separate contaminant forms of interest from the solids which would be retained within the disposal area. As a practical consideration, studies should be conducted to determine the length of time that samples may be stored prior to leaching and the storage conditions. It has been found that freezing or drying of solids can alter their tendency to release or sorb contaminants. Lee *et al.* (Ref. 17-3) found it satisfactory to store their sediments in the dark at 4°C for a maximum period of about two weeks. They felt that longer storage periods would have resulted in sediments and associated contaminants' being sufficiently changed to substantially alter test results.

It is likely that as the various factors discussed above are evaluated for their impact on leaching test results and are defined for a variety of solid waste disposal sites and materials, other factors will be identified as important, constraints will be placed on use of the test procedure, and some factors will be found to be unimportant in affecting results. However, until sufficient field studies are conducted to do this sorting, each of these factors must be carefully evaluated for its appropriateness for the disposal site of concern and the material of concern.

Ham *et al.* (Ref. 17-7) and Thompson (Ref. 17-8) conducted laboratory studies of leaching tests for industrial wastes. Ham *et al.* evaluated three different batch test procedures (as opposed to column leaching tests) in an effort to define which would most closely represent an appropriate "standard" test, both from practical and technical points of view. They stressed the importance of proper interpretation of the results of any leaching test based on contaminant, landfill, and organism characteristics, and pointed out that ". . . a certain concentration of a given parameter in the test leachate should not be taken to indicate that the waste is hazardous in the landfill." Ham *et al.* provided a discussion of merits and drawbacks of each of the three tests evaluated in terms of properly mimicking solid waste disposal sites, interpretability, and the ease of conducting the procedure. They indicated that because landfills (municipal) produce acidic leachates comparable in pH and buffering capacity to the synthetic leachate leaching solution used in one type of test, it would be more appropriate to use such a solution than deionized water or acetate buffer (pH 4.5) also evaluated. As pointed out by Ham *et al.*, while the liquid-to-solid ratio is important, it is generally arbitrarily assigned; they suggested that ratios of 1:10 to 1:4 are reasonable. While tests were run for 24 or 48 hr periods, they pointed out that these leachings may simulate years of *in situ* leaching, during which time the chemical may be substantially altered. The number of elutions and agitation techniques were also evaluated by Ham *et al.*; they indicated that, although they are also ultimately rather arbitrarily assigned, they should be based on chemical characteristics. While stressing the importance of proper interpretation of leaching test results, Ham *et al.* provided little explicit guidance and had conducted no field work to attempt to correlate the results of their work with actual leachate generation in the field. They suggested, however, that this type of leaching test would likely represent worst-case conditions.

Thompson (Ref. 17-8) used a modified elutriate test, similar to the one developed by the U.S. EPA and Corps of Engineers for evaluating contaminant release from dredged sediment, to compare chemical (mostly heavy metals) leaching from industrial sludges both raw and chemically stabilized by a variety of techniques. As part of his modifications to the dredged sediment elutriate test, he apparently conducted some studies on the effects of various test conditions to optimize the release of contaminants in his test. He indicated that a 1:8 solid-to-liquid ratio (on a weight basis) appeared to be appropriate, and that a contact time of 30 min provided as much contaminant release as "longer" periods. He did not apparently evaluate the effect of not controlling redox conditions. As Thompson indicated, there are a number of problems in using this test in its current form to say anything about the hazard associated with solid wastes disposed of on land. The field verification portion of Thompson's overall study has apparently (according to Thompson) been recently completed but no comments were made about those results. It should be noted, as indicated previously, that the elutriate test for dredged sediments has been extensively evaluated by Lee *et al.* (Ref. 17-3) in the laboratory and in the field for its ability to properly assess potential contaminant release during open water dredged sediment disposal. These studies indicated that the dredged sediment elutriate test adequately predicts sediment-associated contaminant release during open water dredged sediment disposal.

The leaching procedure, as indicated above, is a technique of imitating, in the laboratory, the mobilization of solids-associated contaminants within a solid waste disposal site. Once an estimation is made of the forms and concentrations of the contaminants that are likely to migrate out of a particular disposal site, it is possible to assess what concentrations and forms of a contaminant may enter an aquatic system.

17-3.2 Transport and Transformation from Disposal Site to Aquatic System

Lee and Jones (Ref. 17-9) and Lee *et al.* (Ref. 17-6) have presented detailed guidance on the factors to be evaluated in determining the environmental chemistry-fate of a material upon entry into an aquatic system and on developing a tiered hazard assessment approach for estimating the potential impact of a contaminant or discharge on an aquatic system. Before that hazard assessment approach can be applied for solid wastes, an evaluation must be made of the transport and transformation of the contaminants in the solid waste leachate *en route* to an aquatic system, in order to derive the expected concentrations and forms of the contaminants of interest entering the aquatic system. Figure 17-1 is a generalized, schematic contaminant chemistry-fate model alluding to many of the reactions into which a contaminant in a solid waste leachate could enter, all of which must be considered. The major types of reactions that can occur in such a system are acid-base, precipitation, complexation, oxidation-reduction, hydrolysis, photolysis, gas transfer, biochemically mediated reactions – biotransformation, and sorption. Movement over the surface, through the unsaturated soil, in groundwater, into roots of vegetation, as well as volatilization followed

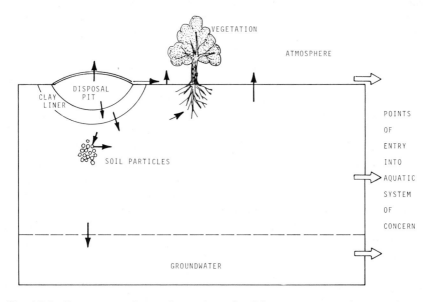

Fig. 17-1. Transport and transformation of solid waste-associated contaminants en route to aquatic systems.

by atmospheric transport are the major modes of transport of dissolved and particulate forms of contaminants. The models to describe the transport and transformation processes must be formulated on a site-specific basis. Of particular importance in model development are reactions with solid phases, such as sorption on soil particles. It has been repeatedly demonstrated that sorption tends to greatly detoxify contaminants or render them unavailable to aquatic organisms.

Another potentially important reaction is the sorption of contaminants on hydrous metal oxides, such as aluminum hydrous oxide and especially iron hydrous oxide. Many solid waste disposal site leachates are anoxic and therefore are likely to contain high concentrations of ferrous iron. Upon contact with oxygen, the iron will be oxidized and will precipitate as ferric hydroxide. Ferric hydroxide is an efficient scavenger for many trace contaminants, which tends to reduce the availability of many contaminants to aquatic organisms.

One of the potentially most important aspects of a contaminant chemistry-fate model for the system described in Fig. 17-1 is the interaction between the contaminant leached from the solid waste and the disposal pit liner and/or the soils of the area. It has been found by Green *et al.* (Ref. 17-1) that a number of organic solvents can affect the characteristics (most notably, permeability) of clay liners frequently used in industrial and some municipal disposal pits. Some solvents, such as carbon tetrachloride and xylene, caused clays to shrink, allowing the fluid to run through the liner in channels. However, when these solvents were mixed with others, such as water or acetone, the clays did not crack. It is important to evaluate this situation at each disposal site and for each of the wastes being disposed of, as the presence of certain organic solvents could markedly affect the transport of these and other contaminants present with them.

It is important to emphasize that one should not assume that the concentrations and forms of a contaminant in a leachate of a laboratory leaching test, such as those proposed by the U.S. EPA or ASTM Committee D-19.12, will be the concentrations and forms of the contaminants that enter a water body near a land disposal site. There is a wide variety of physical, chemical, and biological reactions, and dilution that tend to greatly reduce the concentrations of available forms of contaminants that would enter a water body compared to what would be found in the disposal site leachate. It is evident that the initial proposal of the U.S. EPA that is, to apply a factor of 0.1 for dilution and other reactions to the laboratory leachate contaminant concentration and then judge the "hazard" by comparing this number with the U.S. EPA Red Book criterion, is inappropriate. It may overestimate or underestimate the hazard to aquatic systems of the solid waste; in the experience of the authors, it is more likely that it will greatly overestimate the potential hazard.

The following discussion of how this information is used in an aquatic hazard assessment was adapted largely from Lee and Jones (Ref. 17-9) and Lee *et al.* (Ref. 17-6).

17-3.3 Characteristics of Aquatic Hazard Assessment

As indicated previously, an environmental hazard assessment for chemicals entering aquatic systems is built upon two basic components: aquatic toxicology and
environmental chemistry-fate of the chemical in the aquatic system. The characteristics of each are discussed below.

Aquatic Toxicology. The aquatic toxicology portion of hazard assessment provides
information on the response of aquatic organisms to concentrations and forms
of chemicals and durations of exposure that may be encountered in the environment. The objective is to produce a concentration of available forms (or dilution)-
duration of exposure- "no effect" relationship, such as the general case shown in
Fig. 17-2. All chemicals affect aquatic organisms in accord with relationships of
this type, in which the concentration of available forms of the contaminant can
be increased significantly without harming the organisms, provided that the duration of the exposure of the organisms to the chemical is sufficiently short. Further, for all chemicals, there is a chronic exposure safe concentration for available
forms, below which no known impact of the chemical has been found. At concentrations of contaminants below the chronic safe concentrations, changing the
duration of exposure has no effect on toxicity of the chemical to aquatic life.

It is important to emphasize that the response of various organisms to various
parts of the available form-duration of exposure- "no effect" relationship shown

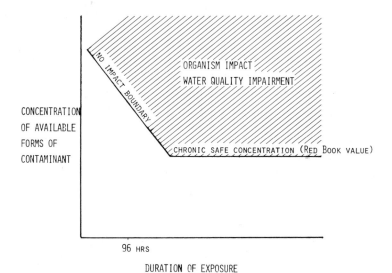

**Fig. 17-2. Generalized relationship among concentration of available forms, dura-
tion of organism exposure, and no impact level.** *Source*: Lee *et al.* (Ref. 17-6).

in Fig. 17-2 will vary. In the high concentration of available form–short duration situation, such as on the left side of the diagram, the effects that are typically
noted are those of acute lethality, while on the right side of the diagram, where
the concentration of available forms has an impact on the aquatic organisms, the
effects of the chemical on the organisms are primarily those associated with impairment of rates of growth or reproduction, alteration of behavioral patterns, etc.
Every meaningful environmental hazard assessment should include the development, even in a rudimentary way, of a concentration-duration of exposure to
available forms– "no effect" relationship shown in Fig. 17-2. It is evident that
environmental chemistry will play a dominant role in developing this relationship
because this is specific not only to a type of chemical, but also to each form of a
chemical that exists within aquatic systems. A free "aquo" species of a chemical
will show a different toxicological behavior in general, then a complexed, sorbed,
or otherwise transformed species.

Although organism toxicity is often the focal point for impact of chemicals
on aquatic systems, there is a variety of other concerns that must be considered
in evaluating the environmental impact of a chemical. Among these is the bioconcentration of the chemical, that is, its effects on higher trophic levels, including man, that would use these organisms as a food source.

Other areas of concern for chemicals in aquatic systems that should be evaluated
under the biological testing for hazard assessment include taste and odor production for domestic and industrial water supplies, stimulation of growth of excessive amounts of certain organisms such as algae by aquatic plant nutrients, as
well as a whole host of aesthetic effects such as color, turbidity, and floating debris.

Environmental Chemistry–Fate in Aquatic Systems. Environmental chemistry-
fate for aquatic systems considers for all modes of input, including solid waste
disposal sites, the transport and transformation of the chemical of concern and
its potentially significant transformation products, from its point of entry to the
aquatic system to its final disposition in the water of concern or its exit from the
system.   Environmental chemistry-fate also considers the chemical processes
that influence the form(s) (chemical and physical) of the contaminant and its
transformation products in each of the major components of the environment. For
example, for the aquatic environment, it is necessary to evaluate the forms and
concentrations of each form in true solution (dissolved), associated with particulate matter (such as erosional materials and organic detritus that are suspended
in the water column and deposited in the sediments), and contained upon or
within aquatic organisms.   There is also the potential for some highly volatile
compounds to be transported to or from gas bubbles in the water column. Few
chemicals are completely conservative (nonreactive) in the environment, i.e.,
whose concentrations in the environment change only as a result of physical
processes of dilution and dispersion. Most chemicals undergo a variety of trans-

formations, the majority of which can have a pronounced effect on the environmental concentrations of the toxic forms of the contaminant.

Major types of reactions that commonly occur in the aqueous environment were previously cited. Sorption reactions can be divided into several categories. One is biotic uptake of the chemical where associated transport and transformation within the aquatic organism must also be considered. Generally included within biotic uptake are those reactions that take place outside of the cell or organism involving extracellular enzymes. Another category of sorption is abiotic sorption, association of the contaminant with particulate matter present in aquatic systems such as clay particles, detrital minerals, organic detritus, and iron and aluminum hydrous oxides. A special case of absorption involves the uptake and release of contaminants from the aquatic system to the atmosphere.

One of the potentially important aspects of an environmental chemistry-fate model for aquatic systems is the interaction between the dissolved contaminant and suspended and deposited sediments. Many of the chemicals having the greatest potential hazard to the environment are highly hydrophobic and, therefore, tend to become strongly attached to particulate matter within the water column and within the sediments. Some of the most important forms of this particulate matter as discussed previously are the hydrous metal oxides of iron and aluminum. Lee (Ref. 17-10, 17-11) has discussed the role of iron hydroxide in influencing the behavior of chemical contaminants in aquatic systems.

Each of the above-mentioned transformations can be described by chemical thermodynamics and kinetics, whereby a position of equilibrium (thermodynamics) is obtained for a particular environmental system. The position of equilibrium is governed by the characteristics of the system, such as temperature, light, mixing-turbulence, suspended solids, and a variety of chemical properties, such as pH, gross and individual organic content, redox (oxidation-reduction) conditions, and, for air-water transfer, Henry's constant. The equilibrium positions are generally described by mass law relationships involving a thermodynamic or quasithermodynamic equilibrium constant.

While many chemical reactions proceed rapidly, being essentially instantaneous from an environmental impact point of view, there are many reactions that are significant in aquatic and other environmental systems, for which the rates are sufficiently slow to require consideration of their chemical kinetic properties. For these types of reactions it is necessary to develop a chemical kinetic-rate expression, a differential equation with a rate constant multiplied by the activities of the chemical species that are involved in limiting the rate of reaction.

Figure 17-3 presents a generalized model describing the environmental chemistry for a chemcial contaminant in an aquatic system. To mathematically represent this model requires the development of a series of differential equations that, when solved simultaneously, provide a description of the distribution of the chemical in various forms and in the various parts of the environment that can interact with aquatic systems.

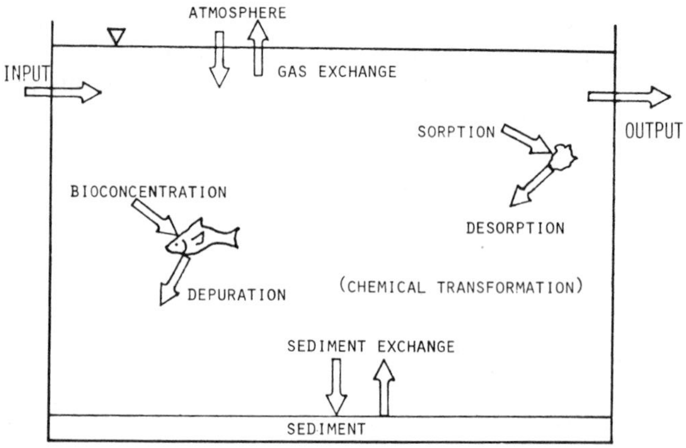

$$\frac{D(\text{AVAIL. FORM})}{DT} = K_1(\text{GAS EXCHANGE}) + K_2(\text{BIOCONCENTRATION}) +$$
$$K_3(\text{SORPTION}) + K_4(\text{CHEMICAL TRANSFORMATIONS}) +$$
$$K_5(\text{ETC.}) \ldots$$

Fig. 17-3. Environmental chemistry-fate model. *Source*: **Lee** *et al*. **(Ref. 17-6).**

It should be emphasized that, for large aquatic systems, there exists, between the point of entry of the contaminant and its ultimate fate, a concentration gradient, that is governed by physical processes of advection and mixing. Therefore, Fig. 17-3 should be modified in general to show this component of the environmental chemistry-fate model that, as noted above, can be the most important component describing the behavior of many chemicals in aquatic systems.

The model described in Fig. 17-3 is generally termed a dynamic model in which there is an attempt to develop differential equations that describe the overall processes involved. The modeler must determine which pathways are significant in transporting and transforming a particular contaminant within the water body of concern such that it can cause degradation of the water quality (beneficial use) of the water. The prescribed model is based primarily on the characteristics of the chemical and how it interacts with environmental compartments rather than being based on a particular system. Other approaches taken for environmental chemistry-fate modeling and their shortcomings are discussed by Lee *et al.* (Ref. 17-6).

A key component of any environmental chemistry-fate modeling effort is the proper verification of the model. No model should be used to make management decisions unless it has been verified; that is, it must have been demonstrated to predict, with adequate reliability, the concentration of the contaminant of interest in various environmental compartments under altered load conditions, without having had the rate constants retuned.

The outcome of the environmental chemistry-fate modeling should be an estimate of the expected concentrations of the potentially hazardous forms of the contaminant of concern in the various aquatic environmental components. This, combined with information on the expected residence times of various types of organisms in the compartments, can be used to estimate the durations of exposure of organisms in each component to the available forms of the contaminant. During the process of making an assessment of the hazard associated with the input of solid waste disposal site-derived contaminants, the predicted concentration-duration of exposure coupling is compared with the "no effect concentration-duration of exposure" relationship developed for the chemical in question in the toxicological portion of the hazard assessment. Figure 17-4 shows schematically a variety of the possible couplings and their relationship to the area of "impact." The hatched area in this figure represents the area shown in Fig. 17-2 in which the combination of duration of exposure and concentrations of available forms of a contaminant is sufficient to have an adverse effect on aquatic organisms and/or water quality. The numbered curves (1–5) show results that could be obtained through environmental chemistry-fate modeling, where a combination of dilution and chemical reactivity bring about a certain concentration–duration of exposure relationship. Curve 1 represents that coupling typical of spill situations, where there is toxicity for a short time associated with the point of entry before any reactions or dilution takes place. This might also be the situation associated with the mixing zone for a particular discharge.

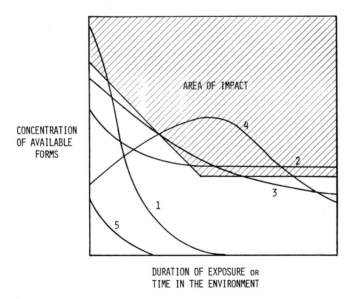

Fig. 17-4. Typical concentration of available forms—duration of exposure relationships in aquatic systems. *Source*: Lee *et al.* (Ref. 17-6).

Curve 2 in Fig. 17-4 is a case where the chemical does not show any acute toxicity at levels that are normally found in the environment but is chronically toxic either to the organism or to higher forms that may use the organism as food. PCBs, DDT, and mercury would all fall into this category. Because of the bioaccumulation of some of these types of chemicals within the higher trophic level fish, there is a potential for harm to man and other animals that use fish as a source of food.

Curve 3 represents the type of situation that might be associated with municipal wastewater discharges that contain ammonia, where for short durations of exposure there is no impact. However, for some receiving waters there is a sufficiently large intermediate zone at some distance from the point of discharge where there would be toxicity to fish that reside in the area. Eventually the ammonia would be oxidized or diluted to nontoxic levels farther down the stream from this zone.

Curve 4 is representative of the situation where there is a transformation of the chemical added to the system that causes it to be more toxic as it goes downstream. Eventually it is either diluted or detoxified through other reactions. An example of this type of situation is one involving the addition of a complexed heavy metal to the environment where the complex is biodegradable, releasing the toxic form of the heavy metal at some distance downstream in sufficient concentrations to be toxic to aquatic life in that region.

Curve 5 is the case that exists for most chemicals for which there is sufficient treatment or controlled use so that there is no toxicity associated with it, either acute or chronic. It is important to emphasize that while the various conditions shown in Fig. 17-4 are represented by smooth curves, for any real systems there are fluctuations about the mean concentration that can be of significance in affecting water quality.

The contaminant input from solid waste disposal sites can follow any of the patterns shown in Fig. 17-4. It is expected that if there is toxicity from this source, it will be localized near the point of entry, since it is not likely that the rate of input and concentrations of contaminants would be sufficient to cause widespread contamination. An exception to this would be contaminants with high volatility, such as PCBs, which are transported in significant quantities from solid waste disposal sites to the atmosphere and then to aquatic systems through precipitation and dustfall. It is expected that solid waste disposal sites would contribute greater amounts of contaminants during wet periods than dry periods. These seasons would also in general be the periods in which the greatest amount of dilution would be available in receiving waters.

It is important in assessing the significance to water quality of contaminants associated with solid waste to focus on the amounts and forms of contaminants that enter the first water body in which there is an aquatic resource recognized to be of beneficial use to man, e.g., a fishable stream or lake, rather than on an intermittent flow system in which there is no significant resource for man. High concentrations of contaminants in soils, groundwater, or surface runoff immediately adjacent to the solid waste disposal contaminant input may not be of significant harm to aquatic life, especially fisheries in receiving waters where there is concern about fisheries or other aspects of water quality.

For existing solid waste disposal sites or where the same kinds of wastes will be placed in similar geological formations adjacent to existing sites, it may be most cost-effective and reliable to install a series of monitoring wells and sample them and surface runoff in order to assess the likelihood of environmental "contamination" and the impact on groundwater quality. This type of assessment should be conducted in accord with the hazard assessment approach outlined by Lee et al. (Ref. 17-6). While their approach specifically discusses application to surface waters, it can be adapted to assessing the hazards to municipal and agricultural users of groundwater supplies. In making this evaluation, it is important to consider the possible transformation and dilution between the point of sample collection and point of water quality concern. As discussed by Lee (Ref. 17-12), it is important to take a much more conservative approach toward groundwater contamination than generally needs to be taken for surface waters. Once groundwaters have been contaminated to the point at which their beneficial use in impaired, they are generally much more difficult to restore than surface waters.

17-3.4 Tiers of an Aquatic Hazard Assessment

A hazard assessment should be conducted in a series of levels or tiers of increasing sophistication and detail, with a decision point at the end of each tier. In each tier an estimate is made of some aspect of the aquatic toxicity of the chemical and also of the expected environmental concentrations (environmental chemistry-fate) of the chemical. These two components are compared to make a decision regarding the potential aquatic environmental impact. Decision choices at the end of each tier's testing could be (1) to not allow disposal of the particular solid waste in the place and manner evaluated because of excessive expected hazard, (2) to restrict such disposal or require treatment of waste to reduce expected environmental hazard to an acceptable level, (3) to proceed with disposal as evaluated, as the expected hazard is acceptable, or (4) to continue testing to better define the potential environmental impact. As the more sophisticated tests tend to be much more expensive, decision to conduct addition testing tiers must be weighed against (1) the cost of assuming a significant hazard does in fact exist and thus of providing the treatment or finding alternative disposal necessary to reduce the environmental hazard, and (2) against the costs, both social and economic, of not allowing production of certain materials.

As successive tiers of testing are conducted, precision and accuracy of the estimates of the contaminant's toxicology, and the environmental concentrations of the contaminant and its transformation products as well as the reliability of the decision made should be improving. Figure 17-5 illustrates this concept.

Because of the tiered structure of this hazard assessment approach, the testing done in the early tiers can detect both those contaminants that would be highly hazardous as a result of their estimated environmental concentration being considerably above the "no-effect" level estimated through the aquatic toxicology testing, as well as those contaminants for which the expected concentrations are well below "no-effect" levels. The situations of greatest concern are those in which there is some overlap in confidence levels (as shown in Fig. 17-5) or a relatively small difference between the estimated environmental concentration and the maximum "no-effect" concentration. Under these conditions continued testing is probably needed, as the reliability of the estimates of both the toxicological properties and environmental concentrations at the lower tiers are usually quite rough. However, as shown in Fig. 17-5, with more sophisticated testing in higher level tiers, it is possible to refine these error bars and thereby more clearly discern whether or not there is a potential impact in a particular system. While there may be considerable controversy regarding an appropriate magnitide of separation between the estimated environmental concentration of available forms and the "no-effect" level, this would typically be on the order of a factor of 10 to 100. There are some site-specific considerations that should be given in establishing the margin of safety that should be used in this type of evaluation.

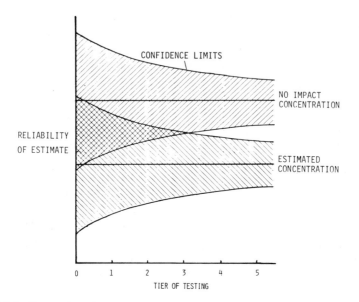

Fig. 17-5. Examples of increasing reliability with increasing testing. *Source*: Cairns *et al*; (Ref. 17-5).

The main reason for conducting a hazard assessment in tiers is that it is generally less expensive, more technically appropriate, and equally protective to evaluate the potential environmental impact to the degree necessary to make decisions on the degree of contaminant control required based on actual expected impact, rather than to treat routinely for worst case conditions in which it is assumed that the toxicity associated with a particular effluent or source extends for considerable time and distance from the point of discharge or that any concentration of contaminants above U.S. EPA water quality criteria or state water quality standards equivalent to these criteria represents a deterioration of water quality. This latter approach is the ultraconservative approach that, while providing environmental protection, will certainly result in the needless expenditure of large amounts of money for pollution control with little or no additional improvement in water quality beyond that provided using the hazard assessment approach. In the 1980s hundreds of millions to billions of dollars will be spent for pollution control in order to comply with the provisions of RCRA, Public Law 92-500, and the 1977 Amendments of this Clean Water Act. It is essential that all funds spent in the name of water pollution control be directed toward controlling water quality problems. The tiered hazard assessment approach provides a technically valid basis for ensuring that the expenditures made in this area provide the opportunity to result in the greatest possible improvement in water quailty for funds expended.

Table 17-1 presents a summary of the types of toxicological and environmental chemistry–fate information that are needed for a generalized tiered hazard assessment program that could be used to assess the significance of the input of contaminants derived from solid waste disposal sites or any other source of contam-

**TABLE 17-1. Environmental Hazard Assessment
for Aquatic Systems**

Tier 1	Bioassay — 96 hr LC_{50} Dilution — Worst-case based on rate of use and input
Tier 2a	Short term bioassay at other exposure durations Measure existing concentrations in environment
Tier 2b	Fish egg-fry and *Daphnia* toxicity test Bioconcentration — octanol–water partition
Tier 3	Develop environmental chemistry–fate model
Tier 4	Conduct field studies, evaluate actual impact Determine chemical species
Tier 5	Evaluate significance of impact on water quality; compare to cost of control program
	Submit results for societal review and decision.

Source: Lee *et al.* (Ref. 17-6)

inant for aquatic systems. Further information on testing requirements of each tier and the use of this information is provided by Lee *et al.* (Ref. 17-6).

17-4 CONCLUSIONS AND ACKNOWLEDGMENT

For a variety of reasons discussed in this chapter, the U.S. EPA's original approach for assessing the aquatic environmental hazard associated with solid wastes disposed of on land by conducting a single leaching test and comparing contaminant concentrations in the leachate multiplied by 0.1 to the U.S. EPA Red Book criteria, is not technically valid. The release of contaminants from solid wastes depends on many factors: characteristics of the chemical of concern, of other chemicals present, and of the particular disposal site. A site-specific leaching test must be conducted for each system of concern in order to estimate the concentrations and forms of contaminants likely to leave a waste disposal site. The studies conducted on the dredged sediment elutriate test can provide considerable insight into the test conditions that need to be evaluated for their appropriateness for representing solid waste disposal sites. Field studies must also be conducted in conjunction with the development of a laboratory leaching test, as was done for dredged sediment disposal, to verify its ability to predict contaminant leaching from solid wastes.

In interpreting the results of leaching tests, it is important that consideration be given to the behavior of the contaminants within the disposal site, as well as en route to a water body of concern. A contaminant can be greatly detoxified, or made less available to aquatic organisms, by many of the physical, chemical, and biological transformations that it undergoes in these systems.

Recently, a hazard assessment approach has been developed for aquatic systems that can be used in conjunction with the information on transport and transformation of the contaminant between the disposal site and a water body of concern, to assess the aquatic environmental hazard of contaminants associated with solid wastes disposed of on land. The investment of time and money in conducting a hazard assessment will be highly cost-effective in that it will provide a technically valid basis by which an environmentally protective solid waste contaminant management program can be designed, avoiding overdesign to cover worst-case conditions.

Support for preparation of this chapter was provided by the Department of Civil Engineering at Colorado State University, the Colorado State University Experiment Station, and EnviroQual Consultants and Laboratories, all of Fort Collins, Colorado.

REFERENCES

17-1. Green, W.J., Lee, G.F., and Jones, R.A., "Impact of Organic Solvents on the Integrity of Clay Liners for Industrial Waste Disposal Pits: Implications for Groundwater Contamination", Final Report to US EPA-Ada, Ada, OK, June, 1979.

17-2. Lee, G.F., "Comments on December 18, 1978 *Federal Register*, 'Guidelines for Hazardous Waste Management,' " Department of Civil Engineering, Environmental Engineering Program, Occasional Paper No. 39, Colorado State University, Fort Collins, CO, March (1979).

17-3. Lee, G.F., Jones, R.A., Saleh, F.Y., Mariani, G.M., Homer, D.H., Butler, J.S., and Bandyopadhyay, P., *Evaluation of the Elutriate Test as a Method of Predicting Contaminant Release during Open Water Disposal of Dredged Sediment and Environmental Impact of Open Water Dredged Material Disposal, Vol. II: Data Report*, Technical Report D-78-45, U.S. Army Engineer WES, Vicksburg, MS, August, 1978.

17-4. Jones, R.A., and Lee, G.F., *Evaluation of the Elutriate Test as a Method of Predicting Contaminant Release during Open Water Disposal of Dredged Sediment and Environmental Impact of Open Water Dredged Material Disposal, Vol. I: Discussion*, Technical Report D-78-45, U.S. Army Engineer WES, Vicksburg, MS (1978).

17-5. Cairns, J., Dickson, K.L., and Maki, A.W. (Eds.), *Estimating the Hazard of Chemical Substances to Aquatic Life*, ASTM STP 657, American Society for Testing & Materials, Philadelphia, PA, 1978.

17-6. Lee, G.F., Jones, R.A., and Newbry, B.W., "An Environmental Hazard Assessment Approach for Water Quality Management," Department of Civil Engineering, Environmental Engineering Program, Occasional Paper No. 46, Colorado State University, Fort Collins, CO, November, 1979.

17-7. Ham, R.K., Anderson, M.A., Stegmann, R., Stanforth, R., *Comparison of Three Waste Leaching Tests*, EPA 600/8-79-001, US Environmental Protection Agency (EPA), Cincinnati, OH, 1979.

17-8. Thompson, D.W., *Elutriate Test Evaluation of Chemically Stabilized Waste Materials*, EPA 600/2-79-154, US EPA-Cincinnati, OH, 1979.

17-9. Lee, G.F., and Jones, R.A., "The Role of Environmental Chemistry–Fate Modeling in Environmental Hazard Assessment: An Overview," Presented at ASTM Symposium on Aquatic Toxicology, Chicago, October, 1979, to be published in symposium proceedings.

17-10. Lee, G.F., "Factors Affecting the Transfer of Materials between Water and Sediments," University of Wisconsin, Eutrophication Information Program, Literature Review No. 1, 1970.

17-11. Lee, G.F., "Role of Hydrous Metal Oxides in the Transport of Heavy Metals in the Environment," Proc. Symposium on Transport of Heavy Metals in the Environment, *Progress in Water Technology* 17, pp. 137–147, 1975.

17-12. Lee, G.F., "Potential Problems of Land Application of Domestic Wastewaters," *Land Treatment and Disposal of Municipal and Industrial Wastes*, Sanks, R.L., and Takaski Asano, (Eds.), Ann Arbor Science Publishers, Ann Arbor, MI, pp. 179–192, 1976.

Index